International
Environmental
Policy

International Environmental Policy

From the Twentieth to the Twenty-First Century

Third Edition

Lynton Keith Caldwell

Revised and Updated with the Assistance

of Paul Stanley Weiland

DUKE UNIVERSITY PRESS   Durham and London 1996

© 1996 Duke University Press
All rights reserved
Printed in the United States of America on acid-free paper ♾
Typeset in Janson Text by Tseng Information Systems, Inc.
Library of Congress Cataloging-in-Publication Data appear
on the last printed page of this book.

Contents

Acknowledgments

The first and second editions of *International Environmental Policy* acknowledged the special assistance of many individuals and institutions. Those expressions of appreciation still stand but are not repeated in this edition. To acknowledge the help of particular institutions is not enough: I would like specifically to identify the persons to whom I owe a special debt of gratitude. In the Indiana University library system the following provided extraordinary assistance: in the Government Publications Department of the Main Library, Alice Wickizer, Marian Shaban, and Linda Kelsey; in the reference section of the library of the School of Law, Keith Buckley, Ralph Gaebler, and Marianna Mason; in the reference section of the Research Collection Ann Bristow and staff; and Michael Parrish in the library of the Schools of Business and Public and Environmental Affairs. I am also grateful to the Center for Urban Policy and the Environment for making possible the assistance of Paul Weiland in the preparation of this edition. His help was invaluable, the more so in view of the significance of life and work in the world's great cities on environmental problems and policies of nations. The assistance of a retired faculty research grant from the Office of Research and Graduate School was also greatly appreciated.

From Washington I received helpful information from Susan Fletcher of the Congressional Research Service, and from Charles E. Myers of the Office of Polar Programs, National Science Foundation.

Finally, I must acknowledge with appreciation the invitation of Reynolds Smith of the Duke University Press to prepare this third edition. Without his initiative, interest, and encouragement this comprehensive and complex task would not have been undertaken. I am also greatly appreciative of the thorough and perceptive copyediting of the text. Author and readers owe a debt of gratitude to Maura High for smoothing transitions and clarifying ambiguities in the original typescript and to Jean Brady for managing the production.

In the thirteen years since the first edition of this work appeared, the scope and substance of international environmental policy have increased exponentially. The field has now nearly outgrown the competence of an individual scholar. Most books in this comprehensive field are now either multiauthored or more specialized in scope. In the future a multiauthored team effort will most likely be required to deal with the material covered in this edition.

Defense of Earth in a Changing World:
An Introduction

A case may be made that the primary condition in any enduring civilization is a stable relationship between humans and their environment (e.g., by Jacques Ellul in *La Technique* (*The Technological Society* 1964).[1] The international environmental movement may be understood in effect, if not explicitly in intent, as a quest for this relationship, even though its underlying and ultimate purpose is often obscured by concern over specific contemporary issues. The movement is an expression of fundamental change in human perceptions of life on earth. More rapidly than in most historical transitions, evidence of the change has appeared at every level of social organization—local, regional, national, and international. A development of great importance for the future has been emerging, but its significance is not widely understood. Sociologist Robert Nisbet has surmised "that when the history of the twentieth century is finally written, the single most important social movement of the period will be judged to be environmentalism."[2]

To understand what has been happening, one needs to know the why and how of this development. Although not addressed in this book, the influence of transnational religions and popular ethics should not be overlooked, for perceptions of relationships between man and nature are fundamental to policies and behaviors relating to the environment. A new, ecological view of the role of humans on earth has been emerging, one that departs from the traditional perception of human dominion over nature and moves toward a more realistic appreciation of humanity's interrelationship with the biosphere. This new view has led to action in which scientific knowledge, lessons of experience, and ethical judgments have been united in public policies and international agreements. Nevertheless, traditional views persist, although now partially offset by ecologically informed trends in ethics and theology.

This book is a comprehensive survey of the worldwide movement for protection of the human environment, with emphasis on intergovernmental agreements and institutional arrangements. Although the growing role of nongovernmental organizations in international policy is considered, less attention has been given to the politics and ideology of the popular environmental movement. Emphasis is upon what has already become, or is well on the way to becoming, public policy. The book is essentially a history and chronology of international cooperation on environmental issues. It describes the expanding dimensions of international environmental policy and its status in the closing decade of the twentieth century.

To understand adequately the significance of international environmental policy in the present and the probable future, its origins and evolution need to be understood. The present can always be better understood against a background of the past.[3] The worldwide explosion of environmental concern that occurred in the 1970s and its recurrence in the late 1980s did not happen without causes and antecedents that influenced the substance and direction of international action. The book thus provides a record of historical events of continuing intellectual and historical significance. Two such events, the Stockholm Conference of 1972 and the Rio de Janeiro Conference of 1992, and their legacies form integrative themes throughout this book. The significance of these events will, I believe, grow with the passing of time; their character as turning points will increasingly be understood. But not all events of relevance or importance have been reported here — to have included all events of significance would have unduly lengthened the text and obscured with detail the main currents of action.

The far-reaching scope of this subject, examined in chronological depth, precludes detailed analytic treatment of most policy issues. For the latter purpose a series of volumes would be required. For many topics a large and growing analytic literature already exists.[4] This book undertakes to provide an overview of international environmental relations, with no more detail than necessary to record the processes through which various environmental policies emerged. The text accordingly is written at a relatively high level of generality, but some readers may seek more detailed information regarding specific environmental issues. For this reason the work is extensively documented; the notes supplementing the text provide guidance to sources of further information. Multiple sources have been provided for many citations, as not all readers will have access to the same publications. Earlier published work has also been cited where it has made a significant contribution to the historical record.

Several technical points require comment. In a book written and published in the United States it is logical that American preferences in spelling be followed. British spelling, used in United Nations documents, is followed when it appears in official titles and quoted passages. Thus, two different spellings of program/programme may appear in the same sentence. Measurements pose a similar problem. Metric measures have been used except where the cited data are in the English system. And to protect the reader from drowning in an alphabet soup of acronyms for international agencies and programs, full names will be repeated at intervals.

Regretfully, but bending to convention, the term "Third World" is frequently used throughout this volume in reference to so-called developing or less-developed countries. This nomenclature, and the equally flawed "North-South" dichotomy, does not adequately identify the differences that distinguish the less industrialized from the more industrialized nations. Nor does it reflect the great diversities among the "Third World" nations that in some cases are more significant than their commonalities. But to propose and defend a more appropriate terminology would further complicate and extend an already complex and extended text. Instead, minimal explanation of these word usages is provided at those places in the discussion where clarification might help the reader. The terms human, humankind, humanity, people, and man are used interchangeably throughout the text. They are generic terms for the human species used alternatively for reasons of style, but always with fidelity to quoted passages.

The author does not bend the contents of the book to fit a theme, yet a unifying concept emerges from the historical evidence. It is the transformation of institutions through social learning in response to the findings of science and the perceived impacts of environmental change. The environmental movement exemplifies learning by large numbers of people from experience made explicit and understandable through science. The movement belongs to a larger transformation in human social thought, which may be likened to a second Copernican revolution. The first revolution removed the earth from the center of the universe; the second removes humanity from the center of the biosphere. The human species is indeed the dominant resident-shaper of environmental change, but only people bound to prescientific theologies (and not all of them) believe that the earth and its biosphere were created for man's exclusive benefit.

Attitudes are changing, but there is nonetheless a lag between the new perception and the assumptions, practices, and institutional policies expressing traditional views of relationships between humanity and the rest of nature. Thus, the environmental movement is transitional between those perceptions and policies that have been widely prevalent in human affairs, and those new beliefs and commitments that are exemplified by the reports, declarations, laws, treaties, and programs described in this volume. While the transition has moved rapidly relative to historical change, many years may yet pass before its implications are fully realized.

Meanwhile, the goals of environmental protection are pursued in a world still unhappily characterized by mutually antagonistic blocs among and within nations. Underlying the political differences are ethnic and ideological conflicts, which at times have threatened to wreck international cooperation of any kind. There are students of international affairs who argue that a political unification of the world is necessary to match its ultimate biospheric unity and to contain the parochial tendencies of humankind that lead to international conflict. But the prospect of a world of politically compatible nations in the foreseeable future seems remote, and the prospect of a unified world social order more distant yet.

This negative assessment is offset to some extent by human adaptability. Nations and people cooperate when convinced that their interests will be served by cooperation. Social learning leading toward an integrated view of humans in the biosphere rationalizes international cooperation on numerous environmental issues, regardless of differences in other respects. Many environmental issues threatening the biosphere today cannot safely be set aside until political, social, and economic antagonisms among nations are resolved. But people who collectively dislike one another can work together when faced with a common threat. This is the politics of antagonistic cooperation—perhaps the only strategy realistically available to defend the earth against human egoism, aggression, and lack of foresight.

Yet, as the transition between the twentieth and the twenty-first centuries occurs, a more cautiously hopeful prospect appears possible. At the close of the 1980s major political reorientations occurred in many countries. An active environmental movement appeared to be growing in Brazil—a country that led Third World resistance to international environmental policies at the 1972 United Nations Conference in Stockholm, yet hosted the 1992 United Nations Conference on Environment and Development. East European states and groups that were mute within the former Soviet Union were demanding environmental as well as political and economic reforms. In July 1989 the political chiefs of seven principal industrial democracies declared their commitment to international environmental cooperation. Issues such as global climate change, disintegration of stratospheric ozone, and long-range transboundary transport of pollutants (including radioactive fallout) appeared to be prompting—at least in principle—the willingness of governments to cooperate for their mutual protection.

A structure of international environmental cooperation has been emerging and developing that may provide a bridge to more inclusive international relationships in the future. Nations that have never worked together before are doing so now in a variety of cooperative efforts, such as the UNEP Oceans and Coastal Areas Programme and the supranational policymaking of the European Union. The North American Free Trade Agreement has been supplemented by a lateral environmental cooperation commission. Common purpose precedes common action, and the broadening understanding of the relationship of environment to economic and social issues may lead to expanded perceptions of common interest. And conversely, international collaboration on common economic problems has led, as in Europe and among the Southeast Asian (ASEAN) governments, to cooperation on environmental matters. Thus the tenor of this book, while not optimistic, does offer some hope. It is hope based upon the demonstrated ability of humans to learn and to apply their learning in practical ways even under conditions of stress and conflict.

Inevitably affected by the dynamics of science, environmental policy should be understood as a developing process. A report confined to the state of environmental policy at a given time has appropriate uses, but it cannot reflect this dynamic, and thus rapidly becomes "dated." However, to understand the significance of environmental policy today, it is necessary to know the circumstances that led to the present situation. Scientists studying soil deterioration or the changing chemical balance in the atmosphere need to know the historical progression of biogeochemical interactions that explain what has happened and why—thus enabling them to more reliably project future probabilities and alternative responses. Similarly the rationale underlying the way in which environmental issues and problems are dealt with in international affairs is best understood in historical (i.e., evolutionary) terms. This book is written so that it may be of continuing value as a record of the rise and development of an international environmental movement. Its emphasis, on emergent processes rather than upon contemporaneous events, is meant to give it continuing utility even when future events displace current arrangements.

For this reason the following chapters will frequently describe the chronological sequence of emerging issues and, so far as feasible, be specific with respect to dates and critical events. Not only is the timing of a development often helpful in explaining its outcome, but dates at specific points of reference are helpful to readers who may wish to learn more about a particular incident. A chronology of significant events is provided in Appendix C.

The scope of the subject of this book is so comprehensive that illustrations of issues and problems must be carefully limited and selective. It is impossible to tell the reader everything that is important to the subject, although opinions and examples have been drawn from nearly every nation. The text is extensively documented to facilitate follow-up, but the large amount of relevant literature and the space limitations of a single volume, preclude (with rare exceptions) citations to publications in languages other than English. However, a high percentage of the world's literature on environmental policy and science has been published in English. This includes periodicals published by the United Nations agencies and nongovernmental international organizations, which are often the best sources of current information in the field.

What Are the Critical Issues?

Environmental policy tends to be focused, at least superficially, on specific and concrete emergencies or events. Environmental issues are numerous and disparate; but public attention is drawn to immediate, comprehensible concerns that are more readily publicized than the larger, more complex developments of which they are manifestations. It is easier to attract public attention to the effects of acid rain, the plight of fur seals, or pollution from pesticides than to the more complex, yet no less fundamental problems of biogeochemical cycles or species-habitat relationships.

Critical environmental issues may therefore be evaluated differently in science and in public affairs. With the growth of scientific understanding of processes and relationships in the environment, public policymakers are beginning to appreciate the larger dimensions of environmental problems. It is also more convenient for governments and international organizations to deal by statute or treaty with a few larger issues than with many lesser ones. It is simpler to have a general treaty for the protection of marine mammals (even with subsections for particular species) than to negotiate separate treaties for whales, porpoises, walruses, and manatees. A treaty to protect all migratory birds is more easily administered than treaties for each species.

Critical issues as categorized by science do not always correspond to the perceptions and priorities prevailing in governments and international agencies. An issue becomes prospectively critical to a government if it becomes sufficiently critical to its political constituency. Moreover, the way in which an environmental issue or problem arises in public affairs and the way in which the public perceives it may differ significantly from the way the same set of circumstances would be understood from the viewpoint of the relevant sciences. For this reason a book like the United Nations Environment Programme (UNEP) report, *The World Environment 1972–1982*, while appropriately organized in a large part under categories into which scientists divide their study of environmental problems (e.g., atmosphere, terrestrial biota, population), also reflects the concerns of UNEP's governmental constituents (e.g., health, settlements, industry, and tourism).

This book is organized upon a different rationale. The emergence of international environmental policy as well as its implementing arrangements, rather than the state of the environment, form its organizing principle. National legislation and international treaty making seldom proceed along lines of systematic scientific relatedness. Yet scientific aspects of critical environmental issues are built into the structure of the book because perceptions derived from science are gaining in recognition and influence upon policy.

There has been a growth of understanding that the goals of policy cannot be achieved solely by attack upon apparent and immediate environmental problems, for they are often manifestations of deeper environmental disorders requiring systematic analysis and explication that the sciences may (or may not) provide. And because the chapters to follow are organized primarily around the emergence of popular and political perceptions of environmental problems, a summary of critical issues as defined through science will be useful. Regardless of popular understanding, these issues are fundamental to the course of international environmental policy development.

There is no universally accepted way to categorize critical environmental issues; many are interactive or overlapping. The issues are seldom mutually exclusive; each one embodies some aspect of others. Classifications adopted by various organizations may reflect particular intentions or responsibilities — as exemplified by the topical categories of the UNEP Earthwatch program. There is less argument about which issues are environmentally significant than about which among them are the more critical. Every categorization of environmental issues is to some extent arbitrary in the sense that other schema may be equally valid — although not necessarily contradictory. The following enumeration of critical issues lists twelve broad categories of environmental significance. The length of time before disastrous or irreversible effects occur is a principal criterion for criticality.

Critical at Present: Immediate action necessary if the threat is to be countered:

1. Genetic loss (threatened extinction of presently endangered species).
2. Ecosystem disruption (massive loss of habitat, genetic material, quality of life, and regenerative capabilities — marine as well as terrestrial).
3. Overpopulation by humans (a critical factor in most environmental issues and requiring early corrective action to counter already escalating ecological and economic impoverishment and social conflict — almost every environmental problem and the prevalence of poverty could be reduced by stabilization of population).
4. Deforestation and overgrazing (many of the above effects, as well as decimation of forest-dwelling peoples, soil deterioration including erosion, laterization, flooding, siltation, and possible reduction of atmospheric oxygen).
5. Contamination of the environment — air, water, soil, and biota (by industrial toxicants including radioactive materials, photochemical reactions, e.g., smog, and by particulates, e.g., dust).
6. Degrading and depletion of fresh water (caused by many of the above threats, eutrophication or acidification of lakes and streams, exhaustion or contamination of groundwater and aquifers, and destruction of wetlands).

Becoming Critical: Prompt response needed, but a lead time of some years allows solutions to be found before disaster occurs. A systematic search for solutions should begin now:

7. Unsustainable assumptions and trends that could lead to ecological and economic collapse. Persistence of ideologies inconsistent with life on earth (e.g., ever-expanding consumption and material growth).

8. Deterioration and erosion of top soil (especially disastrous in tropical countries and closely related to overpopulation, deforestation, and desertification).
9. Climate change and deterioration of atmospheric quality, sea-level rise caused by global warming, disruption of stratospheric ozone layer, precipitation of acidic and other contaminants, and impairment of atmospheric clarity by industrial particulates and dust.
10. Sources and uses of energy (progressive reduction in existing sources, and dangers from their environmental effects make this issue ultimately critical; no early adequate solutions are apparent).
11. Disruption of biogeochemical cycles (relates to all of the above; a combination of destructive trends could—at least in theory—break the linkages that permit regenerative capabilities of the biosphere to function, resulting in massive extinction of life on earth).
12. Maintenance of the built environment and loss of cultural heritage in arts and architecture (the large environment-shaping public works of modern society, e.g., dams, canals, highways, nuclear reactors, and other large structures, entail costs projected indefinitely into the future, and regardless of their continuing utility—which may not always be assured—encumber the future allocation of societies' resources).

Except as offset by conservation practices or technological innovation, all of the foregoing issues become more critical with expanding human populations. Some might also be worsened temporarily by a sudden collapse of world populations. Restoration of damaged environments would become difficult if numbers declined to the minimum required for subsistence. The maintenance of a modern physical infrastructure might prove an impossible burden for a greatly reduced population. The concept of optimal population deserves serious consideration. The present danger is clearly one of excessive population pressure upon a finite environment. The populous world as presently organized is now being increasingly, but perhaps too slowly, perceived as unsustainable. The questions of whether or how to achieve a stable or steady ecological-economic state will certainly pose a major policy problem in the twenty-first century.

All of the processes, developments, and conditions that comprise these issues affect relationships among nations. Some problems affect all nations, whereas others may have largely regional significance (at least in their primary effects). But the ramifications of every major issue have worldwide implications. In this sense they may be called global, but there are few ways to act upon them globally. They must be attacked at their origins. Except for the international commons of the high seas, outer space, and the unique case of Antarctica, the issues arise within national frontiers. However, among their other inequalities, nations are unequally able or willing to manage their affairs so as to avoid the emergence of critical environmental problems. Nor are there many ways under the present disposition of national power and politics that nations may peacefully coerce one another into environmentally prudent policies. Embargoes are uncertain possibilities.

The basic assumption upon which the foregoing statement of concepts and approaches rests is that the environmental movement, as it spreads worldwide, is a manifestation of a major historical change of state, or discontinuity. Eric Ashby has described the so-called environmental crisis of our time as a climacteric.[5] It marks the end of that half-millennium of exuberant, exploitive expansion of (chiefly) Western society that we have called modern times. Symbolic boundary markers might be 1492 to 1992. People well informed, and even many poorly informed, are beginning to understand that the course pursued by modern society during the past five hundred years can no longer be continued indefinitely. In more than a poetic sense, the threats to future life exemplified by global climate change, ozone depletion, toxic contamination, and environmentally catalyzed disease represent a backlash of nature against the improvident optimism born of human ingenuity. There may remain great opportunities for human achievement, but the price of their realization includes an informed restraint upon the human uses of the earth. The ultimate goal of environmental policy is a sustainable relationship between humankind and the life-supporting systems of the earth.

At the close of the twentieth century two collateral movements, notably in the United States, forecast what may become dividing issues in the twenty-first century. They are antienvironmentalism, based chiefly on the perceived necessity for perpetual growth and on an individualistic and possessive view of property rights, and—in contrast—a broader movement for an environmentally and economically sustainable future. The property rights movement has been characteristically American and parochial and would seem to have no great significance for international environmental policy other than a deterrence to cooperation by the United States on certain international issues (e.g., biological diversity). The movement for sustainable development, however, could strengthen international environmental policy and action by integrating ecological and economic theory in the concept of natural economics.

The sustainability concept is not only realistic in its economic perspective but also implies a moral and prudential commitment. When most effectively articulated it could resolve the dichotomy between so-called realistic and idealistic views of humanity's relationship to the earth and to future generations of life on earth. For this to happen the concept of sustainable development will need clarification and its utility will need to be tested in practical application. Even in its present ambiguous form the concept does have utility as a goal and a framework for assessing and perhaps redefining the concept of development.

The encompassing environmental policy question of today is, what happens now? How will the broad range of commitments to protect the environment be put into effect, when, and by whom? How will the costs of environmental efforts be allocated? What trade-offs will be necessary to assure that important environmental needs are met with regard to future generations as well as to those now living? Most important of all, however, is a moral question: By what principles of ethics or moral conduct is human behavior to be judged in relation to the environment?

Twenty-five years ago, in response to an invitation from the Scientists' Institute for Public Information, I described the "environmental crisis" as more profoundly "a crisis of will and rationality—of intelligence and moral character." This "crisis" is not in the environment, which is ever-changing, but in human behavior as it affects the environment. It comes down to a choice among values. And as Ernest Becker put it, "[T]he fundamental question of values in any culture can be phrased in simple terms—what kind of control over what kind of environment?" [6]

I believe that it is hardly possible to overemphasize the need for humanity to find, through its most basic assumptions about life and nature, a truly sustainable relationship with the earth. I therefore conclude this introduction by repeating a statement I made in the UNEP publication for the United Nations Environmental Sabbath/Earth Rest Day, 1–3 June 1990:

> The environmental crisis is an outward manifestation of a crisis of mind and spirit. There could be no greater misconception of its meaning than to believe it to be concerned only with endangered wildlife, human-made ugliness, and pollution. These are part of it, but more importantly, the crisis is concerned with the kind of creatures we are and what our species must become in order to survive. [7]

1
Comprehending the Environment

Before there could be an international environmental policy, there had to be nation-states and a better understanding of the relationships between human society and the planetary environment. Comprehension of an environment that interacts with the behavior of people and nations took shape slowly over many decades and centuries. Major advances in science and science-based technology were necessary before the human environment could be realistically defined and comprehended. There are still many people who not only fail to comprehend the significance of the environment but also deny that it has significance beyond its utility to serve humankind's material needs. Nevertheless, there is sufficient awareness of the importance of the environment to human life in all respects, including our ethical and spiritual values, that governments have begun to develop, however imperfectly, laws and institutions for safeguarding its viability and hence protecting and enhancing human life itself.

Humanity lives in two realities. The abiding reality is that of the *earth* — the planet — independent of man and his works; the other reality — the transient reality — is that of the *world*, which is a creation of the human mind. The *earth* and its biosphere form a grand synthesis of complex interactive systems within systems, organic and inorganic, animate and inanimate. The *world* is the way humanity understands and has organized its occupancy of the *earth:* an expression of imagination and purpose materialized through exploration, invention, labor, and violence. Oceans, islands, species, and ecosystems are integral parts of the earth, but the world is not integrated — its cultures and their values do not comprise a unity. All living people may be of one species, but their cultures are diverse. Physically, humans belong to the earth, yet intellectually they may transcend it — a dangerous liberty when dissociated from regard for the necessities of life on earth. It is an arrogant conceit that whatever we can imagine, we can one day do.

The so-called environmental crisis of the modern world derives from this physical and intellectual duality. Unlike the environmental disasters encountered by prehistoric and primitive people (e.g., ice ages), the modern crisis is largely man-made—a consequence of the failure of human insight and ingenuity to foresee and prevent the ill effects of human purpose and action. Yet this shortcoming of perception does not appear to be inherent in human mentality; more likely, it is evidence of unsubstantiated assumptions and beliefs in human culture. Some individuals have been able to foresee and forebear. The remedy for humanity's failure to assess the needs for continued life on earth lies in the exercise of the human capacity to observe, to learn, and to apply, with action and restraint appropriate to the circumstances.

The constructive achievements of civilization have demonstrated humankind's ability to learn from experience. The environmental movement described in this book is, from one viewpoint, a history of such social learning—a process joining experience, information, interpretation, communication, and action.[1] Learning occurs when there is congruence between what is believed in the mind and what actually happens on earth. People may recognize the danger to their own futures in soil erosion, deforestation, species extinction, and environmental pollution when "seeing is believing." When they do *not* see, they may not believe. Yet people nonetheless do believe things that the experience of life on earth does not confirm; perception may be "virtual reality."

Humanity's environmental relationships thus comprise a profound problem in human psychology. The problem, in essence, is the reconciliation of the human view of life on earth, expressed through culture, with nature's ultimate ground rules for continuing existence. It has been a problem addressed by anthropologists, behavioral scientists, philosophers, and theologians. It has now also become a problem for national policymakers, educators, and politicians. These political efforts to restrain and redirect human behavior toward a continuing coexistence with the rest of the biosphere provide the subject matter of this book. It is, in effect, a study in the history of social learning.

The growth of the environmental movement to international and global proportions has been a historical development, which contemporaries cannot easily evaluate. The development has been without clear precedent—but with antecedents, as are recounted in the chapters to follow. Understood in its full context, it may be seen by subsequent generations as a major change of state in human affairs—an awakening of modern society to a new awareness of the human predicament on earth.

Science has played an indispensable role in this enlarging of perception. The environmental movement has itself been powerfully affected by the consequences of science misused to the detriment of the living world, and even more importantly by what advancing science is revealing about the structure and processes of nature. These two effects of science upon society explain the emphasis throughout this book on the organization and undertaking of environment-related scientific inquiry. Traditional society could rely upon experience for policy guidance to a degree wholly unsafe today. It may be that most of humankind neither comprehends nor trusts science. Nevertheless, its findings and technological applications indirectly as well as directly permeate modern culture. Science clearly influences the policies of governments and international agreements. Its ability to produce substances and effects previously unknown on earth means that policy must now anticipate experience.

This need to foresee the probable consequences of human activities has generated new techniques for monitoring environmental change and for assessing the impact of present and proposed actions upon the natural and social (i.e., cultural) environment. Comparable analytic techniques have been developed for innovations directly affecting health and safety. In addition, an "earthwatch" has necessitated the formation of new linkages and relationships among scientific associations, national governments, and international and nongovernmental organizations. Few people, including scientists, are aware of the extent, diversity, and complexity of the emergent environmental science, as it is addressed especially through the International Council of Scientific Unions and the United Nations specialized agencies. In its degree of scope and complexity this global network of scientific investigation is something new in the world. There is little reason to doubt that it will continue to affect the future of international environmental policy. However, before describing the emergence of the environment as an issue for public policy, it is necessary to take a closer look at the term "environment" and its implications.

The Meaning of "Environment"

As commonly used, "environment" usually means surroundings. In fact, the term is more complex. In one of the better dictionary definitions, "environment" is "whatever encompasses; specifically the external and internal conditions affecting the existence, growth, and welfare of organisms."[2] Thus "environment" includes both that which environs and whatever is environed—in this book more precisely the living world or biosphere including the human species. Thus "environment" most accurately denotes the relationship between the environing and the environed. The more we learn about environmental relationships, the more complex and numerous they appear. The entities or forces that comprise an external environment do so only in relation to other entities and the forces on which they impact. Objects "out there" are environment only in relation to whatever they surround or environ.

A French physiologist, Claude Bernard, recognized the complex relational character of environment a century and a half ago when he distinguished the *milieu extérieur* outside the human body from the *milieu intérieur* comprising the body and its interior organs.[3] The human body, and all living things, are in continuing interaction with this exterior environment, which conditions their existence but which to some extent they modify. Indeed, James E. Lovelock in his book *Gaia: A New Look at Life on Earth* (1979) develops a hypothesis that the advent of life on earth has in fact modified the physical attributes of the planet, making it more hospitable for life. Clearly, "environment" should not be understood as "just those things out there" but the interactive totality that comprises the planet, its biosphere, the individual species and organisms that live in it, and the human habitat and infrastructure.

For millennia humans took their environmental relationships for granted, adapting to external change when necessary. The "discovery" of the planet and the envelope of life that surrounds it occurred in relatively recent times, notably through advances in technologies of measurement, navigation, and observation. These developments permitted the voyages of exploration and discovery undertaken by Western Europeans and contributed to the advancement of sciences descriptive of the earth. And so began the conscious effort of humans to understand their environment and very often to modify or attempt to modify it to suit human purposes. But comprehension of the environment is still incomplete. Modern society has not yet learned how to achieve a sustainable relationship with its external environment.

Discovering the Biosphere

The roughly five hundred years between the European discovery of America in 1492 and the landing of the Apollo XI astronauts on the moon will surely appear in retrospect as a distinct and decisive era in the history of man and the earth. In our times, this half-millennium is called modern—whatever name future eras may give it. The earth can never again be what it was when the era began, nor can prospects for the era to come be forecast by precedents that have given reliable predictions in the past.[4] Nevertheless, cause–effect circumstances that have occurred in the past may recur in the future. To this extent and in a limited way history may be instructive.

"In the twentieth century, man, for the first time in the history of the Earth, knew and embraced the whole biosphere, completing the geographic map of the planet Earth, and colonized its whole surface. Mankind became a single totality in the life of the Earth."[5] Thus the Russian scientist V. I. Vernadsky in 1938 summarized the end of a process of discovery that began at least five thousand years earlier when humans began to leave behind records of their impressions and descriptions of the natural world.

At the beginning of modern times, large areas of the world had no permanent human settlements. The major areas of human habitation were isolated and had developed distinctive cultures. Farming and herding relied largely upon the behavior of natural systems, modified only marginally by public works for water supply, flood control, and irrigation. Today large urban concentrations are absolutely dependent for survival on the continuous operation of artificial systems. Without a steady flow of electricity and fossil fuels, millions of present-day people could not exist. As the human population has grown, the world's peoples have become increasingly homogenized physically and culturally. Nearly all major "premodern" cultures have been extinguished or acculturated by the dominant civilization that we call "modern."

The modern age has been characterized not only by an explosive increase in human population but also of knowledge, especially in technology. Through technology, the impact per human individual upon the biosphere has increased exponentially, accelerating toward the end of the twentieth century. Distinctive among the many forms of human dominion is the nation-state, which has been the characteristic structure for extending human preemption of the earth. It was developed in Europe and accompanied the expansion of European peoples into the Americas, into South Africa and Australia, and across northern Asia to the Pacific Ocean.

The unifying and distinguishing work of this era has been the human preemption and discovery of the biosphere. This is a simple way of stating a complex paradox: the biosphere was occupied and its exploitation well advanced before its true nature—vulnerable and finite—was even vaguely perceived. Before A.D. 1500 knowledge of the nature of the earth or its relationship to the rest of the universe was very limited, and much of what was believed was wrong. By the end of the era, humanity had won an experiential knowledge of the earth and its place in space, and had gathered many clues as to its evolution over time.

The discovery of the biosphere in the latter half of the twentieth century has come none too soon for the survival of human civilization. By the late 1960s, it was becoming evident that the uncontrolled impact of human activity upon the biosphere could not long continue without endangering the basis of life itself. Although opinions differed about the imminence of danger and the prospects for avoiding it, few who read the evidence could discount the potential catastrophe foreshadowed by existing trends.

To understand the changes in attitudes and institutions required for the defense of the earth, it may be useful to trace the discovery of the biosphere as an evolved living system with tolerances and limitations that human exactions cannot exceed without risking or, in some cases, causing disaster.

Locating and Measuring the Earth

To the best of our knowledge, humans alone among the animals are aware of themselves in relation to their environment and able to make objective observations to discover their place in it. The search to locate themselves and guide their journeys has been a major factor in the development of many sciences and technologies. The list includes astronomy, geography, geology, navigation, astronautics, cartography, and surveying.

The first records of our extending knowledge of our environment are maps. The oldest ones known were found on Sumerian and Babylonian clay tablets in the Tigris-Euphrates Valley.[6] The Greeks seem to have been the earliest to try to ascertain the shape of the world: first they thought the earth was circular and, subsequently, an elliptical plane. Pythagoras (c. 532 B.C.) and Aristotle (384-322 B.C.) appear to have believed that the earth was a sphere, but the first globe that we know of appears to have been made by a Greek named Crates of Mallus in about 145 B.C. The science of geodesy may be said to have been founded by Eratosthenes of Alexandria (c. 276-194 B.C.), the first man known to have measured the size of the earth. Relatively accurate measurements were also made by the Arabs, but Eratosthenes' calculations were not substantially improved upon until A.D. 1615 when the Dutch scientist Willebrord Snell measured the earth by triangulation.

The ancient geographer, Ptolemy of Alexandria, author of a geography in eight books (c. A.D. 150), established the concepts of the world that prevailed into early modern times. Maps of the fifteenth and sixteenth centuries generally followed the Ptolemaic projections, and, in 1492 in Nuremberg, Martin Behaim constructed one of the first modern globes following the Ptolemaic concepts.

Toward the end of the Middle Ages, collections and translations of the astronomical and geographical works of Greek and Arab scholars were made available through the *Sphaerarmundi* by an Englishman named John of Holywood, writing under the name of Sacrobosca (c. 1250); and by the eighteen *Libros del saber de astronomía* compiled under the direction of the Spanish king, Alfonso X, el Sabio (c. 1284), an enormous work containing the translation into Spanish of all the contemporary theoretical knowledge of astronomy and descriptions of such scientific instruments as astrolabes, quadrants, armils, and clocks.[7]

The great European voyages of discovery during the latter half of the fifteenth century and after spurred the development of techniques of location and measurement. Successful navigation on the high seas depended on them, and between 1500 and 1600 the sciences of astronomical navigation and cartography developed rapidly. In 1569 the Flemish cosmographer Gerard Mercator published his famous projection, which was further refined by the English mathematician Edward Wright in 1599; it is still regarded by some as "the most useful map of the world in the practice of navigation."[8] Nevertheless, the problem of maritime navigation continued to be so great that in 1713 the English government appointed a special Commission for the Discovery of Longitude at Sea and offered substantial rewards for useful inventions.

The invention of the marine chronometer by John Harrison followed in 1735, and he was eventually (1773) awarded the equivalent of one hundred thousand dollars for his instrument.

The extended voyages of Christopher Columbus, Vasco da Gama, and Ferdinand Magellan began a process of discovering, describing, and mapping the surface of the earth that continued to the end of the nineteenth century. In 1891 a geographical conference in Berne received a proposal to construct an accurate map of the world on a uniform scale. A committee was appointed to pursue the project, and it reported consecutively to geographical congresses held in London in 1895, in Berlin in 1899, and in Washington in 1904. Finally a special conference convened in London in November 1909 managed to develop proposed standards for the map and rules to govern its production. These proposals were accepted by the geographical congress meeting in 1913. A general international conference, at which thirty-four nations were represented, was held in Paris of that year and accepted the project of the Carte du Monde au Millionième on a scale of approximately 1 inch to 15.8 miles.[9]

Exploration of the sea floors required further technology and instrumentation. The scientific voyage of H.M.S. *Challenger* around the world (1872–1876) provided the first comprehensive survey of the physical and biological conditions of the oceans.[10] Subsequently, the laying of submarine telegraph cables required accurate measurements of the ocean depths and added to knowledge of submarine conditions.

The discovery, description, and measurement of the surface of the earth were matched by advances in knowledge of the relationship of the earth to the cosmos. Modern astronomy began with the theories of Nicolaus Copernicus (1473-1543), who held that the planets together with the earth revolved in circular orbits around the sun. Advanced by the work of Galileo, Huygens, and Kepler among others, the dynamics of the universe were first satisfactorily formulated by Isaac Newton (1642-1727) in his *Philosophiae naturalis principia mathematica* (1687). Terrestrial mechanics as formulated by Newton and modified and refined by his successors remained a dominant explanation of the behavior of the universe until Albert Einstein advanced his theories of relativity in the early twentieth century. In the twentieth century, the measurement and description of the universe was extended to the galaxy, which encompasses the solar system to which earth belongs. But the scope of twentieth-century astronomy rapidly transcended the limits of our immediate galaxy to a presently incalculable, but seemingly immense, number of external galaxies or island universes, extending indefinitely into space. Exploration of deep space was advanced by the discovery of K. Jansky in 1931 that radio waves, apparently from the farthest reaches of the universe, can be received and interpreted on earth. Giant radio telescopes are now augmenting our knowledge of the universe.

Thus, by the latter third of the twentieth century, people were not only able to locate themselves at whatever point they might happen to be on the surface of the earth, but also to a degree undreamt of at the beginning of the modern era, they were able to locate the planet with some accuracy in the nearer regions of a universe of incomprehensible size. This process of discovery involved the interaction of virtually all of the physical sciences, discovery in any one of which was contingent upon discovery in others.

Before the mid-twentieth century, different geodetic systems were in use in a number of different countries. Since then, a world system of control points and coordinates has been devised with the assistance of the International Union of Geodesy and Geophysics and the United Nations. A major step toward understanding the physical earth was the International Geophysical Year (IGY), 1957-58, sponsored by the International Council of Scientific Unions and the World Meteorological Organization.[11] One consequence of the IGY was the Antarctic Treaty of 1959, which established the Antarctic continent as the world's first international scientific reserve and set a precedent for the UN Treaty on the Peaceful Uses of Outer Space.[12] Scientific exploration of the earth was greatly advanced by innovations in instrumentation, such as radar, sonar, and remote sensing by satellite, and the development of new methods of photogrammetry, which enormously increased the speed and accuracy of determining the configuration and topographical properties of the physical world, of its climate and oceanographic conditions, and of the extent and condition of its vegetation.

Indeed, in recent years instrumentation of all kinds, essential to obtaining an accurate picture of the earth in context, has increased extraordinarily: telescopes, refracting, reflecting, and radio; microscopes of several varieties, including the electronic microscope; cyclotrons and other accelerators in high-energy physics; chromatographs for the analysis of chemical substances; the instrumentation for space exploration—the list of scientific and technical developments could fill many pages. The very mass of the material and its frequently specialized character have made difficult its synthesis into any coherent picture, especially for the public at large. It has facilitated our exploitation of parts of Earth's resources, but has also fragmented knowledge that might have enabled us to act prudently towards Earth seen as a whole.

Interdependencies of the Living World

In 1807 the naturalist Alexander von Humboldt wrote, "In the great chain of causes and effects no thing and no activity should be regarded in isolation."[13] This interconnectedness of the living world had been recognized over the centuries, but not until the twentieth century did the terminology to designate the specific, systematic interconnections of the natural world come into general use. In 1867 Ernst Haeckel had put forward the word "ecology" to designate the study of living systems in relation to their environment, but, like "biosphere," it was slow to find common usage.[14] The term "ecosystem" does not appear to have been used commonly before an essay by A. G. Tansley published in 1935 in the journal *Ecology*.[15] The ecosystem has also been known by other names, notably, "biogeocoenose," especially in the Russian literatures. It means a definable or bounded system of complex and dynamic biological and physical relationships that vary greatly in size and complexity from the minute or simple to the very large and infinitely complex. The term "ecosphere" has been used to summarize the totality of living systems that envelop the earth and is synonymous with "biosphere."[16]

In the course of discovering the interdependencies of the living planet, organisms of which it was comprised were located and described. Taxonomy and systematics, the description and classification of species, were thus major concerns of biological science in the eighteenth and greater part of the nineteenth centuries, a work particularly associated with the name of the Swedish botanist, Carl von Linne (Linnaeus, 1707–78). Exploration of the continents and the seas and the collection of plant and animal specimens laid foundations for the geography of plants and animals and for more sophisticated understandings of habitat requirements and competition among species.

The distribution of plants and animals was discovered to be neither random nor static.[17] The reasons that a particular species was found to be where it was proved often to be complex. Spatial locations were found frequently to be related to biological dependencies, of which symbiosis, parasitism, and territoriality represented special cases. At any given time the network of interdependencies in the living world was found to be in a state of approximate, although dynamic, equilibrium. This homeostatic state was subject to change through forces acting not only in the physical environment external to organisms but through genetic changes in the organisms themselves. The consequences of this process of change were discovered to result in the evolution of the species, and theoretical mechanics of this process were described by Charles Darwin in 1859 in *The Origin of Species*, by Alfred Russell Wallace in 1870 in *Contributions to a Theory of Natural Selection*, and by the science of genetics after 1900.

The transplanting of species into areas in which they had not naturally occurred, if it did not fail, frequently had disruptive and calamitous results. With advances in the sciences it was learned that the homogenizing and impoverishing of the ecosystems of the earth was a measurable consequence of human interference with natural interdependencies. But nowhere in the expansion of populations, and especially of European populations in modern times, did an ecological awareness or an ecologically oriented policy guide the behavior of explorers, exploiters, and settlers. By mid-twentieth century, however, the disastrous record of untested and unguided human intervention had been

well documented, and there was a growing awareness of the dangers of unin-
formed disruptions of natural systems. Nevertheless, individual and institutionalized human behavior was slow to catch up with human understanding.

Less readily understood than the interconnectedness of things in space was
their interconnectedness in time.[18] The theory of evolution dealt with intervals of time far greater than the experience of any human individual and beyond the comprehension of most of them. Yet, in part because of the work of
Albert Einstein (1879–1955) showing the relativity of time and space, the significance of time in human affairs was changing. Past expectations in relation
to time were becoming less and less reliable as guides for expectations in the
future. Cultural change, based heavily upon innovation in science and technology, was accelerating throughout the nineteenth and twentieth centuries.
This artificial speeding-up of history not only contributed to a disruptive impact upon the natural world, but also created tensions and discontinuities in
personal lives and in society.

The biological and behavioral sciences identified chronological sequences,
periodicities, and interdependencies among organisms that must be respected
to avoid harmful consequences. For example, because of an almost universal
failure to appreciate the significance of the processes of exponential growth,
societies failed to take timely action to prevent the catastrophic explosion of
human population in the last half of the twentieth century. The necessities
for lead time, the inevitability of ramifications, and the effects of time lag had
become concepts essential to managing human behavior in relation to the
biosphere.

The lesson that future constraints and opportunities depend upon what is
done in the present is as old as the fable of the ant and the grasshopper. Before the mid-twentieth century this counsel of foresight and prudence was
not generally applied to the irreversible processes of life in all of its many aspects, but merely to the relatively narrow confines of economic policy. Similarly, the truism that the world is so made that no act occurs in isolation has
only slowly influenced the behavior of individuals or of societies. In 1969 it
was still necessary for Garrett Hardin to reiterate the aphorism, "You can't
do just one thing." [19]

The discovery of the interdependencies of the living world revealed, in
ever-sharpening profile, how anthropocentric and unecological modern behavior has been. There was a tacit assumption that culture had displaced
evolutionary biology, and that human history was totally disassociated from
that of other species. Such notions were challenged by writers like Morley
Roberts (*Biopolitics*, 1938), and by Robert Ardrey, whose book, *The Territorial Imperative* (1966), suggested that the principle of dominance of territory,
widely established among other living species, was also a powerful force in
human affairs, even though disguised by cultural rationalization.

There had been less reluctance to recognize relationships between the development of civilization and the climate, land forms, and resources of the physical world.[20] Environmental theories of history and human development figured prominently among historians and geographers, and indeed often had been pushed beyond the bounds of demonstrable evidence. Reaction against such overenthusiastic environmentalism often led others to unduly depreciate the influences of the physical environment.[21] A result was a lack of attention to the undercutting of the human life support system by overpopulation and by an expanding commercial and industrial civilization.

Curiously, the pinnacle of technological effort, space exploration, led in the 1960s to a new appreciation of the interdependencies of the biosphere. Space travel required technologists to devise a minimum personal artificial ecosystem, the spaceship, and to do this they were required to learn how people must accommodate to those interdependencies that they cannot change. The effort to discover, through space biology and medicine, what was required for human survival beyond the limits of the earth's biosphere, inevitably clarified and emphasized the conditions necessary to life on earth. So the voyages into space had an effect similar to that of the sea voyages of preceding centuries — they added cumulatively to the process of discovering the true nature of the earth and human interactions with it.

The discovery of the biosphere inevitably involved a process of self-discovery. As our profound and often destructive impact upon the earth became more obvious, the need to know more about human behavior became ever more evident. The seeming growth of aberrant behavior among individuals and societies strongly suggested that the human adjustments required by man-made changes in the modern world might be exceeding the ability of many individuals to accommodate them.[22]

A variety of physiological and sociological impacts resulting from the increasingly rapid changes have been summarized by Alvin Toffler in the term "future shock."[23] The factors that cause future shock also often seem to be, upon examination, factors in environmental degradation and pollution, and in the impairment of human health through environmental stress and the contamination of the atmosphere, water, and food chains. The acceleration of history that was altering the face of the earth and the human condition was also bringing us face to face with the parameters of our own existence and the question of the survival of the natural world.

Limitations of the Living World

In 1913 a Harvard University biochemist, Lawrence J. Henderson, published a book, *The Fitness of the Environment*,[24] which undertook to explore the full range of physicochemical conditions under which living matter could exist. It examined the inorganic basis of life and the planetary conditions permitting genesis of life. Henderson showed that the requirements of living matter are conversely, its limitations. Some of Henderson's analysis has been modified as a result of advances in physical chemistry, but the book did give a concise and systematic formulation to an old and continuing question. In the concluding chapter Henderson observed:

> A half century has passed since Darwin wrote "The Origin of the Species," and once again, but with a new aspect, the relation between life and the environment presents itself as an unexplained phenomenon. The problem is now far different from what it was before, for adaptation has won a secure position among the greatest of natural processes, a position from which we may suppose it is certainly never to be dislodged; natural selection is its instrument, even if, as many think, not the only one. Yet natural selection does not mold the organism; the environment it changes only secondarily, without truly altering the primary quality of environmental fitness. This latter component of fitness, antecedent to adaptations, a natural result of the properties of matter and the characteristics of energy in the course of cosmic evolution, is as yet no wise accounted for. . . .
>
> There is, in truth, not one chance in countless millions of millions that the many unique properties of carbon, hydrogen, and oxygen, and especially of their stable compounds water and carbonic acid, which chiefly make up the atmosphere of a new planet, should simultaneously occur in the three elements otherwise than through the operation of a natural law which somehow connects them together. There is no greater probability that these unique properties should be without due cause uniquely favorable to the organic mechanism. There are no mere accidents; an explanation is to seek. It must be admitted, however, that no explanation is at hand.[25]

Henderson's studies pushed him toward a philosophical conclusion very similar to the later evolutionary interpretation by Pierre Teilhard de Chardin in *The Phenomena of Man* (1955).[26] In the concluding paragraph of his study Henderson wrote, "The properties of matter and the course of cosmic evolution are now seen to be intimately related to the structure of the living being and to its activities; they become, therefore, far more important in biology than has been previously suspected. For the whole evolutionary process, both cosmic and organic, is a unity and the biologist may now rightly regard the universe in its very essence as biocentric."[27] Is it also teleological, does it evolve toward some predictable state, as Teilhard appears to have believed? Prevailing opinion in the sciences does not accept a teleological explanation of evolution as necessary or demonstrable. Nevertheless, there are aspects of evolution and ecology that strongly suggest purposiveness. Perhaps our concepts are as yet insufficiently refined to handle the question of teleology in biological processes.[28]

Thirty-five years after first publication of *The Fitness of the Environment*, Harold S. Blum explored the relationships between the second law of thermodynamics and organic evolution in *Time's Arrow and Evolution* (1951),[29] drawing heavily upon Henderson's thesis. In his concluding observations Blum reaffirmed in general Henderson's position that the evolution and present condition of life as it has developed on the earth is an extremely improbable phenomenon, although the enormous size of the universe permits the statistical probability of other planets capable of sustaining life. Blum concluded, "If we think, however, of the delicate balance of conditions our earth enjoys, and to what extent chance has entered repeatedly into biological evolution, it seems that the probability of evolving a series of living organisms closely resembling those we know on earth may be a relatively small number. This becomes poignantly evident when we think of all the chance events concerning the evolution of the human brain—which occurred only once on our planet."[30]

Even more true today than when it was written is Blum's observation:

> And perhaps for this reason alone, this life-stuff is something to be cherished as our proper heritage. To be guarded from destruction by, say the activity of man, a species of living system that has risen to power and dominance through the development of a certain special property, intelligence. Such a development—vastly exceeding that of any other species—has apparently given this particular system the ability to determine its own destiny to a certain extent. Yet at the moment there are all too many signs that man lacks the ability to exercise the control over his own activities that may be necessary for survival.[31]

The very success of modern society in modifying natural environments, in rearranging and augmenting them for human purposes, has encouraged the euphoric assumption that there are no insurmountable barriers to our ability to transcend the limits of our environment. In his book, *The Human Use of the Earth* (1960), Philip L. Wagner observes, "There are necessary natural limitations to the security, stability, and success of artificial environments." The survival of all artificial systems depends upon the stability and the reliability of the biosphere. But as Wagner observes, "Societal limitations likewise assert themselves."[32] Human disruptions of artificial systems can be as destructive to human life and welfare as natural hazards. Economic depressions, epidemic diseases, and political revolutions can impede essential operations of an artificial system and thus endanger the welfare and survival of a city, a nation, or indeed of humanity. Finally, Wagner points to technical limitations on these artificial systems.

Modern industrial societies, for example, are heavily dependent upon a continuing supply of electrical energy and of water. Disruption of the man-made and maintained systems not only stops the functioning of the systems as a whole, but indirectly affects all other systems upon which the continuing supply of water or electrical energy depends. No great concentrations of population in modern cities could long survive the breakdown of their basic life-support systems. Thus the need to maintain the artificial systems imposes a rigorous limitation on the priorities and actions of governments and peoples for whose support unmanaged natural systems, of water for example, are no longer adequate. This constraining necessity now includes the overwhelming majority of people in the industrialized countries of the modern world.

Thus, as we have seen, the requirements of the living world are a reverse way of expressing the limitations of the living world. There are limits beyond which societies cannot go in altering their relationships with the natural world. But in many areas of life the location of the limits is uncertain. Controversies over the use of nuclear energy, of supersonic transport, or of genetics, fertilizers, and pesticides to increase farm crops illustrate a very large number of public questions in which contention over public policy has developed because the outcome of the proposed action has not been certain. Where irreversible consequences may be involved or where the costs of reversibility are so great that it is doubtful that society will pay the price, the course of wisdom would appear to be one of caution and the exploration of alternatives.

Arthur C. Clarke's precepts for ecological prudence are of continuing relevance.[33] They are (1) Do not attempt the unforeseeable, and (2) Do not commit the irrevocable. Regard for the parameters of human existence has never been strong in the dominating philosophies of Western society. If the West, along with all the peoples and nations of the earth, is to take effective measures to preserve humanity's life-support base in the biosphere, some means must be found to include the concept of limitations among the assumptions, attitudes, and institutions of modern people. Far from restraining human energy and creativity, the development of a realistic assessment of the costs and dangers of transgressing the limitations of the living world would further advance applied science and technology toward improving human health, welfare, and happiness.

The Biosphere: Object of Policy

Two developments were needed before an international environmental movement could occur. First, environmental policy had to become legitimized at the national level: governments had to be persuaded to take cognizance of environmental issues as regular and official concerns. Nations cannot easily cooperate collectively upon matters with which they have had no experience individually—national policy has to anticipate, if it does not actually precede, international action. Second, an integrative concept of universal applicability had to emerge. The traditional conservation concept could not fully meet this requirement because its focus was seldom enlarged beyond national needs and economic considerations. Its concern historically had been with resources, defined by human culture and technology in relation to national interest, rather than with conservation of the global environment. Before a comprehensive international environmental movement could occur, the interactive, life-sustaining processes of the biosphere had to be perceived as a concern common to all humankind.

Awareness of the need for a public policy toward the biosphere has grown gradually as the concept of the biosphere itself has been built up by successive enlargements of scientific knowledge.[34] Edward Suess (1831–1914), an Austrian geologist, has been credited with first use of the term "biosphere" in the concluding chapter of a small book entitled *Die Entstehung der Alpen* (1875); there he introduced the "biosphere" in a description of the concentric layers enveloping the earth. The development of the term in modern scientific thought is particularly associated with the work of the Russian mineralogist and forerunner of modern biogeochemistry, V. I. Vernadsky (1863–1945). Although Vernadsky attributed the concept to the French scientist Lamarck (1744–1829), the term gained currency through the publication of Vernadsky's book, *Biosfera*, in Leningrad in 1926, and a French edition, *La Biosphère*, published in Paris in 1929.[35]

Vernadsky's ideas appear to have grown out of his study of biogeochemical phenomena, maturing during the period of World War I and first outlined in 1922–23 in lectures at the Sorbonne in Paris. The essential elements of Vernadsky's concept of the biosphere have been stated in an abstract of a paper published in translation in 1945 in the *American Scientist*.[36] Vernadsky distinguished the biosphere as the area or domain of life, a region where the prevailing conditions are such that incoming solar radiation can produce the geochemical changes necessary for life to occur. The biosphere includes the whole atmospheric troposphere, the hydrosphere or oceans, and the lithosphere—a thin layer in the continental regions extending three kilometers or more below the surface of the earth.

Vernadsky identified sixteen propositions distinguishing living from inert material. Among them are the following: The processes of living natural bodies are not reversible in time, and the vast majority of living organisms change their forms in the process of evolution. The processes of changing growth in living matter tend to increase the free energy of the biosphere, and the number of chemical compounds produced by living organisms probably reaches many millions, whereas the number of different kinds of chemical compounds in inert bodies is limited to a few thousand. New living natural bodies are born only from preexisting ones, and have a common nature in their cellular morphology, substance, and reproductive capacity. All living matter is, therefore, ultimately genetically connected throughout the course of geologic time. Living organisms arise and exist only in the biosphere, and only as discrete bodies; no entry of life into the biosphere from cosmic space has ever been observed.

These propositions were not novel, but upon them Vernadsky based a summation which had never before been so concisely and pointedly stated:

> In everyday life one used to speak of man as an individual, living and moving freely about our planet, freely building up his history. Until recently the historians and the students of the humanities, and to a certain extent even the biologists, consciously failed to reckon with the natural laws of the biosphere, the only terrestrial envelope where life can exist. Basically man cannot be separated from it; it is only now that this indissolubility begins to appear clearly and in precise terms before us. He is geologically connected with its material and energetic structure. Actually no living organism exists on earth in a state of freedom. All organisms are connected indissolubly and uninterruptedly, first of all through nutrition and respiration, with the circumambient material and energetic medium. Outside it they cannot exist in a natural condition.[37]

Apparently influenced by the French mathematician and Bergsonian philosopher Édouard LeRoy (1870–1954) and the French Jesuit, geologist, and paleontologist Pierre Teilhard de Chardin (1881–1955), Vernadsky adopted the concept of the noosphere as the state toward which the biosphere is now evolving. The noosphere, or realm of thought, declared Vernadsky, "is a new geological phenomenon on our planet. In it for the first time man becomes a large-scale geologic force. He can and must rebuild the providence of his life by his work and thought, rebuild it radically in comparison with the past."[38]

As people propagate the noosphere, they extend and transform the biosphere. Vernadsky observed prophetically that:

> Chemically, the face of our planet, the biosphere, is being sharply changed by man, consciously and even more so, unconsciously. The aerial envelope of the land as well as all its natural waters are changed both physically and chemically by man. In the twentieth century, as a result of the growth of human civilization, the seas and the parts of the oceans closest to shore become changed more and more markedly. Man now must take more and more measures to preserve for future generations the wealth of the seas which so far have belonged to nobody. Besides this, new species and races of animals and plants are being created by man. Fairy tale dreams appear possible in the future: man is striving to emerge

beyond the boundaries of his planet into cosmic space. And he probably will do so.[39]

Vernadsky followed his colleague the geologist A. P. Pavlov (1854-1929) in saying that we had entered the *anthropogenic era of geologic time*, "that man, under our very eyes, is becoming a mighty and ever-growing geological force." Nevertheless, individuals were bound by a seemingly infinite number of ties to the biosphere and were, except as their existence was modifiable and modified by their thought and effort, subject to its physical limitations. Within the parameters of the natural world, their mere presence, as well as their deliberate and inadvertent impact, transforms its properties and conditions. There thus arises, as Vernadsky concluded, "the problem of the *reconstruction of the biosphere in the interest of freely thinking humanity as a single totality*." The problem was how to adapt the biosphere to people's needs and desires without impairing its viability. In discovering the nature of the biosphere humanity was creating and simultaneously discovering the noosphere, which, declared Vernadsky, is "this new state of the biosphere, which we approach without our noticing it."[40]

Although Vernadsky developed the scientific concept of the biosphere, and posed the problem which makes it an object of policy, popular awareness of the biosphere may owe more to the writings of Pierre Teilhard de Chardin, whose principal work, *Le Phénomène Humain* [*The Phenomenon of Man*] (1938) took a poetic and metaphysical approach to the concept of the biosphere, which is evident in the following passage describing the fundamental unity of living matter:

> However tenuous it was, the first veil of organised matter spread over the earth could neither have established nor maintained itself without some network of influences and exchanges which made it a biologically *cohesive* whole. From its origin, the cellular nebula necessarily represented, despite its internal multiplicity, a sort of diffuse super-organism. Not merely a *foam of lives* but, to a certain extent, itself a *living film*. A simple reappearance, after all, in more advanced form and on a higher level of those much older conditions which we have already seen presiding over the birth and equilibrium of the first polymerised substances on the surface of the early earth. A simple prelude, too, to the much more advanced evolutionary solidarity, so marked in the higher forms of life, whose existence obliges us increasingly to admit the strictly organic nature of the links which unite them in a single whole at the heart of the biosphere.[41]

This concept of the biosphere as a self-directing entity has been developed further by the British scientist James E. Lovelock in his book *Gaia: A New Look at Life on Earth*. Lovelock takes the concept of the biosphere as a "diffuse super-organism" and makes explicit the way in which "the Earth's living matter, air, oceans, and land surface form a complex system which . . . has the capacity to keep our planet a fit place for life."[42] Gaia, the Greek term for Mother Earth, is used by Lovelock as shorthand for the hypothesis "that the biosphere is a self-regulating entity with the capacity to keep our planet healthy by controlling the chemical and physical environment." Gaia, like the name of a ship, is merely a convenient way of indicating that the materials with which it was made "when specifically designed and assembled may achieve a composite identity with its own characteristic signature, as distinct from being the mere sum of its parts."

The relevance of the foregoing theories of the biosphere to environmental policy may be summarized as follows. First, the biosphere concept, like the theory of organic evolution, continues to be developed, elaborated, and modified by scientific investigation. Second, all interpretations stress the interrelatedness of the elements of the biosphere, forming a single, vast biogeochemical system. Third, all theories find humankind to be a consciously altering and directing agent in the biosphere. And fourth, the extent of humanity's innate ability to reconstruct the biosphere and the ultimate consequences of human impact are still areas of uncertainty; the corresponding extent of the resisting, recuperative, and regenerative powers of Gaia remain largely unknown. Initial skepticism among scientists regarding the Gaia hypothesis have been to some extent modified by Lovelock's further explications and by additional scientific inquiry.[43] The International Geosphere-Biosphere Programme, initiated in 1986 by the International Council of Scientific Unions, appears to have been influenced, at least in part, by a desire to test elements of the Gaia proposition. A catalytic factor in this international scientific effort was the development of forecasts of global climate change attributed to human activities.

To resolve the challenge that Vernadsky posed regarding man's transformation of the biosphere, the biosphere must first become a subject of social concern and then an object of public policy. The first public recognition of the biosphere as an object of international policy came with the resolution adopted in November 1966 by the General Conference of UNESCO at its fourteenth session. Pursuant to this resolution, an Inter-governmental Conference of Experts on the Scientific Basis for Rational Use and Conservation of the Resources of the Biosphere was convened in Paris, September 4–13, 1968. This gathering, known for convenience as the Biosphere Conference, was organized by UNESCO with assistance from the United Nations, the Food and Agriculture Organization, the World Health Organization, the International Union for the Conservation of Nature and Natural Resources, and the International Biological Programme. The significance of this conference and its recommendations for institutional development will be considered in subsequent chapters of this volume. Here the conference is mentioned as evidence of an awareness of the loss of environmental quality that had been occurring throughout the world.

Evolving Concepts

We have seen that the discovery of the biosphere was a process of conceptual growth. Throughout the first five thousand years of human history our knowledge of where we were in the cosmos and the true nature of our relationships with our environment remained substantially unchanged. There were, of course, cultural differences, instances of perceptive insight by philosophers, and a remarkably accurate measurement of the circumference of the earth by Eratosthenes. Yet at the time of the European discovery of America, our knowledge of the physical world had advanced little beyond that extant at the time of the Alexandrian mathematician and geographer, Ptolemy.

As Buckminster Fuller has pointed out,[44] the discovery of the true nature of the existential world has been a product of physical exploration and scientific investigation, of the mapping of the land masses, and of navigation on the sea, in the air, and finally of outer space beyond the earth. The great age of geographic discovery has lasted nearly five hundred years and may be nearing its close. There is much yet to be learned about the earth, but much of that knowledge would appear to be incremental to what is already known. Through unmanned spacecraft the mysteries of the other planets are being probed; but what will be learned is unlikely fundamentally to change our present understanding of humanity's relationship to the biosphere.

Our knowledge of the biophysical world must ultimately be tested, confirmed, and systematized through science. The electron microscope, the seismograph, the spectroscope, the radiotelescope, the gas chromatograph, and many other instruments have supplemented direct observation in discovering the natural parameters of human existence. Yet this age of discovery and of science has thus far occupied less than ten percent of a human history that extends backward in time for at least five millennia. Humanity's ethical systems, religions, and many of its political ideas were born and developed in this prescientific age. The rich heritage of the past has been carried into the modern era, but like all wealth, this legacy that we call civilization brings with it burdens. Among the burdens of the past are those ways of perceiving and dealing with the natural world that are fundamentally inconsistent with the possibility of sustaining long-term relationships with it.

It may seem strange that so radical a departure from historical views of man-environment relationships as are implicit in our present knowledge of the biosphere should not have had more profound effects upon human attitudes and institutions. Perhaps it is our long and intimate association with the physical world that has caused us to be indifferent to it or negligent toward it. Knowledge concerning the biosphere has been the product of findings and theories in nearly every science. Scientific theories regarding the biosphere, supported by sufficient evidence, may in time be accepted as conventional wisdom and thereafter be taken for granted. But the discovery of the biosphere, unlike the splitting of the atom, has not been a dramatic event that can be located precisely in time. Perhaps that is why many intelligent individuals cannot understand how the incremental accumulation of knowledge about the biosphere may add up to an interpretation of humanity's environmental relationships that holds absolutely fundamental and revolutionary implications for the future of human behavior patterns and institutions.

As one may not see a forest for the trees, very large ideas may be lost in the specific arguments of which they are composed. The human mind, moreover, tends to see what it wants to see or what it has been trained to see. The notion of inexorable limitations has not been congenial to modern societies. Even when the finiteness of the world is admitted intellectually, the concept will often be rejected emotionally. The traditionalist will feel that God will intervene somehow, and the unimpressed scientist deep in his own speciality will merely ask, "So what?"

But as more is learned about the grand-scale cycles of energy and chemical processes that sustain the biosphere, processes that people can disrupt but not control, fewer skeptical questions are voiced. A more pertinent question has been asked by George Woodwell: "How much of the energy that runs the biosphere can be diverted to the support of a single species: man?"[45] The answer to that question will be found, with or without searching for it, in the years ahead. But it is not sufficient that scientists alone understand the requirements of a self-regenerative biosphere—there must be a general recognition that there are requirements that societies cannot modify.

Beliefs, attitudes, and behaviors are in some measure the product of experience. If our behavior and institutions are to be reorganized consistent with the requirements of the biosphere, human experience must somehow be modified to make those requirements a part of society's operating assumptions. Something must be added to human experience that did not exist in the prescientific past and is consistent with what is known today about the natural world, something that confirms this knowledge with great emotional power. Knowledge to support such a change of attitude exists and is growing, but a dramatic event may be needed to symbolize the new concept.

Such an event may well have already occurred. The astronauts of Apollo were the first men to ever see the earth whole. The emotional impact of their photographs of the blue earth, alone in the blackness of space, was obviously very great and was shared by millions throughout the world. How this view of the earth from outer space, like a fragile glass sphere, will affect the future behavior of peoples and governments is uncertain, but it represents a powerful and moving event that had not occurred before in all human experience. The symbolic power of the image is suggested by its extensive use on the covers of books, on posters, in advertising, and on postage stamps.

When one considers that it is only in the present generation that people have seen the biosphere objectified in space, and that a body of knowledge and ideas has been brought together in its name, we may have some hope that human beings will behave differently toward it in the future than they have in the past. Upon the fulfillment of this hope, the future of human society and perhaps of life itself depends.

2
Growth of International Concern

How did it happen that in the latter half of the twentieth century a mass movement for environmental protection, remediation, and preservation arose in
several countries, converged, and spread worldwide? Several factors contributed to this international convergence of concern. There were extraordinary
advances in science, new and powerful technologies, expansion and globalization of information, communication, and transportation and unprecedented
economic growth in the so-called developed or industrialized world. And the
growth of affluence was accompanied for many people by greater leisure and
more opportunity to experience the world around them. A consequence was
demand for a higher standard of environmental quality and an awareness of
environmental deterioration.

National environmental concern first appeared in the United States, United
Kingdom, Canada, France, Germany, Switzerland, the Netherlands, and
Scandinavia. But before a popular movement could become international, it
first had to become national. National governments and many if not most of
their people, had to see an issue of concern that could only be addressed effectively through cooperation with another country or countries. But national
responses to these issues were compromised by the doctrine of national sovereignty and narrow perceptions of national interest.

Since modern nations emerged, their governments have sought to satisfy
mutual needs or interests through formal agreements, commonly in the form
of treaties. As noted elsewhere in this volume, boundary rivers and lakes, migratory animals, and protection from the spread of plant and animal disease
were early subjects for international negotiation. While military and commercial treaties were more numerous and generally easier to enforce, agreements
regarding natural resources and environment usually imposed obligations or
restrictions on action within the territorial limits of the parties, and were thus
seen as modifications of or limitations to the doctrine of national sovereignty.
One might argue that in relinquishing certain freedoms in the use of resources
and the environment, a nation lost part of its sovereignty. But unless a nation
had control over its resources and environment, it would not be competent

to enter into negotiations regarding them. This difficulty was avoided in the concept of "merged sovereignty," wherein nations combined their respective rights of unilateral control to obtain advantage that could not be obtained, or attained as effectively, through mutually exclusive unilateral action.

International environmental concerns spread with the growth of cross-boundary transactions and effects, such as commerce and the migration of people across national boundaries, which have often had both direct and indirect environmental consequences. Commerce entails exchanges of information and ideas in addition to goods and services. The North American Free Trade Agreement (NAFTA) could not have been ratified in the Congress of the United States without a paralleling environmental agreement between Canada, the United States, and Mexico. Similarly, the West European quest for economic integration has resulted in extensive environmental controls proposed and overseen by the Commission of the European Union.

International cooperation is also required to control or suppress air and water pollution. Such cooperation requires effective control over environmentally damaging practices in the countries in which the pollution originates. The remedial or preventative costs fall chiefly on the polluting countries. But the benefits are enjoyed by countries freed from external contamination—for example, eastern Canada should no longer receive as much acidic deposition from coal-fired plants in the United States nor Scandinavia as much acidic fallout from Germany and the United Kingdom. This is known as the polluter-pays principle. Some problems have more diffuse origins. Thinning of the stratospheric ozone layer that shields the earth from harmful solar radiation has been caused primarily by chlorofluorocarbon emissions from the developed industrial countries. Similarly, increasing atmospheric levels of carbon dioxide and other compounds that are emitted chiefly in the developed countries could cause a rise in the earth's temperature and melting of the polar icecaps with resulting rising sea levels. All nations would benefit by prevention of these effects, but the cooperation of all nations is required to obtain this protection.

Institutionalizing International Cooperation

For international cooperation to be effective, some formal commitment is necessary. The intent to cooperate needs to be institutionalized—to be given a material expression. In international affairs, institutionalization is usually based on a treaty or less formal written agreement. For major and continuing cooperative programs, such as international environmental programs, the intent to cooperate is institutionalized in a variety of ways, all designed to provide implementation of agreements intended for continuity and relative permanence.

Significant steps in the development of international cooperation were taken after World War II with the establishment of the UN Educational, Scientific, and Cultural Organization (UNESCO) in 1945, the International Union for Conservation of Nature and Natural Resources (IUCN) in 1948, and the Biosphere Conference of 1968.[1] Then, in 1972, the United Nations Conference on the Human Environment extended international consideration to the social, political, and economic factors that influence man-environment relationships, a move that led to the establishment of international institutions with special responsibility for these matters. It may be premature to state that nations now generally recognize the need for protective management of human impact upon the biosphere, but practical necessities and cumulating scientific evidence are forcing often reluctant governments toward environmental cooperation.

The Biosphere Conference of 1968, as indicated in Chapter 1, appears to have been the first major international meeting concerned with the global environment, as distinguished from natural resources. Earlier congresses and conferences concerned themselves with specific physical sciences and biological issues and ignored ecology.[2] Two direct outcomes of the Biosphere Conference were Man and the Biosphere (MAB) (an international cooperative program) and the designation of biosphere reserves described in Chapter 11.

Summarizing some recurring themes of the conference, the final report declared that "although some of the changes in the environment have been taking place for decades or longer, they seem to have reached a threshold of criticalness, as in the case of air, soil, and water pollution in industrial countries: these problems are now producing concern and a popular demand for correction."[3]

"Parallel with this concern," the report continued, "is the realization that ways of developing and using natural resources must be changed from single purpose efforts, both public and private, with little regard for attendant consequences, to other uses of resources and wider social goals." The report emphasized that human exploitation of the earth "must give way to recognition that the biosphere is a system all of which is widely affected by action on any part of it." The problem, the report declared, was not one for science alone. A further consequence of a "new awareness that man is a key factor in the biosphere" was that "natural science and technology alone are inadequate for modern solutions to resource management problems; one must also consider social sciences, in particular politics and public administration, economics, law, sociology and psychology, for, after all, it is resources as considered by man with which we are concerned."

In the Biosphere Conference, the traditional and conventional concept of resources coexisted with the emergent idea of the biosphere: the conference by definition dealt with "Resources of the Biosphere." But not all assumptions of the older economistic and newer ecologic paradigms are necessarily mutually exclusive, and the Biosphere Conference emphasized the importance of their harmonization. It initiated an international effort toward a new policy synthesis.

From 1968 onward, the concept of the biosphere as an evolved, integrated, planetary, life-supporting system was implicit even when not explicitly stated in declarations of international environmental policy, such as those issued by the experts assembled by the United Nations Associations of the Nordic Countries in December 1969 and the "Declaration on the Management of the Natural Environment of Europe" by the European Conservation Conference in 1970. As of 1 January 1970, the term "biosphere" was incorporated into the public laws of the United States in the Preamble of the National Environmental Policy Act of 1969, which declared that it is a part of national policy to "promote efforts which will prevent or eliminate damage to the environment and the biosphere."[4]

In an article summarizing the findings of an international panel of experts assembled for the Biosphere Conference, René Dubos described the role of man in reshaping the biosphere through the extension of the noosphere:

> Planning the future demands an ecological attitude based on the assumption that man will continuously bring about evolutionary changes through his creative potentialities. The constant interaction between man and the environment inevitably implies continuous alterations of both—alterations which should always remain within the constraints imposed by the laws of Nature and the unchangeable biological and mental characteristics of man's nature.[5]

National Basis of Cooperation

Although it is not possible to recount here the national movements for conservation and environmental protection that preceded international action, their indispensable role as precursors of international cooperation must be acknowledged. The delegates who went to Stockholm in 1972 and to Rio de Janeiro in 1992 did not do so wholly on the initiative of their governments. There were the official delegates of national governments and the unofficial delegates representing nongovernmental organizations attending NGO forums. Especially in the leading industrial democracies—Canada, France, Japan, Sweden, Germany, the United Kingdom, and the United States—organized and active citizen groups had pressured their governments for legislative and administrative action to protect and restore the quality of environment. Many of these groups communicated with one another across national borders and formed, or participated in, international organizations.

During the 1960s a crescendo of public concern, first expressed in the popular press, culminated by the end of the decade in national environmental laws and policies. Warnings of adverse consequences from misused technology, overstressed ecosystems, wasted resources, and contaminated environments had been current for several decades, but they had little visible impact upon public attitudes or government policies. In the early sixties, however, a number of books appeared that had an immediate and obvious influence on public opinion. The best known of these was Rachel Carson's *Silent Spring* (1962), which aroused public apprehension over the new chemical pesticide technology and prompted investigation at the highest political levels in the United States and several other countries. Representative of a larger literature were Stewart Udall's *The Quiet Crisis* (United States, 1963), Jean Dorst's *Before Nature Dies* (France, 1965), Rolf Edberg's *On the Shred of a Cloud* (Sweden, 1966), and Max Nicholson's *The Environmental Revolution* (United Kingdom, 1970). These and other writings developing similar themes found a public receptivity not previously evident. The authors were corroborating the direct experience by increasing numbers of people of deteriorating environmental conditions.

In some respects environmental conditions were better than they had been a generation earlier: city streets were cleaner, water-borne disease less common, bituminous coal smoke banned or reduced. But the new dangers exemplified by DDT, atomic radiation, photochemical smog, and massive destruction of natural ecosystems posed more serious threats to the environment. Once the biosphere concept was accepted, human impact upon the environment took on new significance. Whereas an earlier generation had worried about the health effects of coal smoke in the air over London and Pittsburgh, informed people in the 1960s became concerned about burning coal for energy and the resulting acidic fallout from stack emissions and changes in the carbon dioxide balance in the earth's atmosphere. The possible consequences of irreversible changes in world climate and ecosystems created new problems for international environmental policymaking.

In rapid succession, during the 1960s, enlightened public environmental awareness expanded from local to national and from national to international and worldwide horizons. For many issues it became clear that national action alone could not successfully address national environmental problems. For example, African states could not protect endangered wildlife if lucrative markets for skins and trophies were open in Europe, America, and the Far East; the geographically concentrated industrial states of Western Europe could not unilaterally prevent transboundary air and water pollution; no nation alone could protect itself against the fallout from the atmospheric testing of atomic devices or prevent the chemical contamination or depletion of fisheries in the oceans and regional seas.

Belief in the possibilities of worldwide disaster became common, although far from universal. Studies like the *Limits to Growth* report to the Club of

Rome (1972) reinforced the opinions of those who believed that major re-structuring of industrial economies was necessary to prevent their ultimate collapse. These intimations of disaster were initially viewed with skepticism, not only by economists and politicians in the industrial world but also by political elites in so-called Third World or developing countries. Third World presence at Stockholm in 1972 was prompted as much by fear that develop-ment aid might be sacrificed to ecological values as by a genuine concern for biospheric environmental protection. Yet the germ of environmental concern was carried home by some delegates to Stockholm (and to related interna-tional gatherings) and began to spread among the better-educated members of Third World societies.

Throughout the balance of this book it should be assumed that articulate environmental movements at national levels are pressing governments to act. It would certainly be relevant to relate the international policy positions of various national governments to their domestic policies and circumstances. To do this, however, would require another book. In a few instances, such as the attitude of Brazil toward ecosystem preservation or the United States position on the Law of the Sea treaty, explanation seems necessary. But for the most part attention will be focused upon what nations have done—with less attention given to why they have done it.

Early Cooperative Efforts

The earliest conventional forms of international environment-related co-operation were established to clarify jurisdiction and manage the uses of international waters, especially boundary lakes and rivers. In time, interna-tional river basin commissions took on environmental protection responsi-bilities and were the subject of extensive recommendations at the UN Con-ference at Stockholm. The more successful transboundary institutions, such as the International Joint Commission of Canada and the United States, have afforded examples of cooperative arrangements for environmental purposes.

Early efforts toward international cooperation on environmental matters concerned an aspect of the biosphere visibly threatened by the growth of human populations and industrialism—wild nature, particularly migratory wildlife. Migratory birds and animals were an early focus of attention because no nation alone could ensure their protection. National legislation being insufficient, international treaties were invoked to extend protection across national boundaries so that protective measures might correspond to the mi-gratory range of species. The national basis for international cooperation lay in the inability of national governments to protect environmental resources unilaterally beyond their jurisdictional boundaries, resources that were of concern to their people.

Migratory Wildlife

With few exceptions migratory wildlife was the focus of treaties that dealt explicitly with nature protection, negotiated prior to the 1954 London Convention on Oil Pollution. Moreover, these treaties were largely concerned with the direct killing of birds and game animals for sport or commerce, and afforded them little or no protection from indirect threats such as environmental pollution or loss of habitat. The sovereign signatories were careful to avoid committing themselves to undertake any international obligation to establish sanctuaries, reserves, or national parks within their own territories, although sometimes they recognized the need for protection of habitat. The language of the treaties did not go beyond such guarded phrases as "explore the possibility of," and "give consideration to," or even more cautiously to "consider the possibility of establishing."

With the exception of the Fur Seal Convention of 1911, these earlier treaties appear to have been indifferently effective. They were important as commitments of national governments to the idea of the conservation of nature. They were expressive of the emergence of a new public attitude toward nature and the human environment, but the extent to which they retarded the general attrition of environmental quality and of the earth's wildlife was very limited. In almost every instance of early international protective efforts, protracted, complex, and desultory negotiations were required before even minimal agreements could be formalized. And these frequently were qualified by reservations that reduced their already limited effectiveness.

International cooperation resulted initially from efforts by groups of private citizens to persuade their governments jointly to protect certain migratory and endangered species whose ranges overreached national boundaries.[6] A case in point is the European Convention Concerning the Conservation of Birds Useful to Agriculture. Signed by eleven nations at Paris on 10 March 1902, this treaty culminated an effort of almost thirty-five years, beginning with a declaration by an assembly of German farmers and foresters in 1868 addressed to the Foreign Office of Austria-Hungary requesting its aid in obtaining an international treaty for the protection of birds and animals useful to agriculture and forestry. During the following year Italy, Switzerland, and France responded favorably to the idea, which at that stage was to establish a net of bilateral agreements.

In 1872 the Swiss proposed an international regulatory commission, but other sovereign states of Europe were not interested.[7] The issue was discussed at International Ornithological Congresses in 1873, 1891, and 1900, and an International Ornithological Committee that formulated a treaty proposal was set up at Vienna in 1884. In 1887 a joint Austro-Italian declaration committed the respective governments to "strict and comprehensive legislation" for bird protection, but the declaration was not implemented. It did, however, afford a basis for discussion at conferences in Paris in 1895 and again in 1902, at which time the international convention was finally drafted.[8] The inadequacies of the European treaty for wildlife protection soon became apparent, but not until 1979 was the new Convention on the Conservation of European Wildlife and Natural Habitats drafted. Sponsored by the Council of Europe and signed by fifteen states (19 September 1979 at Berne, Switzerland), this treaty was ratified by the European Community and entered into force for its members as of 1 September 1982.[9]

Protection of migratory birds and animals in North America was less complicated than in Europe, political boundaries and cultural differences being less numerous. The 1916 Canadian-American Treaty for the Protection of Migratory Birds has been described as "simply a game law, imposing closed season and limitations on commerce."[10] A practical reason for the negotiation of this treaty (technically with the United Kingdom, which then represented Canada in foreign affairs) was the apparent constitutional inability of the federal government of the United States to protect wildlife through statutory legislation. An international treaty carrying the full force of statuory law enabled the United States to enact implementing legislation that established a Migratory Bird Commission and to begin a system of permanent wildlife refuges that today totals thirty million acres.

Efforts to extend protection to include Central and South American countries were begun pursuant to a resolution of the United States Senate in 1920; however, not until 1936 was the Treaty for the Protection of Migratory Birds and Game Mammals negotiated with Mexico.[11] In 1940, sponsored by the Pan American Union, the Convention on Nature Protection and Wild Life Preservation in the Western Hemisphere committed the signatories to establish parks, reserves, nature monuments, and wilderness areas; to safeguard wildlife habitat; and to "adopt appropriate measures for the protection of migratory birds of economic or aesthetic value or to prevent the threatened extinction of any given species."[12]

The treaty became effective in May 1942, following ratification by five of the twenty-one signatory nations. The language of the treaty represented at least a step toward inter-American cooperation for the quality of the environment. The signatories declared their purpose "to protect and preserve in their natural habitat representatives of all species and genera of their native fauna and flora, including migratory birds, in sufficient numbers and over areas extensive enough to ensure them from becoming extinct through any agency within man's control" and to "protect and preserve scenery of extraordinary beauty, unusual and striking geological formations, regions and natural objects of esthetic, historic or scientific value, and areas characterized by primitive conditions in those cases covered by this Convention." In an effort to help implement this convention, an Inter-American Conference on Conservation of Renewable Natural Resources was convened in Denver, Colorado, 7–20 September 1948. Among several important papers presented at this conference, one by Harold J. Coolidge, executive secretary of the Pacific Science Board of the U.S. National Research Council, provided one of the few contemporary summaries of the developing international approach to nature protection.[13]

As with virtually all treaties for wildlife protection, however, the Inter-American Treaty is basically an exhortation, imposing no binding commitments. Article VII permits "insofar as the respective governments may see fit, the rational utilization of migratory birds for the purpose of sport as well as for food, commerce and industry, and for scientific study and investigation." And the treaty, "totally neglected by the Pan American Union," which was charged with its implementation, has hardly restrained massive development projects or the attrition of the environment by rural populations that in effect destroy the habitat of highly localized plant and animal species.[14] In recent years the Organization of American States has shown some interest in environmental matters, but the subject does not appear to be of high priority. Nevertheless, Victor H. Martínez, vice president of Argentina in 1987, proposed to the organization's legal committee a system for monitoring the implementation of environmental treaties among states in the Western Hemisphere.[15]

The African Convention Relative to the Preservation of Flora and Fauna in Their Natural State (London, 1933)[16] marked a further advance in international efforts toward wildlife protection, particularly in its more specific prohibitions regarding the hunting and harassment of wildlife and in the protection of habitat. Article XXI of the treaty provided that "each contracting government shall furnish to the government of the United Kingdom information as to the measures taken for the purpose of carrying out the provisions of the preceding articles." The United Kingdom agreed to communicate all information so furnished to all the governments "which signed or acceded through the Convention." Thus the United Kingdom acted as a kind of agent or clearinghouse for the signatory powers.

The London Convention of 1933 was, of course, an agreement among the European states governing pre–World War II Africa. It logically followed that with political independence of the African continent a new treaty was needed. The African Convention for Conservation of Nature and Natural Resources was approved and signed in Algiers in September 1968 by representatives of thirty-eight member states of the Organization of African Unity.[17] The broader conceptual base of the new African agreement to "ensure the conservation, utilization, and development of soil, water, flora, and the fauna resources in accordance with scientific principles and with due regard for the best interest of the people" is reflective of the transition from the earlier preservationist view to the broader scope of environmental policy.

As of 1995, rapid growth of human populations and rampant poaching of wildlife threatened to nullify the good intentions underlying the treaty of 1968. And for many forms of wildlife, a treaty limited to a single continent has proved to be insufficient to protect roving or migratory wildlife. In June 1979 a global treaty for the conservation of migratory wildlife was signed at Bonn and went into force in 1983, with twenty-two countries plus the European Community adhering.[18] But thus far the impact of the treaty is uncertain. Its provisions are outlined in Chapter 11.

Living Resources of the Sea

In the case of international cooperation to protect migratory wildlife, nations were to fulfill their commitments almost wholly within their own territorial jurisdiction. Few if any international institutional arrangements or agencies were called for beyond secretariats to receive information and issue reports. In the marine environment, beyond national boundaries, matters were different. Here nations enlarged their experience in developing international institutional frameworks for resource and environmental problems of mutual concern.

The record of this experience is widely varied, showing notable successes, recurring failures, and slow—even reluctant—learning from errors of the past. International issues in the marine environment have historically been primarily economic and military. Early environment-related issues arose primarily out of conflicts over the harvest of living resources in the sea—notably fish, seals, and whales. National concerns had more to do with the conservation and management of stocks and division of the take than with concern for ecological values or the quality of life on earth. Yet the narrow economistic premises of governments in these matters have gradually been modified by growing biospheric and ecologic awareness. Prior to the Stockholm Conference, the Food and Agriculture Organization held two major conferences (1955 and 1958) on the living resources of the sea and sponsored an international convention on their conservation. These efforts had little effect on the marine life that they were intended to conserve.

International conservation efforts have few success stories, but one undertaking that achieved its purpose—but not without criticism—has been the protection and management of the fur seal herd in the Bering Sea. Eighty percent of all fur seals breed on the Pribilof Islands (which are under the jurisdiction of the United States) but travel widely throughout the North Pacific. By the early twentieth century, indiscriminate hunting threatened their survival. The Convention between the United States and Other Powers Providing for the Preservation and Protection of Fur Seals, signed 7 July 1911, and periodically revised and extended, demonstrates that it is possible for humans to be restrained in the exploitation and management of a perishable resource.[19] Although the administration of the treaty has not been without contention (notably Japanese agitation over alleged depredations of the fur seals on commercial fisheries, and objections among Americans to killing of the "surplus" males for fur) the signatory nations—the United States, Canada, Russia, and Japan—have observed its terms and have based their management of the fur seal herd on a biologically sound, scientific research program.[20]

In 1970 a new threat to the seal herd was discovered. In the course of searching for DDT residues, marine biologists in the United States found concentrations of mercury in the livers of Alaskan seals at levels considered toxic to humans. The seals spend the greater part of their lives in the open ocean at least fifty to one hundred miles from polluted coastal areas, but mercury is concentrated in the aquatic food chain, being absorbed from the water by plankton and other small animals that are in turn eaten by fish and finally by the seals. Following further investigation, the United States Food and Drug Administration ordered withdrawn from the market ten thousand iron supplement pills made from the liver of fur seals, and the United States Public Health Service began preliminary examinations of the native Aleut population in the Pribilofs, to whom the fur seal has been an important source of meat.

The threat from mercury illustrates how swiftly and unobtrusively manmade ecological change can overtake and neutralize erstwhile protective arrangements. The Fur Seal Convention of 1911 was effective in protecting the seals from the direct impact of man; it was sufficient to control the taking of seals on the islands and in the surrounding seas. But the new threat could not be so localized, nor was it necessarily restricted to industrial activity among the nations party to the treaty. The implications of global contamination by DDT and mercury were becoming increasingly clear. International protective efforts, voluntary or coercive, that could not obtain change in the industrial practices of nations could not be relied upon to protect people or wildlife from oceanic or airborne pollution.

If the conservation of the fur seals can be considered one of the more successful episodes in the history of international environmental protection, efforts to protect the great whales have surely been one of the more frustrating. Following the rapid decline of whales in the North Atlantic, the first step toward international control of whaling was taken in 1920 with the establishment of the International Bureau for Whaling Statistics at Sandefjord in Norway. Cooperation with the bureau was purely voluntary, all countries participating in whaling being urged to send complete data on their operations to this center.

Meanwhile, through the League of Nations an attempt was made to secure multipartite agreement to a program of conservation for whales. The Convention for the Regulation of Whaling was finally signed at Geneva in September 1931.[21] Twenty-four nations ratified or adhered to the treaty by December 1935, but Japan and the Soviet Union, which were to play major roles toward the extermination of the whales, were not parties to it.

Obligations of the signatories were minimal. The treaty, which for all practical purposes was unenforceable, was supplemented by an agreement concluded by nine nations in 1937 and by an amending protocol signed in 1938 so that the 1931 treaty would become more than a mere statement of good intentions.

World War II brought a temporary reprieve from the virtually unchecked rapacity of the whalers. Following the war, at the initiative of the United States, a new International Whaling Conference was held in Washington in 1946 and the International Whaling Commission (IWC) was established.[22] A code for the whaling industry was formulated and the commission was empowered to amend it without the necessity of further formal conferences. The IWC conformed to the structure that had been established for international fisheries commissions, consisting of one member of each contracting government with one vote for each member. Ordinary decisions were to be taken by simple majority, with changes in the code requiring a three-quarter member majority. Although authorized to administer regulations regarding open and closed waters, periods, methods, and intensity of whaling, including the maximum catch in any one season, the commission had no power to restrict the number or nationality of factory ships or whaling stations ashore, or to allocate specific quotas to any one nation.

The record of the International Whaling Commission during the three decades following its establishment was largely a history of its inability to overcome the studied and stubborn defense of the short-range interests of the whaling industry. The commission's ineffectualness was compounded by its frequent disregard of the findings and recommendations of its scientific advisors.[23] Although the preamble of the 1946 treaty refers to "the interest of the nations of the world in safeguarding for future generations the great natural resources represented by whale stocks," the document is filled with qualifications that have been interpreted to legitimize the destructive and wasteful methods of operation.

At the 1972 Stockholm Conference, the whale became a symbol of mankind's antiecological behavior. At Skarpnack, an abandoned airport at the edge of Stockholm, a tent city was set up sheltering about a thousand young people, where a Celebration of the Whale took place followed by a Whale March into Stockholm. Led by the American delegation, the United Nations conference recommended a ten-year moratorium on commercial whaling. This recommendation, although unenforceable, was intended to pressure the

IWC into more effective protective action. A decade was to elapse, however, before a moratorium was adopted by the IWC.

Why did international cooperation and institutional arrangements succeed in saving the fur seals and initially fail to protect the whales? The circumstances differed greatly. The fur seals were a single species with highly predictable behavior and localized breeding habitat under control of the United States. Each of the four treaty states benefited more from long-term conservation than from short-term overexploitation. But there were many species of whales ranging widely throughout the oceans, their behaviors poorly understood, their numbers difficult to estimate, and management of stocks complicated by numerous factors of politics, economics, and geography. The weakness of the IWC as a protective agency was that, like the international fisheries commissions, its voting members represented in most instances the industry that it was intended to police. Moreover, it was established to regulate whaling, not to prevent the taking of whales. The continuing efforts of the international environmental movement to secure an IWC moratorium on whaling will be recounted in Chapter 9.

International endeavors to protect marine fisheries have also had a history of frustration. Conservation efforts have suffered from the same fundamental weakness: international arrangements and policies have been largely controlled by the industries that would have to be restrained if protective measures were to become effective. In practice, declared conservation objectives have repeatedly given way to division of the spoils.

Protection and management of marine fisheries have been attempted largely through numerous bilateral and multilateral arrangements. The institutional device most frequently employed has been an international commission or council organized for a particular region of the sea or species of fish. But the limited success of these bodies in the international management of fisheries has suggested to some observers the need for a more adequate jurisdictional arrangement for their conservation and has provided argument for a world regime for the oceans and has moved nations closer to accepting the recommendations regarding world resources that were brought before the Stockholm Conference and the United Nations Conferences on the Law of the Sea.

Two approaches have been taken toward international conservation: cooperative research and regulation—research, being easier to organize and less susceptible to political controversy, preceded regulation by several decades. International efforts to conserve and manage marine fisheries began because of the decline of North Sea and North Atlantic fisheries late in the nineteenth century, where diminishing returns were first felt. The first effort was the International Council for the Exploration of the Sea (ICES), established following organizational meetings in Stockholm in 1899 and in Christiania (Oslo) in 1901.[24] Since its inaugural meeting in 1902, the headquarters of the council have been in Copenhagen; its membership was drawn initially from the northwest European states, with Portugal, Spain, and Italy subsequently joining and cooperative relations being established with the United States, Canada, and Japan. The council has collected and published data supplied by its members and has engaged in cooperative and planned research through committees organized both by species of fish and by geographic area. Within the limit of its activities it has been generally considered a success, affording a model for similar research efforts which followed.

In 1919 the International Commission for the Scientific Exploration of the Mediterranean Sea was established; it was supplemented in 1949 by the General Fisheries Council for the Mediterranean (GFCM). Sponsored by the Food and Agriculture Organization, GFCM became operational in 1952 upon ratification of its charter by five governments.[25] From 1920 to 1938 the North American Council on Fisheries Investigation laid the groundwork for the International Convention for the Northwest Atlantic Fisheries. Following a general plan to encourage the formation of regional research councils for the scientific exploration of the sea in parts of the world not served by such bodies, the Food and Agriculture Organization in 1948 sponsored establishment of the Indo-Pacific Fisheries Council (IPFC) for the "development and utilization of living aquatic resources." This process of organizing the fisheries of the world on a regional basis has continued until nearly every part of the sea in which fishing is carried on is now covered by an international council or commission.[26] The effectiveness of these fisheries arrangements has been handicapped by the movements of fish in the sea. Articles 61–70 of the United Nations Law of the Sea Convention define the respective rights of nations in relation to living resources in their exclusive economic zones. The convention initially proposed no new law for fishing on the high seas, but recommended international cooperation with respect to highly migratory species, including the establishment of new regional organizations. In 1995, however, a Draft Agreement was adopted by the United Nations for the Implementation of the Provisions of the Law of the Sea of 10 December 1982 Relating to the Conservation and Management of Straddling Fish Stocks and Highly Migratory Fish Stocks.[27] The more important of the fisheries' regional bodies are listed in Appendix A. But their ability to prevent overfishing remains in doubt.

Research may provide an informational basis for conservation, but it does not of itself conserve. While international cooperative efforts in fisheries research have enjoyed moderate success, difficulties arise when regulatory measures for the conservation of fisheries are proposed. The historical experience has been that freedom of the seas and national rights are invoked to cover the immediate economic interest of the fisheries industries, as in the Cod Wars (1952–76) between Iceland and Great Britain.[28] Nevertheless, repeated international efforts have been made to establish a rational, biologically sustainable policy for world fisheries.

The first international effort to establish a world fisheries policy originated in the League of Nations, which requested the International Council for the Exploration of the Sea to create a committee of jurists to consider "whether it is possible to establish by way of international agreement rules regarding the exploitation of the products of the sea." On 29 January 1926 J. L. Suárez of Argentina, reporter for the committee, submitted his "Report on the Exploitation of the Products of the Sea."[29]

This farsighted document urged a biological basis for policy, with primary attention to the preservation of economically valuable species rather than to the traditional concern of the parties with reciprocal commercial rights. It proposed international regulation of the continental shelf to a depth of two hundred meters and provided for a rotation of zones of fishing to avoid overexploitation, for closed seasons, for the protection of the young, for standardization of methods of capture, and for the fullest possible use of all animals taken. But the report was not implemented and had no immediate consequence. Subsequent consideration under committees of the League of Nations and the International Council for the Exploration of the Sea led eventually only to the whaling treaty of 1931, which, as we have seen, proved largely ineffectual.

In 1955 the United Nations tried where the League failed. At the request of the General Assembly the International Technical Conference on the Conservation of the Living Resources of the Sea was held in Rome at the headquarters of the Food and Agriculture Organization. The meeting was to assist the International Law Commission in the preparation of draft articles on certain basic aspects of the international regulation of fisheries. The conference, as its name implied, was concerned almost entirely with technical problems involved in the conservation of fisheries. Matters of a legal or political nature were not discussed, although these had been the areas in which the fundamental obstacles to conservation of marine resources were to be found. For more than two decades the United Nations attempted to deal with the legal and political issues concerning marine life as part of a series of Conferences on the Law of the Sea (UNCLOS), first convening in 1958. The oceans, presenting a very large and special set of problems, were not fully considered by the delegates to the Stockholm Conference, who deferred major issues to the Third United Nations Conference on the Law of the Sea, which began its sessions in 1973.

With the exception of the North Pacific Fur Seal Convention, a plausible case could be made that, had none of the international conservation agreements negotiated prior to 1970 been consummated, the state of fisheries and world wildlife generally would not have been significantly different. Perhaps a better case might be made for the contention that while the practical effect of the treaties was often disappointing, they represented commitments in principle toward which governments in practice might gradually be persuaded to move. Where public attitudes favored a conserving pattern of behavior, as in the United States, Canada, and northwestern Europe, the treaties were, after the fashion of the day, generally observed; conversely, where the commitments were not supported by the traditional culture or public opinion, ratified treaties were often ignored or given no more than token administration.

The history of the growth of international efforts to protect migratory wildlife, marine mammals, and fisheries in the years preceding the 1968 Biosphere Conference and the Stockholm Conference illustrates the lack of a biospheric or broadly ecological view of environmental protection. In the absence of continuing institutional implementation, treaties proved to be more symbolic than substantive. Yet the cumulative effect of these and other efforts prepared the way for more effectively concerted and directed international action.

3
The Road to Stockholm

The United Nations Conference on the Human Environment, which met in Stockholm in 1972, marked a watershed in international relations. It legitimized environmental policy as a universal concern among nations, and so opened a place for environmental issues on many national agendas where previously they had been unrecognized. In comparison with other UN conferences, the tangible results of Stockholm have been substantial. Yet the growth of international environmental cooperation during the 1970s and thereafter is an aspect of a larger social transition, expressed notably in the 1992 Conference on Environment and Development in Rio de Janeiro. It is an expression of a changing view of mankind's relationship to the earth. Analysts of public opinion have called this view of human life on earth the "new environmental paradigm"; evidences of its reality are found in surveys of values and attitudes from many parts of the world.[1]

The term *paradigm* here represents people's assumptions regarding how the world works. For example, the change from the belief that the sun, moon, and stars revolved around the earth to the Copernican view of earth's place in the solar system was a paradigm shift. The change marked by Stockholm is from the view of an earth unlimited in abundance and created for man's exclusive use to a concept of the earth as a domain of life or biosphere for which mankind is a temporary resident custodian. The older modern view saw this planet as a storehouse of resources to be freely developed for human use. Today, an ecological view sees it as an ultimately unified system of living species and interactive, regenerative biogeochemical processes that may supply human needs as long as the system's rules are respected.

Paradigm shifts do not occur overnight or evenly among societies. The uneven rate of acceptance of this conceptual change creates difficulties for governments and other established institutions. The validity of traditional policies is questioned, but the legitimacy of the new assumptions may not yet be fully accepted. Public and international policies reflect the transition and its uncertainties through halfway, contradictory, or unimplemented measures.

However, even though retrograde developments continue to occur, the movement toward a new relationship between human society and the biosphere is clearly demonstrable.

The emergence of an international movement for environmental protection was evident when people, whose numbers and status were sufficient to be politically significant, became conscious of the common stake of humanity in protection of the biosphere and moved to persuade governments to act in conformity with this awareness. As in most civic and international causes, the critical mass of people concerned about environmental protection coalesces around the more educated, sensitive, and influential members of society. Among these are cosmopolitan individuals who have a worldview of human affairs and some understanding of the interrelationships that form the complex systems of earth and world. Such individuals are most numerous and influential in nations that are more highly advanced in science and technology. During the course of the twentieth century a sufficient number of people became sufficiently concerned about the danger to humanity's common stake in the biosphere to induce international action. This movement was spontaneous and diverse, converging into a truly global effort only in the years immediately preceding the Stockholm Conference.

Progress along the road to Stockholm was neither straightforward nor evenly paced. At the beginning of the twentieth century, neither environment as an integrative ecological concept nor the biosphere as the planetary life-support system was an object of public international concern. International efforts, so far as they focused upon resource conservation, did so largely although not exclusively for economic and strategic reasons. As noted in Chapter 2, there had been efforts to conserve various flora and fauna. These efforts characteristically entailed exchanges of scientific information and cooperation in scientific exploration and research. Only in the light of experience and with considerable caution did nations agree to cooperate in protecting migratory wildlife. Exclusive sovereignty over national resources was —and is still—almost universally asserted, but it became increasingly modified in practice as nations began to merge their sovereignty in order to safeguard birds, fish, and marine mammals that might otherwise be lost to all. From these beginnings, other modifications of sovereignty in practice have followed.

The first steps on the road to Stockholm were taken at least a century before the United Nations Conference on the Human Environment, when in 1872 the Swiss government proposed (unsuccessfully) an international regulatory commission to protect migratory birds in Europe. And even earlier, some governments had taken conservation initiatives, mainly for national economic reasons, to protect forests, inland waters, mineral deposits, and preferred forms of wildlife.

International Congresses

The instrument primarily chosen for initiating international cooperation was the nongovernmental international congress or conference. International scientific congresses began to be held with increasing frequency after the mid-nineteenth century. Botanical, ornithological, and zoological congresses were held regularly, but, although international in participation, they were convened primarily for exchange of information and were seldom concerned with intergovernmental relations. Early in the twentieth century, the format adopted for nature protection was the congress. Among the more important meetings were the International Congress for the Protection of Nature (Paris, 1909), the International Congress on the Protection of Flora, Fauna, and Natural Sites and Monuments (Paris, 1923), the International Congress for Study and Protection of Birds (Geneva, 1927), and the Second International Congress for the Protection of Nature (Paris, 1931). A Consultative Commission for the International Protection of Nature was established at Berne in 1913 with signatories from seventeen European countries. Activation of the commission was prevented by World War I, but although the commission was never actually functional, it retained legal status in the years between the two world wars. It was, in form, the first intergovernmental agency established for the comprehensive protection of nature.

Following World War II, the convening of international conferences for various aspects of nature protection was resumed. During the first half of the twentieth century, governments hosted congresses in which they did not officially participate. But concomitant with expanding government responsibilities at the national level after World War II, governments and intergovernmental organizations increasingly became involved in conservation and, subsequently, in environmental conferences. In July 1947 the Swiss League for the Protection of Nature sponsored a conference at Brunnen attended by delegates and observers from twenty-four countries and nine international organizations. This meeting adopted a draft constitution establishing a provisional International Union for Protection of Nature that, after a follow-up conference called by UNESCO and held in Fountainebleau on 1–7 October 1948, became the basic charter for the International Union for Conservation of Nature and Natural Resources (IUCN), the principal nongovernmental international organization for environmental protection in which governments, nevertheless, have been participants.[2]

Paralleling a transition from almost wholly nongovernmental conservation initiatives to governmental involvement was the coalescing of diverse interests in conservation and nature protection into a more comprehensive international environmental structure. This process appears first to have occurred in Western Europe, including the United Kingdom, and in the United States.

During the 1960s and 1970s an international network of nongovernmental organizations developed and began to effect a transition from traditional and narrowly limited conservation policies toward a comprehensive and open-ended concern for quality of the environment.

When one compares the assumptions, agenda, and emphasis of the United Nations Scientific Conference on the Conservation and Utilization of Resources held 17 August–6 September 1949 at Lake Success, New York, with those of the United Nations Conference on the Human Environment held in 1972 in Stockholm, it is clear that changes in attitudes and institutional relationships had occurred regarding our relationship to the earth and to our use of its resources.

UN Conservation Conference — 1949

Judged in relation to needs, the United Nations Scientific Conference on the Conservation and Utilization of Resources in 1949 had few tangible results. Yet the effort was well intentioned. According to Dr. Harold J. Coolidge, who because of his deep involvement in international conservation matters was in position to know, "[T]he origin of this conference goes back to Franklin D. Roosevelt," who felt that, after the end of World War II, the United States should call such a conference.[3] Official initiative for the conference came in a letter, dated 4 September 1946, from his successor, Harry S. Truman, to the United States representative on the Economic and Social Council (ECOSOC) of the United Nations. The letter from the president declared his belief "that a congress composed of engineers, resource technicians, economists, and other experts in the fields of physical and social science would offer the most desirable method of presenting and considering the definite problems now involved in the resource field."[4] The idea was accepted by ECOSOC, which, however, adopted a narrowed interpretation of its purpose when it asked the secretary-general of the United Nations to proceed with arrangements. In the ECOSOC resolution of 11 February 1948 (109 VI) the secretary-general was reminded that "the task of the conference is to be limited to an exchange of experience in the techniques of the conservation and utilization of resources." Environment had not yet emerged as a policy concept.

Although the conference could not escape its limitations of concept and agenda, some participants understood what its functions should have been. Prominent among these was Fairfield Osborn, president of the New York Zoological Society and the Conservation Foundation, who in an opening address declared that "this world meeting on resources is a sign of the evolution of human society" and observed that "despite the growing evidences of cooperation between nations, it is unlikely that this world meeting would be taking place, even now, if it were not for conditions of obvious increasing seriousness." Extending this theme, he observed that

> Within the last century startling developments of a worldwide nature have taken place. These changes have been so rapid, in some cases so violent, that adjustments to them, socially as well as materially, have barely kept pace. Unquestionably the greatest factor of change is the explosive upsurge in population in virtually all countries resulting in a doubling of the world population within the last century, or an increase of more than a thousand million people. . . .
>
> This doubling of the Earth's population in merely four generations has been accompanied, as we are all well aware, by an almost fantastic series of inventions which have brought about what may be described as the *second* industrial revolution. The consequence has been that the drain upon the Earth's resources has increased not upon a mathematical scale related to population growth, but upon a geometrical scale related to greater numbers of people demanding a greater variety of products from an infinitely more complex industrial system.[5]

Osborn summarized his message in language that, had it been acted upon by the conference, could have contributed greatly to the welfare of humanity and the preservation of the earth. Unfortunately, a quarter of a century was still to elapse before the United Nations arrived at the standpoint from which Osborn's assessment of needs began. He said:

> In the light of experience, and in these terms, conservation becomes a political and administrative problem, an educational, even a social, cultural, and ethical problem. Therefore, it is not one with which the scientists or technologists can deal single-handed. Further, conservation in the sense that it implies the wide use and equitable distribution of the Earth's resources, offers a point of synthesis for international cooperation for which the world is waiting.[6]

The lack of readiness to receive or to understand Fairfield Osborn's message was illustrated by the address, immediately following, of Dr. Colin G. Clark, lecturer at Cambridge University and economic advisor to the government of Queensland, Australia. While Dr. Clark recognized that "man has proved himself capable of the most appalling misuse of natural resources under certain circumstances" and that the "world rate of population growth has accelerated since 1920," he saw no problem in population increase, confining his analysis to an optimistic appraisal of the ability of nations to enlarge food supplies. He concluded that "the claim that in countries where fertility is high it should be artificially reduced is thus groundless economically."[7]

The agenda of the conference was divided by categories of resources, including minerals, fuel and energy, water, forests, land, and wildlife and fish. Although the importance of ecological knowledge as a guide to action was mentioned in several papers and addresses at the conference, at no point was a truly holistic view taken of the relation among population, resources, and environment. However, an International Technical Conference on the Protection of Nature, sponsored jointly by the International Union for Conservation of Nature and Natural Resources and UNESCO, met at Lake Success (22 August–1 September 1949), coincident with the Resources Conference and to some extent supplementary to the United Nations effort.[8] This meeting may be seen in retrospect as a forerunner of the UNESCO-sponsored Biosphere Conference two decades later.

The United Nations Conference on the Application of Science and Technology for the Benefit of the Less-Developed Areas, held in Geneva on 4–20 February 1963, was in many ways a replay, with elaborations and refinements, of the Lake Success conference of 1949.[9] Although separated by fifteen years, almost no change in orientation could be detected between these two conferences. During this same period, and for several years thereafter, a number of scientific conferences relating to energy resources were held under United Nations auspices, notably one on new sources of energy at Rome in 1961 and four on peaceful uses of atomic energy, which were convened at Geneva in 1955, 1958, 1964, and 1971.[10] Without question, these conferences were important for the exchange of views and the marshaling of information about human use of the resources of the earth. But they failed to ask the questions that the growing stress of human demands upon the earth should have suggested, or to bring under consideration not only the full range of human needs and values, but the basis for the welfare and survival of the biosphere. The strongly economic and technological bias of the conferences, along with the uncritical optimism that generally prevailed, precluded any realistic assessment of trends that were rapidly worsening the human environment and the prospects of its improvement.

The Biosphere Conference — 1968

A major advance in international environmental policy formation and a landmark in recognition of the relations of humans to the natural world was the Intergovernmental Conference of Experts on a Scientific Basis for a Rational Use and Conservation of the Resources of the Biosphere, held in Paris in September 1968. More conveniently termed "the Biosphere Conference," it was sponsored by a number of international agencies under the leadership of UNESCO. The conference marked the arrival of the biosphere concept as an object of international policy deliberation.[11]

The Biosphere Conference did not consider inorganic resources except as they provided for the support of plant and animal life; man-made environments, particularly those of human settlements, received cursory attention. But within its context the conference examined a broad spectrum of issues with an emphasis on ecological considerations. Unlike earlier conferences, this one was not content with an exchange of views and experience but rather adopted twenty recommendations for future action by the participating governments, by the United Nations system, and especially by UNESCO. In the concluding paragraphs of its final report the conference summed up its assessment of man-made environment relationships and their political implications:

> Until this point in history the nations of the world have lacked considered, comprehensive policies for managing the environment. . . . Although changes have been taking place for a long time, they seem to have reached a threshold recently that has made the public aware of them. This awareness is leading to concern, to the recognition that to a large degree, man now has the capability and responsibility to determine and guide the future of his environment, and to the beginnings of national and international corrective action. . . . It has become clear, however, that earnest and bold departures from the past will have to be taken nationally and internationally if significant progress is to be made.[12]

Convergent Efforts toward Action

By the mid-1960s changes in the structure of international cooperation seemed probable—first, as a result of the growth of worldwide concern with the general state of the environment and, second, as a result of international efforts on behalf of the environment, which could not succeed without institutional innovation at the international level. Indications of the developing concern and changing emphasis of nations toward the global environment may be seen in comparing the resolutions on the human environment of the Biosphere Conference and of the General Assembly of the United Nations with the agenda and proceedings of the United Nations Scientific Conference of 1949 on the Conservation and Utilization of Resources. The twenty years between the resources (1949) and the biosphere (1968) conferences mark as fundamental a change in perceptions of international responsibilities for the global environment as any that have occurred since the establishment of permanent international organizations.

A process of conceptual change was occurring in the quarter century between the end of World War II and the preparations for the United Nations Stockholm Conference in 1972. But this development was not generally apparent as late as 1963, when Pierre Auger wrote concerning the problem of ecological balance and its disruption by human intervention that "no clear doctrine has emerged as yet, such as a synthesis accompanied by an appropriate evaluation of the conditions in the near future."[13] And even though the conceptual basis for international action did not yet appear adequate to needs and realities, events in the 1960s provided evidence of growing recognition that the intellectual and institutional capabilities of nations needed to be mobilized to arrest the worsening condition of the biosphere.

This recognition of need for environmental protection first appeared at local and national levels where the effects of environmental deterioration were directly experienced and where governments had competence to take unilateral action. Although governments may formally agree (e.g., by treaties) to policies that not all of them adhere to at home (e.g., protection of wetlands and wildlife), their international commitments establish goals or standards that their own citizens may prompt them to honor. Among nations, as among persons, trends and fashions may be contagious. In the years immediately preceding Stockholm, environmental protection had become a politically significant issue in nearly every developed state, and in many of them efforts toward international cooperation were already under way.

For several of the major industrial countries, positions taken later at Stockholm on certain issues were influenced by ecological politics at home. By 1972 several leading states had enacted new and significant environmental legislation or had reorganized ministries and departments for environmental protection purposes. Environment had become a major public issue in Canada, Japan, the Netherlands, Sweden, the United Kingdom, and the United States; and in these countries (among others) the issue was both domestic and international.

In the United States, the White House Conference on Natural Beauty of 24–25 May 1965 was a temporarily influential consciousness-raising event.[14] Although President Lyndon B. Johnson generally failed to give official support to the impetus toward environmental protection generated by the conference, the event was a significant step toward adoption of a national environmental policy for the United States. The conference is perhaps best understood as an exercise in social learning in which a diversity of interests in environmental policy were brought together to focus upon agreed problems. How much of this experience carried over to the American role in the United Nations Conference of 1972 is conjectural. But the 1965 White House Conference may have prepared the way for the role that the United States played at Stockholm.

The subsequent White House Conference on International Cooperation (29 November–1 December 1965) recognized the connection between international development programs and an impending crisis of the biosphere. A report prepared for this meeting by the National Citizens Commission recommended establishment of a number of new international agencies and programs for environmental and ecological protection. These included a specialized UN agency for marine resources, the World Institute of Research Analysis, the UN-sponsored International River Basin Commission, and the Trust for the World Heritage in Natural and Scenic Areas and Historic Sites.[15]

Of more immediate significance was passage of the National Environmental Policy Act (NEPA) of 1969, which committed the United States government and its agencies to a policy of international cooperation in environmental affairs.[16] Section 102(f) of the act declared that the Congress authorizes and directs that all agencies of the federal government shall "recognize the worldwide and long-range character of environmental problems and, where consistent with the foreign policy of the United States, lend appropriate support to initiatives, resolutions, and programs designed to maximize international cooperation in anticipating and preventing a decline in the quality of mankind's world environment." An international obligation was acknowledged in the preamble, extending the scope of national concern beyond national boundaries "to promote efforts which will prevent or eliminate damage to the environment and biosphere and stimulate the health and welfare of man."

In an address at Glassboro (New Jersey) State College on 4 June 1968, President Lyndon B. Johnson called for Soviet-American cooperation in the formation, with other nations, of an International Council on the Human Environment.[17] On 22 July 1968, in an essay entitled "Thoughts on Progress, Peaceful Coexistence and Intellectual Freedom," Russian academician A. D. Sakharov entered a strong and widely circulated plea for cooperation between the Soviet Union and the United States in coping with the global effects of environmental pollution and deterioration. The primary significance of these proposals was that they were made—that they could assume a comprehending and sympathetic audience of sufficient size and influence to justify the risk and effort involved.

The resolution of the United Nations General Assembly (3 December 1968) calling for a world conference on the human environment catalyzed or stimulated a great variety of efforts—scientific, academic, civic, and political—on behalf of international action toward protection of the biosphere. Illustrative of this activity was the identification of environmental problems as a major social concern of the NATO Committee on Challenges of Modern Society (CCMS), the European Conservation Year (1970) under the sponsorship of the Council of Europe, and the Economic Commission for Europe Symposium on Problems Relating to Environment held in Prague in May 1971.[18] In addition, publications, seminars, and other activities were undertaken, including an International Conference on the Environmental Future, cosponsored by the government of Finland, the Finnish National Commission for UNESCO, and the Jyväskylä Arts Festival in midsummer 1971.[19]

It was evident, however, that not all national governments and not all people would agree that environmental protection should receive a high priority on the agenda for international action. Moreover, among those who agreed that an environmental crisis had been reached, there were differences of viewpoint regarding its causes and what should be done about them. Thus the convergence of concern was accompanied by a divergence of opinion, particularly among the socialist and Third World countries.

Because environmental concern was strongest among the more developed nations, suspicions arose among Third World or less-developed countries that the movement concealed a neoimperialist scheme to retard their economic growth and to keep them subservient suppliers of underpriced raw materials and consumers of the industrial output of North America, Western Europe, and Japan. Socialist ideologists did not resist the temptation to encourage this apprehension, or to blame capitalism and imperialism for the environmental degradation of the world. These differences became explicit at the Stockholm Conference. Yet even in the then socialist and Third World countries, there were individuals and groups, particularly among scientists, who were persuaded that the issue was important to all mankind and that their countries had an important stake in the preservation of the biosphere.

There is emulation among nations as among people. Rhetoric of resentment among Third World nations coexisted with conspicuous and too often uncritical emulation of the industrialized First World. Not all Third World leaders favored replicating the West, and for some of them the harsh impact of Western industrialism on the environment was something to be avoided. For example, Julius Nyerere of Tanzania rejected both Western and Soviet-style industrialization, favoring an indigenous, essentially rural economy. But during the years immediately preceding the Stockholm Conference, the example of new environmental laws and agencies established in France, Sweden, the United Kingdom, and the United States (among other countries) and the request of the United Nations Preparatory Commission for status reports from all countries on environmental policy, made possession of an environmental policy a status symbol—evidence that a nation belonged among the more advanced or advancing states of the world and not among the backward nations.

Preparations for Stockholm

Official initiation of the conference began with a letter dated 20 May 1968 from the permanent representative of Sweden to the secretary-general of the United Nations "on the question of convening an international conference on the problems of human environment."[20] The possibility of such a conference had been introduced by the Swedish delegation to the plenary session of the UN General Assembly on 13 December 1967.[21] The Swedish proposal was referred to ECOSOC for consideration, and the UN Secretariat prepared a short paper outlining the activities of the United Nations organizations and programs relevant to the human environment.[22] On 30 July 1968, ECOSOC adopted Resolution 1346 (XLV), which requested that the United Nations proceed with plans for a conference. The resolution stated that the council:

1. Recommends that the General Assembly, at its twenty-third session, consider . . . the desirability of convening a United Nations conference on problems of the human environment, taking into consideration, inter alia, the views expressed during the forty-fifth session of the Economic and Social Council . . . and the results of the Intergovernmental Conference of Experts on the Scientific Basis for Rational Use and Conservation of the Resources of the Biosphere [Biosphere Conference of 1968]. . . .
2. Considers that, in order to assure the success of a conference, should its convening be decided by the General Assembly, detailed and careful preparations would be necessary. . . .
3. Proposes to the General Assembly that it include in the agenda of its twenty-third session an item entitled "The problems of human environment."[23]

On 3 December 1968 the UN General Assembly took up agenda item 91, "The problems of the human environment," and after generally supportive discussion adopted Draft Resolution 2398 (XXIII) without opposition, thereby setting in motion the preparatory efforts leading to the conference to be held in Stockholm in the summer of 1972.

The resolution of 3 December 1968 seems certain to be a conceptual milestone in the history of the relationship between humans and their environment, marking a worldwide recognition that, in the words of the resolution, there was "need for intensified action at the national, regional, and international level in order to limit and, where possible, to eliminate the impairment of the human environment and . . . to protect and improve the natural surroundings in the interest of man."[24] Thus the question whether or to what extent nations would surmount their differences to meet this need was placed on the agenda of world politics.

The objectives of the conference stated in the resolution were "to provide a framework for comprehensive consideration within the United Nations of the problems of the human environment in order to focus the attention of Governments and public opinion on the importance and urgency of this question and also to identify those aspects of it that can only, or at best be solved through international cooperation and agreement." The General Assembly requested the secretary-general, in consultation with the Advisory Committee on the Application of Science and Technology to Development (ACASTD), to submit through the Economic and Social Council to the General Assembly a report concerning the nature, scope, and progress of work being done in the field of the human environment, the principal environmental problems facing developed and developing countries, the time and methods necessary to prepare for the conference, a possible date and place for its convening, and finally the range of financial implications for the United Nations.

On 26 May 1969, the secretary-general submitted his report to ECOSOC, then convened for its forty-seventh session.[25] This report, following the recommendations of the ECOSOC resolution on 30 July 1968, surveyed the main problems of human–environment relationships, reviewed the nature, scope, and progress of research and preventive measures, and set forth the considerations required in the planning of the conference. The secretary-general also reported the invitation of the government of Sweden to host the conference in Stockholm in June 1972. Following the endorsement of ECOSOC, the report was brought before the General Assembly, which by Resolution 2581 (XXIV) on 15 December 1969 established the Preparatory Committee (PREPCOM) whose work would be essential to the success of the Stockholm Conference.[26]

The twenty-seven-nation Preparatory Committee, together with alternates and advisors, was too large actually to organize the conference. So a special staff was appointed under the direction of Maurice Strong, an official of the government of Canada, who was also designated as secretary-general of the conference. Under Strong's clear-focused and energetic leadership, preparations for Stockholm proceeded with a thoroughness unknown to previous international conferences. The two-week interval at Stockholm, 5–17 June 1972, could be sufficient for the representatives to consider and adopt agreed statements only if the issues involved had been thoroughly considered by national representatives prior to the conference. The symbolic value of the Stockholm Conference could be very great, but its practical accomplishments would heavily depend upon the preparatory work preceding it. The important agreements among nations would have to be achieved prior to Stockholm.

A number of intergovernmental working groups were organized to develop proposals and bases for agreement on the major items of an agenda. One group was to prepare a draft Declaration on the Human Environment (membership identical to the full PREPCOM), and there were additional groups on marine pollution, soils, conservation, and monitoring and surveillance.[27]

The first session of the Preparatory Committee was held at the United Nations headquarters in New York, 10–20 March 1970. The committee agreed that its main task would be to assist the secretary-general of the conference in the selection of topics and headings for the conference as well as in the formulation of ideas, suggestions, and proposals with regard to the content of its program. The discussion revealed substantial consensus on the characteristics and seriousness of environmental problems, but the complex diversity of ways in which these problems had arisen in various countries was seen as complicating the protective task.

At its second session, meeting in Geneva, 8–19 February 1971, the Preparatory Committee considered a proposed agenda for the conference culminating in the adoption and signature of a Declaration on the Human Environment. The proposed agenda consisted of six main subjects, consideration of which was divided among three principal committees and which were subsequently the principal working divisions of the Stockholm Conference:

1. Planning and Management of Human Settlements for Environmental Quality
2. Environmental Aspects of Natural Resources Management
3. Identification and Control of Pollutants and Nuisances of Broad International Significance
4. Educational, Informational, Social, and Cultural Aspects of Environmental Issues
5. Development and Environment
6. International Organizational Implications of Action Proposals

The conference preparations inspired a large number of associated or collateral meetings, seminars, and conferences intended to feed into the deliberations at Stockholm. Some of these were initiated or endorsed by the PREPCOM and the secretary-general; others were independently sponsored by national governments and nongovernmental organizations.

On the international scene, the International Council of Scientific Unions (ICSU) and the International Union for Conservation of Nature and Natural Resources (IUCN) were especially active in efforts to influence the conference agenda. In the United States, nongovernmental organizations were busy developing position papers and arousing public interest. For example, the Commission to Study the Organization of Peace published a report on the United Nations and the human environment; the National Research Council Environmental Studies Board prepared at the request of the Department of State a report on institutional arrangements for environmental cooperation; and the National Wildlife Federation made plans to hold an international symposium in Stockholm during the time of the United Nations conference.[28]

One of the more important preconference meetings was the PREPCOM-sponsored study on environment and development undertaken by a panel of twenty-seven experts in economics, development planning, banking, social research, and ecology. Meeting at Founex, Switzerland, in June 1971 the panel concluded that "the kind of environmental problems that are of importance in developing countries are those that can be overcome by the process of development itself."[29] Founex helped to alleviate some of the Third World misgivings concerning their developmental aspirations. It also set terms for the debate over relative priorities of ecology and economics that has continued to the present, reconciled in some measure by the concept of sustainable development.

Working through the United Nations regional economic commissions, seminars were held in August and September 1971 in Asia (Bangkok), Africa (Addis Ababa), the Middle East (Beirut), and Latin America (Mexico City).[30] Emphasis was on the environment–development needs and priorities of developing countries, and the Founex report was a major source document.

Among other gatherings preparatory to Stockholm was a world meeting of scientists which took place simultaneously with the Twelfth Pacific Science Congress meeting in Canberra, Australia. Sponsored by the Scientific Committee on Problems of the Environment (SCOPE) of the International Council of Scientific Unions in cooperation with the United Nations Advisory Committee on the Application of Science and Technology to Development, the Canberra meeting was designed especially to involve scientists from developing or Third World countries and to provide advice to the conference secretariat.[31] Additional scientific input to the conference was provided by a meeting of experts on meteorology and climatology convened in Stockholm in midsummer 1971 to consider the impact of human activities on possible future changes in the atmosphere.[32]

A listing of all official, quasi-official, and unofficial meetings preparatory to the Stockholm Conference would be very extensive and impossible to compile without risk of significant omissions. In addition to international gatherings, meetings were organized within countries in which environmental awareness had become high. For example, in April and May of 1972 a Canadian preparatory committee for the conference sponsored public regional consultations in eleven Canadian cities. Similar efforts were mounted by the United Nations Associations of the Nordic Countries. Among very numerous preconference meetings in the United States were a series of public hearings sponsored by the secretary of state's Citizen Advisory Committee.

This description of the preparatory phases of the Stockholm Conference omits more detail than it includes. But it is sufficient to make clear the extensive and intensive activity preceding the conference in assembling the information, disseminating the concepts, clarifying and reconciling (where possible) the differences that were to make up the substance of the deliberations at Stockholm. The consequences of this preconference activity may have had a more extended and lasting significance than did the actual conference itself. The basic papers and reports prepared for Stockholm and the collateral meetings amassed an immense amount of information concerning the state of the world environment. Exchanges of viewpoints among the participants in meetings and study groups were very widespread and often took the form of dialogues between scientists, citizen groups, and representatives of their governments. Certainly with respect to the state of the earth and humanity's environmental relationships, the years 1968 to 1972 witnessed a worldwide raising of consciousness for which there appears to have been no precedent.

This is not to say that tangible accomplishments immediately followed the raising of consciousness. Setting goals, negotiating agreements, and developing organizational arrangements required time. But one early result of massively increased awareness of threats to the world environment was a corresponding increase in mass dissatisfaction in many countries with the ways in which technology was being applied. This dissatisfaction was most apparent among the better-educated, upper-middle class in the developed countries. Among the youth it tended to be expressed in hostility toward the priorities and behaviors of governments and corporate enterprise. These attitudes were reinforced by the ecologically destructive and socially disastrous course of the war in Southeast Asia—and were readily seized upon by Marxist and other anti-established-order groups to attack government, business, and conventional lifestyles in the developed countries.

A preview of protest at Stockholm could have been obtained from the World Youth Conference held in Hamilton, Ontario, in late August 1971 with sponsorship of the conference secretariat, UNESCO, IUCN, and the Environic Foundation International. The notion of Secretary-General Strong that "the greatest potential influence of youth would be in seeking to influence their own governments" was plausible.[33] But the effect of left-leaning, militant youth seizing control of the Hamilton conference to denounce capitalism, imperialism, racism, and war in an accusatory spirit aimed at the United States in particular and developed countries in general was not calculated to influence many governments in a positive way.

The PREPCOM held its third session in New York in September 1971 with the draft Declaration on the Human Environment being a major item on the agenda. On 22 December 1970, the secretary-general had invited governments to comment on the possible form and contents of a draft declaration that one of the intergovernmental working groups had been assigned the task of preparing for consideration by the full Preparatory Committee. There was substantial agreement among committee members that the declaration should be a document of universally recognized fundamental principles, recommended for action by individuals, national governments, and the international community. Preconference approval of the declaration was accomplished at the Fourth Session of the PREPCOM, 6–10 March 1972, at which final arrangements for the conference were ratified.

The preparatory work had been comprehensive and thorough—four years and the efforts of hundreds of people in nearly all parts of the world having been enlisted in the action. The scene now shifted to Stockholm where the Swedish government and the conference secretariat were already engaged in completing the extensive physical arrangements that a meeting of this character required, especially in view of the political, social, and ideological tensions associated with the environmental issue in its international dimension.

4

The Stockholm Conference

The United Nations Conference on the Human Environment marked the culmination of efforts to place the protection of the biosphere on the official agenda of international policy and law.[1] Specific aspects of the environment had been the objects of international negotiations and arrangements, but the concept of the collective responsibility of nations for the quality and protection of the earth as a whole did not gain political recognition until the years immediately preceding the Stockholm Conference. Stockholm enlarged and facilitated means toward international action previously limited by inadequate perception of environmental issues and by restrictive concepts of national sovereignty and international interest. In effect, nation-states joined together their sovereignty and jurisdiction to resolve collectively issues that previously would have been definable only within the limits of particular national jurisdictions.

As previously noted, international conferences have become major instruments in the development of international policy. They have been both governmental and nongovernmental: nongovernmental meetings often immediately preceding, or coterminous with, those strictly governmental. Conferences affecting environmental policy have often mixed scientific and political considerations—their outcomes being political, but often informed and modified to some extent by scientific evidence.

Positive elements in these conferences have been (1) the stimulation of awareness of issues affecting all or most nations, (2) the opportunity to air grievances and reveal hidden tensions, and (3) obtaining agreement among nation-states sufficient to afford a basis for cooperative action, including research and institutional arrangements.

Negative elements have been (1) opportunities for inflammatory rhetoric and distortion of issues for purposes of propaganda, (2) a tendency to compromise issues to a point of inaction, and (3) uncertainty regarding the ability of governments to honor conference commitments. Particularly for the last reason, some institutionalized follow-up has been essential to the success of most international conferences. Establishment of the United Nations Environment Programme and Environment Fund was indispensable to a productive consequence of the Stockholm Conference. Unwilling to recommend a major new organization for the environment having specialized agency status, the Stockholm Conference proposed a special body within the UN Secretariat to stimulate, assist, and coordinate the international protective efforts. This institution, established by resolution of the General Assembly, was the United Nations Environment Programme (UNEP). At its inception in 1973, there was danger in its being ignored by the old-line UN agencies, and that the so-called north–south tensions among the developed and developing nations would polarize UNEP or dissipate its very limited resources. But with the passing of time, UNEP appears to have established itself within the UN system and has to some extent alleviated north–south tensions in which environment and development appeared to represent conflicting values. It provides an integrative mechanism through which a large number of separate efforts — intergovernmental and nongovernmental, national and regional — can be reinforced and interrelated. Thus UNEP functions less as a unilateral effort than as a coordinative, interactive mechanism among many actors on behalf of the encompassing goal of international environmental protection.

Divergent Viewpoints

On 5 June 1972, after more than two years of intensive preparation, the United Nations Conference on the Human Environment opened its first plenary session at the Royal Opera House in Stockholm.[2] It was immediately confronted by a division that may yet frustrate the achievement of its objectives. Two conflicting viewpoints were present. From the perspective of the first, the primary concern of the conference was the human impact on the biophysical environment with emphasis on control of pollution and conservation of resources. The second viewpoint held social and economic development (as perceived by the viewer) as the real issue. To bridge these differences, the concept was advanced that environmental protection was an essential element of social and economic development, and as one of the participants in building the bridge observed, "Once development had become a dimension of the problem, it quickly gained prominence."[3]

In addressing the conference, Olaf Palme, prime minister of Sweden, declared that his government attached "the greatest importance to the stress laid in the Declaration upon the need for development." He found it "an inescapable fact that each individual in the industrialized countries draws, on the average, thirty times more heavily on the limited resources of the earth than his fellow man in the developing countries." He concluded that "these simple facts inevitably raise the question of equality, of more equal distribution between countries and within countries."[4] Palme's highly selective examples of ecological and economic exploitation by the industrialized countries, particularly by the United States, set the tone of much of the debate in plenary sessions of the conference. Third World statements, often paralleled by First World mea culpas, reiterated the thesis that the foremost environmental problem of the world was Third World poverty caused almost wholly by the exploitive practices of the developed nations.

Indira Gandhi, prime minister of India, found poverty the greatest polluter. As she declared, "Many of the advanced countries of today have reached their present affluence by their domination over other races and countries, the exploitation of their own masses and own natural resources. They got a head start through sheer ruthlessness, undisturbed by feelings of compassion or by abstract theories of freedom, equality, or justice."[5] Mrs. Gandhi's reading of history was widely shared among Third World representatives and expressed most violently by the Maoist spokesman of the People's Republic of China. Even the more moderate Third World delegates such as Helena Z. Benitaz of the Philippines emphasized the injustice theme, asserting that "[a] past age of domination has left in many countries of the so-called Third World . . . stunted and malformed economies perpetuating to this day the poverty of blighted, stagnant and benighted rural communities."[6]

That these ideological attacks did not wreck the conference at the outset owes much to the diplomatic skills of the conference secretary-general, Maurice Strong, and the forbearance of the delegations from the developed nations. But international ideological differences, surmounted for the time being, were not removed. They remained to be confronted subsequently by the General Assembly and the new United Nations Environment Programme. Ameliorative and evasive strategies adopted by the developed nations postponed the day of actual confrontation, but they also bought time for common interests and more rational consideration of differences to emerge.

For the 114 governments represented at Stockholm to have agreed generally on a declaration of principles and an action plan was a remarkable accomplishment. Agreement may have been facilitated by the voluntary absence of the Soviet Union and its East European allies in protest over the failure of the German Democratic Republic to be accepted as a conference participant. The East German exclusion was primarily a consequence of the formula adopted by the United Nations regarding attendance at United Nations conferences.[7] Soviet citizens, however, participated in the conference as official members of the conference staff.

How Stockholm Was Different

The accomplishments of United Nations conferences have generally not been impressive. Their outputs have tended to be heavily rhetorical. Characteristically their resolutions are ambiguous compromises among conflicting ideologies, and in the years after Stockholm they have tended to become vehicles for the complaints of the developing against the developed nations.

Rhetoric flourished at Stockholm, but the United Nations Conference on the Human Environment differed from other United Nations conferences in its initiation of a sequence of positive measures that have translated published resolutions into actual accomplishments. This positive outcome has been the principal distinguishing feature of the Stockholm Conference, and was the result of four factors that at least in degree made this conference different from the others.

First, the conference from its preparatory stage was action-oriented; it was intended by its managers to lead to positive results and not merely to statements of principles.

Second, preparations for the conference were extensive and thorough, with sufficient time to obtain agreements and to resolve or manage the more dangerous political differences. Accommodation among political viewpoints did not necessarily imply agreement, but rather that respective parties understood their differences and were able to find compromises that would avoid disruption of the conference.

Third, popular interest and support reinforced the sense of the necessity for the conference and its action orientation, even though their direct influence upon the delegates at Stockholm was not great. The presence of representatives of nongovernmental organizations (NGOs) has become common at all United Nations conferences, but the number, variety, and attendance at Stockholm was exceptional. A variety of unofficial gatherings took place coterminous with the conference.[8] The Swedish government provided accommodations for the Environment Forum some distance removed from the conference site and opened a deactivated airport near the end of a subway line as a campground. Other gatherings, inclining to the political left, linked environmental considerations to denunciation of the war in Southeast Asia. The so-called Peoples' Forum and Dai Dong were near the outer limits of left-leaning environmentalism.

Partly as a result of the politicizing of the Environment Forum by the radical left, many of the NGOs met separately, having daily briefings from Barbara Ward and others who were officially involved in the conference. Anthropologist Margaret Mead by invitation addressed a plenary session presenting a statement of principles adopted by the NGOs.[9] Although moderate in tone, as contrasted with some of the oratory at the Environment Forum, the NGO principles embodied idealistic concepts of the use of political means to achieve environmental and sociological objectives: their basic but vaguely defined goals were "social justice and redistribution." The NGO report contained many prescriptions with which nearly all informed students of environmental policy would agree. But the report was long on "musts" and "shoulds" and very short on "hows." Among partisans of the left at the Environment Forum, however, priorities were clear. The "party line" reported by one American observer was that "[i]t is useless to talk about controlling pollution until all governments become socialistic, poverty is abolished, and social justice reigns."[10]

Although the unofficial inputs may have had little direct influence upon the deliberations of the conference, they nonetheless reflected many of the political movements and pressures that had brought the official delegates to Stockholm. Their presence was a well-advertised reminder of the hopes and expectations of peoples in many parts of the world for a positive and constructive outcome of the conference.

The information media—using photographic film, radio, television, cheap printing, and low international air fares—made possible a degree of communication, shared purpose, and a visible presence that had never previously characterized high-level international conferences. The unofficial assembly of ecologically concerned youth, radicals, scientists, and conservationists from around the world was not entirely facetiously described as "Woodstockholm" —"a ritual celebration of an emotional commitment to a new orientation toward life and the earth."[11]

Fourth and finally, the success of the conference in achieving a positive outcome owed much to skillful management before, during, and after it. The leadership of the conference secretary-general, Maurice Strong, was consistently directed toward holding the collective effort together and focusing its deliberations on positive outcomes. The continuity of his coordinative role both in the preparatory phases, at Stockholm, and subsequently as the first executive director of the United Nations Environment Programme (UNEP) must surely be counted as a major factor in its success.

Assessment of Achievement

The formal output of the conference consisted of the Declaration on the Human Environment, the Declaration of Principles, Recommendations for Action (109 in number), and the Resolution on Institutional and Financial Arrangements. There were minor resolutions on World Environment Day, nuclear weapons tests, and a Second United Nations Conference on the Human Environment.

In relation to what might reasonably have been hoped, Stockholm may be accounted a success. It avoided foundering on antagonisms born of Third World resentment over First World "injustice." The strategy for this avoidance was incorporation of environmental protection into the Third World's development priorities. Yet this First World concession introduced a new environmental element into the conventional interpretation of development.

The development concept was thus enlarged, and delegates were exposed to evidence that many social and economic problems had environmental connections. Assessments of the accomplishments of Stockholm have varied. But the more informed judgment appears to be that "[i]n general, this experience was technically similar, but perhaps politically superior, to any success and legitimacy that the United Nations can claim in its history of international undertakings."[12]

In retrospect, the primary accomplishment of the United Nations Conference on the Human Environment was the identification and legitimization of the biosphere as an object of national and international policy. Its resolutions provided standards for environment-related acts of government, which even regimes indifferent to environmental values felt obliged to acknowledge as evidence for a status of progressive "modernity."

The infusion of newer international law was not dramatic, but trends in effect before Stockholm relating to marine pollution, transboundary air and water pollution, and protection of endangered species, among others, were reinforced by the Stockholm resolutions. And the proceedings of the conference itself exemplified the process by which positive or codified international lawmaking is increasingly being developed.

Putting aside until later the question of whether Stockholm initiated actions that may eventually lead to a coherent political order for the biosphere, one may identify four ways in which novelty characterized the influence of Stockholm on international relations. There were significant elements of innovation in (1) the redefinition of international issues, (2) the rationale for cooperation, (3) the approach to international responsibility, and (4) the conceptualization of international organizational relationships.

Redefinition of International Issues

As noted in Chapter 3, the conference agenda at Stockholm was divided into six main subject areas:

1. Planning and Management of Human Settlements for Environmental Quality
2. Environmental Aspects of Natural Resources Management
3. Identification and Control of Pollutants and Nuisances of Broad International Significance
4. Educational Informational, Social, and Cultural Aspects of Environmental Issues
5. Development and Environment
6. International Organizational Implications of Action Proposals

From these agenda items, 109 recommendations (or rather sets of recommendations) were addressed to governments, intergovernmental agencies, and nongovernmental organizations. Most of the numbered recommendations consisted of several parts, so that the total number of actions recommended greatly exceeded the number of formal resolutions. For example, Recommendation 20 was concerned with strengthening machinery for the international acquisition of knowledge and transfer of experience on soil capabilities and the degradation, conservation, and restoration of soils; it consisted of at least ten specific lines of action. Recommendation 86 to national governments regarding marine pollution included six separate provisions. Recommendation 51 on international river basins comprised at least thirteen subdivisions.

One apparent reason for the extended character of these and other recommendations was the complexity and political novelty of the issues. Few of them could be defined or dealt with exclusively within given disciplinary or sectoral contexts. In essence, the problems of the human environment were multidisciplinary, and had to be approached with more complex strategies than those employed historically in setting forth the mutual obligations of nations through treaties or by customary law.

This delineation of issues revealed the state of international consensus (or its absence) on problems of the human environment. Aside from educational and organizational matters, the conference recognized four general sets of issues: (1) human settlements, (2) natural resources, (3) pollution, and (4) the conflict or balance between development and environment. Both the Preparatory Committee and the conference addressed issues that national governments realistically could, or would, address. Thus matters that most nations preferred to avoid, such as the dynamics of human population, mismanagement of economic development programs, and the indirect threat of militant ideologies to the global environment, were not major agenda items. The racial policies of South Africa received a prominent place in the conference's declaration of principles, but the oppression of minorities by Third World governments was politely ignored.

The conference tended to focus upon symptoms more often than upon causes—yet, being action-oriented, it could not profitably extend its agenda beyond areas in which nations were willing to act. Nevertheless, in addressing symptoms the conference was sometimes forced to examine causes. It has been a truism of politics that probing causes seldom leads to consensus, and that politically sensitive issues are often best addressed obliquely. At Stockholm the need for research into environmental problems was emphasized, but responsibility for finding ways to pay the costs of research was left to the governments and international agencies.

Rationale for Cooperation

There was conflict between science advisers and foreign policy advisers at Stockholm, reflecting differing assumptions regarding the bases and priorities of international cooperation.[13] These differences had been anticipated before the conference convened and were never wholly overcome. A third source of advice came from nonscientific NGOs whose outlooks differed from both the diplomats and the scientists.[14]

Many of the scientists and more of the nongovernmental participants in the Stockholm Environment Forum called for the institutionalization of new supranational loyalties to the planet and to humanity as a species. This viewpoint implied a positive international law enforceable directly upon individuals and especially on international business enterprise. It also implied a subordination of national sovereignty to some form of international jurisdiction—an eventuality for which no government seemed prepared and which was not even considered at the conference.

There were specific issues on which government delegates were not prepared to compromise their views of national interest for the protection of the biosphere. Nevertheless, the improbable outcome of Stockholm was the extent of international consensus on the broad—but sometimes detailed—provisions of the Action Plan. National objections or abstentions with respect to particular recommendations of the conference were exceptional.

Foreign office views of environmental policy tended to guard traditionally perceived national interest from sacrifice to "idealistic causes." For example, France was unwilling to stop the atmospheric testing of nuclear weapons; Japan was unwilling to observe Recommendation 33 calling for a ten-year moratorium on commercial whaling; the United States was unwilling to accept the principle of "additionality" that would have required an increase in its foreign aid budget to cover the additional costs imposed by environmental protection measures on development projects. Nevertheless, there was general acknowledgment that all nations had a common interest in the preservation of the biosphere, and that unilateral protective action could not be made effective for a wide range of environment-related issues. Thus, as with the rationale behind the International Treaty on the Peaceful Use of Outer Space (1966), the impetus for cooperation at Stockholm was a common threat commonly perceived.

The developing nations differed from the industrial states with respect to priorities; their positions at Stockholm reflected very different circumstances with respect to the character and influence of domestic public opinion. In most developing countries public opinion, to the extent that it existed in relation to environmental issues, was inchoate and inarticulate. Few organized and influential citizen groups or independent and critical presses were pushing their governments toward environmental protection measures at home or at Stockholm. Even so, in some developing countries a small but disproportionately influential elite in government and the universities was aware of world trends and of a national stake in international environmental protection measures.

This opinion was not public in the conventional Western sense, but it was nevertheless essential in moving governments. Its existence was demonstrated in 1971 by a letter of inquiry sent by Francesco di Castri (then a professor in Chile) to more than one hundred scientists in developing countries.[15] Responses indicated that concern over environmental deterioration did exist in many Asian, African, and Latin American countries, and that unqualified commitment to economic growth and industrialization regardless of ecological or social cost was not as general among their elites as many skeptics in the industrialized countries had believed. Nevertheless, international differences regarding the substance and significance of environmental problems had been built into national policy positions and could have defeated cooperative efforts at Stockholm.

In the main, the predicted confrontation at Stockholm between developed and developing nations over the impact of environmental protection on development was effectively contained. This again was largely due to the political skill of Secretary-General Maurice F. Strong, who had formerly headed the international development program of the government of Canada, and to the thorough preparatory meetings in which conflicting perspectives on environment and development had been clarified and, in some instances, reconciled. These preparatory activities, described in Chapter 4, provided elements of a common foundation in knowledge for the representatives of the 114 nations participating in the Stockholm Conference. The existence of a foundation of fact and perception did not mean that a new edifice of international policy or law could be readily erected upon it. But without this common base, action toward a new international environmental order could never begin. The Stockholm Declaration of Principles (supplementary to the Declaration on the Human Environment) provided a rationale for international cooperation for environmental protection. Such a beginning had to occur before general international cooperation could follow.

Enlarging National Responsibilities

Although the Stockholm Conference was conducted according to the traditional protocol of international conferences, the importance of the nongovernmental input to the conference at national and international levels, and the presence of the unofficial gatherings at Stockholm provided, de facto, a broader scope for interaction than is characteristic of strictly intergovernmental deliberations. The presence of representatives of many nongovernmental organizations may not have directly influenced the official delegates at Stockholm, but it provided information feedback to their home populations, their governments, and their news media. This broadened participation affected not only the substance of conference rhetoric, but influenced public perceptions of the conference. The agenda and the substance of the resolutions had of course been fixed before Stockholm. But nongovernmental influence had been present throughout the preparatory phase.

Classic international law is the law of nations—not of individuals or nongovernmental organizations. But the Stockholm resolutions were directed not only to national governments, but also to "peoples," to international agencies, and to governments collectively with respect to issues requiring their collective action. Were the Stockholm recommendations and resolutions anticipatory of an emergent transnational environmental law—a law *for* nations as contrasted with a law *of* nations? A case might be made that environmental declarations, resolutions, and agreements since Stockholm have pointed in this direction. For cases in point, see Appendix E, Environmental Soft Law.

Illustrations of implied transnational issues may be drawn from Recommendations 86 to 94 dealing with marine pollution, and from the reference in the Declaration on the Human Environment to the "common international realm," which by implication included the high seas, the seabed, and the ocean floor, and which was already subject to action by the UN General Assembly and the Third United Nations Conference on the Law of the Sea. The Stockholm recommendations at times exceeded the readiness of nations to act upon principle, a situation paralleled by the difficulties of the Conference on the Law of the Sea in reaching agreement on fundamental propositions that contained restrictive implications for national policies. The possibility that principles of international law might be broadened in the future by emergencies requiring assumption of responsibility by national governments indicates a need for closer attention here to what was actually declared.

The preamble to the Declaration on the Human Environment stated that there is "the need for a common outlook and for common principles to inspire and guide the peoples of the world in the preservation and enhancement of the human environment." Paragraph 7 of the declaration specified the locus of responsibility for achieving the objectives of the conference:

> To achieve this environmental goal will demand the acceptance of responsibility by citizens and communities and by enterprises and institutions at every level, all sharing equitably in common efforts. . . . Local and national governments will bear the greatest burden for large-scale environmental policy and action within their jurisdictions. International co-operation is also needed in order to raise resources to support the developing countries in carrying out their responsibilities in this field. A

growing class of environmental problems, because they are regional or global in extent or because they affect the common international realm, will require extensive cooperation among nations and action by international organizations in the common interest. The Conference calls upon Governments and peoples to exert common efforts for the preservation and improvement of the human environment, for the benefit of all the people and for their posterity.[16]

This provision did not appear to change the subjects of international law, but to the extent that individuals and the variety of organizations included in paragraph 7 interact in carrying out new efforts and programs transcending national boundaries, there could be a logical extension of international law to those activities and relationships. In the long run, modification both of conventional international law and of national sovereignty could be significant, and this trend was continued in Agenda 21 in 1992 by the UN Conference in Rio.

Principle 1 of the Declaration of Principles was addressed to the rights and responsibilities of individuals, with the implication (elsewhere made explicit in Recommendations 95, 96, and 97 pertaining to educational, informational, social, and cultural aspects of environmental problems) that national governments with the assistance of international agencies should enable their people to become informed on environmental issues. The substance of numerous recommendations was what governments should do in relation to their own people rather than, as in traditional international law, what a nation-state should or should not do in relation to other nation-states.

An issue with more solid implications for possible changes in international responsibilities was the extent to which the developed or rich nations were obliged to assist the developing or poor nations in reconciling their development efforts with environmental quality objectives. Two aspects of this issue took shape at Stockholm and remain points of controversy in the post-Stockholm period. These aspects were expressed in Recommendations 103, "compensation," and 107, "additionality." [17]

The principle of compensation belongs to a set of propositions later formalized by the UN General Assembly in the Declaration of a New International Economic Order. Paragraph (b) of Stockholm Recommendation 103 stated the compensation issue in essence: "That where environmental concerns lead to restrictions on trade, or to stricter environmental standards with negative effects on exports, particularly from developing countries, appropriate measures for compensation should be worked out within the framework of existing contractual and institutional arrangements."

This recommendation was debated at length with the developed nations generally in opposition. The United States rejected the proposition that nations should be compensated for declines in their export earnings that could, in fact, be brought about by causes not necessarily fairly attributed to the environmental policies of other nations. Other industrialized countries — France, Italy, and the United Kingdom — opposed the provision less on principle than because they felt it to be unworkable.

The principle of additionality was formalized in several of the pre-Stockholm meetings of the regional economic commissions and in the Preparatory Committee, and subsequent to Stockholm, by Resolution 3002 (XXVII) of the General Assembly (15 December 1972). Its clearest expression at Stockholm was in Recommendation 107, which declared that "[e]nvironmental problems should not affect the flow of assistance to developing countries, and that this flow should be adequate to meet the additional requirements of such countries," and Recommendation 109, which stated that "[i]t should be further ensured that the preoccupation of developed countries with their own environmental problems should not affect the flow of assistance . . . and that this flow should be adequate to meet the *additional* environmental requirements of such [developing] countries."

More specifically, the impact of this concept was that existing development funds should not be diverted to environmental quality purposes, and that funds to carry out the recommendations of the Stockholm Conference should be in addition to those now allocated to developmental purposes. Both the compensation and additionality principles represent efforts to found a new relationship of rights and obligations among nations — in this instance, between developed and developing nations.

With respect to an established principle of international law that a state must pay compensation for injury to another state caused by activities originating in its own territory, the Conference, through Principles 21 and 22 of the Declaration, took a reinforcing position. Principle 22, however, moved beyond custom, stipulating that "states shall co-operate to develop further the international law regarding the liability and compensation for the victims of pollution and other environmental damage caused by activities within the jurisdiction or control of such states to areas beyond the limits of national jurisdiction." This principle could not be easily implemented. Where the origin of acid precipitation is traced to several states, including states most injured, it is difficult to fix proportionate responsibility. Many states may contribute in some measure to the problem, and the precise source of the damage in any one state may be impossible to ascertain. Governments may be unwilling to make compensatory payments, and there is no way to compel them to do so. For example, the Soviet Union did not pay for the harm caused in neighboring countries by the Chernobyl accident in 1984.[18]

In summation, a legacy of Stockholm was an enlarged and reinforced concept of national environmental responsibility that had prospective bearing upon the future of international political, legal, and organizational relationships.

International Institutional Arrangements

In a review of the first year following Stockholm, Maurice Strong emphasized the organizational logic of the Stockholm recommendations, which he said called for "a drastically new concept of management." [19] Conceding that for many purposes the hierarchical bureaucratic structure of governments and international organizations had worked well in the past, he observed that this form of organization "has made it difficult to perceive — and even more difficult to deal with — complex environmental cause-and-effect relationships that transcend disciplinary and institutional boundaries." Continuing, he declared that "the environment cannot be sectoralized. It is a system of interacting relationships that extends through all sectors of activity, and to manage these relationships requires an integrative approach for which present structures were not designed." This means, he said, "that lines of communication and decision making must be given much greater horizontal and trans-sectoral dimensions than are provided for in existing structures," and that "the new patterns of organization in an era of societal management must be based on a multitude of centers of information and of energy and power, linked together within a system in which they can interact with each other."

These concepts reflect not only the logic of the leadership at Stockholm and the subsequent strategy for the United Nations Environment Programme, they may also indicate the probable direction of social organization in the future and hold far-reaching implications for the structure, the subject matter, and the processes of international law and relations.

The concept of sovereignty is problematic to the organizational issue. Although the Stockholm Declaration of Principles speaks of sovereign rights of states to exploit their own resources, the total effect of the document is to modify the exercise of sovereignty. The traditional unidimensional view of sovereignty appears superficially to be inconsistent with Maurice Strong's view of the multidimensional requirements of planetary environmental protection. However, Strong has interpreted the use of sovereignty in a positive and innovative manner. Nations may merge their sovereignty. He writes:

> But the development of new international machinery to deal with the complex problems of an increasingly interdependent technological civilization will not come about through the surrender of sovereignty by national governments but only by the purposeful exercise of that sovereignty. It is only when nations find themselves incapable of exercising their sovereignty effectively or advantageously on a unilateral basis that they will agree — reluctantly — to exercise it collectively by agreement with other nations. It is seldom that nations enter into arrangements which restrict their ability to exercise their sovereignty until circumstances compel them to do so.

Since Stockholm the concept of merged sovereignties has been exemplified by the European Union and by the UNEP Oceans and Coastal Areas Programme, which now includes the formerly separate Regional Seas Programme, in which nations have developed common institutions to cope with shared problems. This development simultaneously decentralizes responsibility on a global basis and centralizes it on a regional basis. The localized character of many environmental problems has made logical the development of localized institutions to deal with them. A consequence is heterarchy, or decentralized structuring of institutions, as contrasted with a global hierarchy administered by an encompassing international agency. To assist the formation and coordination of this "heterarchial" network has been a principal task of the United Nations Environment Programme. The difficulties and hazards are all too apparent. As Maurice Strong has said, the effort will require "a degree of enlightened political will on the part of the peoples and nations of the world that is without precedent in human history."

Before adjournment, the conference passed a separate set of resolutions establishing institutional arrangements for implementing the UN Environment Programme and setting up an Environment Fund. The Governing Council, which was to oversee the implementation of the Action Plan (i.e., the 109 resolutions of the conference rearranged according to function), as originally conceived was to be composed of fifty-four members elected for three-year terms on the basis of equitable geographic distribution. The council was to act largely as a coordinating body among existing United Nations organizations: "To promote international cooperation in the environmental field and to recommend, as appropriate, policies to this end; [and] to provide general policy guidance for the direction and coordination of environmental programs within the UN system." [20] Other resolutions specified the functions of the Council and the Secretariat. In addition, the conference recommended that the United Nations establish a voluntary environmental fund of $100 million for a five-year period and called for member contributions.

Action by the General Assembly

The Report of the Conference on the Human Environment was sent to the Economic and Social Council, which took note of the report and sent it on to the General Assembly for action. On 15 December 1972, upon recommendation of the Second Committee (Economic and Financial), the Twenty-seventh General Assembly adopted eleven resolutions related to the environment and the Stockholm Conference.[21]

By Resolution 2997, the General Assembly of the United Nations accepted the recommendations of the Stockholm Conference on the institutional and financial arrangements for international environmental cooperation with one exception. At the insistence of the developing nations, the size of the Governing Council was expanded from fifty-four to fifty-eight nations—the additional seats to accommodate greater representation from Asia.

Under General Assembly Resolution 3004, the developing nations were able to change the location of the UNEP secretariat from what had been believed to be its probable site in Geneva to Nairobi, Kenya. Initiative for this action was provided by a group of recently independent African states and was supported by most of the developing nations, moved more by resentment that no major UN facility had been located in a developing nation than by their concern for the environment cause.[22] The resolution passed by a wide majority; ninety-three states in favor, thirty abstentions (mostly the developed nations), and one opposed (the United States). The move was seen as a clear indication of the control that the developing nations proposed to exert in UNEP affairs.

The location of the secretariat in Nairobi has complicated both UNEP's coordination with other UN bodies and its contact with governments, the scientific community, the environment movement, and the information media. More important, the Third World's tactical victory on priorities for UNEP threatened to jeopardize financial and political support from the developed countries.

Finally, General Assembly Resolution 3002, paragraph 4, dealt directly with the question of additionality. In this resolution, entitled "Development and the Environment," the General Assembly recognized that funds available for the environment would be scarce in relation to need, but nevertheless supported the claims of the developing countries for additional finance. In the operative section of the resolution the General Assembly recommended "respect for the principle that resources for environmental programmes both within and outside the United Nations System, be additional to the present level and projected growth of resources contemplated . . . to be made available for programmes directly related to developmental assistance." While Resolution 3002 passed unopposed, considerable opposition had been voiced by the developed nations against paragraph 4 in the meeting of the General Assembly's Second (Economic and Financial) Committee.

To the extent that the developing nations were able to agree among themselves on the issues of environment and development, they were able to have a decisive influence on events. During the preconference preparation period they were able to obtain a definition of environmental problems that reflected their interests, and to guarantee consideration by the Stockholm Conference on the issues of compensation and additionality. Because of their relative unity of outlook and because of the need to include them in any agreement, the developing states were able to win at Stockholm a broadened interpretation of compensation and to have legitimized the principle of additionality. At the

Twenty-seventh General Assembly, they were able to marshal their numerical strength to have their views officially adopted.

This consolidating of political interests never fully materialized at Stockholm, but became effective in the Twenty-seventh Session of the General Assembly in the establishment of UNEP. The General Assembly accepted the Stockholm recommendations, with the developing countries predominating in the establishment of UNEP and the Governing Council. And the General Assembly instructed the council to give special consideration in the formulation of programs and priorities that might assist in accelerating the economic development of developing countries without environmental disruption.

The acceptance by the General Assembly of the Declaration of Principles and Recommendations for Action of the Stockholm Conference, however, did not guarantee nor make automatic the implementation of the Stockholm recommendations. The recommendations did not carry the full force of international agreements and did not impose binding obligations on the participating nations. Implementation of the Stockholm Recommendations for Action depended upon the perceived interests and the cooperation capabilities of sovereign nation-states.

In Retrospect

On balance, what difference did Stockholm make? Did the results justify the efforts? The answer depends substantially upon expectations. It is easy to expect too much from international conferences, especially when issues are subject to differing interpretations, and compliance with recommendations is voluntary. The personal interests of governing officials and the influence of powerful economic and political forces can marginalize the best of intentions. The principal achievement of Stockholm was that it occurred, and concluded with a rhetorical show of unanimity and without divisive conflict. It put environment on the international agenda of nations and caused many states to adopt, at least pro forma, a domestic environmental policy. It legitimized an issue that a decade earlier had not been recognized.

Rhetoric is a necessary attribute of evolving policy. Ideas and propositions must be voiced and disseminated before they obtain tangible expression. Chapter 5 will consider the legacy of Stockholm in greater detail. Suffice it to say here that without the impetus of the 1972 UN Conference, the international environmental developments of the 1970s and 1980s would probably not have occurred. Nations must become aware of issues before they can or will act upon them. Stockholm had this consciousness-raising effect. It provided a precedent for international collaboration that guided the behavior of national governments on a variety of environmental issues over the succeeding twenty years, culminating in the Rio Conference of 1992.

5
Post-Stockholm Assessment

The two decades following Stockholm were characterized by a great advance in the international acceptance of global environmental policy.[1] Especially significant was a change of attitude in much of the so-called Third World and the collapse of the Soviet Union, which greatly diminished the frustrating effects of Cold War rivalry.[2] Environment and development were no longer seen as necessarily adversarial. By 1992 nations were ready to move beyond Stockholm in addressing global problems of the environment. Of course, much of this advance was rhetorical (i.e., paper commitments that some signatories were not ready to honor). Even so, declared intention is usually precedent to action. The significance of Stockholm and Rio may be assessed with more factual evidence in 2022 than is possible today.

The United Nations Environment Programme

The principal accomplishments of the Stockholm Conference were twofold: the official recognition of the environment as a subject of general international concern and the institutionalization of that concept in the United Nations Environment Programme (UNEP).[3] As with many new institutions, UNEP came into existence with certain policies predetermined. The direction that it would take could be surmised from the composition of its Governing Council. It was moreover born into an environment of international political contention and socioeconomic instability that would most certainly influence the scope and character of its activities.

Implementing the recommendation of Stockholm for a United Nations Environment Programme, the UN General Assembly on 15 December 1972, through Resolution 2997 (XXVII), established the necessary institutional and financial arrangements.[4] The UNEP secretariat was to be small and serve as a focal point for environment-related activities within the United Nations system. At its head was an executive director (also an under-secretary-general of the United Nations), elected by the General Assembly rather than by the

Governing Council. In addition, an Environment Coordination Board, representing the United Nations agencies with environment-related programs, functioned for several years under the chairmanship of executive director. After 1977 its functions were absorbed by the United Nations Administrative Committee on Coordination, to which it had reported.

The policies of UNEP are set by the Governing Council, which receives reports and recommendations from the executive director. Its membership, first set at twenty-seven by the PREPCOM, was increased to fifty-eight by the General Assembly, thus giving the developing nations a clear majority, but also complicating its agenda-setting functions. From the viewpoint of some observers in developed countries, UNEP appeared to be one more forum for the voicing of frustrations of former colonial states and yet another channel for development assistance. These misgivings have been partially relieved as the Third World majority, under leadership from the UNEP secretariat, has come to recognize the seriousness of environmental issues. And the politics of UNEP has been somewhat eased as the developed countries have sought their international environmental objectives through other organizations, such as NATO, OECD, and the European Union, leaving UNEP to concern itself largely with environmental problems of the Third World or those that require a North–South dialogue. The disintegration of the Soviet Union and revelation of its severe environmental mismanagement relieved the UN of the too often frustrating effects of Cold War maneuvering by the USSR and the United States.

Launching UNEP

It would not serve the purposes of this book to highlight each session of the Governing Council, but the first three sessions (1973–75) were formative and merit special attention because they set the course of UNEP to the present date. The First Session of the Governing Council met in Geneva from 12–22 June 1973.[5] It was at this session that the consolidation of the political interests and numerical strength of the developing nations enabled them to shape the policies of UNEP. Development-related priorities predominated. For example, one of the prime interests of the developed nations was Earthwatch, a worldwide monitoring program to detect significant changes in critical environmental conditions, whose establishment had been an explicit recommendation of the Stockholm Conference. But the International Referral System (now INFOTERRA), an integral part of the Earthwatch program for exchange of information, suffered an initial downgrading at the hands of the representatives from the developing nations. Instead of being put into operation immediately, the secretariat of UNEP was empowered to initiate only a pilot phase.

The Geneva session of the Governing Council was to determine what part of the Action Plan would be the initial focus of UNEP's activities. The developing nations viewed the proposed priority areas, as put forward by Maurice Strong, as placing excessive emphasis upon environmental problems caused by industrial states, rejected the proposal, appointed their own drafting group, and substituted their own priorities. To avoid an open clash, a compromise was reached. The priority areas adopted were (1) human settlements, (2) land, water, and desertification, (3) education, training, assistance, and information, (4) trade, economics, and the transfer of technology, (5) oceans, (6) conservation of nature, wildlife, and genetic resources, and (7) energy.

Compensation for alleged disruptions of exports and demands for additional funding to meet environmental standards were also very much on the agenda of the developing nations at the First Session of the Governing Council. In a report on priority area 4 (trade, economics, and the transfer of technology) the Governing Council majority asserted the developed nations' obligation (1) to ensure that environmental measures adopted by their governments not create barriers to trade, especially with the developing nations, and (2) to examine ways to offset negative trade effects resulting from environmental measures taken by developed nations. Paragraph 5 of the report succinctly summed up the position of the Governing Council on additionality, calling upon UNEP to take steps in collaboration with other appropriate agencies to encourage developed countries to make increased capital assistance available to developing countries so that extra costs of introducing environmentally sound technologies by them would be covered.

The Second Session, held in Nairobi (10–23 March 1974) was described as harmonious, particularly when compared with the First Session. The council reaffirmed its previously designated priority areas and assigned "functional" priority to the creation of Earthwatch and its major components, the Global Environmental Monitoring System (GEMS) and the International Referral System (INFOTERRA). Executive Director Strong, in his subsequent report to the Twenty-ninth Session of the General Assembly, explained that the information referral and monitoring systems were to be developed with emphasis on technical assistance and training to meet the needs of developing countries, thus ensuring that they would be able to derive full benefits from these projects.

The Second Session of the council also saw what may have been the first limited application of the concept of additionality during the decision to institute Earthwatch. Those monitoring stations that would be, or were already, established on the territory of industrialized nations would be paid for by them. In developing nations, whose participation in the network would be needed for its effectiveness, but which lacked the means for implementation, external assistance would be provided. The Environment Fund was seen as the logical source of this additional support. Thus, if the developed nations wanted the monitoring system, it would be necessary to subsidize establishment of the monitoring stations in the developing states.

The Third Session of the Governing Council (17 April to 2 May 1975) adopted over twenty-five major decisions that were consistent with previously established objectives and priorities. The developing nations maintained their dominance, but the continued harmony begun at the Second Session suggested a balance had been struck between the interests of the developed and developing nations. The only apparent discordant note was the warning which Maurice Strong seemed to be directing toward the developed nations. In his opening statement to the council, the executive director declared that short-term economic interests were threatening the world's future. He went on to say, "[A]s long as measures to protect and improve the environment are seen largely in terms of 'added costs' while impairment of our basic environmental capital is not counted as a cost, economic incentives will continue to seem to run counter to environmental interests."[6]

During these initial sessions of UNEP no consensus was reached on the responsibilities of the developed nations for Third World environmental problems. Third World leadership often took positions that, to outside observers in the developed countries, appeared to be contradictory. Foreign economic enterprise was both sought and resented. Industries based in developed countries contributed to the environmental problems of the Third World through timbering, mining, trade in endangered species, tourism, and monocrop agriculture. But these activities were often welcomed as bringing certain economic benefits. Complaints that the industrialized countries were robbing developing countries of their natural resources seemed inconsistent with the Third World objection that the developed nations were substituting synthetics (e.g., plastics) for natural products from the Third World, such as copper.

Program Development

By the Governing Council's Third Session, the United Nations Environment Programme appeared to be launched upon a course of action consistent with the Stockholm recommendations. Its accomplishments are recounted from year-to-year in the introductory reports of the executive director to the annual sessions of the Governing Council.[7] The developing nations, having made their points about (1) the emphasis on development as a vehicle for raising the quality of the environment and (2) the location of the main office of UNEP, were prepared generally to implement some of the priorities urged by the environmental scientists.

Caution is nonetheless needed in estimating the readiness and future willingness of national governments to cooperate for environmental protection where costs—monetary or political—are involved, and where no immediate threat to the welfare of the nation is perceived. The test of international cooperation for the protection of the planet will become more severe as the noose of world needs of ever increasing populations tightens upon the limited supply of land, fresh water, and other natural resources. Meanwhile, in examining the functions and procedures of UNEP, it ought not be forgotten that UNEP is a political organization. It answers to a constituency that hitherto has never placed environmental quality near the top of its scale of priorities. Now, however, environment appears to be an ascending international priority.

Stated in the bland and abstract language that characterizes diplomatic descriptions of international arrangements, the basic functions of UNEP may be classified as the dissemination of information, the cultivation of understanding, and collaboration with the environmental programs of other agencies. The United Nations Environment Programme also undertakes initiatives in collaboration with national governments and nongovernmental organizations, for example, the IUCN. In actual practice, UNEP's functions, assigned by General Assembly Resolution 2997 (XXVII) of December 1972, are often mixed. Its role is best described as "catalytic"; its program initiatives are largely carried on by other organizations, with UNEP providing functions of coordination, information, and reporting. Yet it possesses an intangible influence inherent in the importance of its mission and in the perception that international agencies and governments have of that mission in relation to their own assessments of hazards to the environment. To the extent that governments face common environmental problems, are disturbed by environmental threats beyond their jurisdiction, or need to harmonize environmental policies relating to international trade, UNEP has a significant role to play.

Influence is not easily achieved by any international agency, including the World Bank, which has more benefits to confer than does UNEP. With the decline of old-fashioned nationalism, constructive, self-restraining international cooperation is more easily implemented by the First World countries of Europe and North America than it is by newly nationalistic states of the Third World, many of which are unable to exercise responsible internal control. Yet, in the decade following Stockholm, Third World leadership appeared to broaden its perception of the causes of domestic problems, some of which were acknowledged to be of local origin. Many were increasingly seen to be environmental.

Cumulating evidence from environmental science was demonstrating that not all Third World disasters could be laid at the door of the developed world. Nevertheless the inordinate demand of the industrial states for timber, minerals, and agricultural cash crops for export (e.g., soy beans, cotton, and beef) have had deleterious effects ecologically and sociologically on many developing nations. Deforestation in Nepal has affected flooding in Bangladesh; up-river dams and irrigation projects in many Third World countries have had measurable adverse impacts upon their downriver neighbors. Desertification however has been more often a consequence of indigenous behavior patterns, but possibly made more harmful by misplaced international technical assistance (e.g., in the African Sahel). To some extent off-setting ill-conceived economic developments, UNEP has provided a forum acceptable to the Third World countries for examining their mutual problems free from suspicion of solutions imposed by developed countries. Thus, UNEP's role in cultivating understanding has increased its influence on national attitudes and policies.

UNEP may influence national action through the providing of benefits, but not all such benefits are discretionary, and some are provided without condition. Distinction should be made between grants to specific countries or organizations, and expenditures for services generally and routinely available. Earthwatch is an example of the latter.[8] This program, mandated by the Stockholm Conference, provides a continuing assessment of the state of the global environment. Its four parts include (1) review and evaluation of environmental conditions to identify gaps in knowledge and need for action, (2) research on environmental problems, (3) monitoring of certain environmental variables, and (4) exchange of information among scientists and governments. Many of its elements are described in subsequent chapters of this book; many of its findings have been reported in a compendious report by UNEP—*The World Environment 1972-1982*—edited by three members of UNEP's Senior Scientific Advisory Board.[9]

The monitoring function is provided through the Global Environmental Monitoring System (GEMS), planned with the assistance of the International Council of Scientific Unions.[10] Two major elements of the exchange of information function are the International Referral System (IRS), now called INFOTERRA, and the International Register of Potentially Toxic Chemicals (IRPTC).[11] INFOTERRA is in effect a switchboard or clearinghouse that not only provides information from UNEP data sources, but puts inquirers in touch with intergovernmental or scientific organizations that possess the required information. The system comprises a global network of 107 government-designated, national focal points for identifying and registering sources of environmental information. INFOTERRA publishes the *International Directory of Sources*, drawing upon existing national and international information services and systems and distributes this information throughout the network. A single central data bank for all the world's environmental information needs would require an undertaking of magnitude and complexity straining the present capacity of information technology.

Nevertheless, technology may be brought to answer the need. UNEP has created GRID, the Global Resource Information Data Base, the principal data management program within GEMS. GRID is a distributed system with centers in Nairobi, Geneva, and Bangkok and uses Geographic Information System (GIS) computer-based technology. UNEP's programs require the acquisition and collation of immense amounts of data on a continuing real time basis. This is the challenge that GRID faces and will probably meet—although time, training, and resources will be needed. Electronic technology, Internet, World Wide Web, and the rapid development of computer literacy promises an unprecedented flow of communication and information relative to environmental issues.

Other benefits provided by UNEP, which may be either of a conditional or nonconditional character, are advice and mediation relating to environmental policies and issues. In performing these services UNEP seeks, where relevant, to coordinate its action with other international assistance programs, especially with the United Nations Development Programme (UNDP). Such joint or coordinated action among United Nations agencies assisting planning and development by national governments is especially needed in reconciling economic, engineering, and ecological objectives.

Ecodevelopment, a condensed expression for ecologically sound development, was a major theme of UNEP throughout the decade of the 1970s. The term is not susceptible to precise definition. To describe it as development consistent with ecological principles and conserving of environmental values identifies its general meaning, but raises more questions than it answers. The term "ecodevelopment" has been largely replaced by "sustainable development," notably since the report of the Brundtland Commission in 1987. Possibly the most explicit official statement on ecodevelopment was the "Cocoyoc Declaration," adopted by the participants in the UNEP/UNCTAD Symposium on Patterns of Resource Use, Environment and Development Strategies held in Cocoyoc, Mexico, 8–12 October 1974.[12] Such policy-shaping documents provide UNEP with the rationale needed to justify decisions regarding allocation of assistance among appeals for aid.

Discretionary grants, although limited, are nevertheless important as inducers of cooperation. This leverage in international policy matters derives largely from use of the Environment Fund. UNEP's resources, while small in relation to total monetary transfers among international agencies and national governments, may yet be critical in particular instances. The UNEP contribution to a national development or natural resources project can represent the difference between the initiation or postponement of a project. Nearly three-quarters of the Environment Fund has been allocated through other organizations in the United Nations system. UNEP may assist or participate in projects administered by other organizations where environmental considerations are important, or it may cooperate on equal footing with another agency through the mechanics of joint programming. Within the

United Nations system UNEP has been associated most frequently with the Food and Agriculture Organization (FAO), the United Nations Educational, Scientific, and Cultural Organization (UNESCO), the United Nations Habitat and Human Relations Foundation (UNHHRF), the World Health Organization (WHO), and also with several of the United Nations special bodies, notably the United Nations Development Programme (UNDP).

UNEP's role in the promotion of environmental priorities and international cooperation is played in a variety of ways, sometimes through sponsoring international agreements that cumulatively may shape new international environmental law and behavior. Among the agreements that UNEP has promoted have been the Convention on International Trade in Endangered Species (adopted March 1973, effective July 1975), the Convention on the Prevention of Marine Pollution by Dumping of Wastes and Other Matter (open for signature, December 1972, in force, September 1975), and the conventions and protocols negotiated in the Regional Seas Programme, now incorporated under Oceans and Coastal Areas. The conferences and negotiations leading to these arrangements and guiding their implementation are among the more effective instruments for the propagation of environmental understanding. Published reports of UNEP-sponsored activities, such as the Regional Seas Programme, (e.g., *The Siren*) extend the informational base for understanding beyond immediate participants to a worldwide audience.

The traditional United Nations vehicle for enlarging understanding and shaping attitudes has been the international conference.[13] UNEP has been charged by the General Assembly with major responsibilities for United Nations conferences on housing and human settlements (Vancouver, 1976), on water problems (Mar del Plata, 1977), and on desertification (Nairobi, 1977).[14] Other United Nations agencies are also involved in international conferences and may play leading roles as initiators and organizers. These occasions may also afford opportunity for UNEP to influence outcomes, as for example in the UN Conference on Tropical Timber, 14–31 March 1983, but UNEP does not appear to have significantly affected the results of the United Nations conferences on food (Rome, 1974) and population (Bucharest, 1974) or influenced the Third United Nations Conference on the Law of the Sea (1973–1982). But influences need not be direct in order to have effect, and it seems plausible that UNEP's work has effected a broadened perspective on population-environment relationships as in the 1994 International Conference on Population in Cairo.

Although UNEP is not itself a party to international treaties, it has endeavored to enlarge and strengthen international environmental law in principle as well as through specific practice. Since 1977 it has sponsored a working group of experts to consider international legal responsibilities (liability and compensation) relating to environmental damage. Its activities in this area complement the missions of other UN agencies—for example, the International Maritime Organization (IMO) on liability for pollution in international shipping and the International Atomic Energy Agency (IAEA) regarding radioactive contamination.

UNEP finds another channel of influence through NGOs concerned with the natural sciences and the environment. In recent decades these organizations have generally grown in numbers and strength, many of them sending representatives to Nairobi or keeping in regular communication with UNEP personnel. The Environment Liaison Centre in Nairobi helps UNEP to maintain these relationships, and some NGOs, notably the North American environmental organizations, have developed arrangements to regularize communication with Nairobi on a continuing basis.[15]

Principal collaborators with UNEP outside the formal structure of intergovernmental organizations are the International Council of Scientific Unions (ICSU), the International Union for Conservation of Nature and Natural Resources (IUCN), and the World Wildlife Fund for Nature (WWF). It is inaccurate to describe either ICSU or IUCN as strictly nongovernmental. Government representation is present in both organizations and affords channels of communication with national authorities that are often less formal and possibly more direct than those available through the official United Nations system. Through ICSU and IUCN, UNEP may communicate informally with national ministries and departments, as well as with a broad range of nongovernmental organizations within nations.

Over the years, UNEP's influence has quietly expanded to a degree scarcely comprehended by the world's news media. Its financial resources have never been adequate to the tasks confronting it, and a very major goal of international environmental policy should be to provide UNEP with the funds to pursue the objectives to which nearly all governments and all informed people agree. If this goal could be more readily achieved by raising the Governing Council of UNEP to the status of ECOSOC or the Security Council, such a move would be eminently justified.

Operational Strategies of UNEP

To accomplish its objectives, UNEP developed a tri-level program strategy suggestive of a similar one employed in the pre-Stockholm Preparatory Committee. Level 1 consists of two parts: an annual report entitled *State of the Environment* and a review of priority areas at national, regional, and international levels. Level 2 moves toward action, linking a conceptual approach to an operational strategy, specifying the action needed to address the problems and priorities identified at level 1. The third level comprises those activities to be supported, usually only in part, from the Environment Fund. Optimal use of the fund requires its being regarded as leverage to promote action that might not otherwise occur or be carried out effectively.

To optimize further the use of available resources and to bring UNEP to the scene of action, the executive director proposed establishment of Programme Activities Centres (PACs) to the Second Session of the Governing Council. The PACs were to be adapted to ongoing activities varying in size, structure, and duration, and were intended as foci for networks of cooperating institutions—intergovernmental and nongovernmental—able to mobilize resources and avoid undesirable centralization of UNEP initiatives in Nairobi. These units have now materialized in the form of regional centers and program units such as GEMS and INFOTERRA. In addition to the UNEP Liaison Office at UN headquarters in New York, regional offices have been established in Geneva for Europe, in Nairobi for Africa, in Bangkok for Asia and the Pacific, in Mexico City for Latin America, in Beirut for West Asia, and in Washington for the United States.

The practical necessity for combining the resources of UNEP with those of other United Nations agencies led to another operational strategy called joint programming. As defined by UNEP in the language of United Nations officialese, joint programming is the process of mutual identification by two or more bodies of program concepts, objectives, and activities that are relevant to the aims of those bodies. Through exchange of information and consultation at the time of program formulation, joint programming establishes those areas and issues that are of mutual interest and thereby provides a basis for a shared program of work. In effect this should enable different United Nations bodies simultaneously to attack various aspects of a common objective in a coordinated manner. Joint programming is difficult to accomplish, and yet it is a logical and probably necessary way to deal with the multidimensional problems of the human environments that transcend the scope of any single operational program.

UNEP's Growing Status

A difficulty inherent in UNEP's situation has been its relationship with older United Nations agencies. At the time of UNEP's establishment, the specialized agencies and special bodies were resistant to UNEP's role of coordination and leadership—hard to play within the United Nations system and additionally difficult in a novel and often poorly understood area of policy. Within the United Nations secretariat, UNEP has had the closest relations with UNDP, but in the past UNDP has showed little interest in examining the environmental consequences of the development projects that it funds. Today it appears more receptive.

UNEP did not initially have much influence on international lending institutions, including the World Bank and the Regional Development Banks. Only recently has the World Bank had any significant interaction with UNEP—the bank's environment review process having predated establishment of UNEP. The World Bank has not been uncooperative with the United Nations Environment Programme, but it has not looked to UNEP for guidance. However, the program would surely benefit from closer relationships with the banks. With assistance from the Canadian government, UNEP funded a study by the International Institute for Environment and Development entitled *Banking on the Biosphere* (1977), which described environmental policies of the financing agencies. A declaration in 1980 by UNEP, UNDP, the World Bank, and several regional development institutions promised the incorporation of environmental considerations in development policies and may provide a rationale for closer collaboration in the future. Implementing this declaration has not come easily, but relations between UNEP and the Committee of International Development Institutions on the Environment (CIDIE), which was established to carry out the intent of the declaration, have grown closer. The secretariat of CIDIE is located at UNEP headquarters, and its activities are featured in the annual report of the executive director of UNEP.[16]

Pressure on the World Bank through the NGOs and through the United States Congress, as well as the election of a new president of the bank, Barber Conable, in 1986, ushered in a new relationship between the Bank and UNEP. The sustainable development concept has become a common denominator to the relationship. The UN General Assembly and the national governments are committed to sustainable development, and the UN system must follow suit.

There has been significant increase in the environmental emphasis of those specialized agencies that have environment-related concerns, such as FAO, UNESCO, WHO, and WMO. The activities of WMO are almost wholly environment-related, but the traditional focus of this agency is on science rather than policy, giving it little occasion for competition or conflict with UNEP. From the viewpoint of the specialized agencies, UNEP is more often perceived as representing a special sectoral interest to which appropriate attention should be paid, rather than as a coordinating body intended to ensure the presence of environmental sensitivity throughout the United Nations system. Nevertheless, through joint programming and consultative functions UNEP may, in time, and with assistance from other sources—notably, nongovernmental organizations—attain a greater measure of influence.

It is important to the objectives of the Stockholm Declaration and Action Plan and the Rio Agenda 21 that UNEP play an effective role of leadership and coordination. The Stockholm Action Plan distributed functions among the specialized agencies, NGOs, and national governments. The UN system is itself highly disjunctive, its bureaucratic interests and the inclinations of its scientists being generally mission-centered. As one student of the difficulties of adapting conventional UN behavior to the Stockholm agenda observed, "If there were to be a spur to systematic effort and coherent programming— from within the system—it would have to come through the initiatives of the UNEP." [17]

As of 1996, my evaluation of UNEP differs significantly from that offered in the first edition of this book. At that time, the limitations and difficulties encountered by UNEP received greater weight than I would give them today. The achievements and influence of UNEP have advanced further and more rapidly than most observers could have foreseen at the beginning of the 1980s.

Two factors largely account for this progressive change: first, the energetic and persuasive leadership of UNEP's executive director, Mostafa Kamal Tolba; second, the coming-of-age of the international environmental movement. The first executive director, Maurice Strong, had the difficult task of establishing the "legitimacy" of UNEP among suspicious Third World countries and "turf-sensitive" specialized agencies of the United Nations. UNEP also was handicapped by the apparent indifference of some developed countries, which believed that its role in providing assistance to Third World countries should be supported, as long as it "didn't cost too much." These difficulties were on the way to being overcome under the leadership of Maurice Strong, although the pace of progress was at times discouraging.

With the Report of the World Commission on Environment and Development (Brundtland Commission) in 1987, and the growing concern in the UN General Assembly over environmental issues, the status of international environmental policy visibly heightened. In his 1988 annual report Mostafa Tolba summarized the dramatic change that had occurred during the decade. In his first annual report in 1982, Tolba described the year as "disappointing" in that "commitments and resources failed to match the words and declarations of the commemorative session of the Governing Council ten years after the Stockholm Conference." But in his annual report six years later he could characterize 1988 as the year in which environment became a top item on the world's political agenda. From 8 to 10 July 1988 in Oslo, at the invitation of Norwegian prime minister Gro Harlem Brundtland and UN secretary-general Javier Peréz de Cuellar, twenty-two directors of UN agencies and programs met together for the first time to plan and coordinate their programs for environmentally sustainable development. Significantly, the forty-three volumes of documents on the earlier Brundtland Commission's 1983–86 investigation were to be housed at UNEP headquarters, available to governments, the public, and researchers. It appeared that the UN system might be beginning to affect environmental policy, as intended by the Stockholm Conference.

One need not be prophetic to foresee an enhanced role for UNEP in the future. The cursory description of its activities necessitated by the limitations of this book should nonetheless be sufficient to indicate the scope of its responsibilities and the importance of its functions. The greatly increased activities of NGOs, especially their growth in Third World countries, provide powerful reinforcement for UNEP's programs. Of all the United Nations agencies, UNEP may have a broader, more diverse, and more directly involved constituency than any other. The clientele of the specialized agencies, e.g., FAO, WHO, WMO, ICAO, and ILO, tend to be specialized professionals; UNESCO has a large and varied constituency, but much of it also tends to be specialized, e.g., professional educators, scientists, or artists. It is perhaps not pushing rhetoric too far to describe UNEP as the general public's UN agency. As the environmental NGOs grow in strength in their respective countries and learn to collaborate effectively through the Environmental Liaison Center and the IUCN (World Conservation Union), they will (and do) constitute powerful forces for policy to which governments and the older international agencies will give serious attention. The presence of UNEP within the UN system has already influenced the character of the General Assembly, helping to move it from a forum for disaffection and acrimony toward true international cooperation on issues of common concern to people everywhere.

Post-Stockholm Assessment

Perceived national interest, rather than the Stockholm Declaration of Principles or the interventions of UNEP, has been the prime mover of international environmental cooperation. Nevertheless, the Stockholm Conference is a major landmark in the effort of nations collectively to protect their life-support base on earth. The balance sheet concept is not a feasible way to evaluate the consequences of the Stockholm Conference. The conference and UNEP are parts of a larger global development that might have occurred even if the United Nations had never existed. Implementation of many of the conference recommendations will require time measured by decades, reinforced by Agenda 21, which was adopted by the 1992 UN conference in Rio de Janeiro.

Separate roles have been played by other international organizations, both intergovernmental, such as the Organization for Economic Cooperation and Development (OECD) and the European Union, and nongovernmental, such as ICSU and IUCN. But UNEP as activator of the Stockholm Action Plan has given the international environmental movement a universality, a legitimacy, and an acceptability in Third World countries that under the circumstances could hardly have been obtained otherwise. Maurice Strong's assessment one year after Stockholm is still valid:

> Environmental actions taken to date are still of fairly marginal significance compared with those yet to be confronted. The difficult choices—

about the imbalance created by man's activities, about equity in the use of common resources, about the sharing of power both within national societies and internationally, about the fundamental purposes of growth and the sharing of its benefits as well as its costs—remain to be made.[18]

The nations at Nairobi have demonstrated the practicality of limited action within defined areas of agreement—for example, adoption of the Global Environmental Monitoring System within the Earthwatch program. As previously noted, the Nairobi commitments had largely to do with arrangements for information and assistance. Few of the priority topics implied possible interference with or reorientation of national priorities. Yet even this latter possibility was present in passage by the Governing Council of a resolution asking the executive director to prepare proposals for cooperation in the field of the environment concerning natural resources shared by two or more states.

National governments determine if, how, and when the Stockholm recommendations are implemented. Post-Stockholm efforts to extend international law by conventional methods were initially slow to yield positive results. Yet the cumulative international environmental agreements consumated between 1972 and 1992 are impressive and unprecedented. The NGOs urged the UNEP Governing Council to push the ratification of four major conventions negotiated during or after the Stockholm Conference, although the marine pollution treaties were more a responsibility of the International Maritime Organization than of UNEP. The four were as follows:

—Convention on the Protection of the World Cultural and Natural Heritage, Paris, 23 November 1972
—Convention on the Prevention of Marine Pollution by Dumping of Wastes and Other Matter, London, 29 December 1972
—Convention on International Trade in Endangered Species of Wild Fauna and Flora (CITES), Washington, D.C., 3 March 1973
—Convention on the Prevention of Pollution from Ships (MARPOL), London, 2 November 1973

These treaties were negotiated with considerable difficulty, but also with substantial international support. But once open for ratification, the nations were not in a hurry to act—low priority rather than domestic opposition being the usual retarding factor.

A decade after Stockholm, three of these treaties were at least technically in effect, but the Convention on the Prevention of Pollution from Ships (MARPOL) did not come into force until 2 October 1983, one year after it had received the required number of ratifications. It has been described as "the most important and comprehensive treaty to fight marine pollution."[19] Only moderate success was achieved with several additional treaties, some of which were negotiated prior to Stockholm or apart from UNEP. For example:

— Convention on Wetlands of International Importance Especially as Water-
fowl Habitat, Ramsar, Iran, 2 February 1971
— International Convention on the Establishment of an International Fund
for Compensation for Oil Pollution Damage, Brussels, 18 December 1971
— Convention on the Prohibition of Military or Any Other Hostile Use of
Environmental Modification Techniques, New York, 10 December 1976
— Convention on the Conservation of Antarctic Marine Living Resources,
Canberra, Australia, 20 May 1980

During the latter half of the 1980s, however, international environmental
treaty-making gained momentum. Two protocols to the 1979 Convention on
Long-Range Transboundary Air Pollution were negotiated during the following years:

1985   Protocol on the Reduction of Sulphur Emissions or Their Transboundary Fluxes by at Least 30 Percent
1988   Protocol Concerning the Control of Emissions of Nitrogen Oxides
or Their Transboundary Fluxes

UN conventions of global significance were signed between 1985 and 1990:

1985   Vienna Convention for the Protection of the Ozone Layer
1987   Montreal Protocol on Substances That Deplete the Ozone Layer
1986   Convention on Early Notification of a Nuclear Accident
1986   Convention on Assistance in the Case of a Nuclear Accident or
        Radiological Emergency
1989   Convention on Control of Transboundary Movements of Hazard-
        ous Wastes and Their Disposal
1991   Convention on Environmental Impact Assessment in a Transboun-
        dary Context
1992   Framework Convention on Climate Change
1993   Convention on Biological Diversity

In addition, several older conventions were gradually receiving ratifications.
The 1982 Law of the Sea (LOS) treaty received the required number of ratifications to put it into effect. The United States adhered to most of its provisions but refused to ratify. A 1995 amendment to the LOS appears to have
made it generally acceptable. A 1988 Convention on the Regulation of Antarctic Mineral Resource Activities intended to open the protected continent
to mineral exploitation and backed by the Bush administration in the United
States, was blocked by Australia and France. Several proposed treaties were
being formulated by 1990. Although they could not be directly attributable to
Stockholm, the conventions on the conservation of biological diversity and
on global climate change prepared for signature in 1992 at the UN Confer-

ence on Environment and Development were clearly within the Stockholm legacy. Other proposals, e.g., on forest principles, did not attain treaty status.

A growing problem in the post-Stockholm period, although not new in principle, has been the settlement of international environmental disputes when the parties in the first instance are private or corporate rather than governmental. International litigation based on treaties can be an extended process and depends upon governments assuming the claims of nongovernmental parties. Only national governments have standing before the World Court. The growth of multinational corporate enterprise has added to the complexities of litigation, as for example in fixing responsibility for pollution by tankers at sea having multinational private ownership.

One consequence of these developments and evidence of the maturing of environmental politics has been a growth of mediation as a method for the resolution of international environmental disputes. In the United States, mediation has been used with some success in domestic environmental controversies. And the prospects for more extended uses of this technique have led to the establishment of at least one private enterprise effort in the Washington, D.C.-based firm, Environmental Mediation International.[20]

The politicizing of the international environmental movement following Stockholm was to some extent inevitable. Obviously, important interests and values are affected by changes or innovations in environmental policies. There is evident in many countries a disenchantment with government and international organizations, in part because of failure to honor commitments or to get things done without seemingly interminable maneuvering and delay. Further, governments and some international agencies are mistrusted because of environmentally destructive policies that they have historically pursued. Nevertheless, given sufficient time it appears that rational consensus can—not necessarily will—be achieved in international bodies and a conceptual basis established upon which more effective environmental action can be taken.

The Strategy for the Third United Nations Development Decade, adopted 23 October 1980 by the General Assembly, moved further than previous UN declarations in recognizing the interrelationships among development, environment, population, health, and resources. The Assembly document declared:

> It is essential to avoid environmental degradation and give future generations the benefit of a sound environment. There is a need to ensure an economic development process which is environmentally sustainable over the long run and which protects the ecological balance. Determined efforts must be made to prevent deforestation, erosion, soil degradation and desertification. International cooperation in environmental protection should be increased.[21]

The difficulty of accurately attributing influences prevents a meaningful assessment of the effects of Stockholm in other than very general terms. A summation of principal trends and developments in the post-Stockholm decade relating to the Action Plan was prepared by the staff of UNEP for the Governing Council's Session of a Special Character, 10–18 May 1982. By the end of the decade some positive changes in international environmental policy were demonstrable, and assessments of progress were undertaken by a number of individuals.[22]

Tenth-Anniversary Session

By Resolution 35/75 of 5 December 1980, the UN General Assembly decided to commemorate the tenth anniversary of the United Nations Conference on the Human Environment by asking the Governing Council of UNEP to convene a Session of a Special Character to assess the impact of the Stockholm meeting of 1972 on the world environment. This session, open to all member states of the United Nations, met in Nairobi 10–18 May 1982. Two major tasks shaped its agenda: (1) review of the extent to which the objectives of the 1972 Action Plan had been realized, and (2) recommendations with respect to future environmental trends, perspectives, action, and international cooperation to be addressed by UNEP during the succeeding decade.

The details of the tenth-anniversary assessment are too numerous to report here. They have been published in two UNEP documents that sum up the achievements and disappointments of the post-Stockholm decade and estimate the course of future development: *The Environment in 1982: Retrospect and Prospect* and *The World Environment 1972–1982: A Report by the United Nations Environment Programme.*

The latter volume contains a comprehensive survey of the condition of the world environment as of 1982 with analyses of major research efforts, problems, and trends. The shorter preceding document, prepared for the Session of a Special Character, provides an assessment of the overall achievement of international environmental policy in the post-Stockholm decade. The following statement on "The Over-All Achievement," slightly paraphrased from *The Environment in 1982*, provides a concise yet specific summation of progress under the Action Plan:

> [T]he United Nations Conference on the Human Environment was a powerful force for change, which led to two major achievements of a general nature:
> (a) Awareness of the significance of the environment and of the implications of environmental change increased substantially at the policy-making level (Governments and legislatures) and among the public at large;

(b) New environmental programmes were created at all levels (international, regional and national), and existing programmes were intensified, extended and accelerated.

. . . Following were the major achievements in terms of the functional components of the Action Plan, i.e. assessment, management, and supporting measures. In environmental assessment, GEMS (Global Environmental Monitoring System), though incomplete, is operating and expanding in a systematic way; INFOTERRA (International Referral System) is functioning, although it should be more widely and frequently used; IRPTC (International Register of Potentially Toxic Chemicals) is established as a world center for information on toxic chemicals; IPCS (International Programme on Chemical Safety) is providing environmental dose-response data for an increasing number of substances; and the status of stratospheric ozone is under active investigation.

In environmental management, a number of international and regional conventions have been concluded, but there have been delays in ratification and shortcomings in enforcement; action plans have been adopted to combat desertification, to improve water supply and management, and to improve human settlements, but their implementation has been distressingly slow; the World Conservation Strategy has been well received; and there has been some progress in conceptualizing the objectives and procedures for environmental management, with what has been worked out now being tested. The tasks comprising supporting measures, i.e. environmental education and training, technical assistance and public information, are immense. While useful efforts have been made, e.g. the environmental education programme and the inclusion of environmental training programmes in the activities of cooperating agencies, more remains to be done. The Regional Seas Programme, now a part of Oceans and Coastal Areas, is unique in combining assessment and management activities along with supporting measures.

The sector-by-sector review of implementation of the Stockholm Action Plan and the functional highlights noted above, suggest a mixed record of achievement. An over-all assessment is that fair progress has been made in implementing some of the elements of the Action Plan, and that there has been a significant impact on the public and on Governments. For other elements, progress has been very slow. However, these conclusions must be tentative.[23]

The declaration adopted by the Session of a Special Character on 18 May 1982 stated that "the Action Plan has only been partially implemented and the results cannot be considered as satisfactory." This assessment did not evaluate all international environmental action since Stockholm. A comprehensive review of post-Stockholm developments would include action by national governments and separate environment related activities of the specialized agencies and NGOs. Yet, even with these additions the extent of the gap between conservatively estimated needs for protective action and the readiness of governments to act afforded little ground for optimism.

With respect to the second major task of the Session of a Special Character — an estimate of the environmental future — the prospect as reported to UNEP was not bright. Offsetting the good-to-fair assessment of achievements under the Action Plan were the areas in which no progress could be reported. The achievements were largely in monitoring, information exchange, research, and consciousness raising — none of which crossed economic or development interests or required a major reorientation of policy and administration at national levels. Governments had been reluctant to commit themselves to substantive action except in those infrequent cases of urgent popular demand. In brief, the inference appeared to be that the state of the environment would worsen, but UNEP would be able to monitor the where and why of its decline.

Yet, as the UNEP assessment declared, conclusions had to be tentative and await the results of the 1992 UN Conference on Environment and Development. The Stockholm resolutions were a charter for a revolution in national and international values and behavior. It would be unrealistic to expect that, in no more than a decade of post-Stockholm effort, the priorities of national governments would be redirected and the erosive effects of overpopulation, overconsumption, poverty, socioeconomic disturbances, war, and reluctance to commit today's resources to preventing tomorrow's problems would be overcome. Nevertheless, significant changes had occurred at national levels during the decade, and these may be more reliable indicators for the future of international environmental policies than the actual implementation of the Action Plan. In a world of nation-states, the ultimate results of any international effort depend upon what national governments actually do. What they do depends upon what their public officials and constituent peoples desire. Changes in popular perspectives on human–environment relationships could thus be precursors of impending changes in national policies.

By the end of the 1980s there was evidence that such a change in popular attitudes was occurring and that governments sometimes reluctantly were beginning to take heed. On 9 May 1989, UNEP reported the results of the first worldwide survey on public opinion regarding the environment. Subsequently reported in UNEP's magazine, *Our Planet* (1, no. 2/3, 1989), the results of the fourteen-country survey by Louis Harris and Associates confirmed what other studies had shown: worldwide and widespread popular concern about the environment. In 1992 a Gallup twenty-four-nation survey (the Health of the Planet) and a 1988–89 Harris and Associates sixteen-nation survey (Public and Leadership Attitudes to the Environment in Four Continents) confirmed the continuation of earlier trends.[24]

World Charter for Nature

The complicated and often paradoxical course of international environmental policy development is illustrated by the history of the United Nations World Charter for Nature. This document, adopted by the General Assembly on 28 October 1982, illustrates the widespread verbal acceptance of the principles enunciated at Stockholm and the practical difficulty of making these principles operational in a world of sovereign and antagonistic states. Nevertheless, the adoption of the charter is important; the symbolic act is often a necessary precursor of implementing action. Future history may find the charter to be a landmark in the evolution of transnational affairs.

The idea of a charter was initiated at the Twelfth General Assembly of the International Union for Conservation of Nature and Natural Resources (IUCN) meeting at Kinshasa, Zaire, in September 1975. In an address to this gathering, President Mobutu Sese Seko proposed that the IUCN develop a charter for nature based on the proposition that "all human conduct affecting nature must be guided and judged."[25] This proposal was accepted by the IUCN, and an international group of experts was designated to prepare a draft charter to serve as a code of conduct for managing nature and natural resources. Following several revisions, the draft was presented to President Mobutu whose endorsement made it a matter of national significance and pride.

On 2 June 1980, the permanent representative of Zaire to the United Nations proposed to the secretary-general that the IUCN charter be included on the agenda of the Thirty-fifth Session of the General Assembly. On 30 October 1980, the General Assembly by resolution invited member states to communicate their views and observations to the secretary-general and on 9 March 1981, the executive director of UNEP on behalf of the secretary-general addressed a letter to all member governments soliciting their response before a deadline of 16 September 1981. Meanwhile representatives of "nonaligned countries" (presumably opting out of the U.S.-U.S.S.R. Cold War), meeting in New York in October 1980 designated the charter as a priority item for General Assembly consideration. And in June 1981, the Council of Ministers of the Organization of African Unity strongly recommended adoption of the charter.[26] Clearly, a large number of developing nations regarded this charter, initiated by one of their own members, as an important point of protocol. Fifty states replied to the secretary-general's letter—including Australia, Canada, France, Japan, and the Soviet Union. The United States did not reply, but the attitude of the incoming Reagan administration toward the United Nations and environmental issues was known to be negative. Lack of response reflected the personal bias of President Reagan and was in no way an expression of American public opinion.

On 28 October 1982, the World Charter for Nature came before the General Assembly for final action and was adopted by a majority of 111 with 18 abstentions and the United States casting the single negative vote.[27] Among the abstaining delegations were representatives of the eight signatory states to the Treaty of Amazonian Cooperation—Bolivia, Colombia, Ecuador, Guyana, Peru, Surinam, and Venezuela—led by Brazil. South American nations, and notably Brazil at Stockholm, had consistently taken a strongly nationalistic position with respect to permanent and exclusive sovereignty over their natural resources. But other Third World countries that would hardly have agreed to the language of the charter had it been proposed at Stockholm in 1972 overwhelmingly endorsed it in New York in 1982. And the United States, which had been in the forefront of international environmental policy development in 1972, now cast the sole dissenting vote, paralleled by a 146 to 1 vote on a General Assembly resolution to protect people from imported products harmful to health and environment.[28]

The mere record of voting on these UN resolutions provides no adequate indication of the actual position of the member states. Two considerations in particular weighed heavily in the outcomes. First, the World Charter for Nature, with its sponsor a developing, nonaligned nation, was able to elicit Third World solidarity behind the initiative. The pre-1989 former Soviet bloc found the charter a costless and convenient way to demonstrate fraternity with the Third World and to demonstrate the isolation of the United States in the General Assembly. Informal inquiry among pro-charter delegations caused some observers to infer that the Third World vote was a noncommittal way to please President Mobutu and the delegation from Zaire. The resolution imposed no obligations upon them, and its implementation need not be taken seriously. The second consideration, the reasons for the negative vote of the United States, illustrates the vulnerability of international environmental policy to the domestic politics of nations.

Following the endorsement of the initial proposal for the charter in 1980, which the United States supported, a major reversal of political priorities had occurred in Washington. The election of Ronald Reagan to the presidency in November 1980 brought into power an administration which sharply downgraded environmental issues on its scale of priorities, and showed a diminished regard for the United Nations as a useful instrument of international cooperation. Not only was the Reagan administration unwilling to spend money for new international environmental activities, but it proposed to eliminate or cut back severely on programs previously supported, such as UNEP. Moreover, the Reagan administration tended to act on the assumption that environmental policies and regulations unnecessarily burdened the private economic sector and retarded economic growth.

Members of the United States delegation to the United Nations endeavored to make the point that the negative votes of the United States should not be taken as rejections of the purposes of the resolutions, but rather resulted from their government's decision to oppose proposals with new financial implications regardless of merit. In the case of the World Charter for Nature, however, the apparent indifference of the Reagan administration to environmental affairs and the low priority assigned to policy decisions on such matters as the charter left the U.S. Department of State unable to obtain timely action on its policy position. In a sharp retort to a request by the United States to postpone a vote on the charter until consensus could be achieved, the representative from Zaire declared that "it is not our fault if the United States delegation has not been in position for the past three years to submit its comments, which could have been taken into consideration by the Ad Hoc Croup of Experts."[29]

Given the present status and conventions of international law and relations, the language of the charter was idealistic. On all twenty-four points, national action was required in the imperative and mandatory language of "shall." For example, point 14 stated, "The principles set forth in the present Charter shall be reflected in the law and practice of each state as well as at the international level." Point 24 concluded the charter with a moral precept that could hardly be made operational by the most sympathetic of governments. "Each person," it declared, "has a duty to act in accordance with the provisions of the present Charter; acting individually, in association with others or through participation in the political process, each person shall strive to ensure that the objectives and requirements of the present Charter are met."

What does this episode of the World Charter for Nature signify regarding the state of international environmental policy? Evaluations may differ, but I regard the charter as a significant symbolic expression of a hope among nations to achieve a more harmonious and sustainable relationship between humanity and the rest of the biosphere—between mankind and earth. The charter declares a standard of ethical conduct that governments today are more willing to profess than to practice. If the hortatory language of the charter does not reflect the behavioral realities of the world today, it does declare standards for evaluating the behaviors of people and nations and goals toward which efforts should be directed. Symbolic expression is an important element in the development of public and international policy; it sets the context within which operational strategies and programs are developed. The World Charter for Nature might be regarded as the decalogue of the International Environmental Movement and the World Conservation Strategy as its expression in practice. In historical retrospect, the World Charter for Nature may be seen as an event more significant than was apparent at the time of its adoption.

1992 United Nations Conference on Environment and Development

On 22 December 1989 the General Assembly of the United Nations by Resolution 228 voted to accept an invitation from the government of Brazil to hold a major conference on environment and development in that country in 1992.[30] The theme and focus of this conference had already been set by the report of the World Commission on Environment and Development (Brundtland Commission). The prospect of this conference in Brazil signifies a significant change of attitude in Brazil, which twenty years earlier at the Stockholm Conference had led the opposition to international environmental restrictions. Concern both within and without Brazil about destruction of the tropical rainforest in the Amazon basin may have been a factor toward willingness to consider international environmental agreements formerly regarded as prejudicial to national sovereignty.

In 1989 two groups of Latin American and Caribbean states, through the Declaration of Brazilia (31 March) and the Declaration of Manaus (6 May), although reiterating their insistence on sovereignty over resources, indicated their willingness to participate in environmental protection measures if linked to development and to relief from some of the burden of foreign debt.[31] Fervent nationalism hostile to international environmental cooperation continues, however, to be cultivated by large landowning and military interests in Latin American countries and elsewhere in the Third World. Since 1981 the United States has relinquished its former leadership role in international environmental affairs and for the present appears to be one of the less cooperative states in the shaping of environmental policy. In any case it would be difficult for the president to lead a neutral to negative Congress. Even so, there appears to be a growth of environmental concern among the better-informed sectors of the population in all countries. This trend, and environmental conditions imposed by international lending and technical assistance institutions, have forced a grudging acceptance of developmental restraint on behalf of ecological integrity and sustainability.

Assessment and Prognosis

Perhaps the best informed assessment of global environmental protection during the first two post-Stockholm decades was made by the executive director of UNEP in his annual report for 1989:

> If 1988 was "year in which 'environment' became a top item on the world's political agenda" (as I characterized it in my 1988 Annual Report), then—from the environmental standpoint—1989 was certainly "the year of decision."
>
> The year 1989 witnessed an unprecedented number of environmental conferences, meetings, decisions, resolutions, conventions, declarations, communiqués and statements. On the very first day of the year, the Montreal Protocol on Substances that Deplete the Ozone Layer entered into force, only some 15 months after the signing of that Protocol. In

early March the government of the United Kingdom, with the support of UNEP, hosted a "Saving the Ozone Layer Conference" in London. Later in March, at a meeting of Heads of States and Governments in the Netherlands, the Hague Declaration on the Environment was adopted. March also saw the adoption and signing of the Basel Convention on the Transboundary Movements of Hazardous Wastes and Their Disposal at an international conference organized by UNEP and hosted by the Government of Switzerland. April witnessed the first meeting of the Contracting Parties of the Vienna Convention and of the Montreal Protocol in Helsinki, and May saw the adoption by 81 governments of the Helsinki Declaration on the Protection of the Ozone Layer.

But let us ask ourselves a few questions. Is the quality of the air we breathe today better than that of 17 years ago? Is the water we drink safer? Have we halted the advance of deforestation, soil erosion, desertification? Have we made inroads toward the eradication of poverty? Are our oceans and seas cleaner? Is our life-sustaining atmosphere less threatened? In short, are we living in a cleaner, healthier and more sustainable environment than that of 17 years ago at the time of the Stockholm Conference?

While there have been some gains at the national level in air, land and water pollution clean-up, the sad and discouraging answer to all these questions on the larger plane is "No". The world as a whole is not cleaner or healthier. Deforestation, soil erosion and desertification continue their encroachment unabated. Our atmosphere is threatened not only by the continuing depletion and destruction of the fragile ozone layer, but by the accumulation of greenhouse gases that are expected to start an irrevocable change in our global climate. At the local level, particularly in many of the large urban centres of developing countries, the air is so polluted that it is literally not fit to breathe.

In spite of these gloomy facts, I would be the last to say that our efforts over the past 17 years have been in vain. In this context, the words of the late Sir Winston Churchill about the United Nations come to mind: "The United Nations was set up not to get us to heaven, but only to save us from hell." And so it is with UNEP: while our human environment is not better than it was 17 years ago, it is not as bad as it would have been without our efforts.[32]

There is a larger dimension to the assessment however. If the plethora of rhetoric during the post-Stockholm years had disappointing practical results, it nevertheless represented a new awareness, legitimization, and commitment which in public and international affairs appear necessary precedents to tangible action. The magnitude of the environmental problems and the difficulties in coping with them were almost universally underestimated. The serious and accelerating deterioration of the biosphere occurred over at least two centuries. Restoration of sustainable conditions and remedial action to eliminate cumulative threats may require at least the greater part of the twenty-first century—if necessary tasks are seriously undertaken and exploding populations, poverty, and pandemic disease do not obstruct constructive efforts.

6
Rio de Janeiro and Agenda 21

Although the United Nations 1992 Conference on Environment and Development (UNCED) in Rio de Janeiro was a much larger affair than the 1972 conference at Stockholm, greater emphasis is given to Stockholm in this book. Many more nations, heads of states, NGOs, and journalists came to Rio, but the conference was not a pathbreaker. The Rio Conference and its Agenda 21 were possible only because Stockholm had occurred. Rio built upon and carried forward the work of the Stockholm Conference, which legitimized and initiated *environment* as a focus of international policy. Agenda 21, Rio's "blueprint" for policy in the coming century, had as its starting point the Stockholm recommendations and made them more extensive and detailed. Moreover, the negotiations at Rio, the so-called UNCED process, followed a pattern laid down at Stockholm. In any case, the significance of Rio and Agenda 21 cannot be assessed with assurance until well into the twenty-first century, whereas the more than two decades that elapsed between Stockholm and Rio have left a factual record of the legacy of the Stockholm Conference, which may be assessed.

The UNCED Process

After three years of preparation, the United Nations Conference on Environment and Development (UNCED), representing 178 national governments, concluded two weeks of deliberation at Rio de Janeiro (officially 3–14 June).[1] Perhaps the principal implication of Rio for policymaking is that national and international environmental policies can no longer be compartmentalized. The ultimate unity of the complex biosphere requires that national environmental policies be made with recognition of the nature and scope of their transnational effects.

The principal official agreements achieved at the conference were the Rio Declaration on Environment and Development, a Statement on Forest Principles,[2] and Agenda 21.[3] The declaration was a statement of twenty-seven principles to govern the rights and responsibilities of nations toward the en-

vironment. The Forest Principles were adopted in default of agreement sufficient for a binding treaty. Agenda 21 was an action plan, forty chapters long,
to guide the environment-related policies of government from the present
into the twenty-first century (in the news media, it was often called the Earth
Charter).

The conference was also the occasion for signature of two treaties of policy
significance, the Framework Convention on Climate Change,[4] and the Convention on Biological Diversity.[5] These treaties had been separately negotiated for some months prior to the Rio Conference. They were open for
signature at Rio but were not, strictly speaking, products of the conference.
The Non-Legally Binding Authoritative Statement on Forest Principles was
adopted at Rio, but did not obtain the degree of consensus necessary for
a formal convention. Meeting coterminously with the official conference,
the unofficial Global Forum of an estimated two thousand nongovernmental organizations from 150 countries produced thirty-three "alternate draft
treaties" the future significance of which is conjectural. But they document
a growth in public concern since the 1972 UN Conference on Human Environment and a recognition of national responsibility for protection of the
life-supporting systems of the human environment and the biosphere.

The UNCED process of preparation and negotiation was not new. Similar
activities preceded the 1972 United Nations Conference on the Human Environment and are now becoming an expected way of negotiating major global
agreements. The expectations are for openness and public participation in
defining agendas and negotiating compromises, and are especially demanded
by nongovernmental citizen organizations and the news media.

The UNCED process and its antecedents in the 1972 Stockholm Conference and subsequent treaty negotiations involved large numbers of vol-
unteer organizations and concerned individuals. One reason for the greater
transparency of these international negotiations, compared with less open
conventional diplomatic negotiations, may be that at national levels government officials (e.g., diplomats)—who are duly constituted lawmaking authorities—are more cautious in proposing international agreements than are those
conference delegates—who represent nations but are not necessarily official
policymakers. In addition, the presence and participation of the more influential NGOs from developed countries gave widespread publicity to official
proceedings. International agreements must be returned to national legal authorities for ratification and implementation, and governments must take account of conflicting national interests and capabilities, but they can hardly
help taking notice of the active participation of their citizens in conferences
such as UNCED.

Participation in agenda setting and policy formulation by unofficial representatives of public interests (NGOs) has not been common to the political
ethos of most countries. Only recently has this NGO participation in interna-

tional affairs (as distinguished from lobbying) become conventional in North America and Western Europe. For example, its role has been documented in negotiations leading to the Canadian–United States Water Quality Agreements.[6] It is not surprising that NGOs entered into the debate over the North American Free Trade Agreement (NAFTA).[7] As a condition for accepting NAFTA, the NGOs persuaded the U.S. Congress to support a parallel institution for protection of the continental environment with particular concern for Mexico and the U.S.-Mexico border.[8] This was an instance of citizen involvement in what traditionally would have been an exclusive diplomatic process. The concept of citizen initiative in governmental and international affairs appears to be contagious and has been spreading even into some countries traditionally nondemocratic in governance.

The era of exclusive official diplomacy may be over, yet the influence of NGOs should not be overdrawn. Their role in the UNCED process, although significant, often fell short of their more expansive ambitions; and yet they have attained a legitimate and recognized role in the United Nations system. NGOs are now an established factor in world politics.[9]

The UNCED process, though not unique, identifies and confirms a form of political decision making that seems certain to become more common in international environmental negotiations having transboundary, and especially global implications. The future format of international negotiations cannot be foreseen with assurance, but in an era of unprecedented mobility in information, communication, and transportation, a broadening of public participation in international affairs seems inevitable.

The UNCED Context

If not unique in principle to international policymaking, how was UNCED distinctive in practice? There were at least three contextual factors shaping UNCED's substance and procedures. They characterized not only the conference agreements (e.g., Agenda 21 and the Rio Declaration), but also the draft treaties·and actions taken by the NGO Global Forum. Expressed in simple terms, the factors shaping the conference output were scope, time, and basic assumptions or paradigms, as they influenced beliefs about human-environment relationships. They interrelated to form a comprehensive psychological ambiance within which goals and strategies were adopted. These contextual factors took form in concrete and specific expressions of ethical and ethnic values, economic status, political circumstances, geographic conditions, and expectations for the future. They were also expressed in the influence of science and in the attitudes of both governmental and nongovernmental participants toward the issues of UNCED.

The *scope* dimension grew through a larger understanding of the interrelatedness and ubiquity of environmental problems. International environmental policy was seen to require a transnational-transgovernmental global context. Allowing for differences in interpretation, the "Only One Earth" logo of Stockholm became an operational reality at Rio. The immediate reasons were two. Anxiety over the thinning of the ozone layer and the projected consequences of global climate change brought nations to a recognition that human activities had created problems, the solutions for which required global international action. Although a minority of science skeptics and anti-environmentalists ridiculed the climate issue as a self-serving scam by grant-seeking scientists and exaggerating alarmists, the accumulating evidence had aroused sufficient apprehension to cause governments through WMO and UNEP to convene in 1988 an Intergovernmental Panel on Climate Change and to agree to preparation of the multilateral framework convention which was open for signature at Rio. Although possibly 95 percent of the world's atmospheric scientists concluded that global warming of the earth's atmosphere and oceans was occurring, a skeptical minority dissented.[10] This gave antienvironmentalist politicians reason to argue against any counteraction because "scientists don't agree."

The ozone and climate treaties require national and local implementation. They were salient among a much larger set of issues set forth in Agenda 21 that could be addressed effectively only if all levels of government and nongovernmental sectors cooperated. The clear implication of UNCED was that all levels of government shared responsibility for safeguarding the earth and that actions taken at each level ought to take account of consequences at other levels (e.g., Agenda 21, chapter 37.2 and 37.76d).

This proposition does not fit well with traditional interpretations of inviolable national sovereignty or territorially bounded exclusive legal jurisdiction. Collective transboundary responsibility for planet Earth achieved a degree of confirmation at Rio, hitherto only recognized as a practical necessity in the governance of activities in outer space. It is safe to assume that there will be some rearguard national resistance in the future, but the logic of a planetary scope and a broad spectrum for environmental policy, reinforced by scientific findings, presages a transition to a new era in both policymaking and international relations. But this requires a raising of science literacy almost everywhere.

Behind the permeability of jurisdictional boundaries, and beyond the findings of science, we have noted the expanding role and scope of extragovernmental involvement in policy formation. The term "nongovernmental" (NGO), although technically accurate, is growing less descriptive of these voluntary issue-oriented civil society organizations, which have now become actual or potential participants in governance. The presence of these "civic" groups at Rio and their semiofficial influence on the official delegations were visible aspects of a trend in intergovernmental conferences that is unlikely to be reversed.

Paralleling UNCED was the NGO Global Forum, through which inputs from local, national, and regional organizations were expressed. There has been a growing recognition in the UN system that environmental protection and ecologically informed management of resources require multilevel participation from local to global.[11] A second paralleling event was the formation of the international Business Council on Sustainable Development, which published under the title of *Changing Course* a book of thirty-eight case studies of business initiatives for sustainable development.[12] The International Network for Environmental Management (INEM) organized the main industry event at the Global Forum and launched Industry 21 as an "umbrella" program for UNCED follow-up by INEM's affiliates worldwide.[13] In January 1995 the Business Council on Sustainable Development combined with the World Industry Council for the Environment to form the World Business Council for Sustainable Development.

The actual influence of the NGOs on the Rio deliberation is difficult to assess. But it is reasonable to believe that their presence and influence was felt by many of the official delegates. A more definitive test of their influence will come at the national level in the ratification and implementation of the treaties opened for signature at Rio.

In a study of patterns in the ratification of global environmental treaties, David W. Cook (1990) found that the highest number of ratifications was found in those countries with the greater number of international NGOs, in countries that were dominant in the world economy, and were more open, politically and economically.[14] During the two decades between the conference at Stockholm and Rio, there has been a significant increase in environmental NGOs in less-developed countries, but their influence on national governments varies widely.

Rio confirmed what should already have been evident—that environmental policymaking (both governmental and nongovernmental) must henceforth take account of the scope of cultural receptivity, ethical perception, and social impact outside of official government. Implementation of the comprehensive Agenda 21 of Rio will require an openness and breadth of input in policymaking that would change conventional assumptions about responsibilities for policymaking at all levels of government. Doctrines of exclusively official participation in international environmental policymaking are now on the defensive.

Through scientific methods of analysis and prognosis, we can estimate some trends toward elements of the future with a degree of probability not possible heretofore. In the minds of informed observers, projected consequences of a deteriorating stratospheric ozone system, of global climate change, and of excessive population growth call for policy decisions now to forestall unwanted consequences later. It is possibly too late already to prevent some damaging consequences of these trends.

But most persons, including public officials, are not informed observers of systemic environmental change, many are misinformed, and some have conflicting priorities. The obvious truism that all people now living live in the present and the expectation that few will live through the twenty-first century makes policy a contemporaneous affair. The long-range future is uncertain and seldom presents a compelling need for immediate action. Democracies are notoriously short-range in their policy preferences. The weight of scientific evidence often contradicts conventional "common sense." Yet the Rio conference by implication posed the question, Can modern society learn to shape its policies toward longer-range contingencies, allowing for uncertainties and risks? The International Conference on Environment, Long-Term Governability and Democracy: Twenty-first-Century Prospectives for the Environment, organized by the government of France and a group of scientific and environmental organizations with support from the European Commission, held at the Abbaye de Fontevraud, France, should provide insights into prospects for Agenda 21.

The Rio agreements imply the necessity for a reassessment of time as a consideration in policymaking. Governmental response to international agreements depends heavily upon the role of the NGOs in policy formulation and execution. The work of the United Nations Commission on Sustainable Development, which monitors and reports national performance on international agreements, including those with long-time horizons, suggests that political leadership today is willing now to accept some degree of collective responsibility for the longer-term future.[15]

The third aspect of the Rio context was the rhetorical commitment to a planetary worldview or *paradigm*. Not all delegations were committed to its implications for national policies and responsibilities. Among those professing commitment, not all would honor it in conflict with national priorities — notably economic and demographic. Nevertheless, the commitment to planet Earth, even when largely symbolic, is a significant indicator of changes in perceived political realities. And for many of the delegations, concern for the future of the biosphere was genuine. Geography made a difference among small island states (e.g., in the South Pacific and Caribbean) that might be inundated if rising sea levels followed from global warming.

This concern with the interrelations of national priorities and the biospheric future was primarily but not exclusively noted among the delegates from the more economically and technologically advanced countries. They were also countries in which public opinion and citizen action influenced decisions as to what is desirable and feasible in the international relations of government and business. And in a globalizing international political economy, even governments hitherto indifferent to environmental knowledge and values may have to accede to environmental policies that are gaining acceptance among informed people throughout the world.

That an attitudinal change has been occurring is evidenced by the representation of 178 countries at the UNCED and the personal attendance of 110 heads of state.[16] Attendance by nongovernmental groups (an estimated two thousand) exceeded any previous UN conference, and news media coverage was possibly double that given to the 1972 conference in Stockholm. A member of the United States delegation, Senator (subsequently Vice President) Albert Gore had written a widely publicized book, *Earth in the Balance*, which clearly and forcefully articulated the biospheric paradigm.[17] For an active politician to author a book advocating an international "Marshall Plan" for environmental protection and development would seem to be indicative of changing attitudes and values. But to some observers, Vice President Gore seemed to distance himself from his book, possibly because of presidential ambitions that he believed the book might prejudice.

Even so, the United States, which led at Stockholm in shaping an environmental agenda (with some reservations), was visibly divided at Rio. Ironically, the major antagonist to the United States and Stockholm, Brazil, which voiced the objections of the developing countries to international environmentalism, was the host of the 1992 Rio Conference. In some respects, Rio witnessed a reversal of roles, with the United States often being a negative voice among the nations.

Agenda 21

The most significant product of the Rio Conference was Agenda 21, an extensive and detailed statement of goals and principles to guide the actions of governments at all levels into the twenty-first century. It provided a course of action for attaining environmentally healthy, sustainable, and equitable conditions throughout the world. Agenda 21 was (and is) an action plan, but the actual implementation of the plan was left to the decision-making institutions of society, chiefly governments. The scope, content, and direction of Agenda 21 are best displayed by the following listing of the titles of its forty chapters. There are numerous findings, declarations, and principles under each chapter heading. To explicate the sections and subsections under each of the forty chapters could exceed the volume of this book. But the chapter headings are sufficient to indicate the substance of Agenda 21:

1. Preamble

Section I—Social and Economic Development
2. International Cooperation to Accelerate Sustainable Development in Developing Countries and Related Domestic Policies
3. Combating Poverty
4. Changing Consumption Patterns

Action to consider UNCED recommendations is being taken by a number of national governments. In June 1993 the secretary-general of the UN reported to the newly appointed Commission on Sustainable Development that seventy countries had established committees or commissions to examine UNEP recommendations and more than forty nations had requested assistance from the United Nations Development Programme (UNDP) in preparing national versions of Agenda 21. It seems possible that Agenda 21 will have a reinforcing effect upon other international environmental agreements.

From Perspectives to Policy

As to be expected in multinational sociopolitical developments, some inconsistencies prevailed at Rio and some important issues received what some observers regarded as insufficient attention (i.e., population and energy). UNCED may thus be most realistically seen as a milestone in a policy process begun at least as early as the 1972 UN Conference in Stockholm—or perhaps earlier, in the Biosphere Conference in Paris in 1968. The UNCED process is definitive in that it will be difficult for future international conferences to achieve political legitimacy unless provision is made for openness and for participation of all major parties affected. Closed-session deals by diplomats and expert advisors may have greater difficulty than heretofore in obtaining national ratification. The General Agreement on Tariffs and Trade (GATT) and the North American Free Trade Agreement (NAFTA) negotiations in 1990–91 encountered resistance to ratification partly because of a perceived absence of public input.

Keeping in mind the sociopolitical context of UNCED regarding the scope of issues and of conference participation, the factor of time in the environmental impact of human activities, and the changing paradigm of man's relationship with Earth, we may comprehend more fully its policy implications. But implications are neither prognoses nor predictions. They represent real possibilities, even trends, but unforeseen factors may intervene to alter the future.[18]

With regard to the transboundary, multilevel implications of Agenda 21, UNCED has accentuated and accelerated trends already discernible in international relations. More than ever, policymaking at every level must take account of interactions with other levels—national, regional, provincial, municipal, and international. International policies are usually initiated and must be confirmed at national levels, but in environmental issues the distinction between domestic and international considerations is rapidly becoming blurred. This is nowhere more true than in matters of multinational business and finance.

Agenda 21 depends for its implementation on national action often affecting policies once regarded as wholly domestic (e.g., agriculture, forestry, indigenous people). Developed countries are under growing pressure to assist capability-building in developing countries (Agenda 21, chapter 37). Policies regarding "foreign aid" and international technical assistance will be under pressure to respond to existing realities rather than, as formerly, as expressions of national, largely political, ideologies and abstract economic theories. Developing countries may be expected to resist "conditional" assistance that is not supported by demonstrable findings of environmental science. Some governments may reject "conditions" even when supported by conclusive evidence. But NGOs are now more likely to oppose development projects by donor governments or intergovernmental agencies (e.g., the World Bank) that degrade the environment or disrupt the lives of people.

In the course of time, the ubiquity of many environmental problems (e.g., atmospheric transport of pollutants, oil and chemical spills in boundary waters and international seas, and the transport of hazardous materials) will tend to force a harmonization of relevant policies and laws among all levels of government. The push toward common or comparable standards for environmental policies may be felt most directly by international business. But business interests also are among the promoters of uniformity in environmental laws and regulations, primarily to prevent or to reduce trade restrictions. This effort has usually been directed toward modifying or removing environmental laws and regulations regarded as restrictive to international trade. Here there may be conflict among assumptions, values, and paradigms. Is the business of the world, wholly business (e.g., trade and commerce)? Or are the dimensions of world affairs more inclusive and more fundamental?

Policymakers at national and intergovernmental levels are now facing conflicts generated by the simultaneous growth of the free trade and environmental movements. Conflict is prevalent between advocates of economic and trade priorities and those defending environmental and consumer protection, public health, and preservation of biological species. In principle, these conflicts need not be irreconcilable; in practice, they often are. Foresight into some of the political problems that implementation of the Rio agreements may encounter may be obtained from experience with trade and environment-related treaties or agreements. As has been noted, an illustrative case in point appeared in the *New York Times* of Monday, 14 December 1990. A full-page appeal entitled "SABOTAGE! of American's Health, Food Safety, and Environmental Laws—George Bush and the Secret Side of Free Trade," was endorsed by twenty-one nongovernmental civic action organizations including the Sierra Club, Greenpeace, the Humane Society of the United States and the National Consumer League. The purpose was to arouse citizen opposition to provisions of the North American Free Trade Agreement (NAFTA) and the "Uruguay Round" of the General Agreement on Tariffs and Trade (GATT) that might prevent the United States from barring the importation of environmentally harmful products.[19] This opposition was fueled by the GATT ruling in the tuna-dolphin controversy, in which the enforcement of the U.S. Marine Mammal Protection Act was held to be in violation of GATT agreements and discriminatory against Mexican fishermen, whose nets trapped many dolphin.

Representatives of developing countries have complained since the Stockholm Conference that developed countries used environmental restrictions to prevent importation of their products in order to protect home industries from competition. At Stockholm the developing countries claimed compensation for discrimination based on alleged environmental protection. But countries seeking global trade may also be motivated to raise their environmental standards, especially in relation to exports.

Perhaps the objectives and priorities of free trade can be reconciled with the goals and values implicit in UNCED. Compromise rather than compensation or reconciliation seems more likely, because two different sets of assumptions and two dissimilar paradigms are in confrontation. The Declaration of Rio represents a broad, albeit sometimes indeterminate, public commitment. The trade agreements represent a much smaller, but more explicit set of objectives advanced by interests that have had an "inside track" in the course of national policymaking.

Balance or Synthesis

The grand purpose of UNCED was to reconcile environmental and developmental objectives (i.e., primarily but not exclusively economic) in national and international policymaking. In this effort the objective could be either balance or synthesis, but the implications of the choice between the two are not the same. Balance suggest compromise—so called splitting the difference, or each side giving up something to reach an agreement that is in essence political. The UNCED objectives in the Declaration of Rio, the associated treaties, and Agenda 21 are heavily influenced by scientific findings or hypotheses that do not easily yield to political or economic expediency. This was notably true of section 2 of Agenda 21, "Conservation and Management of Resources for Development."

The UNCED process of compromise and the role of science was more implicit in the formulating of policies to achieve UNCED objectives than is explicit in the actual text of its documents. Given the diversity of nations, agreement on principles or goals is more attainable than on action to be taken. Not all UNCED objectives are supported by scientific evidence, yet in order to realize most of the substantive objectives some measure of scientific information is required. In the absence of science and science-based technologies, many UNCED objectives would not be politically attainable. They would lack indicated operational means, popular credibility, or both.

The perceived necessity for choice between trade, finance, and economic growth on the one hand and the multiplicity of environmental problems of various urgencies on the other confronts governments everywhere. The outcome of policy choice cannot be predicted with assurance, but if the numbers of people affected and the pressures of organized citizen response prove to be politically decisive, trade agreements may not be sustainable unless modified to accommodate environmental, public health, demographic, and perhaps even quality-of-life considerations. Conversely, environmental policies may not be defensible if they are unnecessarily obstructive of generally beneficial economic processes.

As Agenda 21 is implemented by national governments and international organizations, institutional means to attain coordination and consistency among policies will be necessary. Establishment of the UN Commission on Sustainable Development (Agenda 21, chapter 35), and reform within the Global Environmental Facility (associated with the World Bank) indicate a trend toward institutional restructuring. The aphorism that "you can't solve problems through reorganization" may give way to a realization that complex, interrelating sets of problems will not be resolved spontaneously by uncoordinated, unplanned action. In any case, public (and business) policymaking will increasingly be compelled to deal with the complexities and interrelatedness of the problems of the world. And if the risk of policy failure is to be reduced, the linear thinking that has characterized many public policies must give way to multiplex assessment of the probable consequences of proposed action.

Agenda 21 is probably as specific in its forty chapters as any action plan could be when the action must largely take place in more than 150 different countries with different political, social and environmental circumstances. Beyond inevitable compromises, the concept of a holistic future-oriented global perspective was implicit in the documents presented to the delegates at Rio. The prime minister of Japan, Kiichi Miyazawa put the products of Rio in perspective when he declared that: "The joint endeavor to protect the global environment has only just begun." In an opinion editorial for the *Earth Summit Times*, he wrote that:

> The Rio Declaration and other policymaking agreements reached at the Conference on international cooperation in the field of the environment and development constitute an important first step in the effort to attain sustainable development. Even more important, however, is how effectively we translate the political will demonstrated in Rio into action to save our planet. This, I believe, must be the yardstick by which the success of the Conference is to be measured.[20]

To conjecture how each of the forty chapters of Agenda 21 or the treaties on global climate change and biodiversity will specifically affect policymaking would be impractical. The Agenda 21 action plan is nonbinding and may be implemented in different ways in different countries toward a common objective. The treaties, however, *are* legally binding under international law, but the two treaties signed at Rio specify agreed-upon goals and do not impose operational obligations. Framework conventions like these two treaties only declare policies; they will doubtless require protocols in the course of their implementation. Before that happens, much depends on the course of change in popular attitudes and national politics. Public opinion and policies may also be influenced by scientific assessments of probability and risk yet to be made.

Some generalizations are feasible, however, regarding the orientation of policymaking as national and international organizations undertake to implement Rio commitments. The growing influence of environmental, consumer, and health NGOs suggests that it will not be easy for governments to satisfy these constituents with rhetoric. In addition, the environmental trends and problems that shaped the substance of the Rio Conference will not disappear, and on these issues science seldom offers arguments that support inaction.

From the Rio provisions, one may deduce two sets of propositions regarding the legitimate options of policymakers in the future. Although seldom expressed in explicit guidelines, the Rio provisions provide, in principle, a general consensus regarding what governments both ought to do and ought not to do. If put into practice, the obligations implied at Rio would represent a significant reinterpretation of the traditional sovereign right of a nation to do as it pleases with its natural resources and economic assets. Although the Rio Declaration asserted a state's sovereign right to exploit its own resources in accordance with its own policies without harming the environment elsewhere (Principle 2), the assertion contains a significant qualification: there are many acts of government both of commission and of omission that could harm the environment of other states or of areas beyond the limits of national jurisdictions.

The principle of international law that a nation should not use its territory to harm its neighbors has increased in relevance as science has documented the migration of pollutants throughout the biosphere, the effects of massive deforestation, and the consequences of desertification. The principle of sovereignty is now paralleled by the principle of national responsibility for custody and care of its environment, for the effects of its actions on the biosphere, and upon the lives and well-being of people affected by the action or inaction of governments.

Because protective measures are often expressed as positive obligations, the distinction between *do* and *don't* is often a matter of the way in which an issue is presented. Here, the more effective use of environmental impact analysis is likely to reveal hitherto unrecognized causes of environmental degradation and to clarify linkages and relationships among environmental and economic issues. Governments may often protect the environment by ceasing to engage in or to subsidize environmentally damaging activities. Among the preventive measures expressed in Agenda 21 were assessments of impacts upon the environment and the monitoring of environmental effects and changes. These obligations were clearly set out in chapter 35, "Science for Sustainable Development," and in chapter 40, "Information for Decision-Making."

Endorsement of the adoption and improvement of environmental impact assessment and the advancement and dissemination of scientific information appear directly or by implication throughout Agenda 21. Techniques of information gathering and analysis were repeatedly recognized as necessary to the formulation of rational and realizable policies for environmentally and economically sustainable development. More than ever, open communication and adequately informed judgment will be regarded as requisite for acceptable legislative and administrative decisionmaking. This expectation has implications for the education, appointment, and promotion of public officials, for the staff support required, and for relationships between legislative committees, administrative agencies, the worldwide scientific community, and concerned citizens.

An obligation to protect the life-support systems of the planet—atmospheric, terrestrial, and marine—and to conserve biodiversity, were major themes of the UNCED documents. They had been implicit in many prior treaties and declarations, and especially in the 1987 report of the World Commission on Environment and Development (Brundtland Commission), *Our Common Future*.[21] Responsibility for the conditions of human life was emphasized notably in the Rio Declaration and was also prominent in draft treaties prepared by NGOs. A notable departure from the historic practices of governments was a declared obligation in Agenda 21 to respect and protect indigenous peoples (chapter 26).

There was little new in what governments (as distinguished from nations) were obligated to refrain from doing. The historic principle of international law *pacta sunt servanda* (treaties should be obeyed) was, however, made more than a moral obligation in provisions to discourage noncompliance or negligence in treaty obligations. Agenda 21, chapter 38, "International Institutional Arrangements," and chapter 39, "International Legal Instruments and Arrangements," dealt particularly with issues of monitoring, reporting, compliance, and dispute prevention. The UN Commission on Sustainable Development was established by the UN General Assembly in December 1992 to monitor and report on progress by governments and international organizations toward the goals of Agenda 21. Another effective means for obtaining compliance with declared commitments is reporting and publicity: public officials everywhere try to avoid embarrassing disclosures.

In the Global Climate Change and Biodiversity treaties, action by governments would be necessary to prevent or restrict practices that would be harmful to the environment and the biosphere. In the Rio documents, as often happens in international agreements, controversies were frequently avoided or obscured by ambiguous or innocuous choice of words or phrases that could be read in more than one way without offending sensibilities. In many cases, as in the Convention on Global Climate Change, the objectives were strongly worded, whereas the obligations to act and the timetables for action were deliberately left vague. On some issues, representatives from developing countries insisted on the "right rhetoric," even though it was subsequently modified or contradicted by implied obligatory commitments. An example of euphemistic substitution was the use in Agenda 21, chapter 5, of the term "demographic dynamics" instead of "population trends."

In brief, the UNCED agreements moved forward the conditions and processes of national policymaking toward increased reliance on scientific information and analytic methods, with greater involvement of popular representation in decision making, and much broader assessment of the range of consequences of policies and decisions. But moving toward is not being there now. In an opening address to the conference, the secretary-general of the conference, Maurice F. Strong, declared, "The Earth Summit is not an end in itself, but a new beginning. The measures you agree on here will be but first steps on a new pathway to our common future. Thus, the results of this conference will ultimately depend on the credibility and effectiveness of its follow-up."[22]

Among the many appraisals of UNCED's accomplishments, the following observation may be among the more prescient:

Instead of being judged against a single conception of what its outcome should have been, the conference . . . must be judged within the context of a process of increasing attention, sophistication, and effectiveness in the management of environment and development issues. . . . Thus, the important question is not how many treaties were signed or what specific actions were agreed on but, rather, how effectively UNCED contributed to this broader process.[23]

As might be expected, assessments of the results of Rio differ. Unrealistic expectations led some members of the NGOs to regard the conference as disappointing. A cautious appraisal was made by a representative of IUCN, the World Conservation Union: "We will not be able to assess for some years how far all this will result in deep-rooted changes of attitude. In the immediate future UNCED will be measured for its success by the agreements it produced. Here the record is unimpressive."[24]

A conference that brought together representatives of some 178 countries, 110 heads of states, 2,000 NGOs, 8,000 journalists, and thousands of spectators can hardly be expected to present a coherent picture. News coverage was uneven and tended to emphasize controversy rather than accord. Some American delegates and observers were especially critical of U.S. journalism for overemphasizing the negative and contentious aspects of U.S. participation and underrating its positive contribution.

Contemporaneous issues and conflicts aside, UNCED placed a broad set of policy and action imperatives before the community of nations. Principal among these imperatives was a reconciliation or synthesis between environmental protection and sustainable development. Conference objectives, to be realized, force a more comprehensive and fundamental consideration of the purposes and procedures of what has been loosely termed "development." They also require a broader and more sophisticated approach to social and economic policy to achieve sustainable development and environmental protection.

Implementation of the Rio agreements will require reorganization of both national and international structures for the administration of policies requiring interdisciplinary and interjurisdictional coordination. If the goal is to strike a balance between environmental and economic objectives, no great change in the organization of public administration may be necessary. If, however, sustainable development cannot be achieved through a fractionated system, and requires an effective interrelating of policy and administration across jurisdictional boundaries, innovative restructuring of public agencies may be required. Can a new paradigm for politics induce new processes and structures for public administration?[25] The appropriate answer is the cliché "It remains to be seen."

In summary, the Rio Conference proposed an agenda and set normative goals that have major implications *for* national policies and for *how* policies are made, both at national and international levels. Were UNCED an isolated event, skepticism regarding its effect on policymaking might be justified. As a large milestone in a sociopolitical trend under way for more than two decades, it deserves continuing follow-up. If the present world order does not respond voluntarily to self-improvement for its future, adherence to Agenda 21 will ultimately be not so much a matter of choice as of necessity.

7
International Structures for
Environmental Policy

In this chapter and the one that follows the focus will be on international arrangements and institutional structures that have evolved out of efforts among people and national governments to protect their perceived interests in the global environment and its principal regions. These chapters are concerned less with the details of policy, law, or cooperation than with the organizational arrangements through which international efforts take effect. Relationships between purpose and structure will be considered as expressed through international agreements (e.g., treaties) and institutional arrangements.

The primary organizational unit for political and jurisdictional functions in modern times has been the nation-state. But this fact alone is of limited significance because there are many kinds of nations and states, contrasting in almost every respect. There are peoples who regard themselves as nations, yet have no state; there are multinational states; and there are states, such as Vatican City, Andorra, and San Marino, that are not nations.

But the world is not wholly organized around political jurisdictions and governments. A large and complex network of nongovernmental international organizations has developed during the past century. Its diversity is nearly as great as the diversity of human interests, and the ranges in size and influence among NGOs is very great. Some of these organizations have been formed explicitly for environmental protection. Some are federations of scientific and professional associations; still others include philanthropic and religious organizations that are sometimes concerned with quality of life and environmental issues. Moreover, the institutional structure of international environmental cooperation includes not only those intergovernmental or nongovernmental organizations established for environmental protection, but also the more numerous bodies, governmental and nongovernmental, including multinational corporations, that impact upon the global environment and whose cooperation must be obtained in defending the integrity of the earth.

It will be useful to outline the structure within which action occurs before describing, in the next chapter, the various forms of international cooperation for environmental protection. This framework is still evolving and, with the partial exception of the United Nations system and organizational relationships created by treaty, is not built into a fixed pattern by predetermined design. It has emerged in response to diverse challenges; its agencies have sought new forms of international cooperation to cope with newly perceived environmental problems.

This book cannot deal with all of the environment-related functions of the organizations and arrangements considered here.[1] Its purpose is to show how they interact to produce understandings, obligations, and practices that may be called international environmental policy, and may ultimately gain recognition as sources of international law.

But first it is necessary to identify the principal components of the structure of international environmental cooperation. Although individual persons, identified at various places in this book, have played important roles in shaping international environmental policies, the principal actors are organizations. They are national governments, regional and multilateral institutions and agencies established by national governments, cooperative regimes, and also nongovernmental organizations—in some of which governments are nevertheless direct or indirect participants. Binding the structure together is a body of agreements and arrangements, ranging from unofficial expressions of national interest to firm treaty commitments under international law. In the perspective of this volume the uneven progression and often uncertain status of an evolving international environmental law argues for its treatment as the ordering and legitimizing aspect of international environmental arrangements rather than for it to be accorded separate consideration as international law per se.

Organizations for Intergovernmental Cooperation

Many intergovernmental organizations—for example, the European Union—have been established for cooperation on matters including, but not primarily, environmental ones. There are also numerous intergovernmental organizations for particular environment-related purposes. Their functions are as diverse as protection of cultural monuments, management of marine fisheries, and control of desert locusts. Consideration of these specialized environment-related arrangements will mostly be deferred to those chapters in which their missions are described. Similarly, organizations essentially regional in character will be considered in Chapter 8, although they too are part of the inclusive structure of international environmental cooperation.

Political Alliances and Associations

There are a number of multilateral political alliances or associations of states that include environmental matters within the scope of their cooperative relationships. Environment occupies a relatively minor phase on their continuing agenda, but occasionally they become participants in transactions of environmental significance. Although regional considerations of a political character influenced the establishment of several of these organizations, few have retained a strictly regional character. Organizations embracing whole continents (e.g., the Organization of African Unity) or with membership as far flung as Canada, Japan, and New Zealand (i.e., the Organization for Economic Cooperation and Development) can hardly be regarded as "regional" for purposes of environmental cooperation.

The North Atlantic Treaty Organization (NATO) was established in 1949 by the World War II Western Allies for purposes of common defense, but it also included economic, cultural, and, indirectly, environmental affairs among its functions. Its Committee on Challenges of Modern Society (CCMS), established in December 1969, sponsored studies of water pollution control and the national and regional impact of air pollution.[2] CCMS was also concerned with the environmental problems of urban communities, and a Conference on Urban Affairs was convened in Indianapolis, Indiana, in May 1971. Comparative studies were undertaken in Frankfurt am Main, in Ankara, and in St. Louis, Missouri. CCMS activities continue and may gain more promise in NATO if its military mission becomes less urgent. Among its activities have been studies of restoration of contaminated land, air pollution, and estuarine management and seminars on the environmental impact of military land acquisition.

For the historical record, the Soviet-sponsored Council for Mutual Economic Assistance (CMEA/COMECON), like NATO, was primarily political and economic in its inception, but also fostered technical and scientific cooperation and research among its member countries.[3] Its activities impinged upon certain environmental problems, albeit environmental policies per se do not appear to have been among CMEA's higher priorities. Nevertheless, at its Thirty-first Session in June 1977, CMEA adopted an expanded program of environmental cooperation and agreed to cooperate with the Economic Commission for Europe in subject areas such as transboundary air pollution, waste reduction technology/conservation and rational management of water resources, along with protection of flora, fauna, and landscape. In 1979, UNEP announced an agreement with CMEA to "develop and establish contacts in the field of environmental conservation" including exchange of publications, information, and documents. Environmental conservation was an ambiguous priority in the Soviet bloc of states, but the rapprochement with

UNEP appeared to indicate a moderating of a traditionally economistic orientation. The effect of the disintegration of Communist-dominated governments in Eastern Europe during 1989 upon the future of CMEA was unclear at the beginning of the 1990s. What became clear with the fall of Communist-dominated regimes was the abysmal failure of environmental protection in the Soviet Union and the Eastern European states. Defenders of the environment were no match for the overwhelmingly dominant economic and military authorities.

The Organization of African Unity (OAU) and the Organization of American States (OAS) have been primarily concerned with matters other than environmental. However, the OAU did sponsor the new (1968) treaty on the protection of African wildlife, and has also sponsored the College of African Wildlife Management at Mweka in Tanzania. Founded in 1963, the college has been assisted by a number of agencies including the United Nations Special Fund, the British Ministry of Overseas Development, the Ford Foundation, and the African Wildlife Leadership Foundation.[4]

The concern for environment within the OAU appears to have grown with growing appreciation of its relevance to development and to such basic human priorities as food, water, housing, health, and employment. The ecodevelopment concept and the conservation for sustainable growth strategy promoted by IUCN, UNEP, and several of the UN specialized agencies have helped to narrow whatever gap may have existed in the way in which African leadership perceived environment and development.

The OAU Lagos Plan of Action 1980–2000 brought environmental concern directly into socioeconomic planning. Eight priority areas were identified for attention: (1) environmental sanitation and health and safe drinking-water supply, (2) desertification and drought, (3) deforestation and soil degradation, (4) marine pollution and conservation of marine resources, (5) human settlements, (6) mining, (7) air pollution control, (8) environmental education and training, legislation, and information. Organizations such as the International African Institute have helped to bring about a growth of basic environmental awareness reflected in the Lagos plan.[5] And at their Thirty-seventh Session (May–June 1981), the Council of Ministers, with specific reference to the Lagos plan, recommended adoption of the draft World Charter for Nature, which had been proposed to the United Nations General Assembly by an OAU member state, Zaire. However, in view of political differences among many African states and especially given a military dictatorship in Nigeria with a record of hostility to environmental reformers, it is doubtful that much action will follow from the Lagos Plan.

The Organization of American States (OAS) has not shown a major concern for environmental affairs, although it has undertaken environmental assessments of proposed river development projects, assisted by the United

Nations Environment Programme (UNEP). In addition, the OAS (then the Pan American Union) sponsored the 1940 Convention on Nature Protection and Wild Life Preservation in the Western Hemisphere, and has assisted various scientific and technical meetings on environmental matters.[6] Nevertheless, nature and wildlife continue to deteriorate in many Latin American states. The OAS was a signatory to a 1980 declaration by international development-related agencies on environmental policies and procedures relating to economic development. In 1987 Vice President Victor H. Martinez of Argentina introduced into the legal committee of the OAS a proposal for a system to monitor the observance of environmental treaties by the American states. But mere declarations have not often led to action.

Organization for Economic Cooperation and Development (OECD)

The OECD, with headquarters in Paris, originally consisted of countries participating in the Marshall Plan post–World War II reconstruction in Europe, but now includes additional members such as Australia, New Zealand, and Japan, twenty-four members as of 1995.[7] Although the present purpose of the OECD is the promotion of economic growth and the advancement of international trade, it provides an important forum for the study and exchange of information relating to environmental problems and the coordination of policies relating to the environment. The OECD has encouraged movement toward a balance between quantitative and qualitative growth resulting from a growing belief in member countries that "greater emphasis will have to be placed on the complementarity and compatibility of environmental and economic policies." In 1986 OECD published a detailed review of its environmental activities since the early 1970s. The printed record is impressive.[8]

In 1970 the OECD established the Environment Committee at the ministerial level as an expression of its "desire to extend existing sectors to embrace fully the economic implications of environmental problems and specific sectoral work on the environment to wider decisions on economic and social policy."[9] The committee, now directorate, was charged:

— to investigate the problems of preserving or improving man's environment with particular reference to their economic and trade implications;
— to review and confront actions taken or proposed in Member countries in the field of environment together with their economic and trade implications;
— to propose solutions for environmental problems that would as far as possible take account of all relevant factors, including cost effectiveness;
— to ensure that the results of environmental investigations can be effectively utilised in the wider framework of the Organisation's work on economic policy and social development.[10]

In 1975 environmental work within the OECD was organized into four main sectors: environment and energy, environment and industry, urban environment, and land use; and several cross-sectoral projects, e.g., tourism and transfrontier pollution.[11] That year also marked a shift in OECD policies toward a more integrative approach to the environment and an emphasis on policies of prevention rather than cure. Attention was also directed to the development of new management and policy instruments such as environmental impact assessment. The program of work undertaken during the period 1974 to 1978 included projects on power plant siting, environmental impact of off-shore production of oil and gas, urban environmental indicators, application of environmental criteria and standards to land use planning, waste management, and control of specified water pollutants and toxic chemicals.[12]

The movement toward a more integrative approach to the environment was continued at the May 1979 meeting of the environment ministers with a recommendation that member countries prepare periodic national reports on the state of the environment that would be used as a basis for regular publication of an international report on the state of the environment throughout the OECD. The first international report, prepared prior to the meeting, acknowledged that environmental work had previously been concerned primarily with the formation of institutional arrangements and the adoption of laws and regulations for the abatement and control of pollution. The report heralded the emergence of a "second generation" of environmental policies focused on anticipatory approaches to environmental problems, the conservation and management of natural resources, and improvement in the quality of life.[13]

The Declaration on Anticipatory Environmental Policies issued by the environment ministers indicated the direction of work to be accomplished in the 1980s. In addition to asking that "environmental considerations be incorporated at an early stage in all economic and social sectors likely to have significant environmental consequences," the ministers committed themselves to the encouragement of public participation in the preparation of such decisions and the promotion of environmental objectives and awareness in the field of education. Four main themes emerged from this meeting, which form the framework of the current OECD approach to the environment. They are (1) environment and economics, (2) natural resources and the environment, (3) chemicals and the environment, and (4) urban affairs.

Numerous recommendations have been made by the Environment Committee, with implementation left to the discretion of member states in such areas as reduction of emissions of harmful substances such as mercury, sulfur oxides, and particulate matter into the environment; transfrontier pollution; water quality; and the integration, examination, and systematization of environmental and energy policies. Recommendations regarding noise pollution and waste management have also been promulgated.[14]

Because the OECD is primarily a consultative organization, the majority of work undertaken is designed to evaluate and recommend policy options together with guidelines and strategies for their implementation. In April 1981 a Special Session of the Environment Committee on the OECD and Policies for the 1980s to Address Long-Term Environment Issues was held to mark the beginning of the committee's second decade of work.[15] Following this meeting a series of papers was commissioned in response to policy questions arising from the session. These included identification of the most critical issues, the extent of scientific consensus concerning them, the source of these issues in the activities of OECD countries, and opportunities in dealing with them.

The resulting papers, prepared by the OECD secretariat, focused on the "international dimensions of eleven selected issues and on those aspects that could benefit from international cooperation." An annex to this valuable report lists "[h]ighlights of relevant activities underway or planned by international organizations on selected environment and resource issues." [16] The progression of OECD environmental policies and programs over a decade have been recounted in some detail because it exemplifies a growth of understanding of the interrelatedness of economic and environmental issues that was also occurring in other agencies. These developments indicate a broadening of the basis for interagency and international cooperation.

The United Nations System

There is no easy way to describe the environment-related activities of the United Nations system that will be at once comprehensive, coherent, concise, and accurate. To list these activities in detail in a book of general character would not be meaningful for most readers, and would rapidly become dated as programs and policies change. Our concern is less with the operations of the system, than with the emergence of environmental policies in its various components as illustrative of the scope and direction of international environmental concern. However, to understand the roles and relationships of the various components of the United Nations system in environmental policy it is necessary to understand at least the outline of its structure and the functions of its principal parts.[17]

The Charter of the United Nations was signed in San Francisco on 26 June 1945 at the conclusion of the United Nations Conference on Human Organization. It has subsequently been amended in detail, chiefly with regard to membership on its three constituent councils.[18] The environmental policy role of the United Nations has both political and technical aspects, and both are affected by the missions and the organizational histories of the United Nations specialized agencies. These agencies have been established separately, each by its own treaty, thus adding to the complexity of the structure of international environmental policy.

There are significant differences in the legal and political status of the other organizations within the system. Beyond the structure established by the Charter and the treaties, the relationships of governments and of international organizations within the United Nations system depend upon the concurrence of member states. Not all nations participate in all functions of the system. Membership in the United Nations councils and special bodies, such as the Governing Council of the United Nations Environment Programme, is limited and does not include all UN member states.

General Assembly and Secretariat

To the politically literate public, the United Nations is most likely to be identified with the General Assembly. Membership in the General Assembly, which the *World Almanac of 1996* reported as 185, constitutes membership in the United Nations, but not in the specialized agencies, which have separate procedures for admission. Although it is the principal policymaking body of the United Nations system, the General Assembly tends to function at high levels of generality, and its deliberations reflect the ideological disunity of the world.[19]

The United Nations works largely through a structure of committees, commissions, councils, and semiautonomous special bodies. Of the three principal councils, only Economic and Social Affairs (ECOSOC) is directly concerned with environmental policies, although environment-related issues have been brought before the councils on security and trusteeship. For example, in 1953 Syria appealed to the Security Council to stop Israel from diverting Jordan River water in demilitarized territory.[20] And in 1977 a petition from the Pacific islands of Palau (now Belau) to the Trusteeship Council protested the complicity of the United States, the trustee for Palau, in the construction of a superport that the petitioners alleged would "disrupt and adversely affect the land, water and society of Palau."[21]

The Secretariat of the United Nations provides administrative services for the General Assembly and its subdivisions.[22] Its executive officer, the secretary-general, is elected by the General Assembly. Within the Secretariat are departments that serve the constituent councils—for example Political and Security Council Affairs, Trusteeship and Decolonization, and Economic and Social Affairs. The structure of the Secretariat is complex, and the activities of the offices for which it has responsibilities are diverse. The secretary-general has little formal power; however, the fractionalization of power in the General Assembly provides openings for influencing the course of events. Thus, the personal skill of the secretary-general is a major factor in the effectiveness of the office in the United Nations system.

With establishment of the United Nations Environment Programme

(UNEP) in 1972, much of the initiative in United Nations environmental concerns passed to the executive director and Governing Council of UNEP. In the actual exercise of its functions, however, UNEP characteristically acts in concert with UNESCO, FAO, and the nongovernmental IUCN. Nevertheless a UN secretary-general could play a significant role in environmental policy, if so inclined in this direction. During the years preparatory to the United Nations Conference on the Human Environment, Secretary-General U Thant of Burma made a major contribution to international awareness of a worldwide environmental issue of growing urgency.[23] His being a national of a Third World country helped to counterpoise the allegation that the environmental issue was of concern only to countries with advanced technology.

Special Bodies of the United Nations

The greater number of environment-related United Nations activities occur under the authority or overview of the Economic and Social Affairs Council. But relationships between ECOSOC and programs associated with it are sometimes largely pro forma. For example, the United Nations Environment Programme reports annually to the General Assembly through ECOSOC, which then transmits the report with such comments as it deems necessary.

Achieving some measure of coordination among the numerous and separate United Nations organizations is the Administrative Committee on Coordination (ACC), established in 1946 by ECOSOC resolution, and consisting of the secretary-general of the United Nations and the executive heads of the specialized agencies and the International Atomic Energy Agency. This committee meets twice yearly to resolve such problems of coordination as may arise among the various bodies and agencies comprising the United Nations system. Following establishment of UNEP, the Environment Coordination Board (ECB) functioned for several years, but since 1977 its functions have been merged with those of the ACC.

ECOSOC is one of the more powerful elements within the United Nations system. It may establish functional and regional commissions and in various ways promote coordination among the specialized agencies.[24] Among its functional commissions with relevance for environmental policy are Population (1946) and Transnational Corporations (1974). The latter was established to follow up on the 1974 report of the Group of Eminent Persons to Study the Impact of Transnational Corporations on Development of Economic and Social Affairs. The impact of multinational-transnational corporate enterprise upon natural resources and development, particularly in Third World countries, and its involvement in mining, agriculture, lumbering, international resource flows (e.g., oil), superports, and supertankers will be detailed at various places throughout this volume.

Regional economic commissions have been established by ECOSOC and some of them have made contributions to environmental policy.[25] As already noted, the regional commissions, especially the Economic Commission for Europe, contributed significantly to preparations for the Stockholm Conference and continue to give high priority to environmental issues, especially in relation to sustainable development.[26] The commissions with the dates of their establishment are as follows:

—Economic Commission for Africa (ECA), 1968
—Economic Commission for Europe (ECE), 1947
—Economic Commission for Latin America and the Caribbean (ECLAC), 1948
—Economic Commission for Western Asia (ECWA), 1973
—Economic and Social Commission for Asia and the Pacific (ESCAP—before 1974 known as the Economic Commission for Asia and the Far East or ECAFE), 1947

Within ECOSOC there are a number of committees with indirect or incidental bearing on environmental policy issues. For example, ECOSOC, by Resolution 228(X) of 27 February 1950, established criteria for organizations that could be accorded consultative status under Article 71 of the United Nations Charter,[27] the Committee on Nongovernmental Organizations considers and reports upon what consultative status, if any, should be accorded to NGOs. There are numerous nongovernmental groups that desire consultative relationships with ECOSOC and other United Nations agencies, and many of them have environment-related concerns.[28] Increasing pressure for NGO participation was felt at the UN Conference on Environment and Development at Rio de Janeiro in 1992. It appears that some accommodation can be made at UN conferences as distinguished from regular consultative relationships.

ECOSOC Committees on Housing, Building, and Planning (1962), Development Planning (1964), and Natural Resources (1970) have been concerned with environment-related aspects of United Nations development policy. Broader in scope, the Advisory Committee on the Application of Science and Technology to Development (1963) was established following the United Nations Conference on the Application of Science and Technology for the Benefit of Less Developed Areas. The 1979 United Nations Conference on Science and Technology for Development resulted in some changes in previous committee structure, but did not fundamentally change the pattern of organizations.[29] ECOSOC has added population to the commission structure, but the concerns of this council are heavily social—environment plays a minor role.

Closely related to ECOSOC, but attached administratively to the United Nations Secretariat, are a number of semiautonomous bodies that in varying ways and degrees relate to environmental policy, broadly conceived. Those more frequently mentioned in the following chapters are listed as follows by date of establishment:

— United Nations Institute for Training and Research (UNITAR, 1963)
— United Nations Conference on Trade and Development (UNCTAD, 1964)
— United Nations Development Programme (UNDP, 1965)
— United Nations Industrial Development Organization (UNIDO, 1965) became a specialized agency in 1986
— United Nations Environment Programme (UNEP, 1972)
— United Nations Commission on Sustainable Development (UNCSD, 1992)

UNEP is by far the most important agent of environmental policy in the UN system and throughout the world. Since its functions have been described at length in Chapter 5, they will not be repeated here. As appreciation of humanity's stake in the world environment grows, UNEP's resources and functions seem certain to grow as well.

Commission on Sustainable Development

The commission (CSD) was established pursuant to a recommendation by the UN Conference on Environment and Development.[30] It reports to the General Assembly via the Economic and Social Council and its functions are

> To monitor progress in the implementation of Agenda 21 and activities related to the integration of environmental and developmental goals throughout the United Nations system through analysis and evaluation of reports from all relevant organs, organisations, programmes and institutions of the United Nations system dealing with various issues of environment and development, including those related to finance.

> To consider information provided by governments, including, for example, information in the form of periodic communications or national reports regarding the activities they undertake to implement Agenda 21, the problems they face, such as problems related to financial resources and technology transfer, and other environment and development issues they find relevant.

> To review the progress in the implementation of the commitments contained in Agenda 21, including those related to provision of financial resources and transfer of technology.

> To receive and analyze relevant input from competent non-governmental organisations, including the scientific and the private sectors, in the context of the overall implementation of Agenda 21.

To enhance the dialogue within the framework of the United Nations with non-governmental organisations and the independent sector as well as other entities outside the United Nations system.

To consider, where appropriate, information regarding the progress made in the implementation of environmental conventions, which could be made available by the relevant Conferences of Parties.

To provide appropriate recommendations to the General Assembly through the Economic and Social Council on the basis of an integrated consideration of the reports and issues related to the implementation of Agenda 21.

To consider, at an appropriate time, the results of the review to be conducted expeditiously by the Secretary-General of all recommendations of the Conference for capacity-building programmes, information networks, task forces and other mechanisms to support the integration of environment and development at regional and subregional levels.

This commission interacts with other sectors of the UN system and with international financial institutions and other intergovernmental organizations including the industrial, business, and scientific communities.

Global Environment Facility

In May 1991 the GEF became operational on a three-year trial basis, implemented by UNDP, UNEP, and the World Bank.[31] Established pursuant to an initiative by France, supported by Germany and the World Bank, its purpose was to mobilize additional funding for basic issues of environmental concern. Its mission was to assist financing in developing countries to meet four specific challenges to the environment: (1) reduction of global warming gas emissions, (2) protection of the biosphere, (3) protection of international waters, and (4) protection of the ozone layer. A Scientific and Technical Advisory Panel (STAP) was established by the executive director of UNEP to assist implementation of the facility, for example, to provide Criteria for Eligibility and Priorities for Project Selection for GEF funding. In 1994 the GEF was restructured on a permanent basis following a meeting in Geneva, Switzerland in March 1994 and acceptance by seventy-three countries of an Instrument for the Establishment of the Restructured Global Environment Facility. Administration of the Global Environment Facility, a trust fund, and responsibility for investment projects are functions of the World Bank.

World Trade Organization

To this UN group should be added a separate international organization of a more strictly economic character whose decisions nevertheless may have environmental impacts. This is the World Trade Organization, formerly the General Agreement on Tariffs and Trade (GATT), which was established in 1947 in lieu of an International Trade Organization, which failed to obtain intergovernmental agreement. GATT functioned as the secretariat for the Interim Commission of the International Trade Organization, and along with UNCTAD jointly administered the International Trade Center, which provided advisory service to developing countries on exports and marketing activities, many of which have environmental implications. On 15 April 1994, the World Trade Organization was in effect established in the final act of the Uruguay Round of trade negotiations under GATT. The WTO, which became operational on 1 January 1995, replaces GATT and achieves the objective unsuccessfully sought fifty years earlier in the proposed International Trade Organization.[32]

World Court

Associated with the United Nations is the International Court of Justice (the World Court), the successor to the pre-1945 Permanent Court of International Justice. The first court, like the second, was established as part of an effort to create a structure of international organization, in this instance the League of Nations following World War I. The court adjudicates cases submitted to it by states; private persons or corporations have no standing before it. As of 1996 the World Court had not played a significant role in environmental affairs, but with the growing number of environmental treaties it may in time become a factor in international environmental policy.[33]

United Nations Specialized Agencies

A large part of environment-related activities within the United Nations system is administered by the specialized agencies. An important legal distinction is made between these organizations (sixteen as of 1 January 1990) and the special bodies of the United Nations, such as UNEP and UNDP. The former have been established directly by treaty among their member governments, whereas the latter have been established by the General Assembly or by some other subdivision of the United Nations system. The International Atomic Energy Agency (IAEA) is associated with the United Nations and operates like a specialized agency, but is regarded as an independent intergovernmental organization and does not have specialized agency status. The environmental concerns of the IAEA have grown with the spread of nuclear technology in the generating of energy. After the Chernobyl nuclear reactor accident in 1986, the IAEA sponsored two international conventions of environmental significance: the first on Early Notification of a Nuclear Accident, the second on Assistance in the Case of a Nuclear Accident or Radiological Emergency. Since its establishment in 1957 the IAEA has extended its essentially technical considerations to include environmental implications of atomic energy.[34]

Of the sixteen specialized agencies of the United Nations system, less than half are directly involved in environmental protection. Following are listed the more important of these together with their dates of establishment. Like the United Nations special bodies, these organizations are commonly referred to by their acronyms:

—Food and Agriculture Organisation (FAO), 1945
—International Labour Organisation (ILO), 1919
—International Maritime Organisation (IMO), 1958
—United Nations Educational, Scientific, and Cultural Organization (UNESCO), 1945
—World Health Organisation (WHO), 1948
—World Meteorological Organisation (WMO), 1950
—International Bank for Reconstruction and Development (IBRD), 1946
—United Nations Industrial Development Organisation (UNIDO), as special body in 1965, as a specialized agency in 1988

The environment-related functions of these agencies and of several others with minor environmental concerns will be described in greater detail in subsequent chapters; here, their relation to the institutional structure of international environmental cooperation will be summarized. The notes provide references to more detailed accounts of their missions, administrative organization, and activities.[35]

Food and Agriculture Organisation (FAO)

FAO impacts the environment in sometimes contradictory ways. Agricultural expansion accounts for a large part of the destruction of soils, forests, wildlife, and the natural environment. FAO opposes the displacement or destruction of natural ecosystems, but in some instances its assistance may have ecologically questionable consequences. For example, FAO has assisted the Joint Anti-Locust and Anti-Avarian Organisation in West and Central Africa, which is in part intended to destroy "grain-devouring birds." But FAO is also concerned with the prevention of ill effects from the uses of fertilizers, pesticides, and herbicides. FAO was an active participant in the 1954 United Nations Convention on the Prevention of the Pollution of the Sea by Oil, and the following year organized the International Technical Conference on the Conservation of the Living Resources of the Sea.[36] In 1970 it sponsored the Technical Conference on Marine Pollution and Its Effects on Living Resources and Fishing. Other environment-related functions of FAO include preparation of a world soil map, establishment of regional forestry commissions, preparation in cooperation with UNCTAD for the United Nations Conference on Tropical Timber (1983), and technical assistance in land use planning.

International Labour Organisation (ILO)

Environmental activities of the ILO apply primarily to safety and health in the occupational environment.[37] Since 1969 a joint ILO-WHO Committee on Occupational Health has studied problems and standards of airborne pollutants. Illustrative of its activities were the three Meetings of Experts on the Control of Atmospheric Pollution in the Working Environment during 1973–74. Subsequently, ILO convened a Meeting of Experts on Noise and Vibration in the Working Environment. Exposure standards and codes of practice have been developed in these meetings and similar types of programs and projects have continued. A report prepared for the ILO's Tripartite Meeting of Experts in Employment and Training Implications of Environmental Policies in Europe (20 November–5 December 1989) addressed the question "Do environmental policies kill or create jobs?" The report found that few job losses could be attributed to environmental concerns but new environmental protection technologies could require greatly increased training.

International Maritime Organisation (IMO)

Established by treaty in Geneva in 1948, this agency did not come into effect until 17 March 1958 when the treaty had been ratified by twenty-one nations. In 1982 its name was changed from the Intergovernmental Maritime Consultative Organisation (IMCO) under which it had been established to the International Maritime Organisation (IMO). Its purpose has been to facilitate international cooperation on matters of safety and environmental protection in maritime navigation and shipping.[38] Its principal environmental role concerns the pollution of the seas, primarily by oil. IMO's responsibilities are threefold: to prevent pollution, to provide remedies when prevention fails, and to assist the development of jurisdictional powers to prescribe and enforce pollution control standards through intergovernmental cooperation. A major part of IMO's activities has related to marine safety, although prevention of pollution is involved in many safety issues. The policy-shaping entities within IMO are conferences called to conclude or revise treaties and permanent committees that tend to be dominated by national shipping interests.

Pollution prevention measures have proceeded very slowly against resistance from shipping nations. Prevention and remediation of air pollution have been major elements in the UN Regional Seas Programme.

United Nations Educational, Scientific, and Cultural Organisation (UNESCO)

UNESCO relates to international environmental policy primarily in two ways: in education and in scientific research and training.[39] The environmental concerns of UNESCO are extensive and are primarily organized under three sectors: education, natural sciences, and social sciences. Both the education and social sciences sectors have important environmental functions, but the larger number of UNESCO environmental activities fall under the natural sciences sector. An assistant director-general for natural sciences and their application to development has general coordinating responsibility for environmental programs.

UNESCO's programs and its extensive involvement with other agencies will be considered in subsequent chapters, but among the principal areas of UNESCO environmental activities have been the Intergovernmental Programme on Man and the Biosphere; ecology and the integrated study of natural resources; oceanography and marine environment; the International Geological Correlation Programme; the International Hydrological Programme; protection of the cultural and natural heritage; training of personnel for environmental management; development of general environmental education. UNESCO has played a major catalytic role in the mobilization of international concern and effort on behalf of the protection of the biosphere. P. L. DeReeder has characterized this role as centering "on man and his specific interests and values, infused with the idea of the quality of life."[40]

United Nations Industrial Development Organisation (UNIDO)

The environmental interests of UNIDO have not been extensive but are increasing. The principal environment-related activity of UNIDO is the Centre for Genetic Engineering and Biotechnology. *The UNIDO Newsletter*— after August–September 1994 *UNIDO links*—gives attention to industrial safety relating to human health and environment and to matters relating to pesticides, agricultural fertilizers, and energy development.

World Health Organization (WHO)

The influence of environmental factors on health makes inevitable a WHO interest in a broad range of environment-related activities. And yet the effects of its activities on the quality of life and the environment are mixed. A highly successful world campaign against malaria in the 1950s may have been a factor in the explosion of human populations in tropical countries, with consequent problems of food supply, housing, and education, as well as destructive exploitation of the natural environment.

WHO has organized some of its activities on a regional basis with offices at Brazzaville for Africa; Alexandria for the eastern Mediterranean; Copenhagen for Europe; New Delhi for Southeast Asia; Manila for the western Pacific; and Washington for the Americas. The regional office for the Americas also serves as headquarters of the Pan American Health Organisation (PAHO), which, established in 1902 as the International Sanitary Bureau, predates WHO by nearly half a century.[41]

The WHO system of associated organizations is further elaborated on a subregional and specialized basis. In the Americas, for example, there are eight regional research centers under PAHO sponsorship. Among these and of special environmental significance are the Pan American Zoonoses Center in Buenos Aires (1956), the Pan American Center for Sanitary Engineering and Environmental Sciences in Lima (1968), and the Pan American Center for Human Ecology and Health in Mexico City (1975). The latter research center has worked with PAHO member countries on health effects associated with large scale river basin developments.[42]

World Meteorological Organisation (WMO)

WMO evolved from the International Meteorological Organization established in 1873. It is perhaps the most exclusively scientific agency in the United Nations system.[43] Organized into six regions, WMO operates through eight technical commissions. WMO provides a liaison function and a network for cooperation among nations that is essential to the monitoring and forecasting of changes in weather and climate. It carries on four operating programs, of which the most widely known is the World Weather Watch, a multinational cooperative program to improve the collection of meteorological data on a global scale and to ensure availability of this data to all nations.[44] WMO and the International Council of Scientific Unions have collaborated in the Global Atmospheric Research Programme (GARP), which is organized with a number of subprograms dealing with such phenomena as large-scale motions of the global atmosphere, deep convection systems in the tropics, and exchange processes between the atmosphere and the surface of the sea. Ability to make reliable long-range weather forecasts is one objective of these efforts. More significant for environmental policy have been investigations of the impact of human activities on the atmosphere, particularly with regard to effects upon climate.

World Bank Group and Development Banks

The principal single source of international funding for economic development, and hence an inevitable factor in environmental policy, is the International Bank for Reconstruction and Development (IBRD, or World Bank). It heads a larger group of agencies known as the World Bank Group including the International Finance Corporation (IFC) and the International Development Association (IDA). The group's lending and investing functions for development purposes have brought them directly—although not wholly willingly—into the arena of international environmental policy.[45]

Since 1970 the bank has taken a rhetorically positive although cautious position in relation to ecologically sound development. In 1970 it appointed an environmental adviser to review and evaluate investment projects from the standpoint of their potential effect on the environment. In 1972 the bank published a staff handbook entitled *Environmental, Health and Human Ecological Considerations in Economic Projects*, and the following year it expanded its capabilities in this field through establishment of an Office of Environmental Affairs. Consonant with a new approach to environmental aspects of development, the bank since 1988 has issued *Environment Bulletin: A Bimonthly Newsletter of the World Bank Community* and in September 1989 published *World Bank Support for the Environment: A Progress Report.*

In addition and outside of the World Bank Group, several regional development banks have been established: the European Development Fund of the then European Economic Community (1957), Inter-American Development Bank (1959), African Development Bank (1964), Asian Development Bank (1966), Caribbean Development Bank (1969), and the Arab Bank for Economic Development in Africa (1974). Primarily financial in purpose, lending by these banks affects a range of environmental matters, including land use and settlement, housing, water supply, and sanitation facilities. For example, note *The Environment Program of the Asian Development Bank: Past, Present, and Future*, a report published by the World Bank in April, 1994.

In February 1980, the development banks, in association with the World Bank, UNDP, UNEP, OAS, and the Commission of the European Communities (now Union), formed CIDIE—the Committee of International Development Institutions on the Environment—and jointly adopted the Declaration of Environmental Policies and Procedures Relating to Economic Development.[46] In 1983 the European Investment Bank joined CIDIE, and in 1990 the accession of the UN Food and Agriculture Organisation (FAO) brought the number of multilateral organizations subscribing to CIDIE to sixteen. Thus, institutional means and continuity have been provided to give the declaration operational significance.

After a detailed study of the environmental policies of the World Bank, Philippe G. Le Prestre concluded that their effect has been mixed—sometimes contradictory.[47] The directors and staff of the bank have seen its role as essentially fulfilling the economic and developmental purposes for which it was created. With the assistance of its Office of Environmental Affairs the bank has adopted policies, in principle, to avoid subsidizing environmentally destructive projects. But, in practice, it has often been unable to overcome the resistance of its own economists and national governments to allowing environmental considerations a determining weight in investment decisions. Thus, World Bank policy more often undertakes to blunt the worst effects of environmentally unsound projects than to deny them financial support. Whether in the long run bank policies have been helpful or harmful to international environmental protection cannot be determined without qualification. The bank appears to play an active role in the periodic assessments of environmental aspects of development undertaken by the Committee of International Development Institutions on the Environment (CIDIE), comprising the signatories to the previously noted Declaration of Environmental Policies and Procedures Relating to Economic Development.

In fact, the declaration appears to have uncertain significance. The environmental sensibilities of the spokesmen for the World Bank do not appear to have been shared by the economists and development technicians who have dominated lending policies. The IBRD and the several development banks have continued to subsidize tropical deforestation, ecologically and socially destructive water projects, and monoculture export agriculture. Environmentally concerned people and organizations became increasingly frustrated with the bank's performance. In the United States, public protest at last moved Congress to threaten the bank with funding cuts if real reform was not forthcoming. This message was understood even by the bank's economists, and with the designation of a new, environmentally committed bank president in 1986, the prospects for a change in policy appeared more promising. As of 1989, when CIDIE held its Tenth Session in Manila, there was ground for more optimism regarding its effectiveness. Its secretariat is now at UNEP headquarters in Nairobi and its sixteen member institutions, which account for nearly $40 billion in development financing, have agreed to abide by the 1980 declaration, with emphasis on sustainable development.

United Nations Conference on Trade and Development (UNCTAD)

Since 1964 UNCTAD has functioned as an agency of the General Assembly. It has published a number of reports on environmental issues affecting international trade. Among these are *Combatting Global Warming: Possible Rules, Regulations and Administrative Arrangements for a Global Market in CO_2 Emission Entitlements* (1994), and *Controlling Carbon Dioxide Emissions: The Tradeable Permit System* (1995).

Nongovernmental Organizations (NGOs)

To round out this discussion of institutional structures for environmental policy it is necessary to consider the most diversified and least easily classified components, the NGOs. Simplified for easier comprehension, three classes of NGOs may be distinguished: first, restricted membership organizations with essentially scientific or professional interests, such as the International Council of Scientific Unions and the International Union for Conservation of Nature and Natural Resources (World Conservation Union); second, institutes and centers for purposes of information, education, and consultation, for example, the International Institute for Environment and Development, the Institute for European Environmental Policy, the Foundation for Environmental Conservation, and the World Resources Institute; third, activist advocacy organizations with generally open memberships. This last class—most numerous in North America and Western Europe—includes such groups as the Sierra Club, the National Audubon Society, the Environmental Defense Fund, and the Natural Resources Defense Council in the United States, the Federation Française des Societés de la Protection de la Nature in France, the Natur Schutz Bund and the Bund für Umwelt in Germany, and multinational organizations such as Friends of the Earth and Greenpeace. All three classes have played significant but different roles in environmental policy, although their distinguishing characteristics are less important than their ability to mobilize public opinion.[48]

A small fourth class has more recently emerged and its role is not easily evaluated. This class comprises organizations consisting of or associated with business firms. Examples include the International Environmental Bureau, a division of the International Chamber of Commerce, the World Environment Center in New York, and, since 1995, the World Business Council on Sustainable Development. Most of these organizations are genuinely devoted to environmental quality and protection. Yet some are open to suspicion as possible public relations fronts for industries whose environmental records have hitherto been dubious. Doubts may be resolved by performance.

Nongovernmental groups, since the beginnings of the environmental protection movement, have served a critical function in identifying dangers and problems, informing the public, and in importuning governments and international agencies to action. The NGOs now constitute a worldwide network interacting with governments and international intergovernmental organizations in shaping international environmental policies. It is hardly an exaggeration to suggest that this complex global phenomenon facilitated by electronic communications technology in some respects simulates the noosphere or domain of intellect postulated by V. I. Vernadsky and Teilhard de Chardin.

The legal status of these organizations becomes significant when they enter into relationships with official intergovernmental organizations. The matter is particularly pertinent in relation to the United Nations agencies because many NGOs have obtained representational or consultative status with them, and many more have requested an official affiliation. NGO relationships present problems for United Nations organizations because the UN constituency consists of governments, not people. To clarify relationships and guide official policy a review of UN–NGO relationships, coordinated by the United Nations Institute for Training and Research (UNITAR), was undertaken following the 1975 Schloss Hernstein (Austria) Conference on Non-Governmental Organizations in Social and Economic Development.[49] At its Third and Fourth Sessions (1975, 1976), the Governing Council of UNEP reviewed its linkages with numerous and diverse unofficial groups. To some extent these relationships have been facilitated and coordinated by the NGO initiative in establishing the Environment Liaison Centre in Nairobi. The growth of environmentally concerned NGOs in developing countries during the decades following the 1972 Stockholm Conference has broadened, so that UNEP is no longer dealing only with the NGOs of developed countries.[50]

There is great diversity among NGOs concerned with international environmental affairs. Many of them are constituent members of the IUCN and the Environment Liaison Centre (ELC).[51] The center coordinates a network of more than six thousand organizations and is linked informally and unofficially with UNEP. The center grew out of the collective efforts of the NGOs at Stockholm in 1972 to improve worldwide communication on environmental issues. Among its functions is the promotion of an annual World Environment Day (June 5), development of an environmental education network, and provision of a mechanism for information exchange among NGOs in different countries confronting similar problems. It publishes the bimonthly *Ecoforum*, the quarterly *News Alert*, and periodically a directory of environmentally concerned NGOs. Major NGO sources of current information regarding environmental and sustainable development issues are the *Brundtland Bulletin*, published in Geneva, Switzerland by the Centre for Our Common Future, and the *Earth Report* published by the Earth Council based in San José, Costa Rica.

Also a disseminator of information, but not an open membership organization, is the World Environment Center (WEC)—prior to 1980, the Center for International Environment Information—which was established in New York City in 1974 by the United Nations Association of the United States with support from UNEP. The center publishes the quarterly *Network News*, which reports on corporate environmental activities, the WEC International Environment Development Service (training), and the International Development Forum (exchange of information). The International Environment Bureau of the International Chamber of Commerce is a corporate NGO supported by a group of American and European business firms; it publishes a newsletter periodically.

Among the NGOs with continuing programmatic and operational relationships with the UN system, two require particular mention. They are the International Council of Scientific Unions (ICSU) and the International Union for Conservation of Nature and Natural Resources (IUCN–World Conservation Union). Both organizations were involved in preparations for the two United Nations Conferences on the Environment, and both have been involved in working with UNEP, UNESCO, and other specialized agencies in implementing environmental policies.

International Council of Scientific Unions (ICSU)

ICSU is a federation of federations; its twenty-three constituent unions are umbrella organizations covering a large number of subsidiary associations.[52] For example, there are sixty-eight national members in the International Union of Geodesy and Geophysics (IUGG) and seven international associations relating to geodesy, seismology, meteorology, geomagnetism, oceanography, hydrology, and volcanology. In order to focus the diverse and specialized resources of the unions on major global problems, coordinative mechanisms are needed. The principal mechanisms have been interunion interdisciplinary committees addressing themselves to particular scientific questions and issues.

Of the several types of committees periodically established by ICSU, the continuing scientific committees have been the more significant in relation to an international structure for environmental policy. Among these committees are the Scientific Committee on Oceanic Research (SCOR) formed in 1957, the Scientific Committee on Space Research (COSPAR) in 1958, the Scientific Committee on Antarctic Research (SCAR) in 1958, the Scientific Committee on Water Research (SCOWAR) in 1993, the Scientific Committee for the International Geosphere-Biosphere Programme (SC-IGBP) in 1986, and of special relevance to our concern, the Scientific Committee on Problems of the Environment (SCOPE) in 1969. Special committees have been created for special international cooperative programs, several of which have contributed significantly to the conceptualization and understanding of the biosphere. Among the more important of these for environment-related research have been the Comité Special de L'Année Geophysique Internationale (CSAGI), which planned and coordinated the International Geophysical Year of 1957–58, the Special Committee for the International Biological Programme (SCIBP), which was established in 1963 for the study of large-scale transnational ecosystems and continued its work until 1974, and the Special Committee for the International Decade for Natural Disaster Reduction

(SC-IDNDR) 1990, preparatory to a World Conference on Natural Disaster to be held in 1999.

The ICSU committee most broadly and specifically concerned with international environmental policy has been SCOPE, the Scientific Committee on Problems of the Environment, established in 1969 with two principal objectives:

> To advance knowledge of the influence of humans on their environment, as well as the effects of these environmental changes upon people, their health and their welfare—with particular attention to those influences and effects which are either global or shared by several nations.

> To serve as a non-governmental, interdisciplinary and international council of scientists and as a source of advice for the benefit of governments and inter-governmental and non-governmental bodies with respect to environmental problems.[53]

As of 1995, SCOPE reported thirty-six national members and twenty-two adhering international scientific committees and unions. Its work has been organized within topical areas such as sustainability, biogeochemical cycles, ecotoxicology, global environmental monitoring, man-made lakes as modified ecosystems, risk assessment of environmental hazards, and the state of environmental sciences in developing countries. The actual work of SCOPE is decentralized to focal points in various research institutes and universities, and has been associated with the Monitoring and Assessment Research Centre (MARC) located at Chelsea College, University of London, and co-sponsored by UNEP, WHO, and the Rockefeller Foundation.

SCOPE has made important contributions to scientific understanding of the environment, but in other respects its influence is limited. While it is a source of the knowledge necessary to the formation of international environmental policy, it has not assumed an active advisory role to governments toward that end. Rather, it appears to accept an assumption widely shared among scientists who believe that their public mission is largely fulfilled when scientific studies are made available to governments and international organizations. SCOPE also has no significant relationships with the social and behavioral sciences except in relation to those aspects of geography that relate to social behavior. This does not lessen the quality or significance of its scientific work, but does represent a self-imposed limitation that could reduce its contribution to the actual shaping of international environmental policies.

Of major potential significance is the Special Committee for the International Geosphere-Biosphere Programme established in 1986 "to describe and understand the interactive physical, chemical, and biological processes that regulate the total Earth system, the unique environment it provides for life, the changes that are occurring in this system, and the manner in which they are influenced by human actions."[54] The ongoing findings of this investigation should confirm, refute, modify, and in any case illuminate J. E. Lovelock's Gaia hypothesis.

International Union for Conservation of Nature and Natural Resources (IUCN)/World Conservation Union

IUCN is the nongovernmental organization most consistently and comprehensively involved with the earth's environment. An outgrowth of earlier efforts for the protection of nature, it belongs neither to the ICSU nor UN systems, although it has cooperative relationships with each. Its ongoing activities are reported in the *IUCN Bulletin*. Its initial and basic concerns have been with threats to the quality of the natural environment—particularly to wilderness areas and endangered species. It has investigated circumstances that threaten wildlife and ecosystems, proposed methods by which human-environment problems may best be resolved, and promoted educational measures to further general understanding of the values to be protected in the natural world. Its concerns, however, have gradually broadened with the recognition that its primary objectives can be realized only through the development of ecologically and economically sustainable human societies. This extended and enlarged perspective is exemplified by the World Conservation Strategy, which is considered in the concluding chapter of this book, and of which IUCN was the principal architect.

The unique and important role of the IUCN in international environment policy justifies a more detailed look at its structure. It is a federative membership organization, composed primarily of scientific, professional, environmental, and conservation organizations, and governments, or governmental agencies. As of 1994, 111 nation-states and a total of 678 individual government agencies, national and international nongovernmental organizations, and affiliate members were represented.[55]

IUCN was established in 1948 as the International Union for the Protection of Nature, following an international conference at Fontainebleau sponsored by UNESCO and the government of France. Its name was changed in 1956 in Edinburgh at the Triennial General Assembly to the International Union for Conservation of Nature and Natural Resources, which reflected the broadening of the original concept of the protection of wildlife to include the protection of renewable resources.[56] Over the years this name was found to be awkward and perhaps redundant. The question of the name was raised at the 1981 General Assembly at Christchurch, New Zealand, but no definitive action was taken. Today, however, the IUCN is alternatively designated as the World Conservation Union.

IUCN operates through a number of commissions and committees that specialize in different aspects of the union's work. Standing commissions include Environmental Science, National Parks and Protected Areas, Education and Communication, Species Survival, Environmental Strategy and Planning, and Environmental Law. Union headquarters are in Gland, Switzerland, but its environmental law center is located in Bonn, Germany, and its conservation monitoring centers are in Cambridge and Kew in the United Kingdom. The union provides the secretariat for the Convention on International Trade in Endangered Species, and in 1981 it established a conservation for development center to monitor and assist the progress of ecologically sound and sustainable development throughout the world.

The IUCN convenes a general assembly every third year as a forum for the discussion of international conservation problems and to act upon issues of current importance to the Union. Recent assemblies have been held in Christchurch, New Zealand (1981), Madrid, Spain (1984), San José, Costa Rica (1988), Perth, Australia (1990), and Buenos Aires, Argentina (1994). Associated with the general assemblies are meetings of technical experts on more specialized problems of conservation. Regional meetings sponsored or cosponsored by IUCN have been focused upon the conservation problems of more restricted areas, for example the Pan African Symposium on Conservation of Nature and Natural Resources in Modern African States (Arusha, Tanzania, 1961). A representative listing is provided in the notes.[57]

Separate from, but historically associated with IUCN, is the World Wildlife Fund, now officially designated as the World Wide Fund for Nature (WWF), perhaps best described as a nongovernmental international foundation. Formed in 1961 to provide financial support for wildlife protection generally, it has assisted the IUCN and acts to safeguard the wildlife of the world wherever threatened.[58] Annual World Wildlife Fund appeals in the various countries of the world constitute a voluntary international mobilization of resources toward more effective protection of the wildlife. From its inception, WWF has enjoyed the active support of an exceptional number of persons highly placed in government, science, international society, and public affairs.

Paralleling the IUCN in a particular area has been the older International Council for Bird Preservation (ICBP), which Max Nicholson has described as "the senior international conservation agency," established in London in 1922 under the leadership of T. Gilbert Pearson, an American ornithologist who was also a major force in organizing the National Audubon Society in the United States. The council now shares common headquarters with a federation of other bird protection associations under the title of Bird Life International. The council has been organized in sixty-four national sections as of 1980.[59] It has held several major international conferences and played an important role in negotiating treaties and other international agreements, notably the "United States–Japan Convention for the Protection of Migratory Birds and Their Environment and Birds in Danger of Extinction," signed by both countries in 1972.

The International Legal Order

National and international law are both expressions and tools of international environmental policy. As fields of law they involve considerations of principle and practice that are extraneous to the subject of environmental policy, which is a recent arrival in the field of law. An emergent organizational structure to shape and administer international environmental obligations and relationships implies a corresponding legal framework. But an emergent structure is also by implication in a formative and transitional stage, and new legal concepts associated with this emergence are just as likely to be provisional.

To attempt to deal definitively with the structure and status of international environmental law today would be imprudent. Prior to World War II there was a world legal order dominated by the European powers and the United States. It had evolved over several centuries during which European political control or influence was extended throughout the greater part of the world. Since 1945, with the replacement of colonial regimes by non-European independent states, whatever certainty existed regarding the prevailing norms of international conduct has diminished. Lawful status has been claimed by many developing countries for propositions and principles never accepted by any developed country under the older international law. They generally have treated such concepts as compensation, additionality, and redistribution of wealth throughout the world as obligatory upon the developed states and as constitutive, in effect, of a new international legal order. Today it is unclear how far the old rules of international conduct apply or how widely new rules are accepted.

Environmental Aspects of International Law

International environmental law is not exclusively or even primarily a field of legal practice.[60] There are, of course, international lawyers and litigants, but lawsuits primarily pertaining to environmental issues per se have been relatively infrequent. From the perspective of international policy, environmental law is perhaps best understood as the collective body of agreements among states regarding mutual rights and obligations affecting the environment. It is embodied in conventions among states (treaties) and, to lesser effect, in international declarations, collective principles, opinions of jurists, and generally accepted practices among states. Enforcement of its provisions, customary or specified by treaty, are usually sought through negotiation (e.g., diplomacy) rather than through adjudication. Its boundaries are definable only in broad terms because new scientific findings of international significance and enlarging perceptions of human–biosphere relationships have continually if unevenly expanded its frontiers.

Emergent aspects of international environmental law include those extensions, codifications, or reinterpretations of rights and obligations among nations that have been long accepted as customary international law. An example is the identification of transboundary air pollution as entailing an obligation of a state to prevent the use of its territory to inflict harm upon its neighbors. There are also new principles and obligations derived from formal agreements, usually treaties, regarding subjects hitherto untouched by the law of nations. Examples may be found in the conventions governing international spaces—notably outer space and the deep seabed—where technology has enabled some nations to establish an operational presence beyond territorial jurisdictional claims.

An uncertain source of international law, potential but portentous, may now be found in resolutions and declarations of the General Assembly of the United Nations and of major international conferences such as the United Nations conferences on the environment in 1972 and 1992. Actions of majorities in the General Assembly, in the governing bodies of the specialized agencies, or in conferences within the UN system do not make law in the traditional sense. Yet, the official statements and pronouncements of United Nations bodies, especially when adopted by overwhelming majorities, have the potential of becoming recognized as obligatory rules of national conduct. The barrier to realizing this possibility is that the newer developing states, while most inclined to "legislate" through UN declarations and resolutions, are also jealous of their national sovereignties, and besides, they do not possess the economic, technical, or military capabilities of enforcing their policy preferences upon the greater powers, or of enforcing domestically environmental protection policies to which they have officially adhered. The legal influence of UN declarations and resolutions depends greatly upon the degree of actual consensus and the extent to which those pronouncements can be applied in practice. The World Charter for Nature, for example, elicited a high degree of at least formal consensus, but it provided no means to make its precepts operational.

A process leading toward international lawmaking has now become established. The first step is a proposal to ECOSOC or the General Assembly to establish a committee or to convene an international conference to consider a major issue of international concern. Then preparatory committees, working groups, regional meetings, and symposia are organized and opinions of governments and of nongovernmental organizations invited. Usually, but not invariably, preconference preparation assures substantial agreement on official conference action. In the end it is not the action of an international conference or of the General Assembly that confers legal status; it is rather the effective consensus of nations that make it law. But the conferences and (potentially) the deliberations of the General Assembly provide the forums in which consensus building can occur.

A secondary aspect of the United Nations role in the development of international law derives from Article 13 of the UN Charter, which states that the General Assembly "shall initiate studies and make recommendations for the purpose of . . . encouraging the progressive development of international law and its codification." In 1947, the General Assembly established the International Law Commission, which undertakes the "progressive development of international law" through preparation of draft conventions on issues still in a formative stage and through "codification of international law" for topics upon which substantial agreement exists in practice among states. Drafts and recommendations prepared by the commission are reviewed by the Sixth (Legal) Committee of the General Assembly and may be laid before the Assembly for action. An alternative route may be the convening of a United Nations conference to consider an international convention or treaty. The United Nations system thus provides a principal, although not exclusive, structure through which international law is developed and codified. The World Court has not played a significant role in the development of international environmental laws. The UN specialized agencies and IAEA have been far more important actors, and the principal forum for the initiation of new international legal concepts has been the UN General Assembly.[61]

The very circumstances that necessitate the growth of an international legal order for the environment make its development difficult. The traditional structure of international law emerged out of a political configuration far simpler than that now existing. The issues were fewer and more manageable; the forms of political and economic action much less diverse. Even so, through a large number of treaties, a body of positive environmental law has been adopted, notably since the 1960s. In addition there is a growing body of international environmental "soft law" consisting of official declarations, resolutions, recommendations, and principles, which, short of binding law, may nevertheless influence the policies of nations. A selected listing may be found in Appendix E.

As noted elsewhere in this volume, some environment-related agreements among states have been in effect for more than a century. At present the most comprehensive published collection of environmental treaties and other agreements has been the multivolume work by Rüster, Simma, and Bock, *International Protection of the Environment: Treaties and Related Documents* (Oceana, 1975-, 30 vols. and supplements). UNEP has published a compendium of its own legislative authority and a register of international environmental treaties, supplemented periodically to indicate new agreements and new accessions to prior agreements. Collections of the texts of treaties have also been published by the British Institute of International Law and Comparative Law, by UNEP, and by the IUCN Environmental Law Centre in Bonn, and by individual editors.

Environmental law has become a focus of concern among a large number of lawyers, scholars, and public officials. The International Council on Environmental Law (ICEL), formed in 1969, is a network of individuals and organizations professionally concerned with environmental law. It shares common headquarters with the Environmental Law Centre in Bonn. The center possesses what may be the most comprehensive collection of documents on environmental law and policy anywhere in the world (more than 21,000 items).[62] Developed through the initiative of Wolfgang E. Burhenne and Françoise Burhenne-Guilmin, this collection, accessed through a computerized system of retrieval, can provide information regarding provisions of environmental law in all major countries of the world. This service is of obvious advantage to scholars and to governments considering changes or new provisions in their environmental legislation. Opportunities for the training of young environmental lawyers have also been offered by the center.

An important step in the development of international environmental law was taken by a UNEP-sponsored Ad Hoc Meeting of Senior Government Officials Expert in Environmental Law, which convened in Montevideo, Uruguay, from 28 October to 6 November 1981. The purpose of this meeting was "to establish a frame, methods and programme, including global, regional, and national efforts, for the development and periodic review of environmental law."[63] A secondary objective was to contribute to preparation and implementation of the environmental law component of UNEP's medium-term environment program.

The Montevideo meeting specified three subject areas in particular in which guidelines, principles, or agreements should be developed: (1) marine pollution from land-based sources, (2) protection of the stratosphere ozone layer, and (3) transport, handling, and disposal of toxic and dangerous wastes. In addition, eight subject areas were identified for action: (1) international cooperation in environmental emergencies, (2) coastal zone management, (3) soil conservation, (4) transboundary air pollution, (5) international trade in potentially harmful chemicals, (6) protection of rivers and other inland waters against pollution, (7) legal and administrative mechanisms for the prevention and redress of pollution damage, and (8) environmental impact assessment.

The Montevideo meeting adopted a program for "the development and periodic review of environmental law" to be recommended to the Tenth (1983) Session of the UNEP Governing Council. A call for cooperation with

UNEP in the development and implementation of this program was directed to the UN organizations, to national governments, to intergovernmental organizations outside the UN system, and to nongovernmental organizations active in the field of environmental law. Thus the meeting identified the principal components of the emergent structure of environmental policy, law, and cooperation, and proposed the development of a legal system that would give this structure greater consistency, coherence, and effectiveness. To this end, the meeting proposed specific strategies for each of the enumerated subject areas of environmental concern. As of 1995, most of the eight subject areas identified by the 1981 Montevideo meeting have been addressed through the UN system or other multinational agreements.

International Liability for the Environment

The principle of liability for environmental harm has been recognized even by governments that least favor international action in the field of the protection of the environment. There seems to be no doubt about the liability of states for damages which they may cause (e.g., through negligence) to the environment of other states. Principle 21 of the Stockholm Declaration applies not only to this kind of damage, but also to the common spaces, to the environment of areas beyond the limits of national jurisdiction (e.g., the upper atmosphere, the oceans and deep sea bed, outer space, and Antarctica), and the same principle is embodied in Article 21(d) of the World Charter for Nature, and in many marine pollution conventions, including the UN Law of the Sea (1982). However, under the present state of international law, this liability is not regarded as absolute except for nuclear damage or in relation to outer space.[64] Yet even in these cases the acceptance or enforcement of liability is highly uncertain—the transboundary effects of the Chernobyl nuclear explosion or French nuclear tests in the South Pacific are cases in point.[65]

There is an important difference between traditional views of national rights and obligations and the new perspective, which would extend liability for damage done not to a given state or its nationals but to what may be regarded as the common property of all nations. Hence, it is necessary to examine the two different aspects of international liability for environmental damages: the traditional one, which would redress harm suffered by particular states, and new and broader interpretations of legal responsibilities. Each type provides occasions for international controversy.

International environmental law regarding a state's liability for damage caused beyond its borders, although not new in principle, is still in a developmental state, with many points of law unsettled. Advanced technologies allowing human penetration of outer space and the deep sea complicate traditional assumptions. Even so, there are specific treaties between nations from which the principle of liability may be invoked for new or changing technologies as they affect, for example, transfrontier pollution. One of the more familiar is the 1909 Treaty Relating to Boundary Waters and Questions Arising Between the U.S. and Canada, signed by the United States and the United Kingdom (which was still responsible for Canada's foreign affairs). This treaty established the International Joint Commission (IJC), which in its fact-finding role provided the factual basis for resolving the *Trail Smelter Case*, still the only international arbitration on the merits of a pollution dispute. But provisions for arbitration have been written into the Nordic Convention on the Protection of the Environment (1974) and the Convention on the Protection of the Rhine against Chemical Pollution (1976).

In 1935, Trail Smelter, a metal refinery in British Columbia, was found through an IJC inquiry to have discharged sulfurous gases that damaged farm crops across the border in the state of Washington. Mainly on evidence presented by the IJC, an international tribunal comprised of an American, a Canadian, and a Belgian found the Canadian government liable for damages of $350,000. The *Trail Smelter* arbitration set an important precedent in international environmental law on two counts. First, it established the precedent of an international tribunal to investigate a case concerning transboundary pollution damages. Second, the legal principles of the *Trail Smelter* decision were recognized at the Stockholm Conference on the Human Environment in 1972.[66]

The growth in international trade and transport, and the transboundary effects of modern technology, have led to increased attention to liability for environmental damages. Oil spills in coastal waters have caused nations to demand some form of redress from offending polluters; such incidents led to the Convention on Civil Liability for Oil Pollution Damage (Brussels, 1969). But establishing the origins and fixing responsibilities for transboundary pollution of air and water often have been difficult, as demonstrated by the controversies in Europe and North America over liability for acid precipitation. The Convention on Long-Range Transboundary Air Pollution, sponsored by the UN Economic Commission for Europe (ECE) and in force since 1983, attempts to clarify this problem. Scientific investigations of the flow of pollutants through the environment may in time yield findings that will open new subject areas to international law. Formal adjudication under the present world political order may not be the most effective way to resolve international environmental disputes. Nevertheless, proposals have been made to establish a permanent international criminal court competent among other issues to adjudicate crimes against the environment.

The difficulty that the question of liability presents to the building of an international structure of environmental law is illustrated by the controversy over the French atmospheric nuclear tests. From 1966 to 1972, the French government conducted several series of these tests from a base on Mururoa in their South Pacific territory of New Caledonia. In May 1973 it seemed that France was about to engage on a new series, to last until 1975 or later. Australia and New Zealand asked France to put an end to the atmospheric tests, which they contended had resulted in an increase in radiation doses to their populations. They further argued that any radioactive material deposited in their territories constituted a potential danger. France refused to cancel or modify its scheduled test program, and the Australian and New Zealand governments applied to the International Court of Justice, asking it to declare that further atmospheric tests of nuclear weapons were inconsistent with applicable rules of international law, and to order France to abstain from carrying out any such tests.[67] In addition, considering that irreparable damage might be caused by any test that would be carried out before the court's judgment, the Australian government requested the court to order interim measures of protection.

The World Court examined the record on the basis of Article 41 of the Statute of the International Court of Justice, which states: "The court shall have the power to indicate, if it considers that circumstances so require, any provisional measures which ought to be taken to preserve the respective rights of either party." The court could not have specified provisional measures unless the rights claimed in the Australian and New Zealand applications appeared to fall within its jurisdiction. The court avoided addressing the issues; instead it waited until it became apparent that the French would cease nuclear testing in the atmosphere near the end of 1974. The World Court then concluded that France would not conduct further atmospheric nuclear tests in the South Pacific and, the purpose of the applications having been achieved by other means, that the dispute had disappeared. Since the plaintiffs' claims no longer had an objective, the court reasoned that there was nothing to be decided. On 20 December 1974, the court declared its decision not to address the issues of whether it had jurisdiction and whether the plaintiffs' applications were admissible. In 1995 the issue again arose over the intention of France to conduct a series of underground nuclear tests on Mururoa Atoll in the Polynesian South Pacific. Following completion of the tests, the French government declared that it would adhere to the global nuclear test ban treaty. Nevertheless, the tests were widely condemned by other governments and nongovernmental organizations.

The nuclear test cases demonstrate that not all states regard seriously international responsibilities as reflected in Principle 21 of the Stockholm Declaration. It is nevertheless true that the declaration has been an effective catalyst for the development of an environmental ethic for the global community. But there is obviously a difference between the affirmation of principles by national delegations in the euphoria of international assemblies and practical implementation where perceived national interests are involved.[68]

Crimes against the Environment

The concept of crimes against the environment was broached in the early 1990s, notably after the war in the Persian Gulf. The Iraqi dictator, Saddam Hussein, ordered the firing of the Kuwaiti oil wells by the retreating Iraqi army and crude oil was also discharged into the Persian Gulf. This action led in mid-1995 to a proposal before the UN General Assembly to establish a permanent international criminal court. The proposal failed to obtain agreement and was put off to a later date. Operating separately from the International Court of Justice at the Hague, the proposed International Criminal Court would have dealt with genocide, war crimes, crimes against the environment, and other offenses against humanity.[69] The issue has been further advanced by international concern over war crimes committed during the 1992–95 war in Bosnia. The concept of crimes against the environment is not new, however, having been the subject of a study and working paper in 1985 by the Law Reform Commission of Canada, and numerous comments on international environmental damage perpetuated by Iraq in the 1990–91 Persian Gulf War.[70]

Law of the Sea Conferences

The problems and difficulties in developing a generally accepted and consistent body of international law are illustrated by the history of United Nations efforts to codify and extend the law of the sea. The public order of the oceans and the law of the sea have been the subject of a vast literature, which is only partly relevant to environmental concerns and can be no more than acknowledged here.[71] Many of the issues confronted in the UN conferences on the law of the sea belong to the larger category of issues relating to the law of common spaces, which today also pertains to the upper atmosphere, outer space, and Antarctica.

Three Conferences on the Law of the Sea (UNCLOS) have taken place under UN sponsorship: in 1958 (UNCLOS I), in 1960 (UNCLOS II), and from 1973 to 1982 (UNCLOS III). Details of the negotiations, the full range of issues, and the positions of the participating states cannot be recounted within the space available, nor would all of it be relevant to environmental policy. Yet there were important environmental issues involved, if sometimes only by implication. These issues include pollution, territoriality, fisheries, and the deep sea bed; concern here is less with their substance than with what they reveal concerning the development of international environmental policy.

Pollution of the oceans was addressed by UNCLOS I. Three of the four conventions drafted by the conference made reference to pollution control; the Convention on the High Seas makes specific mention of radioactive materials (Article 25).[72] Yet, many aspects of marine pollution were untouched or ineffectively addressed so that subsequent treaties were necessary to deal more specifically with oil pollution, land-based pollution, and ocean dumping.

Territoriality was a major issue in UNCLOS II and continued to be one in UNCLOS III. The Second Conference was compelled to ratify what many states were doing unilaterally, for example, extending national jurisdiction out to two hundred miles beyond the coastline. This partitioning of regional seas and the oceans was inconsistent with rational planning for control of marine fisheries such as tuna. It afforded protection to certain national fishery industries but reduced the scope of international management. Once again the nations found themselves unable to agree on common principles to govern shared spaces and resources.

In 1967 Dr. Arvid Pardo, the United Nations ambassador from Malta, introduced a proposition that thereafter has complicated not only the conference deliberations, but also international law generally. He suggested that the deep sea bed be regarded as "the common heritage of mankind," and that its wealth be used to foster the economic progress of the world's poorer regions. More importantly, he appears to have put the "heritage of mankind" concept on the agenda of world politics.[73] The "common heritage" idea was promptly adopted by representatives of the Third World countries—the so-called Group of 77, now numbering over one hundred—countries that without effort or investment would clearly benefit from this principle.

In UNCLOS III, after years of negotiation, agreements were reached on nearly all points except provisions relating to the exploitation of minerals on the deep sea bed.[74] The government of the United States—the state with the greatest capabilities for deep sea activities—declined to sign the treaty. Under the treaty, the United States would be required to share its technology with other (including competing) nations, and earnings from sea bed mining would be shared with the UN supervisory agency (called "The Enterprise"). Further, operations would be under the agency's control without necessarily having American representation on its governing board.

The long, drawn-out deliberations show how difficult it is to persuade nations to act against their perceived economic interests. A political order of the oceans may result from UNCLOS, but the application of the "common heritage" concept has an uncertain future in a deeply divided world. Yet, after nine years of often contentious labor, the Convention on the Law of the Sea was readied for signature by the representatives of national governments. The convention, if observed in practice, will go far toward establishing a public order of the oceans. Its scope, as summarized on page 1 of the *International Marine Science Newsletter* (no. 32, Summer 1982), deals with almost every human use of the oceans—navigation and overflight, resource exploration and exploitation, conservation and pollution, fishing and shipping. Its 320 articles and nine annexes constitute a guide for states' behavior in the world's oceans, defining maritime zones, laying down rules for drawing boundaries, assigning legal duties and responsibilities and providing machinery for settlement of disputes.

To satisfy objections of the United States and other industrialized countries whose consortia had been investing in deep sea mining technology and in prospecting, a provision authorizing their access to mine sites under the general provisions of the convention was adopted by the conference, but failed to resolve their misgivings. Nevertheless, the majority of provisions were acceptable to nonsignatory states and probably would be observed by them.

On 10 December 1982, the Law of the Sea Convention was signed by representatives of 117 countries convening on the island of Jamaica.[75] The United States, the United Kingdom, France, the German Federal Republic, and Japan initially declined to sign. However, as of 6 May 1983 some 125 states had signed the convention. Yet it seems probable that some time may elapse before the treaty can definitively establish a universal law for the sea. The convention provided that after the initial signing, it would be open to further signature and to ratification for a period of two years. It came into force one year after being ratified by the sixtieth state (16 November 1994). One hundred and fifty-eight states now adhere to the treaty, but the United States has declined to ratify. How the provisions of the treaty will be reconciled with other marine treaties is a matter for future legal scholarship, diplomacy, and adjudication. Yet it is clear that a world order for the ocean environment is emerging. Discussion of the oceans as a subject of international policy will continue in Chapter 9.

Steps toward a World Order of the Environment

In the concluding paragraph of her book *World Public Order of the Environment* (1979), Jan Schneider observes that the progressive development of international law depends upon the reorientation of fundamental assumptions underlying international community expectations.[76] But the community of nations is so called primarily because of common occupancy of the earth, and not because of many shared assumptions or expectations. There are large and basic areas of human experience and value where no consensus exists—and some in which overt hostility is the norm.

This book assumes the apparent paradox of a political world deeply divided on a planet that is a complex ecological unity. The well-being and survival of life on this planet now depends upon human behavior that will prevent the disruption or destruction of this ecological unity. We now know that ordinary human activities expanding exponentially can destroy the earth as surely as can violence. Yet, it is not utopian to believe that the environmental concerns of nations may induce their cooperation more rapidly than have the more conventional issues of international relations such as armaments, monetary exchange, trade, investment, and human rights. The proliferation of environmental conventions since the mid-1960s is evidence of environmental concern among nations and of their willingness, at least in principle, to accept mutual obligations for the benefit of others as well as themselves if they do not regard the price as too high.

The movement to establish international principles or standards for the internal conduct of national governments (e.g., human rights) is a step toward broadening the scope of international law. It may not be a great step from international standards for a government's conduct in relation to its people to standards for management of its environment and resources. But international sanctions are not now effective in these latter areas, although increasing numbers of informed and concerned persons throughout the world, associated through the NGOs, form a new force in shaping international policy and law.[77] If the several environmental treaties adopted during the post-Stockholm decades are effectively administered, domestic law and policy in many countries will be changed or developed to conform to international law.

It may be that institutional changes to accommodate changing attitudes toward humanity and the biosphere will occur primarily at the national level. States do not appear ready to relinquish their powers of governance to international agencies. Yet their citizens may in time desire a broader spectrum of environmental rights from international law than are presently available under national law. Individuals need not be direct subjects of international law to be beneficiaries of it. Evolution of environmental policy through the OECD, the European Union, and numerous United Nations cooperative programs shows the progressive interlinking of national and international policy and law.

Experience supports the opinion of Professor Richard B. Bilder that "it is unlikely that the International Court or other judicial agencies will, in the near future, play an important role in environmental dispute management."[78] It may be that if international environmental controversies involving legal obligations increase in number, recourse to arbitration or adjudication might become more frequent. But many environmental conflicts are not easily reduced to judicial questions. Various forms of intergovernmental fact finding, problem solving, negotiation, and mediation may offer more practicable approaches to the resolution of international environmental disputes.

The clarification and refinement of new environment-related legal concepts will be an important part of creating a more effective world order of the environment in the decades ahead. I share the view of Cyril Black and Richard Falk who write: "There are many problems that call for institutional management of a global character, but we do not believe that it lies within the capacity of the present system of states to alter to this extent the attitudes, values, and sources of support necessary for such a transformation within the near future."[79] But I also believe that socioecological catastrophe over large areas of the earth could change this assessment.

The effectiveness of international treaties and other agreements varies widely from subject to subject and country to country.[80] Unfortunately many are ignored or poorly enforced by the signatories. International monitoring of observance might strengthen their utility, but some states would regard this as an affront to their sovereignty. And so a politics of adjustment, accommodation, and even of antagonistic cooperation may be the most that can presently be achieved. Meanwhile the structure of environmental policy, law, and cooperation is building incrementally, aided perhaps by a growing perception in almost all countries that there are policy areas of common human concern that people and governments can address regardless of differences in other respects. Politics and coordinate action, in addition to formal or positive law, may offer the most promising route toward an ecologically more orderly world.

8

Transnational Regimes and Regional Agreements

Some environmental problems are global in effect and require the cooperation of all or many nation states. These global problems pertain to the atmosphere, the high seas, and the electromagnetic environment (e.g., global climate change and depletion of the stratospheric ozone layer). There are also international agreements between nations, primarily for purposes other than environmental (e.g., economic and political) that are neither universal nor regional (in the conventional sense of "region"). For example, the OAS and the OAU are continental in scope, NAFTA is semicontinental, and OECD and NATO are discontinuous groups of nations.

The forms of organizations are usually, or initially, adapted to the particular problem for which they were established. Some, as noted in Chapter 7, have added environmental considerations to other, usually primary, concerns. A growing group of international agreements now addresses environmental problems as primary concerns. Some are global in effect but regional in extent (e.g., tropical rainforests). Consistent categorization of these forms of cooperation is impossible. Although some environmental problems may be approached on a regional basis, many are distributed to an extent that defies conventional bounded political jurisdiction (e.g., migratory wildlife, desertification, epidemic disease). Solutions to these kinds of situations have been sought through a variety of international arrangements. Many are geographically bound regional organizations; others, less formal, are called "regimes."

Regimes are defined in different ways by different scholars.[1] The term as used here refers to agreements—usually single-purpose ones bounded by the problem rather than by territorial jurisdiction. Regimes are not governments, but some of them do undertake limited administrative functions, as in control of the desert locust. Characteristically, regimes are agreements between states to abide by common rules or policies respecting, for example, actions in Antarctica, discharge of chlorofluorocarbons into the atmosphere, or regulating oceanic fisheries and the taking of marine mammals. They are seldom hierarchical in structure, having minimal, if any, organizational structure. No clear or firm line separates all regimes from regional organizations, although well-defined examples of each may be distinguished. Unlike regimes, most regional arrangements operate within the jurisdictional boundaries of member states. In practice, however, international or transnational arrangements are too numerous and diverse for any categorical classification to be particularly useful. Utility lies in the appropriateness of the arrangement in relation to the problem.

There are also various ways of defining regions, and there are numerous ways in which organizations may be regarded as "regional." As noted in Chapter 7, regional members of CIDIE, the Committee of International Development Institutions on the Environment have environmental programs (e.g., the Asian Development Bank). Common resources shared by two or more states may lead to a regional arrangement, such as the territorial reserves in Bolivia and Peru for the protection of the endangered vicuna or the agreements between Canada and the United States for the management of the Great Lakes.[2] Common problems may suggest regional arrangements, as in the South Asia Cooperative Environment Program (SACEP) established by the governments of Afghanistan, Bhutan, Bangladesh, Maldives, India, Iran, Nepal, Pakistan, and Sri Lanka. But their effectiveness is often short-lived and minimal. The Association of Southeast Asian Nations (ASEAN), however, appears durable and has environmental cooperative agreements. Negotiations have been under way for a Northeast Asia Environmental Regime. Six countries are involved—China, North Korea, South Korea, Mongolia, the Russian Federation, and Japan. The rationale for the regime and early negotiations, encouraged by the United Nations Commission for Asia and the Pacific (ESCAP), UNEP, and UNDP, were reported in the journal *International Environmental Affairs*, vol. 6 (Fall 1994).

Mutual need to deal with a common problem may lead to the establishment of common administration or at least coordinate action. For example, regional projects have been initiated by the United Nations and by the specialized agencies, such as the campaign against the desert locust in Africa and western Asia, begun in 1952 by the Food and Agriculture Organization (FAO).[3] In 1960, the United Nations Development Programme (UNDP) joined this effort with financial aid. A network of nineteen international research stations has been established, and information concerning the influence of weather on the behavior of the locust has been provided by the World Meteorological Organization (WMO). A series of international treaties and cooperative arrangements provide structure for the effort. In 1955 the International African Migratory Locust Organization (OICMA) was created by the colonial administrations of Belgium, France, and the United Kingdom. Following the independence of many African states new treaties have established locust control organizations in central, south, east, and west Africa. A commission for controlling the desert locust in southwest Asia has also been established with FAO cooperation.[4]

There is no mutually exclusive way to categorize organizational arrangements among nations for dealing with environmental problems. Many factors influence the scope and structure of organizations, especially at the regional level. For example, the circumstances that led to regional treaties for Antarctica differ greatly from those shaping international fisheries commissions. In this chapter, three important aspects of the implementation of international environmental policy on regional bases will be considered. They are international river basins, European regional cooperation including agreements regarding the North and Baltic seas, and UNEP's Oceans and Coastal Areas Programme (formerly Regional Seas Programme).

International Rivers and Lakes

A large number of intergovernmental agreements regarding international lakes and rivers have been negotiated over the years. Their legal status has been described in reports by the International Law Commission to the General Assembly of the United Nations and in a collection of case studies and essays published by the Hague Academy of International Law.[5] Most of these agreements concern jurisdictional boundaries or conditions of transit, but some have environmental aspects. There are more than two hundred separate river basins shared by two or more countries, and, with growing demand on water supplies for competitive agricultural, industrial, environmental, and domestic purposes almost everywhere, access to these sources is becoming a major matter of political concern.[6]

Among the oldest multilateral arrangements that have acquired in some sense an environmental significance are European commissions governing the rivers Rhine and Danube.[7] The Commission for the River Rhine was established in 1815 at the Congress of Vienna but did not become active until after the Treaty of Mannheim in 1868. The Danube Commission was established in 1878. In addition to its primary concern for navigation, it has assisted in flood control activities and is involved in integrated hydroelectric planning and international water pollution control. Over the years, agreements have been expanded and new treaties negotiated. Supplementing this matrix of agreements in the basin of the Upper Rhine, a tripartite commission was established in 1975 by agreement among France, Germany, and Switzerland.[8] Cooperative arrangements also involve private groups and local authorities. Thus, the legal status of international control over the rivers of central Europe has become very complex.

Where rivers are of mutual concern to adjoining nations and riverine economic and environmental relationships are closely conjoined, as with the

Rhine and its tributaries, interlocking arrangements may become necessary. A 1961 French-German agreement established an International Commission for the Protection of the (River) Saar against Pollution. The 1956 Convention on the Canalization of the Moselle (to which the Saar is tributary) had already set the precedent of environment-related compacts among the nations in the Rhine river basin, which in turn led to a protocol among France, Germany, and Luxemburg establishing (in January 1963) an International Commission for the Protection of the Moselle against Pollution. In April 1963, a five-nation International Commission for the Protection of the Rhine (to which the Moselle is a tributary) was formed following negotiations initiated in 1946. Each of the Rhine basic commissions has collaborated with the others, as is necessary to be effective in an interconnecting river system. In 1995 agreements were concluded between Belgium, France, and Luxemburg on Protection of the River Meuse in the Rhine Basin. A separate agreement was reached among the countries in the watershed of the nearby Scheldt.[9]

Most multilateral river commissions follow European prototypes in the primacy of concern for navigation, water use, and protection from floods and other hazards. For example, a nine-member Niger River Commission was established in 1964 by an agreement replacing earlier conventions among the former colonial powers that administered the river. Integrated planning for water resources in the Niger Basin is one of its major responsibilities. Similarly, the Organization for the Development of the Senegal River (Organisation pour la Mise en Valeur du Bassin Fleuve Senegal, OMVS) was formed in 1972 by three West African states (Mali, Mauritania, Senegal). It is concerned only with the economic development of the middle and lower Senegal Basin because early cooperative efforts had broken down owing to objections from the upper-river state of Guinea.[10] A comprehensive action plan for the management of the Zambesi river basin has been developed by the Southern African Development Coordination Conference (SADCC) with assistance from UNEP and the International Institute for Applied Systems Analysis (IIASA).

As international regimes for the joint control of rivers and lakes have developed throughout the world, diverse geographical, economic, and political circumstances have led to a corresponding diversity of institutional structures. The purpose of the newer arrangements in semiarid to arid areas of the world has broadened from the predominantly navigational, and later pollution control, concern of the European river commissions to multipurpose integrated water management. Particularly in regions of water scarcity, competition between consumptive uses—domestic water supply, agriculture, and water-intensive industries—and nonconsumptive uses—fisheries, transportation, and water power—has led to an emphasis upon water conservation.

An interesting and apparently successful example of a multilateral arrangement for water management is the Lake Chad Basin Commission, established in 1964 by the Treaty of Fort Lamy among the adjoining riparian states of Cameroon, Chad, Niger, and Nigeria with two members from each treaty state.[11] The Lake Chad drainage basin covers an area of nearly 2.5 million kilometers in the center of northern Africa, including not only substantial parts of the treaty states but also of the Central African Republic, from which the lake receives more than 40 percent of its total inflow.

The Lake Chad Basin Commission's essential functions are advisory and coordinative. But to facilitate projects funded by treaty and assisted by UNDP, FAO, and UNESCO, the commission has assumed supervisory responsibilities. The executive manager of the commission was designated as comanager of the projects undertaken by the two UN specialized agencies. From an oversight function of the technical assistance projects, commission deliberations have extended to a broader range of issues. In 1972 an Agreement for the Establishment of a Lake Chad Basin Commission Development Fund was accepted by the member states, and the treaty was amended to permit the commission to negotiate external loans. A major accomplishment, further indicative of the growth of commission functions, has been the Agreement on Water Utilization and Conservation in the Lake Chad Basin drafted at Niamey, April 1972. As a result of prolonged drought during the 1980s, Lake Chad suffered serious deterioration. In 1988, under its environmentally sound management of inland waters program (EMINWA), UNEP assisted the commission with preparation of a comprehensive master plan and action program for basin-wide environmental management.

As the African experience suggests, international management of a river basin is vulnerable to political conflict among participants. A case in point is the Mekong River Basin in Southeast Asia. The Committee for Coordination of Investigations on the Lower Mekong Basin was established in 1959 under United Nations auspices as an experimental model for similar efforts in other developing countries. Member states included Laos, Cambodia, South Vietnam, and Thailand.[12] During the first decade of its existence only minimal attention was given to the ecological aspects of its projects and activities. In 1969, however, a team of scientists was assembled to undertake an ecological survey of the Lower Mekong Basin in relation to development planning. The prospect for early effectiveness of this effort was blighted by warfare throughout the region during the 1970s. However, in 1978 three riparian countries, Laos, Thailand, and Vietnam, revived the effort and established an Interim Committee for Coordination of Investigation of the Lower Mekong Basin. Despite political antagonisms among the member countries, the committee is functioning and has adopted an indicative basin plan and work program. It appears to exemplify the politics of antagonistic cooperation. In 1995 the governments of Cambodia, Laos, Thailand, and Vietnam concluded an agreement on the Cooperation for Sustainable Development of the Mekong Basin. So the cooperation appears to be continuing.

A bilateral, permanent Indus Commission with multilateral sponsorship was established in 1960 by India and Pakistan to administer an Indus Basin Fund deposited in the World Bank.[13] But mutual distrust limits its potential for development.

Bilateral and multilateral arrangements for implementing boundary waters agreements have tended to expand from their original purpose (e.g., jurisdiction, navigation, or water diversion) to multipurpose environmental responsibilities. The bilateral International Joint Commission of the United States and Canada (IJC) is a notable case in point. Established initially to oversee the provisions of the Boundary Waters Treaty of 1909, the IJC has gradually become a supervisory agency of the two treaty powers for a range of transboundary environmental issues, but with special emphasis on management of the international waterways of the Great Lakes.[14] Although its jurisdiction is more extensive, the IJC, in relation to the Great Lakes–St. Lawrence drainage system, resembles a comprehensive river basin commission. It lacks administrative authority, however, and policymaking is retained by the governments of Canada and the United States.

During its earlier decades the IJC, like similar international commissions, paid little attention to environmental issues. Since the late 1960s, however, they have been high on its agenda. Although water quality was a consideration in the Boundary Waters Treaty of 1909, not until the surge of environmental protectionism in 1972 did the nations conclude an agreement on Great Lakes water quality specifying quality objectives and means toward their attainments. The 1978 Great Lakes Water Quality Agreement explicitly adopted an ecosystem concept of organization, which included the entire Great Lakes system—the lands surrounding the lakes, the streams flowing into them, and the lakes themselves.[15] The 1978 agreement involved more than water quality: as revised, it embodied recommendations submitted to the IJC by the Great Lakes Research Advisory Board and its Ecosystem Study Committee, and by the Pollution from Land Use Activities Reference Group (PLUARG). This agreement extended the authority of the IJC not only to report to the two federal governments, but to advise and make recommendations concerning a broad range of lake-related issues.

In 1987 a protocol to the 1978 agreement further specified the criteria for water quality and delineated more clearly the responsibilities of the IJC and of the two federal governments. Popular dissatisfaction with the slow pace of policy implementation was dramatized by large and well-organized public attendance at the IJC's biennial review of the water quality agreements. The need for strengthened institutional mechanisms to coordinate, monitor, and perhaps direct the actions of government agencies in the basin was becoming apparent.

Because of their great size and immense economic and ecological importance, the Great Lakes have occupied the greater part of the attention of the

IJC. But other issues, especially in more recent years, have arisen that have also required its attention. Illustrative of these transborder issues are problems of oil pollution in the Puget Sound area, construction of the Skagit-Ross High Dam affecting British Columbia and Washington, the Poplar River Project affecting Saskatchewan and Montana, the Garrison Diversion Unit affecting Manitoba and North Dakota, and the Richelieu River Project involving Quebec and Vermont.[16] Each of these issues involves environmental problems of a localized regional character. In each of them the IJC proved to be an indispensable instrument of international environmental policy. Its methods of careful nonpartisan fact finding, public hearings, and published accounts have won confidence in both of the participating countries.

Of much narrower scope is the International Boundary and Water Commission of the United States and Mexico, concerned primarily with management of the waters (including water quality) of the Rio Grande, and the Colorado and Tijuana rivers. However, new cooperative arrangements have emerged. In 1983 Mexico and the United States ratified the Agreement for the Protection and Improvement of the Environment in the Border Area, and in 1992 they signed the Agreement Regarding the Strengthening of Bilateral Cooperation through the Establishment of a Joint Committee for the Protection and Improvement of the Environment. A binational network of environmental and community nongovernmental organizations from Texas and four northern Mexican states was formed in 1991 to address relationships between natural resources and environmental concerns in the binational area.[17]

Hardly a regional arrangement but perhaps more of a continental regime is the North American Commission for Environmental Cooperation (NACEC) established 1 January 1994 in response to demands for environmental accountability in implementation of the North American Free Trade Agreement (NAFTA). The commission is representative of the trinational governments—Canada, Mexico, and the United States. Its secretariat is located in Montreal. As of this writing it is too early to foresee how its mandate to promote the protection and improvement of the environment and sustainable development in the NAFTA area will work out in practice.[18]

European Regional Cooperation

The environmental concerns of a number of European regional or European-centered organizations, notably the United Nations Economic Commission for Europe, the supraregional OECD, and the several commissions established for the management of European waterways have been described and their work need not be recounted here. Another important agency for regional environmental cooperation in Europe that has not yet been discussed is the European Union, particularly because of its role in European economic affairs. Its significance has grown as it has gradually assumed the character of a functional intergovernmental agency. The Council of Europe is yet another regional organ that has become increasingly involved with environmental concerns. Further, the ecological hazards that confront any major commercially used and technologically exploited waterway have led to cooperative action by the nations that border the North, Black, and Baltic Seas. The problems and their multinational character have been exacerbated as pollution carried by the vast river systems of north and central Europe contaminates the seas.

The European Free Trade Association (EFTA), representing seven countries that were not then members of the European Community, was formed in 1960 chiefly to reduce trade barriers. It established a cooperative relationship with what was then called the European Community and after the joint Declaration of Luxembourg (9 April 1984) gave attention to environmental priorities along with economic considerations.[19] In 1992 (Switzerland and Liechtenstein excepted), the EFTA and the EU combined to form the European Economic Area (EEA) — the world's largest and most comprehensive trading area. Seventeen countries constitute a single market. How this economic block, which became effective in 1994, will affect European environmental policy remains to be seen.

The European Union

Formerly designated as the European Community, now the European Union, this regional organization in some respects resembles an international government. It is an outgrowth of the movement for European unity initiated at the Congress of Europe, meeting at the Hague in 1948. Constituent parts of the European Community were formed separately as the European Coal and Steel Community (1951), European Economic Community (by the Treaty of Rome, 1957), and the European Atomic Energy Community (EURATOM, 1957). In July 1967, these organizations merged to form a single community consisting of the European Parliament, the Council of Ministers, the (administrative) Commission of the Community, and the European Court of Justice.

Transition from community to union was affected through the Treaty of Maastricht, signed in 1992, in effect in 1994. The change of name and status create a slight awkwardness in reporting the history of European environmental policy. In the text that follows, "European Community" or "EC" will be used for actions taken before 1993 when the European Union came into existence. "Union" (EU) will apply for events after the treaty went into effect. European environmental policies under the European Union are complex and evolving. It is only possible here to outline the general arrangements and the principal instruments for coordination and change. For more detailed treatment see the references.[20]

European Union involvement in environmental issues has been a gradual development, reflecting the progression of European opinion from consciousness raising and goal setting, in which the Council of Europe, the Economic Commission for Europe, and OECD played significant roles, to implementation of policies in which considerations of political economy become critical. Writing during the earlier years of international environmental concern, Homer G. Angelo observed, "[a]t this stage in the growth of the Communities, political sensitivities restrain any rapid exercise of control by supranational authority to achieve conservation objectives."[21] Since this observation was made, the role of the Union in regional environmental affairs has expanded greatly. The several European regional political organizations interrelate on environmental policies, but the EU has increasingly taken a lead role in the negotiation of environmental agreements and their implementation.

The EU offers an instructive example of the emergence of environment as a focus for international policy. The regional organizations from which it was formed served almost wholly economic purposes. The successive addition of environmental issues to the agenda shows (1) the interconnectedness of economic and environmental affairs and (2) the growth of environmental awareness among the people of the EU member states. The organizational structure and legal authority of the EU allow it more freedom than other European organizations in transcending parochial national interests in the formulation and implementation of environmental policy.

In 1985 heads of state and government of the then European Community called for the year starting 21 March 1989 to be designated European Year of the Environment (EYE). On 6 March 1986 the EC Council of Ministers agreed upon the following action program:

—to make all citizens aware of environmental pollution
—to integrate environmental policies into the different policies of EC and member states
—to emphasize the European dimension of environmental policy
—to demonstrate the progress already made and the achievements realized by EC environmental policy

Symbolic, but nonetheless politically significant, was a declaration on behalf of environmental protection issued by the EC Council meeting held 2–3 December 1988 on the Greek island of Rhodes. Substantive work of the Union is performed by general directorates of the EU Commission (not to be confused with the Economic Commission for Europe). Various types of environmental activities are distributed within these directorates (e.g., development of pesticide regulations undertaken by the agricultural directorate general). However, as a result of the emphasis placed upon formulation of a comprehensive environmental policy at a 1972 meeting of the heads of state of the EC, a permanent and autonomous environmental unit was created and incorported into its administrative structure.

This unit, the Directorate General for Environment, Consumer Protection, and Nuclear Safety, is currently the functional agency responsible for the development and articulation of Union policy concerning the environment. As such, the environment directorate general not only tries to ensure that the environmental dimension is present in the work of the various general directorates of the Commission, but also independently initiates and executes projects not included within the programs of other directorates general. Its personnel are international civil servants and have competence in both administrative and technical areas. This professional orientation provides an atmosphere conducive to the advocacy of international and environmentalist interests rather than sponsorship of and adherence to individual national policies.

The directorate is a department of the Commission of the European Union. The commission is the administrative arm of the EU. In its role as guardian of the treaties, initiator of policy proposals, and mediator of disputes, it constitutes a forum in which collective community interests take precedence over individual national concerns. The leadership of the EU Commission consists of fourteen commissioners, two each from Germany, the United Kingdom, France, and Italy, and one representative from each of the remaining member states. The supranational character and collegial nature of the EU Commission is also reflected in its simple majority voting rule, making the commission unique among European regional organizations. Furthermore, the commission, representing all member states, has been a signatory to several international environmental protection treaties. Of particular regional significance are the Convention for Prevention of Marine Pollution from Land-based Sources (Paris, 1975); the Convention on the Protection of the Mediterranean Sea against Pollution (Barcelona, 1976); the Conventions on the Protection of the Rhine against Chemical Pollution and against Pollution by Chlorides (Bonn, 1976); and the Convention on the Conservation of European Wildlife and Natural Habitats (Berne, 1982). The commission has also been given authority to negotiate on matters of broader worldwide significance, for example, the nuclear accidents treaties of 1986.

Policy initiatives originating in the Directorate General for Environment, Consumer Protection, and Nuclear Safety are recommended by the EU Commission for approval by the Council of Ministers, which has final decision-making authority. Therefore, even though the policy proposal phase of decision making is characterized by the predominance of European Union interests over national interests, the adoption of policy is dependent upon a consensus of the member states. Once adopted, however, Union policy in the form of regulations, directives, and decisions has the force of law in all member states. The EU Commission has to date promulgated regulations (the most binding decision category) dealing with the establishment of a European Foundation for the Improvement of Living and Working Conditions (1974), the provision of financial support for demonstration projects to exploit alternative energy sources and demonstration projects in the field of energy saving (1978), as well as in other areas. Numerous directives (of somewhat lesser force) have dealt with such topics as the quality of surface waters intended for use as drinking water (1975), the disposal of polychlorinated biphenyls or PCBs (1976), and the protection of migratory birds and their habitats (1978). Decisions involving Union participation in international conventions on the prevention of marine pollution and the establishment of common procedures for the exchange of information in cases of air pollution by certain substances (e.g., sulfur dioxide) have also been issued.

In addition to the above decision categories, the EU Commission also issues recommendations, resolutions, and opinions. Although these actions have no binding force, they serve as indicators of Union-wide attitudes toward policy positions (e.g., the Union has issued a recommendation on the implementation of the "polluter-pays principle," whereby costs of pollution control are to be internalized by the polluting industry). Member states have also entered into an information agreement, in which prior notification of proposed national laws, regulations, and administrative procedures concerning environmental affairs is to be forwarded to the EU Commission. This agreement permits the commission to check that proposed action is in line with Union legislation or to require that action be stayed to allow time to consider if a Union law is required on a particular subject. The commission has authority to enforce national compliance of council decisions. Both the EU Commission and individual member states can bring noncompliance lawsuits before the European Court of Justice; however, sanctions for noncompliance remain unspecified.[22]

Since 1973 the European Community (Union) has undertaken coordinative efforts through five Action Programmes. The first comprehensive environmental program of the European Community was approved by the Council of Ministers on 22 November 1973 and covered the period from 1973 to 1976. The council at that time declared that "improvement in the quality of life and protection of the natural environment are among the fundamental tasks of the Community" and that continuous and balanced expansion of economic activities must necessarily proceed in tandem with an effective campaign to reduce and prevent pollution, leading to a qualitative improvement of the environment.[23] The objectives of the First Action Programme were to

—prevent, reduce, and as far as possible eliminate pollution and nuisances
—maintain a satisfactory ecological balance and ensure the protection of the biosphere

—ensure the sound management of and avoid any exploitation of resources or of nature that cause significant damage to the ecological balance
—guide development in accordance with quality requirements, especially by improving working conditions and the settings of life
—ensure that more account is taken of environmental aspects in town planning and land use
—seek common solutions to environment problems with states outside the European Community, particularly in international organizations

A primary focus for study as reflected in the First Action Programme was the effect of national environmental policies on trade and economic development and the repercussions of such policies on the Common Market (the European Economic Community). Harmonization of such policies at the national level was deemed necessary to avoid nontariff barriers to trade within the Community. In recognition of the fact that the constitutional, legislative, and legal systems of member states differed significantly, provisions of the Action Programme would be progressively implemented until harmonization at the Community level could be effected.

On 17 May 1977 the Council of Ministers approved the Second Community Action Programme[24] for the period from 1977 to 1981. In addition to reaffirming the European Community's commitment to environmental protection as embodied in the First Action Programme, the Second Action Programme represented a reorientation toward "longer-term policies, aimed at promoting a qualitatively superior form of economic growth as a foundation for the future." Reactive and curative measures were to be supplanted by an increased emphasis on prevention of environmental deterioration and on the rational management and conservation of resources and space.

After the adoption of the Second Action Programme in 1977, work proceeded on the projects outlined within its framework. Acts relating to air, water, and noise pollution, chemicals in the environment, and toxic wastes were approved by the Council of Ministers. The Waste Management Committee was established to assist in the formulation of a comprehensive policy in this field. Investigation of the energy potential of wastes and their utilization in agriculture was begun, and action programs on packaging and waste paper were adopted. Studies were undertaken on the recycling of nonferrous metals and the development and ecological management of coastal regions. Additional work was completed toward the creation of a network of pilot schools for environmental education.

In 1979 the Commission of the European Community issued its second report on the state of the environment.[25] This report underlines the Community's support of environmental protection and insists that "there is no need for major conflicts between economic growth and a clean environment."

Rather than contributing to unemployment, the report asserts that the pollution control industry (construction of sewage treatment plants, production of air pollution filters and scrubbers, and the design and installation of treatment plants for toxic wastes) as well as environmental works projects such as reafforestation, recultivation of abandoned land, and the creation of various recreational facilities in the countryside have served as an economic stimulus throughout the regions of the Community.

The 1979 report elaborates upon proposed policy instruments such as environmental impact assessment. Studies have been undertaken by the Community that examine environmental assessment procedures from administrative, legal, and technical perspectives.[26] The EC Commission has been cooperating with individual member states on feasibility studies of this procedure in specific cases, and its work culminated in submission to the council in June 1980 of the Proposal for a Council Directive Concerning the Assessment of the Environmental Effects of Certain Public and Private Projects.[27] Under provisions of the proposal, an environmental impact assessment is to be required for both public and private projects likely to have a significant impact upon the environment.

The Third Community Action Programme (1982–86) was approved by the Council of Ministers on 7 February 1983.[28] This program reaffirmed the commitment of the Community to the environmental quality objectives of the previous action programs and further stated that a "common environmental policy is motivated . . . by the observation that the resources of the environment are the basis of—but also constitute the limits to—further economic and social development and the improvement of living conditions." As a result, the effective management of resources was deemed essential in achieving Community goals. The program also made note of the "socioeconomic context of the 1980s" and indicated that action on behalf of the environment must contribute to the solution of the major problems facing the Community at this time (unemployment, inflation, energy, balance of payments difficulties, and increasing regional disparities). These problems, however, in turn, should not be permitted to dampen efforts at ensuring environmental quality.

The basic thrust of the Third Action Programme was the continuation and completion of the projects initiated in the First and Second Action Programmes. Although work has proceeded on these projects, they remain largely unfulfilled due to the "discrepancy between the scale of the projects and the means available for implementing them."

A Fourth Action Programme for the years 1987–92 was adopted by the EC Commission on 19 October 1987 declaring: "It is no longer seriously contested that environmental policy has a central part to play in the whole corpus of Community policies, and that environmental protection needs to be taken into account as a fundamental factor when economic decisions are taken." As with its predecessors, the Fourth Action Programme was comprehensive, but focused particularly on control of pollution and management of natural resources. In December 1988, the European Council of Ministers meeting at Rhodes adopted a declaration urging the Community to redouble its efforts in the field of environmental policy, and in early 1989 the commission proposed that the Council of Ministers set up a European Environment Agency, coordinating a network to provide technical, scientific, and economic information and to utilize techniques of modeling and forecasting. The Fifth Action Programme 1993–2000 was initiated in May 1993 under the title "Towards Sustainability—A European Programme of Policy and Action in Relation to the Environment and Sustainable Development." Further elaborating the institutional arrangements for the European environment, the European Environmental Agency became operational on 30 October 1993 and is located in Copenhagen. It functions as a research and development facility for the European Union. The full text of the Council of European Communities detailing the functions of the agency is available in the *International Environment Reporter* S-115, 131: 8001 (January 1993).

The Council of Europe

Created by treaty in May 1949 and comprising thirty-four nations as of January 1996 with a combined population exceeding 300 million people, the Council of Europe provides a forum for the expression of European sentiment on environmental among other regional issues. The Council, with headquarters in Strasbourg, consists of a committee of ministers, a consultative assembly that periodically meets in joint session with the European Parliament, and a secretariat. Unlike the European Union, the Council is primarily a deliberative body. It does not carry on major operational functions, rather it serves as a symbol of European identity and sponsors regional agreements, activities, and events.[29]

The Council's freedom from major administrative responsibilities facilitates its initiating and collaborative role in European conservation and environmental affairs. For example, the Council assisted the environment ministers of the European Community in drafting the Berne Convention on the Conservation of European Wildlife and Natural Habitats.[30] Associated with the Council is the European Committee for the Conservation of Nature and Natural Resources established in Strasbourg in 1962 as advisory to the Council's Committee of Ministers. This committee has in this role directed and publicized the environmental activities of the Council. These cover a wide range of topics and are reported in ten languages in the *Newsletter-Nature*, and in the journal *Naturopa*.

The Council was an early agent of environmental policy reform in Europe.

On 2 May 1966, the Committee of Ministers adopted the Programme of Work for the Intergovernmental Activities of the Council of Europe. It included a section on physical environment and resources, the objectives of which were investigation into and adoption of planned action to ensure that the human physical environment shall be balanced, wholesome, and enjoyable, and that Europe's natural resources shall not be wasted, misused, or destroyed.[31] In 1970 the Council sponsored the European Conservation Year, emphasizing development of environmental awareness, and proposing more than thirty guidelines to action. Through the European Information Centre for Nature Conservation the Council carries on a continuing program of education and information exchange. Also interactive with the Council as well as with European governments, regional organizations, and environmental groups generally, is the independent European Environmental Bureau, established in 1974 in Brussels. It engages in policy review, public information and education, and research on specific environmental problems.

A Pan-European Conference of Environment Ministers took place in 1991 at Dobris Castle near Prague in the Czech Republic initiating the Environment for Europe Programme. Assisted by the European Environment Agency and the European Commission a comprehensive report on the state of the European environment was prepared.[32] The Conference of Environment Ministers has been institutionalized, subsequent meetings being held in Lucerne in 1993 and in Sofia in 1995.

The North Sea

Situated amidst highly industrialized and populous countries in northern Europe (the United Kingdom, Belgium, Denmark, France, the Netherlands, Germany, and Norway), this cold and often turbulent body of water presents formidable problems of international control over pollution.[33] The North Sea is not only heavily traversed by ships, but is also a sink for the pollution-laden Rhine and Elbe rivers as well as in the path of fallout from the stack emissions of industrial Britain, Germany, and the Netherlands. On its coasts are several great seaports—Amsterdam, Rotterdam, and Hamburg. The North Sea is a depository for large amounts of domestic, industrial, and agricultural wastes. Toxic metals such as zinc, copper, and nickel have been found concentrated along the coastlines of England and the Continent where aquatic breeding grounds are most vulnerable to damage from these contaminants. Because the existing pattern of coastal currents hinders dilution of such pollutants, accumulation is intensified in these areas. These toxic metals and persistent substances such as salts from potash mines, phosphates from municipal sewage, and agricultural pesticides and fertilizers enter the North Sea via polluted rivers, as well as by means of direct coastal discharge. Furthermore, radioactive material originating in northern France flows into the sea through the English Channel/La Manche. A new threat to the North Sea, however, has arisen from the exploration and exploitation of underwater oil reserves located along the coasts of Scotland and Norway, which complicates the already difficult task of ensuring its environmental protection.

Because of the harsh climate, high winds, strong tidal movement, greater depths, and relative instability of the ocean bottom in the northern portion of the sea, the construction of deep, permanent offshore structures for the drilling and transport of oil involves serious environmental risks.[34] Oil spills originating from structural failure of offshore platforms, pipeline failures, and blowouts in exploratory drilling structures and equipment (e.g., the Norwegian *Ekofsk Bravo* blowout of April 1977 in which 21,300 tons of oil were released into the sea) have already occurred and are becoming increasingly probable as exploration intensifies. Furthermore, such exploration and exploitation generates additional oil tanker traffic in a region in which sea lane congestion is already extremely heavy. The groundings and accidents in nearby waters, involving such vessels as the *Torrey Canyon* and *Amoco Cadiz*, have dramatically highlighted the disastrous environmental effects of oil pollution on the marine life of the area, as well as on the fishing and tourist industries.

The disposal of defunct oil drilling structures is now becoming an environmental problem with economic aspects. Seeking to avoid the costs of salvage, oil companies might prefer to sink the often very large rigs to the bottom of the sea. International environmental organizations such as Greenpeace have opposed this solution on environmental grounds. There are also economic and safety reasons to oppose such action. Salvaging the structural material might be regarded as a cost of the entire operation that the oil companies should bear. The hazard to navigation offers another objection to submergence. A widely publicized controversy in 1995 involving Greenpeace and several European nations, notably Germany, over disposal of the Brent Spar oil rig resulted in withdrawal by Shell U.K. of a proposal to sink the rig.[35]

A number of international agreements including or specifically pertaining to the North Sea area are presently in effect, the most prominent of which are the 1969 Bonn Agreement Concerning the Pollution of the North Sea by Oil among seven North Sea states and Sweden; the 1972 London Convention on the Prevention of Marine Pollution by Dumping of Wastes and Other Matter (global in scope; entered into force in 1975); the 1972 Oslo Convention on Control of Dumping from Ships and Aircraft (regional in scope; entered into force in 1974); and the 1974 Paris Convention for the Prevention of Marine Pollution from Land-based Sources (regional in scope; entered into force in 1978).[36]

The Oslo Convention, which was ratified by all North Sea states, Iceland, and Sweden, covers the North and East Atlantic, the Norwegian Sea, the North Sea, and parts of the Arctic Ocean. This treaty attempts to regulate the deliberate disposal of harmful wastes at sea by means of a total ban on the dumping of substances considered most harmful to the environment (included on a "Black List") and through a system of permits, administered by national authorities, to control the discarding of substances considered less harmful ("Grey List") based upon uniform assessment criteria. Supervision of the implementation of the convention is performed by a Standing Commission composed of representatives from the contracting parties. The commission also collects scientific information concerning the state of the marine environment, compiles dumping records, and oversees the performance of national governments. Enforcement, however, remains the province of national authorities.

Administrative procedures for implementing the London Convention correspond to those of the Oslo treaty. The London Convention, however, is global in scope and applies to the total marine environment with the exception of internal waters of states. It is also more comprehensive in its designation of blacklisted substances (e.g., crude oils, high-level radioactive wastes, biological and chemical "weapons"). Strictly limited supervisory and monitoring activities are performed by the International Maritime Organization (IMO), with contracting parties assuming the major responsibility for collection and dissemination of information relative to the implementation of the convention. Similarly, enforcement measures as well as adaptation and amendment of it remain under the control of national governments.

The Paris Convention covers the same regional area as the Oslo agreement and is similar in institutional structure to both the Oslo and London compacts. Concerned with pollution originating from inland sources, the treaty stipulates that national pollution control measures may never be less stringent than those enunciated by the convention. Although harmonization of policies is called for, the right of governments to enact more stringent environmental requirements is recognized. The convention attempts to regulate discharge of harmful substances into waterways through the establishment of a "Black List" and a "Grey List" similar to those of the previous two agreements. Moreover, the treaty calls for elimination of deliberate transfrontier pollution, whereby states alleviate their own pollution problems by contributing or transferring pollution to neighboring states. Additionally, its definition of land-based sources of marine pollution as "(1) water-courses which enter the sea; (2) pipelines which carry wastes from the coast into the sea; and (3) man-made structures which come under the jurisdiction of contracting parties within the Convention area" is interpreted as encompassing offshore rigs and drilling operations which are, as such, subject to its provisions.

The Paris Convention provided for a commission, PACOM, to oversee implementation as well as to engage in the collection and distribution of information provided through monitoring activities. This commission is authorized to assist in the settlement of disputes and in conciliation among adhering states. Ultimate enforcement and implementation of the provisions of the treaty are responsibilities of the individual national governments.

Mention has previously been made of the earlier regional Agreement Concerning the Pollution of the North Sea by Oil (1969, Bonn), which was formulated in response to the *Torrey Canyon* incident and which attempted to obtain cooperative action on the part of North Sea states threatened by coastal pollution from potential similar disasters.[37] Following discovery of oil beneath the North Sea, eight northwest European states adopted the Convention Relating to Civil Liability for Oil Pollution Damage Resulting from Exploration and Exploitation of Seabed Mineral Resources (1976, London).[38] But the effectiveness of this treaty is impaired by its reliance upon national authorities to formulate principles and specific preventive and protective measures applicable to offshore industry. This convention, and the treaties relating to public law liability (e.g., the International Convention Relating to Intervention on the High Seas in Cases of Oil Pollution Casualties and the so-called private law International Convention on Civil Liability for Oil Pollution Damage — both signed in Brussels in 1969), concern responsibility for environmental damage.

Provisions of the aforementioned agreements could be reinforced or supplemented by two conventions of global applicability. These are the International Convention on the Prevention of Pollution from Ships (MARPOL), in force 2 October 1983, and the Convention on the Law of the Sea, signed on 10 December 1982 by representatives of 117 governments, and in force as of 16 November 1994. Thus the North Sea, in common with the rest of the marine environment discussed in Chapter 9, is covered by a complex, multitiered structure of legal agreements to prevent or control pollution.[39] Nevertheless, marine pollution continues to be a major international problem. Implementation of agreements on a global or even regional scale remains very difficult. To simplify and extend the several North Sea agreements a new Convention for the Protection of the Marine Environment of the North-East Atlantic was signed in 1992 by fifteen states and the European Community. This agreement would embrace and replace the Oslo and Paris Conventions.

The Baltic Sea

Among the more significant regional environmental arrangements undertaken outside of the UN system have been those negotiated for the protection of the Baltic Sea.[40] The Baltic is today one of the earth's major brackish (mildly saline) bodies of water: fresh water is contributed by precipitation and rivers, and salt water is flushed in from the North Sea. Its total area of about 370,000 square kilometers includes the Gulf of Bothnia, the Gulf of Finland, the Baltic proper, the Belts, the Sound, and the Kattegatt, connecting the Baltic with the North Sea. A relatively shallow sea, its average depth is only fifty-five meters.

The natural characteristics that make the Baltic distinctive among the world's seas also make it highly susceptible to deterioration resulting from human activities. It has a limited capacity for water exchange, caused by the narrowness of the linkage with the North Sea and by the large number of sills that form the seabed and basins of the Baltic. As a result, it has been estimated that a complete renewal of the water of the Baltic Sea requires twenty to forty years. Pronounced variations in salinity and temperature between surface and deep water produce horizontal stratification phenomena that further prevent effective mixing, which subsequently impede an exchange of oxygen between its layers; the history of the Baltic has been characterized by intermittent periods of marked deoxygenation of the deep basins. This natural resistance to mixing, as well as the enclosed nature of the sea itself, increases the risk of deterioration in water quality caused by introduction of municipal and industrial wastes. Acceleration of eutrophication from excessive discharge of nutrients into the Baltic, as well as the tendency for toxic substances to remain for long periods in its water, represent a significant threat to its viability.

Similarly, the severe climatic fluctuations characteristic of the Baltic region contribute to vulnerability of its ecosystems. Large portions of the northern Baltic Sea (the Gulf of Bothnia and the Gulf of Finland) experience long periods of ice cover during the winter months, and the cold temperature of the waters increases the possibility of severe environmental damage caused by petroleum spillage and dumping. The cold also impedes the mineralization of organic wastes.

In addition to the natural phenomena that continue to alter the marine environment, human factors play a significant role in effecting ecosystem changes. The Baltic states (Sweden, Finland, Denmark, Germany, Poland, and the Russian Federated Republic, as well as the republics of Estonia, Latvia, and Lithuania) are all highly industrialized and economically developed, engaging in modern intensive agricultural and forestry practices, and in vigorous shipping and navigation. Moreover, the Baltic is used for commercial fishing as well as recreational purposes. Both human and industrial waste products are currently deposited into the Baltic by means of direct emission, indirect discharge via rivers, atmospheric deposition, and dumping and accidental spillage at sea.[41] The combination of eutrophication and the accumulation of toxic substances in both living and nonliving components of its ecosystems contribute to the characterization of the Baltic as one of the world's most severely polluted seas.

Recognizing that unilateral action alone would not be adequate to protect and maintain the unique marine environment of the Baltic, the seven adjacent states on 22 March 1974 signed the Convention on the Protection of the Marine Environment of the Baltic Sea Area (the Helsinki Convention).[42] The treaty, which went into effect on 3 May 1980, includes the area of the Baltic proper, the Gulf of Bothnia, the Gulf of Finland, the Sound, the Belts, and the Kattegatt up to a border drawn between Skagen on the Danish and Goteborg on the Swedish side. Its terms apply to both the international water area (the high seas) and to the territorial seas of the coastal states. Although the sovereignty of states over their own internal waters is assured, and these waters are not included within the convention, Article 4 calls for the voluntary application of its terms to internal waters as well. Between the time of signature and ratification, the treaty's terms were implemented by the Interim Baltic Marine Environment Protection Commission (IC).

Substantive work accomplished by the IC included the coordinated monitoring of the state of the marine environment, a decrease in the levels of DDT and other substances classified as most harmful (a prohibition against the use of DDT is in effect in all Baltic states), cooperation in combating oil residues, and recommendations to establish measures restricting the discharge of other harmful substances into the sea such as constructing reception facilities at Baltic ports and supplying vessels with the necessary equipment for handling such substances.

The IC gained permanent status with the ratification of the Helsinki Convention in 1980 and maintains a secretariat in Helsinki. The Baltic Marine Environment Protection, or Helsinki, Commission (HELCOM) is active on both the administrative and scientific levels to promote cooperation among the Baltic states. In addition to the secretariat, HELCOM is composed of two permanent working groups. The Scientific-Technological Working Group (STWG) concerns itself with water protection technology, the establishment of criteria and standards of water quality, and the monitoring of the marine environment. The Maritime Working Group (MWG) handles problems related to pollution from ships, especially oil pollution resulting from dumping, and coordination of action in the event of accidental spills.

The Convention was described at the time of its adoption as the most comprehensive agreement then in effect to protect the marine environment. It covers discharges from land-based sources, including airborne pollution; discharges from ships, including oil, chemicals, sewage, and garbage; the dumping of wastes; international cooperation to combat spillages of oil and hazardous substances at sea; and the prevention of pollution resulting from the exploration and exploitation of the seabed and its subsoil. By means of the technical and scientific cooperation of participating countries, the Helsinki Convention attempts to stimulate national, municipal, and private efforts to prevent the pollution of the Baltic Sea.

The Convention itself consists of twenty-nine general articles and six specific annexes. Although the treaty does not impinge upon the sovereignty of contracting states, it does provide a framework for cooperation, and Baltic states have demonstrated considerable willingness in implementing its provisions by means of national legislation. Responsibilities of the Helsinki Commission (HELCOM) include overseeing the implementation of the convention, recommending measures pursuant to its purposes, and facilitating information exchange. Decisions within HELCOM are taken unanimously, each state having one vote. The commission, which convenes at least once a year, is responsible for the continued development of scientific and technical cooperation as well as collaboration with other regional and international organizations. Although implementing measures are left to national authorities, a "lead country" is usually designated in cases where more coordinated action is required in the formulation of practical arrangements.

Since explicit mechanisms concerning dispute settlement are incorporated within the convention, contracting parties therefore resort to negotiation or other available legal methods in disagreements arising over its interpretation or application. If agreement cannot be reached using these methods, disputes could, upon common agreement, be placed before an arbitration tribunal or the International Court of Justice (Article 18).

In addition to the Helsinki Convention, the Baltic Sea region is covered by other bilateral and multilateral conventions both regional and global in scope. The International Baltic Sea Fishery Commission (IBSFC) administers the Convention on Fishing and Conservation of the Living Resources in the Baltic Sea and the Belts (the Gdansk Convention).[43] This convention, entered into force on 28 July 1974, promotes the conservation and rational management of the living resources of the Baltic through its recommendations concerning total allowable catches of commercially exploited fish species, fishing technology, national quotas, and stipulations of legal minimum sizes. The Gdansk Convention is reputed to be the first international agreement that attempts to protect all living marine resources, thereby creating a potentially effective mechanism for the prevention of inadvertent overexploitation of these resources (e.g., through disruption of food chains).

A bilateral convention signed by Denmark and Sweden in 1974 and ratified in 1975, the Convention on Protection of the Sound (Oresund) from Pollution, provides guidelines for the purification and treatment of municipal sewage and industrial wastes as well as for the establishment of control measures to prevent pollutants from exceeding recommended levels.[44] More comprehensive marine pollution treaties, previously described (e.g., Oslo, 1974; London, 1975; Paris, 1978) apply to parts of the Baltic area.

In a meeting a month before the seven-nation Helsinki Conference of March 1974, the Nordic Convention for the Protection of the Environment was signed by Denmark, Finland, Norway, and Sweden. Ratified in 1976, the convention calls for each contracting party to place "the environmental protection interests of their neighbors on the same footing as corresponding interests in their own country"[45] and covers the discharge of waste material into lakes, rivers, or the sea in any way such as to constitute an environmental nuisance.

In addition to and, in some cases, in conjunction with research executed under the auspices of the above-mentioned conventions, cooperative scientific investigation of the Baltic is carried out by a number of international and regional organizations with special interests in the area. Among the more prominent of these are the International Council for the Exploration of the Sea (ICES), the Nordic Council (NC) and its various subsidiary councils and committees, and Baltic Marine Biologists (BMB). Investigative research initiated by these organizations includes surveys of the inputs of pollutants into the Baltic, the gathering of baseline data concerning concentrations of toxic substances in marine biota, and basic hydrographical research.[46]

Nordforsk (Scandinavian Council for Applied Research), a regional organization whose membership includes the Nordic countries' research councils, engineering academies, and other scientific and technical bodies, has also been active in the promotion and organization of Nordic as well as international cooperation in the field of environmental protection. Its permanent Secretariat for Environmental Sciences was established in 1970 and is located in Helsinki. In 1968 Nordforsk carried out an inquiry concerning environmental protection in the Scandinavian countries, which has formed the basis for continued activities and research, particularly in the areas of air and water pollution.[47]

UNEP's Regional Seas Programme—Oceans and Coastal Areas

As noted at the outset of this chapter, environmental problems with global implications almost always require localized or site-specific solutions. So it is with the oceans. Just as the regional environmental problems of the North Sea and the Baltic have required concerted action by adjacent nation-states, policy and administrative measures must be applied to specific areas to protect the 72 percent of the earth's surface covered by the seas.

Serious qualitative deterioration has been occurring in semilandlocked bays, gulfs, and seas marginal to the continents. Experience with regional cooperation in most of these areas was insufficient for timely cooperative action. Therefore, UNEP at the Third Session of its Governing Council in 1975 determined that the ocean environment would be one of its "concentration areas." From this decision UNEP played a critical role in sponsoring the Regional Seas Programme.[48] In 1985 the name of the program was changed to Oceans and Coastal Areas (OCA) and its headquarters moved from Geneva to Nairobi. Under an expanded mandate, OCA now includes projects concerned with the living resources of the sea (e.g., fisheries, marine mammals, and aquaculture)—matters also of concern to FAO.

The Regional Seas (OCA) Programme is premised on the belief that countries of each region share a common interest in safeguarding their marine environment and have a mutual need for sustainable utilization of marine resources. Coastal waters, estuaries, and enclosed and semienclosed seas suffer the earliest and most severe degradation from the effects of urbanization, industrialization, agriculture, transportation, and other human activities, while concurrently containing ecosystems that are among the highest in biological productivity (90 percent of the total marine catch coming from such areas). These waters are therefore logical starting points for combating marine pollution. The regional approach allows specific priorities, needs, and capabilities to be taken into account in the formulation of action plans designed to cope with the environmental problems of the oceans as a whole.

By initiating conferences to consider regional maritime problems, UNEP provides opportunity for affected governments to agree upon an action plan tailored to the requirements of a specific area. Active participation of governments in the region is encouraged at the first stages of each program. The development of an action plan is guided by the governments concerned, by the UNEP Governing Council and secretariat, and by United Nations and NGO consultations. Each plan must be formally accepted by the governments before the operational phase can begin.

Although the content of individual action plans vary from region to region in response to prevailing conditions, each plan follows the "framework for action" endorsed at the Stockholm Conference in 1972, which calls for ongoing assessment and monitoring activities to be undertaken by national institutions. Technical assistance by United Nations specialized agencies may be provided. Data gathered from this assessment procedure are evaluated and form the basis for management decisions at the national level. Financial support is initially provided by UNEP, which also sponsors educational efforts and the dissemination of information to increase public awareness of environmental problems within regions.

The assessment component of an action plan includes both scientific and socioeconomic data. Sources, amounts, and behavior of pollutants are investigated as well as the effects of these pollutants on human health and marine ecosystems. Also studied are methods for exploiting living and nonliving natural resources of the region, direct and indirect environmental effects of socioeconomic development activities, and the scientific and technical capabilities of national institutions. This information is then provided to assist policymakers of national governments in integrating environmental concerns and safeguards into development planning and practices to ensure preservation of the environment as well as the sustainable use of natural resources.

Environmental management to protect the seas must begin with environment-affecting activities taking place on land. These include cleaner and more efficient utilization of energy sources, protection of fresh water resources, prevention of soil erosion and desertification, development of tourism without ecological damage, and the mitigation of environmental degradation associated with human settlements. The legal framework for the action plan is usually provided through establishment of regional treaties or conventions, elaborated through and supplemented by specific technical protocols.

Supporting measures, such as technical assistance, training programs, and financial aid, are continued through UNEP and various specialized organizations within the United Nations system. However, because the ultimate objective is that individual regional programs become self-executing, implementation is sought primarily through designated roles for existing national institutions and personnel. Similarly, an increasingly large portion of the financial responsibility for the program is expected to be assumed by the participating governments. Special regional trust funds are among the mechanisms created for this purpose.

The Regional Seas (OCA) Programme involves a comprehensive, progressive assault upon degradation of the marine environment. Since its establishment, the scope of the program has expanded to include twelve regional seas in which action plans are operative or presently under development.[49] These include the Mediterranean, the South Asian seas, the Red Sea and Gulf of Aden, the Kuwait region of the Persian Gulf, the Wider Caribbean region, the West and Central African region, the East Asian seas, the Southeast Pacific, the South Pacific, the East African region, the Northwest Pacific, and the Black Sea. Programs are at different stages of development in each region. A Southwest Atlantic program has been considered, but little action appears to have been taken toward implementation.[50] Over one hundred states currently participate in the program, and many of its regional components appear to have become increasingly self-actuating, showing promise of a continuing future.[51]

The Mediterranean

Nearly enclosed, the Mediterranean Basin is presently bordered by twenty nation-states (Albania, Algeria, Bosnia and Herzegovina, Croatia, Cyprus, Egypt, France, Greece, Israel, Italy, Lebanon, Libya, Malta, Monaco, Morocco, Slovenia, Spain, Syria, Tunisia, Turkey) and the continents of Europe, Africa, and Asia.[52] Its narrow linkages with the North Atlantic at the Straits of Gibraltar and the Black Sea at the Dardanelles inhibit the exchange of water with these adjacent seas.

As a result of the topography as well as the pattern of marine currents, pollution from both coastal discharges and dumping in the open sea tends to accumulate along shorelines. Although the northwestern Mediterranean coastal waters experience the most severe contamination from pollutants, the discharge in the southeastern Mediterranean from the Nile has historically been a significant source of nutrients for fisheries, although now diminished since construction of the Aswan High Dam.

The Mediterranean is the depository for the waste products of 200 million coastal inhabitants. Moreover, the seasonal influx of 100 million tourists attracted by the climate and natural beauty of the area severely taxes the assimilative capacity of its coastal waters. And in addition to being an area of great industrial, agricultural, and commercial activity (especially the northern coastal region), the Mediterranean serves as a transit route for many products, particularly large quantities of petroleum.

The General Fisheries Council for the Mediterranean has estimated that more than 320,000 metric tons of hydrocarbons are discharged yearly into the sea; UNEP officials estimate that an additional ninety tons of DDT and other pesticides are introduced each year. These estimates from the 1970s are approximations intended to give some idea of the volume of pollution. Although remedial efforts are under way, the situation may become worse before it gets better. The threat of thermal pollution, although at present not serious, poses a concern for the future as numerous electric-generating plants are scheduled for installation along the coastline before the year 2000. Beyond this, the Mediterranean is used for recreation, commercial fishing, and aquaculture. Subsurface mining and drilling for oil have potential for causing grave environmental damage.[53]

Before 1972, efforts to protect the Mediterranean had been sporadic and uncoordinated. However, regional concern for the preservation and rational management of this irreplaceable natural resource began to materialize in the wake of the 1972 United Nations Conference on the Human Environment. The Mediterranean states themselves requested UNEP to formulate an action plan to protect the sea, and the Governing Council of UNEP designated the Mediterranean as one of seven endangered marine regions, selecting it as the inaugural project of its Regional Seas Programme.[54] The resulting Mediterranean Action Plan was initially endorsed by sixteen of the Mediterranean states in February 1975 at the Intergovernmental Meeting on the Protection of the Mediterranean (Barcelona I).[55] Currently twenty states and the European Union participate in some phase of the Action Plan. UNEP has provided its secretariat.

The Action Plan is a comprehensive and interdisciplinary attempt to develop and implement substantive programs for the protection of the Mediterranean. It includes four primary activities: (1) integrated planning for the development and management of the resources of the Mediterranean; (2) coordination of research, monitoring, and assessment of levels of pollution, and implementation of protective measures; (3) development of a framework convention and related protocols; and (4) establishment of institutional and financial arrangements to execute the plan.[56]

The Action Plan has evoked other complementary regional attempts at long-range planning, such as the "Blue Plan" adopted in February of 1979, the purpose of which is to reconcile optimum socioeconomic development with the protection of the environment.[57] A critical component of the Action Plan is the Coordinated Mediterranean Pollution Monitoring and Research Programme (MED POL). MED POL projects have included baseline studies of concentrations of hydrocarbons, PCBs, pesticides, and metals in marine waters and organisms, investigation of the transport and synergistic effects of pollutants, and the identification of sources of land-based pollution. Provision for training and equipment to carry on these activities has also been made available.

The recommendations of the Action Plan were further realized in 1976 when sixteen states attending the Conference of Plenipotentiaries of the Coastal States of the Mediterranean Region for the Protection of the Mediterranean Sea (Barcelona II) adopted the Convention for the Protection of the Mediterranean Sea against Pollution and two protocols, the Protocol for the Prevention of Pollution of the Mediterranean Sea by Dumping from Ships and Aircraft and the Protocol Concerning Cooperation in Combating Pollution of the Mediterranean Sea by Oil and Other Harmful Substances in Cases of Emergency.[58] Besides the Mediterranean governments, signatories to the Barcelona Convention included the European Community (EC), the Organization for Economic Cooperation and Development (OECD), the Arab League, the North Atlantic Treaty Organization (NATO), and the UN Economic Commission for Europe (ECE). The convention, which took effect in 1978, calls for the development of additional protocols, procedures, and arrangements "to prevent, abate and combat pollution of the Mediterranean Sea Area and to protect and enhance the marine environment."

Most significant was the endorsement by fifteen states in May 1980 of a third protocol, the Protocol for the Protection of the Mediterranean Sea against Pollution from Land-based Sources.[59] UNEP officials had previously estimated that 85 percent of the pollution entering the Mediterranean originates from such sources. Furthermore, 90 percent of all industrial wastes, municipal sewage, and agricultural pesticides and fertilizers discharged into the Mediterranean is untreated or only inadequately treated. This protocol establishes a list of substances considered highly toxic (e.g., cadmium, mercury, radioactive material), the discharge of which is prohibited, and a second list of less toxic substances (e.g., lead, silver, tin), whose discharge is allowed in only strictly regulated quantities. Control of emissions will be regulated in conformance with established standards based upon determination of acceptable levels of pollution.

Meeting in Cannes in March 1981, sixteen Mediterranean countries and the European Community agreed on a comprehensive three-year program of activities to continue the work already done to protect the sea. A budget of $12 million for the 1981–83 period was agreed upon ($0.5 million contributed by UNEP and the remainder to be contributed by the participating governments and the EC). The budget required large sums to be spent on the Blue Plan's projections of the impact of industrialization, tourism, and transportation on the Mediterranean basin environment; on the Priority Action Programme (PAP), which involves projects relating to aquaculture and solar energy; and on the Regional Oil Combating Centre (ROCC), established in 1976 in Malta, whose usefulness was unanimously reconfirmed by participating governments at the conference. Athens was chosen as the headquarters of a small coordinating group that directs activities of the Mediterranean Action Plan.[60]

In addition to these regulatory efforts, measures were also undertaken to extend protection to Mediterranean forests and animals. In October 1980, UNEP sponsored the Intergovernmental Meeting on Specially Protected Areas in the Mediterranean in cooperation with IUCN, FAO, and UNESCO. Emphasis was on establishing habitat reserves and preparing a protocol on protected areas for submission to the cooperating Mediterranean states.[61] The Protocol Concerning Specially Protected Areas was adopted in Geneva in April 1982 and signed by fourteen of the parties to the convention.

Kuwait Region of the Persian Gulf

The Persian Gulf is a narrow, shallow body of water having only a single outlet to the Indian Ocean through the Strait of Hormuz. The extremely hot, dry climate of the area produces a high rate of evaporation and this, coupled with minimal rainfall and the virtual absence of inflow of freshwater (except for the Shatt-al-Arab Channel), produces a high degree of salinity. The warm and salty water reduces the capacity of the gulf to break up and absorb industrial waste products and urban sewage. These meteorological conditions and the enclosed nature of the Persian Gulf contribute to its characterization as "one of the world's most fragile and endangered seas."[62]

Besides serving as the transit way for over half the oil on the international market, the Persian Gulf region is currently the scene of some of the most rapid and intense development activities in the world. Expanding populations (16 percent per year in the United Arab Emirates) are concentrated in coastal areas where existing sewage and waste treatment facilities are inadequate to meet the needs of increasing populations; in some cases there are no facilities whatsoever.

A greater danger to the marine environment and to human health is posed by the discharge into the gulf of industrial waste (including heavy metals such as mercury). In addition to being the depository of both sewage and industrial waste, the Persian Gulf is also the sole source of drinking water for the region and the principal site for recreational activities. Development efforts of the bordering states of Kuwait, Bahrain, Qatar, and the United Arab Emirates are wholly in the coastal area, while the development efforts of Iran, Iraq, Saudi Arabia, and Oman—countries with extensive inland areas—are nevertheless heavily involved in development along the shore. Large-scale industrial, trade, and development activities including petroleum refineries, shipyards, cement industries, steel mills, and petrochemical complexes are scattered in some twenty urban settlements along the gulf.

The most apparent threat to the marine environment, however, arises from the dense shipping traffic. A 1975 UNEP mission to the region reported that approximately one hundred ships, predominantly tankers, entered the gulf daily through the Strait of Hormuz. The resulting congestion is evident in nearly every port; thirty to forty ships sometimes anchor for long periods at a time without being unloaded, risking large, accidental oil spills either from tanker collision or groundings. Offshore exploration and dredging represent an increasingly significant threat to the marine environment. Severe damage was inflicted by the deliberate release of oil by the Iraqi government of Saddam Hussein in the Persian Gulf War of 1990–91.[63]

In 1975 the Persian Gulf was designated by UNEP as one of seven endangered marine regions and was incorporated into its Regional Seas Programme. A regional conference was held in Kuwait 11–23 April 1978, at which time an action plan as well as a framework convention and protocol were formulated and adopted by the eight gulf states (Bahrain, Iran, Iraq, Kuwait, Oman, Qatar, Saudi Arabia, and the United Arab Emirates).[64] The Kuwait Action Plan (KAP), an attempt at the harmonization of development objectives with environmental protection "for the benefit of present and future generations," provides "a framework for an environmentally sound and comprehensive approach to coastal area development, particularly appropriate for this rapidly developing region."[65]

The Kuwait Regional Convention for Co-operation on the Protection of the Marine Environment from Pollution and the Protocol Concerning Regional Cooperation in Combating Pollution by Oil and Other Harmful Substances in Cases of Emergency became effective on 30 June 1979.[66] The signatory states agreed to "prevent, abate, and combat pollution" resulting from intentional or accidental discharge and dumping of waste materials from ships and aircraft. Discharges from land-based sources, whether entering the gulf directly, by water, or by air, and pollution resulting from exploration, exploitation, and dredging of the seabed, its subsoil, and the continental shelf were also covered in the agreement. The convention called for scientific and technical cooperation, including the designation of a national authority responsible for pollution research and monitoring within the areas under its jurisdiction, and for procedures to determine liability and compensation for damage resulting from violation of the provisions of the convention. Protocols dealing with these issues were to be formulated at a later date.

Article 3 of the protocol now in force provides for the establishment of a Marine Emergency Mutual Aid Centre (MEMAC) whose objectives include information exchange, technological cooperation and training of personnel, and development of national capabilities to handle pollution emergencies. This center, administered with assistance from the International Maritime Organization (IMO), is located in Bahrain. In 1981 the Regional Organization for the Protection of the Marine Environment (ROPME) was formally established. With the establishment of this organization, UNEP's role as coordinator of the Action Plan was ended, although it continues to provide support for development and implementation of the plan.

Although implementation of the Action Plan suffered setbacks due to the outbreak of the Persian Gulf War between Iran and Iraq in 1991 (both are signatories to the convention and protocol), antipollution projects and activities are planned to continue by means of bilateral arrangements between UNEP and individual Persian Gulf state governments. The continuation of hostilities prevented implementation of these agreements in containing and stopping a massive oil spill resulting from Iraq's bombing of oil wells in the gulf.[67] But although the foregoing environmental agreements were effectively nullified by the government of Iraq, ROPME played an important role in rehabilitating the gulf ecosystem following the war.[68]

The Red Sea and Gulf of Aden

The Red Sea, a narrow strip of water extending for 1,930 kilometers southeast from the Suez Canal to the Strait of Bab el Mandeb, separates the coasts of Asia and Africa and has become one of the world's major commercial waterways. The narrow passage through Bab el Mandeb (16 kilometers) makes the Red Sea nearly landlocked and hence vulnerable to accumulating pollution. The Gulf of Aden, by contrast, is an extension of the open sea.[69]

Although a regional convention and Action Plan for the Red Sea and Gulf of Aden were signed in Jeddah in 1976, activities in the area have been until recently primarily confined to the training of marine scientists and enhancing the capacity of national institutions in the field of marine science. However, because pollution levels in the Red Sea have become increasingly serious, efforts to expand the program have begun. In early 1981 a new and more comprehensive Action Plan for the Conservation of the Marine Environment and Development of Coastal Areas in the Red Sea and Gulf of Aden as well as a regional conservation convention and a Protocol on Cooperation for Combating Marine Pollution by Oil and Other Harmful Substances in Cases of Emergency were finalized. The Action Plan and legal agreements were adopted at a Conference of Plenipotentiaries convened by the Arab League Educational, Cultural, and Scientific Organization (ALECSO) in February 1982 in Jeddah, Saudi Arabia. Also planned are agreements and protocols relating to scientific and technical cooperation, the control of pollution from exploration and exploitation of the continental shelf and seabed as well as from land-based sources, and liability and compensation for damage by pollution. The major environmental threat is posed by oil pollution, but degradation of this waterway also results from industrial wastes, domestic sewage, and siltation arising from dredging operations. Although coordination of activities relevant to the region is now handled by ALECSO, joint ALECSO-UNEP programming sessions are convened regularly to discuss development of projects supported by UNEP.

Three countries in the Red Sea and Gulf of Aden region are participants in other action plans as well. Egypt participates in the Mediterranean Action Plan, Saudi Arabia in the Kuwait Action Plan, and Somalia in the East African Action Plan. The Gulf of Aquaba presents a four-nation environmental challenge to Jordan, Saudi Arabia, Israel, and Egypt.[70]

Wider Caribbean Region

Although the Caribbean is more "open" than many regional seas, industrialization, urbanization, and mass tourism are presenting increasing threats to its fragile ecosystems. The extensive coral reefs, mangroves, lagoons, and seagrass beds that shelter most of the Caribbean's aquatic life are essential to the region both economically and ecologically. The siting and construction of heavy industry and petrochemical processing complexes, and the expansion of agricultural activities in coastal areas, endangers the health and productive capacity of these ecosystems. Furthermore, the region is one of the major oil-producing areas of the world in which intensive shipping activities occur. Tourism, which degrades the environment through increasing the demands placed upon the waste-assimilative capacity of the area as well as through the construction of harbors, hotels, marinas, and resort areas, increases the stress upon vulnerable ecological balances.

The cultural, economic, and political differences that have long characterized the region have hindered efforts at intergovernmental cooperation. Regional consciousness is a comparatively recent development. The states and islands of the Caribbean differ in culture, tradition, political structure, economic potential, and ecological character. Experience with self-government has been brief, and the level of the island economies is, with few exceptions, low. External funding and technical assistance are necessary to permit regional initiatives to proceed. However, UNEP, working in conjunction with the Economic Commission for Latin America and the Caribbean (ECLAC), has been successful in achieving agreement among twenty-five of the twenty-seven nations, territories, and dependent islands that comprise the wider Caribbean area (extending from Guyana to Florida and the Gulf of Mexico) for the protection of their shared environment.[71]

A Conference on Environmental Management and Economic Growth in the Smaller Caribbean Islands was held on Barbados in 1979 under the joint sponsorship of the UNEP-ECLAC Caribbean Environment Project, the United States Man and the Biosphere Committee, UNESCO, the Caribbean Development Bank, and the United Nations Department for Economic and Social Affairs, and with the assistance of the United States Agency for International Development and the Caribbean Conservation Association. This conference brought together a large amount of data and opinion relevant not only to the smaller Caribbean islands, but to the wider Caribbean area as well.[72] It was a prelude to the consideration of the Caribbean Environment Project Action Plan then in preparation.

Meeting at Montego Bay, Jamaica, 6–8 April 1981, a UNEP-ECLAC-sponsored conference of regional governments adopted the Caribbean Action Plan comprised of sixty-six environmental projects of common interest.[73] Of twenty-five projects designated as "high priority," eight were selected for immediate action. These included contingency planning for combating oil spills, guidelines for the management of watersheds, improvement of environmental health services, protection of coastal marine resources, renewable energy production, environmental education, and prevention of negative environmental effects of tourism.

Subsequent to this conference, 26–29 August 1981, the meeting of Non-Governmental Caribbean Conservation Organizations on Living Resources Conservation for Sustainable Development in the Wider Caribbean was held under joint IUCN-Caribbean Conservation Association sponsorship at Santo Domingo, Dominican Republic. Paralleling and closely related to the UNEP-sponsored Caribbean Action Plan, the IUCN in 1980 issued its Strategy for the Conservation of Living Marine Resources and Processes, which has been applied to the wider Caribbean region. An aspect of the IUCN-UNEP World Conservation Strategy, this effort was complemented by the Eastern Caribbean Natural Areas Management Programme, assisted by the World Wildlife Fund, the Rockefeller Brothers Fund, the University of Michigan, and the Caribbean Conservation Foundation.

The Montego Bay conference and resulting Action Plan and associated nongovernmental efforts constitute "the greatest collaborative effort of the wider Caribbean to date."[74] These meetings, plans, and strategies represent an attempt to attain the sustainable, nondestructive development of the region. In order to accomplish this goal, the Caribbean Action Plan's immediate objectives are assessment of the state of the environment in the region and assistance to governments in the solution or minimization of environmental problems by means of management of development activities. This effort has, among other ways, been assisted by publication of a *Caribbean Data Atlas* by IUCN and WWF, which is the first step toward comprehensive systems analysis mapping for the region.

A Caribbean Trust Fund has been established, $1.5 million of which is to be contributed by regional governments and $1.38 million to be contributed by UNEP; and a nine-country monitoring committee was formed to work with UNEP in executing the activities of the Action Plan. Additional progress in the region was evidenced when in March 1983 thirteen of the twenty-seven Caribbean nations signed the Convention for the Protection and Development of the Marine Environment of the Wider Caribbean Region in Cartagena, Colombia.[75] This so-called Cartagena Convention is being supplemented by protocols dealing with specific problems outlined in the convention. An initial protocol concerning cooperation in combating oil spills in the wider Caribbean region was adopted at the Cartagena meeting.

West and Central African Region

The waters off the west coast of Africa, especially the Gulf of Guinea, are exposed to oil pollution from ships traversing the offshore corridor from the Indian Ocean to Europe. Health hazards to coastal populations and fisheries are caused by the discharge of domestic sewage and industrial waste into coastal waters. Additional wastes, including pesticides, fertilizers, and sediment runoff are carried into these waters by rivers from inland sources. Extensive damage to beaches, marshes, and lagoons is produced by large-scale land reclamation and coastal engineering projects, such as the construction of harbors. Oil discoveries along the West African coast make pollution control measures even more urgent.

In order to prevent further deterioration of their marine environment, sixteen of the twenty coastal states of West Africa, meeting in Abidjan, Ivory Coast, in March 1981, agreed upon a Convention for Co-operation in the Protection and Development of the Marine and Coastal Environment of the West and Central African Region, which "recognizes the threat to the marine and coastal environment, its ecological equilibrium, resources and legitimate uses posed by pollution and by the absence of an integration of an environmental dimension into the development process."[76] Specifically mentioned in the Abidjan treaty are pollution from discharge or dumping from ships; pollution from land-based sources, such as industrial wastes, agricultural runoff, and sewage; pollution from exploration and exploitation of the seabed and its subsoil, atmospheric pollution; and coastal erosion. Also included in the treaty was a resolution that approves "a right of hot pursuit," under which naval vessels may give chase to oil tankers cleansing their tanks within territorial waters of a coastal state and subsequently moving into the coastal waters of adjacent states.[77]

An accompanying protocol on Cooperation in Combating Pollution in Cases of Emergency pledges signatories to cooperation in the event of pollution emergencies, such as oil or toxic chemical spills. The treaty and protocol entered into force with formal ratification by six of the regional governments. Participants at the Abidjan conference, which was held under the auspices of UNEP, also endorsed an Action Plan for the region. The plan calls for assessments of oil pollution within the region, suspended and dissolved matter in rivers, chemical residues from industrial and agricultural activities, and domestic wastes. UNEP has been delegated secretariat responsibility for the Action Plan, the convention, and the accompanying protocol.

East Asian Seas

The East Asian region is rich in natural resources, uneven in the concentration of its populations and in economic and technological development, and complex in its political relationships. Adoption by its maritime nations of the two-hundred-mile territorial sea principle has added to already existing jurisdictional conflicts. The UNEP Regional Seas Programme has found acceptance primarily in the southern subregion, but China, Taiwan, and Vietnam have claims and interests in the South China Sea and their involvement will be necessary to any region-wide program. Mutual antagonisms have obstructed cooperation among these countries; present initiative rests with the Southeast Asian (ASEAN) group of states.[78]

The East Asian seas region experiences the oil pollution, destruction of coral reefs and mangrove swamps, siltation, threats to public health, fisheries, and tourism now common to other tropical waters. Coastal areas consisting of coral reefs and mangrove swamps represent a particular environmental concern not only because such ecosystems serve as a significant source of food supplies, but also because they contain a number of rare species of flora and fauna unique to this region. Of particular importance are the mangrove ecosystems, which are breeding grounds for marine biota and buffers against natural disasters endemic to the region (e.g., seismic waves and cyclones).

In order to deal effectively with such problems under present political conditions, a subregional approach involving the Association of Southeast Asian Nations (ASEAN) has been adopted. Membership of ASEAN consists of Brunei Darussalam, Indonesia, Malaysia, the Philippines, Singapore, and

Thailand. Ultimate expansion of a regional seas program to include other neighboring states is expected. Preparatory activities relating to regional oil spill contingency planning, toxicity of oil dispersants, land-based pollution sources, river pollution transport, impacts of pollution on mangrove ecosystems and on living aquatic resources, and environmental problems of offshore exploration and exploitation are presently under way.

Following meetings in 1979 and 1980 held in Jakarta, Penang, and Baguio, the ASEAN states established environment priorities a regional seas program and followed by environmental impact assessment, urban water and air quality monitoring, and pollution control technology, all of which had an ultimate relationship to the safeguarding of the Southeast Asian seas.[79] An Action Plan was adopted in Manila by the participating governments in April 1981, but no treaties to implement the plan were negotiated at the time. UNEP was delegated responsibility for secretariat functions of the Action Plan.

Projects of particular significance within the proposed Action Plan include the investigation of the effects of organic effluents on mangrove aquatic communities, bioassays of pollutants on mangrove biota, and the impact of economic exploitation of mangrove forests on fisheries and coastal waters.[80] Adoption of the Action Plan is also expected to enhance the capabilities of national institutions involved in marine research and environmental management.

In January 1980, UNEP sponsored a meeting organized by IMCO (IMO) and Indonesia, and attended also by Malaysia and the Philippines, to control oil pollution in the Celebes (Sulawesi) Sea. An oil spill response plan was adopted following procedures established by the ASEAN countries in 1976.[81]

Environmental cooperation among the ASEAN states is not confined to the Regional Seas Programme. An ASEAN Workshop on Nature Conservation was held on 15–16 September 1980 in Bali attended also by representatives of FAO, UNDP, UNEP, IUCN, and WWF. Agenda items included (1) ASEAN cultural and natural reserves and participation in the World Heritage Convention, (2) endangered species and renewable natural resources, (3) training in management of nature reserves, and (4) an information exchange program.[82] The ASEAN states have taken collective action on environment and development signified by the Singapore Resolution of 1992, taking a common approach to the UN Conference on Environment and Sustainable Development, meeting in Rio de Janeiro, and after Rio by the Bandar Seri Begawon Resolution on Environment and Development adopted by the ASEAN Ministers for the Environment in April 1994.[83]

South and Southwest Pacific Region

During 1974 the South Pacific Commission (SPC) initiated discussions on an Action Plan for the region. The commission, established as early as 1947 by multilateral treaty among the states with territorial jurisdiction in the region, belongs in the structure of international environmental policy.[84] The scope of its communication and coordination functions includes agriculture, environmental planning, health, and marine concerns in the area. Cosponsored by the Commission, UNEP, the South Pacific Bureau for Economic Cooperation (SPEC), and the Economic and Social Commission for Asia and the Pacific (ESCAP), a regional Conference on the Human Environment for the South Pacific took place in Rarotonga, Cook Islands, 8–12 March 1982.[85] An Action Plan was approved at this conference, and a convention and protocols were subsequently negotiated. The South Pacific Commission through the secretariat of the South Pacific Regional Environment Programme (SPREP) was formerly responsible for the technical coordination and implementation of the Action Plan, while overall coordination and guidance was provided by a Coordinating Group consisting of representatives of the sponsoring agencies. The South Pacific Regional Environmental Programme has now separated from the commission.

The Southwest Pacific region includes the east coast of Australia, New Zealand, the Solomon Islands, Papua New Guinea, and Vanuatu. Although the area is relatively lightly populated and generally not highly stressed, interest in the preservation of coral reefs and protection of fisheries has led to the development of an Action Plan.[86]

Southeast Pacific Region

The Southeast Pacific Region extends along the length of the west coast of South America and includes the coastal states of Panama, Colombia, Ecuador, Peru, and Chile. The region is characterized by physical and biological diversity, encompassing tropical, subtropical, temperate, and subantarctic ecosystems. A review of the levels of marine pollution and coastal development problems was prepared by the Permanent Commission for the South Pacific (PCSP) (to be distinguished from the unrelated South Pacific Commission), which acted as coordinator of the program for this region.[87] An Action Plan, regional convention, and agreement on regional cooperation in combating pollution of the Southeast Pacific by hydrocarbons and other harmful substances in cases of emergency were adopted at a conference in Lima, Peru, on 11 November 1981. A trust fund was also established and PCSP designated as secretariat for the action plan, convention, and agreement.

Eastern African and Southwest Atlantic Region

The Governing Council of UNEP at its Eighth Session designated these two regions for inclusion into the Regional Seas Programme. Initial steps toward the formulation of action plans were taken in 1982-83. East African states have agreed on a draft Action Plan for the eastern Indian Ocean. However, by September 1993, the plan had not been activated because only three out of the required six states had ratified the agreement. Exploratory discussions have been undertaken with the Southwest Atlantic countries, which include Argentina, Brazil, and Uruguay, but political problems in the region have complicated the course of negotiations.

The Northwest Pacific and Black Sea Regions

In 1989 the Governing Council of UNEP requested preparation of action plans for areas not yet covered by the Regional Seas Programme. These areas were the Northwest Pacific and the Black Sea.[88] Cooperation in both regions has been retarded by political and military antagonisms. However, in 1992 the Bucharest Convention on the Protection of the Black Sea against Pollution was adopted and was followed by a Ministerial Declaration on Protection of the Black Sea at Odessa on 7 April 1993 and which constitutes an interim action plan. The parties were Bulgaria, Georgia, the Russian Federation, Romania, Turkey, and Ukraine. The declaration contains the essence of an action plan. A secretariat for the convention is located in Istanbul.[89]

International Straits

Certain bounded areas of the seas afford passageways between islands and continents. Although regional in character they differ from the international rivers. They lack the upstream–downstream characteristics that cause river pollution and diversion (e.g., for irrigation) to become environmental policy issues. They are not common spaces because they fall under the jurisdiction of adjacent states. Nevertheless, they are transnational insofar as they are open to innocent passage by ships of many nations. In his treatise on the international law of straits Erik Brüel offers the following distinctions:

> In the light of the elements contained in these and other geographical definitions of straits it appears that a water in order to be a strait in the geographical sense of the word must firstly be a part of the sea, i.e. not artificially created, so that canals, as f. i. the Suez-Canal, the Panama-Canal and the Kiel-Canal, however close their function approaches that of the straits, fall outside the range of conception.
>
> Secondly, the particular water must be a contraction of the sea, i.e. that, at any rate compared with the adjoining waters, it must be of a certain limited width, without it being possible, however, to state and definite measure. Further, the particular water must separate two areas

of land, it being immaterial in this connection whether it separates two continents, one continent and an island, two islands, etc. Finally, in order to be a strait in the geographical sense of the word, that particular water must connect two areas of the sea that otherwise, if the strait be eliminated, were separated—either totally or at least in that particular place—by the territories between which the strait runs. The size of the connected seas is so far unimportant. . . .

Correspondingly, one might equally well divide the straits according to the territories that are separated by them, into "intercontinental" straits i.e. those that separate two continents (Gibraltar, Bosphorus and the Dardanelles, Bab-el-Mandeb, Bering-strait), continental-insular i.e. those that separate an island from a continent (the strait of Calais, Little-Belt) and finally, inter-insular i.e. those that separate two islands (Great-Belt, St. George's Channel).

A last and final classification according to the function of the straits, is the classification between straits of a certain larger interest to the international intercourse and straits of no such interest, which is the classification that is the basis upon which this treatise will determine the question of what straits should occupy a separate legal status.[90]

The international law of straits pertains chiefly to navigation—commercial and military.[91] Being constricted areas, straits are especially vulnerable to pollution from passing ships. The English Channel, for example, has experienced serious oil spills. Canada has enacted legislation to protect the Arctic Northwest Passage from contamination. The Law of the Sea and the several oceanic antipollution conventions apply to the international straits. Nevertheless, the vulnerability of the more heavily used passages indicates a need for close attention in the implementation of international environmental maritime law.

Polar Regions

The Arctic

The foregoing pages provide selective illustration of regional arrangements for the development and implementation of international environmental policy. Other examples could be given, and some are noted elsewhere in this volume. The polar regions are contrasting cases, the Arctic an ocean, the Antarctic a continent. No part of the earth is now immune from transboundary contamination by air and oceanic currents, and land-based pollution is now affecting the Arctic Ocean and its coastal areas and islands. In 1984 the Congress of the United States adopted the Arctic Research and Policy Act (P.L. 98-373) establishing the Arctic Research Commission and the Interagency Arctic Research Policy Committee to develop a national research policy and five-year implementation plan. Commenting on the importance of the region, the Arctic Research Commission declared that "[t]he Arctic Ocean, including its marginal seas, is a truly remarkable component of the global Earth system. Its importance reaches far beyond strategic and economic considerations. Indeed, the Arctic Ocean is a significant element in the complicated machinery that regulates the entire biosphere, our environment for life."[92]

At the invitation of the government of Finland, in September 1989, representatives from eight Arctic countries (Canada, Denmark, Finland, Iceland, Norway, Sweden, the Soviet Union, and the United States) met in Rovaniemi, Finland, for the first Ministerial Conference on the Protection of the Arctic Environment. Follow-up meetings were held in Yellowknife, Canada, in 1990,[93] in Kiruna, Sweden, and Rovaniemi in 1991, where on 14 June at a ministerial-level meeting the Declaration on the Protection of the Arctic Environment was signed. The resulting agreement consists of principles and objectives and summary reports on the Arctic environment. Actions to be undertaken include an Arctic Monitoring Assessment Program (AMAP), and agreements on Emergency Prevention, Preparedness, and Response, and the Exchange of Information on Arctic Flora and Fauna.[94] A regional convention on the conservation of polar bears has been in force since 1976. Participating states have been Canada, Denmark, Norway, the former Soviet Union, and the United States.

Antarctica

Because of their remoteness from populous centers and the severity of their climates, the polar regions of the earth present unique problems of international environmental policy. The north polar region is a landlocked ocean surrounded by territories over which nations have established jurisdiction. The south polar region is a continent over which nations have asserted claims, but none have internationally recognized jurisdiction. Unlike the high seas, the atmosphere, and outer space Antarctica has an internationalized arrangement for governance. The continent and its surrounding waters are "governed" by treaties, which although not establishing a government are sufficiently coherent and authoritative to constitute a regime.[95]

Advanced and only recently developed technologies have been necessary to explore and to occupy the polar regions, even temporarily. Early exploration was often marked by tragedy, as in Captain Robert Scott's unsuccessful effort to return from his 1912 journey to the South Pole. Geophysically the polar regions differ in ways that have significant environmental and political consequences. The north polar region is an ocean, covered during the present geological era with a permanent cap of sea ice, but during extended periods in the geological past it was evidently an open sea. Being essentially landlocked and far more accessible than Antarctica from surrounding continental shores, the north polar region has been treated as an international commons, although Canada and Russia, because of their geographic proximity, exercise substantial political authority over large areas of the Arctic Ocean. For example, in 1970 Canada adopted extensive legislation to protect its Arctic waters from oil pollution.[96]

The circumstances of Antarctica are very different. The region is remote from any routes of commerce and from other continental land masses or population centers. Antarctica is a continent, parts of which are elevated well above sea level forming rugged chains of mountains. Surrounding the continent is an ice pack beyond which lies one of the most turbulent oceans on earth.

These physical factors have made Antarctica adverse to exploration and occupation, and until recently the territorial claims of nations upon the continent were largely symbolic. Although the Antarctic continent is believed to contain significant mineral deposits, and although the waters of the Antarctic have abundant marine life, the principal importance of Antarctica to modern society has been through the opportunities it affords for scientific observation and research. On 1 December 1959, an international treaty proposing to treat Antarctica as an international scientific reserve was signed by the twelve nations involved in research in Antarctica and including the seven states asserting claims to its territory.[97] National claims were not relinquished, but have been suspended during the life of the treaty.

The future of Antarctica and of the treaty is clouded by growing world problems of energy and the availability of scarce minerals. Reviewing developments in the 1970s, David F. Salisbury foresaw international difficulties:

> During the last 15 years, international cooperation in Antarctica has been exemplary. Under the terms of the treaty, nations can inspect one another's bases and must publish the results of their research in scientific journals. . . . But there are signs that the potential wealth of the ice-locked continent is putting an appreciable strain on this cooperation. When a confidential U.S. geological survey estimated that the Antarctic shelf may contain 45 billion barrels of petroleum and 115 trillion cubic feet of natural gas became public, it created an international stir. Since then, the treaty nations at their biannual meetings have begun to discuss how to deal with resource exploitation.[98]

Development of an international environmental policy for the Antarctic region began with the International Geophysical Year (1957–58). Research bases in Antarctica for scientific investigations were established by a number of nations during this period, and the principal impact of human activities has been in the immediate vicinity of these sites. Policies and practices in Antarctica are administered by the Consultative Parties—signatories to the

Antarctic Treaty.[99] However, it appears that not all of the provisions governing activities in Antarctica are faithfully observed, notably with respect to the disposal of waste. In order to implement the treaty, the Consultative Parties adopted in 1964 a set of Agreed Measures for Antarctic Fauna and Flora.[100] F. M. Auburn, in a comprehensive treatment of Antarctic policies, observed that "these arrangements are frequently praised as one of the most comprehensive and successful international instruments for wildlife conservation on land that have yet been negotiated."[101]

Because of uncertainties and disagreements as to the applicability of the Antarctic Treaty of 1959 and the Agreed Measures of 1964 to the surrounding Antarctic seas, two additional conventions have been negotiated. The first, signed in London on 1 June 1972, was the Convention for the Conservation of Antarctic Seals;[102] the second, signed in Canberra on 20 May 1980 and effective on 7 April 1982, was the Convention on the Conservation of Antarctic Marine Living Resources (Southern Ocean Convention).[103] The latter treaty provides for an arbitral tribunal to resolve disputes among the participating parties. But questions have been raised as to whether the Southern Ocean Convention will prove to be an effective conservation measure. A study undertaken by a team of experts associated with the International Institute for Environment and Development, under contract to the IUCN and funded by the World Wildlife Fund, concluded that changes in the treaty were necessary and that "parts of the area should be held in trust for future generations; the . . . convention should be submitted for review by an internationally representative body; a technical and economic committee should be established with links to other international bodies, and preferential quota allocations for Third World countries should be provided for."[104]

A basic concern has been the management and control of fishing for krill. This small but prolific crustacean is the basic component of food chains in the Antarctic or Southern Ocean. The harvesting of krill has become a focal point of international controversy over priorities in Antarctica and adjacent seas. Because krill is a presently abundant, even though inconveniently located, source of protein, it became the object of aggressive exploitation by Russian and Japanese fishing fleets. Poland and Germany also entered the competition, which then raised serious questions regarding the consequences for Antarctic food chains. Krill is the principal diet of the baleen whales (e.g., the great blue whale) whose numbers have been dangerously depleted by overkill. If the krill are severely reduced in numbers to provide food for people who hitherto have been nourished without them, will the recovery of southern whale stocks be prevented even if adequate protection from whalers is eventually obtained?

The question has become highly political. Some conservation and scientific groups would like to restrict krill harvesting and set aside areas of the Antarctic or Southern Ocean in which taking of krill was forbidden. The FAO has undertaken extensive studies on the harvesting of krill and its ecological effects, but it is reported to have withheld publication of findings that displease the krill harvesting governments.[105] The issue was stated summarily by W. Nigel Bonner of the British Antarctic Survey:

> In its simplest form we might say that if we harvest krill at its maximum sustainable yield we cannot expect the whale stocks to recover, for their food base will be removed. Similarly if we opt for harvesting a restored stock of whales, as a means of utilization of the production of the Southern Ocean, we cannot expect simultaneously to maintain a large krill fishery.[106]

This increasing pressure for exploitation of the food resources of the southern oceans led the Scientific Committee on Antarctic Research (SCAR), the Scientific Committee on Oceanic Research (SCOR), FAO, and the International Coordination Group for the Southern Ocean of the Intergovernmental Oceanographic Commission to organize an international cooperative research program under the title of BIOMASS (Biological Investigation of Marine and Antarctic Systems and Stocks).[107] The principal objective of this undertaking is a study of the structure and dynamics of the Antarctic and marine ecosystems. Particular emphasis is upon krill as a key factor in the Antarctic fishery and the food chain of Antarctic animals, although BIOMASS studies other living organisms of the Antarctic Ocean, including seaweeds, birds, fish, squid, seals, and whales.[108]

Whether the present consultative regime for Antarctica will continue to be viable is uncertain. The consultative group of states has been concerned to keep Antarctica free from United Nations intervention. The Third World countries controlling the United Nations General Assembly and Secretariat would expect to share in any division of spoils that might accrue from exploitation of Antarctic resources. And as the 1982 war between Argentina and the United Kingdom over the Falklands/Malvinas demonstrated, national claims in the Antarctic region continue to be taken seriously.

Summarizing the political status of Antarctica, F. M. Auburn concluded that:

> There is no real administration; Consultative Meetings are clumsy and inefficient. Unanimity seriously limits the capacity to reach controversial decisions. The [Antarctic] treaty is inflexible because formal amendment is not practical. *De facto* revision may therefore be expected on the precedent of the changing interpretation of the treaty area and the limitations placed upon the acquisition of Consultative status. It has been pointed out that a number of breaches, some of them being of major significance

have taken place. More would no doubt be revealed if the records of governmental departments and Treaty meetings were publicly available.[109]

Yet in a divided and imperfect world the Antarctica system, if administered under the several treaties and agreed measures, could be regarded as a partial and potential success so far, if the test of success is not pushed too far. Longer-range prospects are uncertain, particularly if a nation should seriously attempt mineral development on the continent, or if attempts to harvest the living resources of the Antarctic seas should impair the viability of its ecosystems.

As of today, prospects for broader internationalization of Antarctica are doubtful; nor, under present political circumstances, could broadened control promise a more rational and restrained administration. If mankind is determined to exploit every feasible niche on earth, Antarctica cannot be held an exception, should the requisite technology become available. And the technology is becoming available, developed especially for the more accessible Arctic regions. Mineral exploitation of Canadian islands in the Arctic Ocean is now going forward; mines are being developed. In early February 1982, Canada's National Energy Board opened hearings on a proposal to transport liquefied natural gas from Arctic islands to southern markets on a year-round basis.[110]

In July of 1981 the Antarctic consortium agreed to develop a regime governing mineral exploitation "as a matter of urgency."[111] So the prospect of Antarctica remaining a natural reserve for science and nature appeared to be fading. The demands of modern industrial society for materials and fuels seem insatiable, and industrialization in the Third World may be expected to increase the demand, with resultant intensive, competitive maneuvering among claimant states for control of prospective resources in Antarctica. Under these circumstances, ICSU's Scientific Committee on Antarctic Research (SCAR) may provide valuable guidance for prospective development activities.

A series of Special Consultative meetings were held during the late 1980s, following these earlier developments. On 2 June 1988 at Wellington, New Zealand, representatives of nineteen Antarctic Treaty Consultative Parties, and of thirteen contracting (nonparty) states adopted the Convention on the Regulation of Antarctic Mineral Resource Activities. This agreement would in effect have legalized development activities in Antarctica, thus annulling its protection as a scientific reserve. But the convention was rejected by two Consultative Party members—Australia and France—so that its provisions were not enacted.

In 1991 the Madrid Protocol to the Antarctic Treaty of 1959 was adopted, providing strengthened and more specific measures for protecting the Antarctic environment. Among other provisions the protocol included a requirement for environmental evaluation and assessment of proposed action affecting the Antarctic environment. As of 1996, however, few signatory states had ratified the protocol.[112]

Regionalism as an Organizing Concept

Regionalism is a multidimensional concept, too complex and extensive to be treated here in other than geopolitical arrangements for environment-related purposes. These international regional agreements described in this chapter illustrate the practical necessity of finding cooperative arrangements most appropriate to coping with particular environmental problems in particular geographical areas. Institutional arrangements to be workable must somehow achieve a consistency corresponding to the physical, ecological, economic, social, and political configurations of participating nations. No set of uniform governmental or intergovernmental structures can accommodate the diversities of nature and culture around the world. Regional arrangements have been established for particular interests that can be regionally bounded. There are circumstances and relationships sufficiently localized in particular areas to make regional cooperation the logical way to deal with common environmental problems. For certain areas of the earth, the environmental problems are not wholly dissimilar (notably in the tropical seas), and certain patterns of cooperation have evolved.

It is too early to estimate the extent to which these regional environmental arrangements will be effective. Will a unifying effect of working together on common environmental problems be sufficient to offset political antagonisms and short-term economic interests among participating states? Indications in the Baltic and Mediterranean areas are encouraging. But hostilities between Iraq and its neighbors in the Persian Gulf among states in the Red Sea and between Argentina and the United Kingdom in the Southwest Atlantic dramatize hazards to international cooperation. Still, nations are only now beginning to develop workable models for collaboration among unequal states; antagonistic cooperation is as yet a largely undeveloped art among nations. But effective regional programs require much more than considerations of international relations—domestic politics are unavoidably involved. There are formidable hazards, especially where short-term national objectives conflict with broader regional interests.

Serious environmental damage has occurred in areas where a sense of regionalism has been relatively weak. The breakup of the Soviet Union and associated territorial disputes have complicated cooperative intentions. An especially difficult and disastrous case is the shrinking and degradation of the Aral Sea in Central Asia. This sea lies between the republics of Uzbekistan and Kazakhstan. Its decline is largely a consequence of the agricultural and economic policies of the former Soviet Union. To save the sea and the region from total disaster, major political and economic changes will be necessary.[113]

The emergence of environmental awareness along with scientific capabilities in developing countries has made possible the degree of cooperation already attained. In none of UNEP's regional seas would the programs under way in 1996 have been possible or even realistically predictable in 1972. Then, for example, the scientific and institutional capability and the governmental receptivity that allowed the ASEAN nations to propose extensive regional cooperation based upon national commitment simply was not available. There are numerous risks to the continuing effectiveness of these regional programs, but where there is evidence that cooperation is possible, there is reason to hope that they may be effective.

9
International Commons: Air, Sea, Outer Space

A globe depicting the political divisions of the world illustrates the most characteristic aspect of its governance: its division into a large number of so-called sovereign nation-states. Until recently, large areas of planetary and extraplanetary space were beyond the reach of national sovereignty, but under economic and political pressures, aided by innovations in technology, nations have claimed extended jurisdictions upwards and outwards from these traditional terrestrial boundaries. But while the actual extent of unclaimed planetary space has greatly shrunk, the legitimacy of many extended territorial claims remains in controversy.

All of the earth's atmosphere and the more than 70 percent of its surface that is covered by the oceans, especially that percentage beneath the deeper parts of the ocean, are much less amenable to political control than are dry land surfaces. The environmental conditions of the polar regions, especially of Antarctica, severely restrict normal human occupancy. Even less amenable to human influence is the magnetosphere and the gravitational field of the earth in outer space. Thus, extensive parts of the planetary environment may be affected by mankind but are not under its effective political jurisdiction.

These areas, over which national jurisdiction is ambiguous or ineffective, are nevertheless of increasing concern to so large a number of nations that they may appropriately be termed "international commons." Indeed, efforts have been made to establish by legal principle that these areas, notably the high seas and deep seabed, are "the heritage of mankind." But as of today, this principle has not been universally accepted as international policy or law. The practical difficulties of extending effective national jurisdiction over these areas have permitted them to be used as international commons—that is, open to use by any nation with the requisite technological capability.

Exploration and temporary residence beneath the sea, on the Antarctic continent, or in outer space, are possible only with the assistance of special technological facilities. Advances in technology have in fact enlarged the area of the earth that may be regarded as "habitable." Nevertheless it has been difficult historically, and continues in some respects to be impossible, for national governments to establish and maintain effective political jurisdiction over these areas which have been identified as an international commons, and have no permanent human populations.

Industrial, aeronautical, and communications technologies have made national preemption over parts of the commons easier, but also more problematic. It is easier for governments commanding advanced air and sea capabilities to maintain surveillance over areas where their sovereignty is asserted, but effective control is limited. Efficient and rapidly deployable systems for exploiting the living resources of the sea (e.g., electronic fishing techniques and floating processing factories) may deplete migratory species of sea animals that move into, out of, and through maritime jurisdiction of many different nation-states. Industrial and agricultural technology creates oceanic and atmospheric pollution that may travel unhindered for hundreds or even thousands of miles in oceanic or atmospheric currents—uncontrollable beyond its point of origin.

Competition and noncooperation remain facts of international relations, but among many nations, threats to the survival of many fish and marine mammals and exposure to transnational pollutants by air and sea have increased a receptivity to collaborative action. Telecommunication across the commons clearly facilitates such international cooperation. Several problems remain, however. It has become evident that the more advanced industrial societies have the power to exploit the international commons to the point of denying access by the less developed world. In addition, ecologically harmful use ultimately hurts all people, impairing the quality of life on earth, foreclosing opportunities, and diminishing wealth. Some forms of international cooperation in protection of the world's commons may be the ultimate requirement of human survival, for example, to stop disintegration of the stratospheric ozone layer. As more is learned about geosphere-biosphere interactions, the need for concerted international cooperation becomes ever more evident.

Atmosphere

Until the advent of large-scale aeronautical technology, air space was essentially beyond political control.[1] After World War II control became more of an issue, as new and advanced weapon systems were designed to use air space as a medium for military action. The problems are not yet acute, for, with the exception of nuclear contamination, conventional air warfare thus far has little relationship to changes in the quality of the atmosphere. Chemical warfare (e.g., toxic gases) could be a localized temporary exception as could the use of military aircraft to break up developing hurricanes through cloud seeding or to utilize weather modification techniques as military strategies. However, environmental and geophysical modification technologies as weapons of war might have uncontrollable long-term effects, not easily reversed.

The atmospheric commons was hardly a matter of national concern until after science-based industrial technology inadvertently created threats to air quality through the burning of fossil fuels and the diffusion of atomic radiation. When Scandinavian lakes acidified as an apparent result of fallout from sulfate emissions of oil- and coal-fired furnaces in Germany and the United Kingdom, and when residents in the United States and Canada were disturbed by reports of measurable fallout from nuclear testing in China, national governments began to be advised that the invisible threats in the atmosphere represented more immediate dangers than the anticipated visible threats of missiles and aircraft. But short of the dubious alternatives of economic or military retaliation, there has been no way for a nation to protect itself unilaterally against impairment of its atmosphere by activities carried on in neighboring countries. Peaceful means for protection exist (e.g., arbitration and adjudication), but international cooperation is necessary for them to work.

A difficulty with management of the atmosphere as an international commons is that, for practical purposes, it is a medium or carrier. It is human action, not the air itself, that protective policy must address. Even states of continental extent, such as the United States and Canada, receive airborne pollutants originating in other countries. Actions that significantly affect atmospheric quality and are amenable to public policy action are land-based in particular nations. But the effects may be felt largely in other nations; thus, international issues arise. It follows that two conditions must be met before international policies for atmospheric protection can be developed.

First, the substance, origin, and behavior of the harmful phenomena must be understood and cause-effect control relationships established. This is largely a function of science aided by technology, as in sensing, analyzing, and monitoring pollutants in the atmosphere and in the discovery of technically and economically feasible methods of their elimination or control. Second, a forum or basis for political negotiations must exist. This may be a bilateral arrangement such as the International Joint Commission of the United States and Canada or a multilateral regional organization such as the European Union. Agreement among all individual states is needed if controlling the increase of carbon dioxide in the atmosphere requires restricting the use of organic fossil fuels—coal in particular.

A potentially major development in international air pollution control is the International Treaty on Long-Range Transboundary Air Pollution, sponsored by the Economic Commission of Europe, which entered into force in March 1983. Canada and the United States have adhered to the treaty.[2] This agreement has had little immediate practical effect. It is essentially an agreement to exchange information, to encourage research, and to consult. It does not reduce the transboundary transport of atmospheric pollutants. Still, it is a first step.

Two types of events in the atmosphere are of international concern. The first are temporarily disruptive effects. The second are continuing alterations of the physical properties of the atmosphere. Both (but chiefly the latter) require policy action by governments or international agencies.

Among the temporarily disruptive effects is the noise created by supersonic flight. The International Civil Aviation Organization (ICAO), headquartered in Montreal, would appear to be a logical agency to supervise international control over noise created by aircraft. Thus far, however, ICAO has not demonstrated a major concern with international policy in this area (although it has been involved with other environmental effects of supersonic flight, e.g., its impact upon the stratosphere). On the other hand, in regions of urban concentration, as in Western Europe, conventional aircraft and motor vehicle traffic crossing national boundaries present continuing problems. The Council of Europe, the European Union, and the OECD have variously addressed noise in the atmosphere as an international concern.[3]

Phenomena not influenced by man, such as particulate emissions from massive volcanic eruptions (e.g., Krakatoa, 1883), may temporarily affect the atmosphere and weather. Such effects may be cause for international concern, but hardly call for action beyond temporary relief measures. More serious in its long-term effects is airborne dust resulting from human misuse of land (e.g., desertification). The burning of fossil fuels (e.g., in automobiles) reduces air quality, and during the Persian Gulf War in 1991, air was polluted over a wide area of Southwest Asia by Iraqi firing of Kuwaiti oil wells.

The distinction between short-term and long-term effects is not always clear. Temporary phenomena, such as heavy emissions of acidifying chemicals into the atmosphere, could bring about the acidification of slow-flushing lakes, continuing long after the causal emissions are abated. More threatening could be the dispersal of radioactive material throughout the atmosphere. Relatively temporary radiation could result in permanently damaging effects. International concern has been aroused sufficiently to have caused 110 nations to adhere to the UN-sponsored Partial Nuclear Test Ban Treaty (1963). But some nations with nuclear capabilities have declined to ratify it, among them China and France. Testing of nuclear devices by France in the South Pacific and by China in Inner Asia have led to international protests and may be regarded as infringements in principle of customary international law.[4]

The accident at the Chernobyl Nuclear Power Plant on 26 April 1986 released a large amount of radioactive material into the atmosphere with serious consequences for the countries of northern and eastern Europe. The international response resulted in two treaties signed within six months of the accident: the Convention on Assistance in the Case of a Nuclear Accident or Radiological Emergency and the Convention on Early Notification of a Nuclear Accident. Many governments and persons pressed claims for compensation for damages suffered, but the U.S.S.R. had never entered into treaties that would require it to acknowledge liability or to accept adjudication. Its breakaway republics do not appear to have accepted liability for damages attributable to the former union.

Human-induced weather modification produces relatively short-term effects, but if it occurs at a particularly critical time and place or recurs repeatedly, it could lead to conflict among nations. Deliberate modification of climate could be far more serious. But there are other causes for concern, especially the inadvertent modification of the atmosphere and weather, for example, by dust and gaseous emissions originating in nature or human activity.[5]

The atmosphere and human impact upon it have now become a major focus of international cooperative scientific research. Present efforts are more concerned with understanding the atmosphere than with its control.[6] Beginning in 1946, a series of investigations undertaken at the General Electric Research Laboratories by Nobel laureate Irving Langmuir and his associates led to techniques of cloud seeding which under certain circumstances could change local weather.[7] The results of efforts to apply these techniques have been mixed, but their potential for misuse has aroused public apprehension.

It seems clear that some forms of weather modification, if successfully undertaken by one country, could affect weather in adjacent or nearby countries. Following the 1971 report of its Committee on Atmospheric Sciences, the National Academy of Sciences recommended that the United States sponsor a resolution in the United Nations dedicating weather modification to peaceful purposes.[8] On 11 July 1973, the U.S. Senate, after an unsuccessful effort in 1972, adopted a resolution calling for international action to prohibit the use of geophysical weapons.[9] In 1974, President Richard Nixon and the general secretary of the Communist Party of the U.S.S.R., Leonid Brezhnev, signed an agreement to find ways to overcome the dangers of environmental modification for military purposes.[10] The Soviet Union subsequently introduced a resolution in the UN General Assembly to prohibit the use of climate modification techniques for other than peaceful purposes, and in 1977 the Convention on Prohibition of Military or Any Other Hostile Use of Environmental Modification Techniques was signed by representatives of fifty-five nations.[11] But until 1980 the United States Senate, reflecting Defense Department opposition, declined to ratify this treaty.[12]

International environmental policy dealing with the atmosphere was initially concerned with deliberate manipulation using science and technology. While knowledge of inadvertent effects of human activities upon global atmospheric quality has been largely at a stage of theory and conjecture, accumulating scientific evidence has pointed toward long-term cumulative consequences of alterations in the atmosphere resulting from ordinary human behavior.[13] These effects have become international issues because actions that could fundamentally change the composition of the atmosphere anywhere, ultimately might change it everywhere. Three potentially critical threats in the global atmospheric environment—ozone depletion, the "greenhouse" effect, and acid precipitation—will now briefly be considered, followed by an account of international efforts to address the uncertainties of the global atmospheric environment scientifically.

Ozone Depletion

The first threat manifests itself in the high-altitude environment of the ozone layer and stratosphere. The danger lies in human-induced chemical changes that could affect solar radiation penetrating the upper atmosphere. When commercial supersonic flight on a large scale was first proposed, there was concern among atmospheric scientists that oxides of nitrogen in the exhaust gases of supersonic planes might unbalance the molecular composition of the stratosphere sufficiently to impair its shielding effect against lethal solar radiation.[14] The apparent uneconomical performance of supersonic transport (SST) has at least temporarily retarded its expansion and so alleviated this anxiety. But a new worry has arisen.

It was discovered that chlorofluorocarbons released from ground levels (in refrigeration systems and as propellants in household aerosols) find their way into the upper atmosphere. Here they react photochemically to reduce ozone concentrations.[15] Stratospheric ozone shields the earth from intensities of ultraviolet radiation that would otherwise threaten life in a number of ways. There is strong evidence linking the incidence of more than one form of skin cancer to excessive exposure to ultraviolet radiation. And it has been surmised that crop yields of several kinds of agricultural plants would be threatened with reduction should a substantial ozone depletion occur. Moreover, the larval forms of several important seafood species might suffer appreciable die-off. In the 1990s an apparent worldwide decline of amphibian species was suspected by some scientists to be caused by sensitivity to increases in ultraviolet radiation.

Studies undertaken in the United States by the National Academy of Sciences have reported that the continued release of chlorofluorocarbons (CFCs) could lead to an average warming of the earth's surface by a few tenths of one degree Celsius before the mid-twenty-first century, as well as effecting changes in the temperature of the stratosphere that in turn would affect ozone chemistry.[16] The inquiries concluded that although there are many uncertainties connected to the effects of fluorocarbons upon the ozone, the evidence is sufficient to justify international concern with the release of these chemicals into the environment, and that the possibility of irreversible damage deserves continuing attention.

As early as 1977, the United States acted under the Clean Air Act Amendments to restrict the use of aerosols containing CFCs. In 1980 the Governing Council of UNEP asked all governments to reduce the production and use of CFCs. As a consequence of growing information and popular concern, the Vienna Convention for the Protection of the Ozone Layer was signed by representatives of twenty countries on 22 March 1985. As with many first steps in international environmental cooperation, the Vienna Convention chiefly provided a basis for more substantial future action by confirming the existence of an international problem and calling for information exchange, monitoring, and research. During succeeding months negotiations continued, and on 16 September 1987 twenty-eight nations signed the Montreal Protocol on Substances That Deplete the Ozone Layer. A schedule for the progressive phaseout of CFCs was adopted. But although only a 50 percent reduction in the use of CFCs was agreed upon, the precedent for positive, measurable action has been a major advance in international environmental policy.

Within four months of the Montreal Protocol going into force, however, it was already out of date and unacceptable to a growing number of nations. At a meeting in Finland in the spring of 1989, representatives of eighty-one nations adopted by consensus the Helsinki Declaration on the Protection of the Ozone Layer (2 May 1989) calling for a total phaseout of production and consumption of CFCs by the year 2000.[17]

Carbon-Dioxide-Balanced Greenhouse Effect

Of greater uncertainty of effects and greater difficulty of control is the continuing increase in atmospheric carbon dioxide (CO_2) as a result of the combustion of fossil fuels.[18] It has been estimated by the Committee on Atmospheric Sciences of the U.S. National Academy of Sciences that were the amount of atmospheric CO_2 to double, the global average temperature might rise as much as two degrees Celsius, a climatologically significant amount. Since this report in 1977 numerous studies and estimates of trends and effects in climate change have been made.[19] As of 1996 there appeared to be general agreement among atmospheric scientists that a trend toward global warming is under way. Differences of opinion are over the rates of change, the likelihood of counteractive forces, and effects in particular geographic areas.

In addition to CO_2, several other gases generated by human activities have been accumulating in the upper atmosphere. They include methane, freon, and CFCs, and their effect upon the climate is to block infrared radiation from the earth into outer space, thus creating a "greenhouse effect" upon the bio-sphere. Global consequences could be severe. Rising sea levels resulting from melting of the polar ice caps could inundate large areas of coastal land, flood-ing such cities as New York, London, Shanghai, and Tokyo. Rainfall patterns would be affected, agriculture gaining in some cases and losing in others. Sur-vival prospects for various plant and animal species would be jeopardized.

There was sufficient evidence of the possibility of major and irreversible changes in the atmosphere from increasing concentrations of CO_2 to make this possibility the subject of several major international investigations. For example, a 1978 workshop on the Global Biochemical Carbon Cycle involv-ing seventy-six scientists from twenty-two countries was hosted by the Inter-national Institute for Applied Systems Analysis and cosponsored by ICSU-SCOPE, UNEP, the German Research Council, the University of Hamburg, and the Shell Petroleum Company. The scientists concluded that, although the consequences of human activities for the concentration of CO_2 in the atmosphere might be less than some of the more pessimistic forecasts, the continued burning of fossil fuels could lead to at least a three- or fourfold in-crease in atmospheric CO_2 concentration. In the United States, two reports issued in late 1983 argued that the warming effect of increased atmospheric CO_2 was inevitable, but they differed in their assessments of the severity of the consequences.[20] Subsequent workshops in Austria and Italy, and additional investigations confirmed and extended these earlier findings and stimulated proposals for a comprehensive international law of the atmosphere.

By the late 1980s the issue had become sufficiently important to cause the World Meteorological Organization and the United Nations Environ-ment Programme to create the Intergovernmental Panel on Climate Change (IPCC) in 1988. The panel was charged with the assessment of scientific information on climate change, including an evaluation of geographic and socioeconomic consequences. In addition, the panel was charged with the task of formulating appropriate response strategies. The report and response strategy released in 1990 by the IPCC scientists received serious attention from governments. Hearings on the issue were held in both houses of the U.S. Congress. The need for international cooperation was recognized, at least in principle, in the Framework Convention on Climate Change signed at the UN Conference on Environment and Development in 1992. But the imple-mentation of corrective measures is fraught with political uncertainties.

Acid Precipitation

The burning of fossil fuels has given rise to another problem of atmospheric quality, hardly less tractable than the problem of atmospheric carbon di-oxide—acid precipitation. Exhausts from motor vehicles, especially nitrogen oxides, and emissions from coal-fired electric-power-generating plants, especially sulfur dioxide, are transformed in the atmosphere to become nitric and sulfuric acids, which combine with particulates and water vapor to precipitate as rain or snow in areas often far removed from their point of origin. Thus, acid deposition has become an international problem.[21] Emissions from the British Isles and northwestern Germany are carried across the sea to Scandinavia, and transnational acidic pollution has become a matter at issue between the United States and Canada, but there are many uncertainties regarding origins and incidence.

It is not presently clear how the international contentions raised by transnational pollution by acid precipitation will be resolved; it would appear that failure to contain the damaging acid emissions in the country of their origin would violate the international obligations of that country to other countries. There are technical remedies or alternatives to the generation of energy through fuels that cause acid precipitation; the difficulty lies in their cost and efficiency. Governments such as the United States and the United Kingdom have been reluctant to impose strict emission standards on older, less efficient factories, especially as the need to rely upon coal as a basic fuel has increased. Nevertheless, here is a case where some nations are allowing their territory to be used for activities which result in serious environmental damage to other countries. And as long as this circumstance exists, the acid precipitation issue will remain a serious and, very likely, a growing source of antagonism in international relations and environmental policy debates in industrial countries. If, as seems likely, an international convention for the atmosphere is negotiated, the acid precipitation problem will surely be covered in its provisions.

An International Policy for the Atmosphere?

Of uncertain consequence, but of potential multiple damage, is the effect of devegetation on land and sea upon atmospheric composition and quality. Overgrazing, marine pollution, urbanization, and deforestation, especially from the cutting of tropical forests, reduce the biomass of oxygen-emitting vegetation.[22] Possible effects range from decrease in atmospheric oxygen to changes in rainfall and increase of soil erosion by wind as well as by rain. Heat and particulates from large cities are known to affect local weather patterns. These and other effects may not be confined to countries where surface disturbance occurs, but may be felt in lands and water downwind. Some scientists believe that the life-supporting capabilities of the ocean could be impaired.[23]

Some atmospheric scientists report that large quantities of dust are carried into the atmosphere as a result of soil tillage in arid and semiarid areas, and of the exposure to wind of particulate matter due to severe overgrazing and deforestation. It is apparent that great amounts of sediment and chemical pollutants reaching the oceans are airborne. Possible effects upon the oxygen-producing green plant life of the sea has aroused concern. Loss of phyto-plankton productivity, which could be a result of severe marine pollution, is at least a potential subject of international protective inquiry. Atmospheric dust has also been identified as a possible factor in desertification.[24]

Interrelationships among these various phenomena are not well under-stood. Nevertheless, studies of the consequences of natural catastrophes such as massive volcanoes as well as man-made disturbances affecting the atmo-sphere are laying a foundation in scientific information that could in the future become the basis for international policies regarding land use in rela-tion to the quality of the atmospheric environment.

International scientific and technical cooperation in the study of atmo-sphere and weather has, of course, been long established. Primarily sponsored by the World Meteorological Organization (WMO) and the International Council of Scientific Unions, a complex series of observations and experi-ments has been undertaken. For details of these various international scien-tific investigations, issues of the *WMO Bulletin* may be consulted.

The most comprehensive of these investigations to date has been the Global Atmospheric Research Programme (GARP), an undertaking following from Resolution 1802 (XVII) by the UN General Assembly on 14 December 1962.[25] The organization of GARP took shape in 1966 and was projected as a single global experiment of twelve months duration intended to provide comple-mentary support to the newly initiated World Weather Watch (WWW). Be-ginning on 1 December 1978 after more than ten years of preparatory work, the earth's atmosphere and weather phenomena were observed and measured from land-based ocean weather stations and by ships, aircraft, and meteoro-logical satellites.

The transnational character of weather and climate have obvious implica-tions for both inadvertent and deliberate modifications of the atmosphere. In 1978, an international workshop found that the present understanding of the dynamics and social impacts of climatic change was as yet too uncertain to provide an adequate basis for policy action.[26] Climate is known to have changed, and to have changed greatly in the past, and it may be assumed that comparable changes will occur in the future. Such changes could greatly alter the relationship among nations, transforming their economic capabilities and their ability to sustain both natural and man-made ecosystems.

The ability to influence fundamental changes of climate is as yet uncertain. But if climate as an aspect of the atmosphere is held to be part of the global commons, it affords a field of national and international environmental policy that seems certain to have a future. The significance of that future was implicit in the remarks of Robert White, former head of the United States National Oceanic and Atmospheric Administration, to the first World Climate Conference (Geneva, 1979). For certain purposes, he said, "we must put climate alongside such global commons as the deep-sea bed and outer space as a concern of mankind for which new international obligations must be derived."[27]

A decade later, climate had become a major concern of international policy. In 1983 the International Council of Scientific Unions initiated the International Geosphere-Biosphere Programme. A series of four meetings held during the 1980s at Villach, Austria, progressively elaborated the scientific aspects of the climate change issue. At conferences at Bellagio, Italy, in 1987 and Toronto, Canada, in 1988, questions regarding the policy implications of global climate change were introduced. During 1988–89 many other colloquia, symposia, and conferences on global climate change were held, and on 3 January 1990 the president of the United States, George Bush, announced a global climate conference to be held in Washington during 1990.

By early 1990, global climate change could be regarded as the single greatest international environmental policy issue. Protection of the ozone layer had moved to a state of implementation. Assessment of the "greenhouse" effect and global warming was a more complex task; there was less agreement concerning the urgency and extent of international response. The president of the United States appeared to back away from earlier protestations of concern. More research was proposed as preferable to costly premature action. However, in 1987 the U.S. Congress passed the Global Climate Protection Act, and in 1992 it ratified the treaty signed earlier at the UN Conference in Rio.

The response elsewhere appeared to be more positive. In 1988 and 1989, the need for international action was emphasized in declarations by the European Community (Rhodes, 1988), by 24 states at the highest political levels (The Hague, 1989), by 123 countries attending the London Conference on Saving the Ozone Layer (5–7 March 1989), and by the 82 countries and the European Community meeting at Helsinki (2 May 1989). The UN General Assembly, in Resolution 43/53 of 6 December 1988 entitled "Protection of the Global Climate for Present and Future Generations of Mankind," strongly urged timely action to be taken within a global framework. In February 1989 an international panel of Legal and Policy Experts on the Protection of the Atmosphere met in Ottawa, Canada, and recommended an international convention or conventions with appropriate protocols to protect the atmosphere and limit the rate of climate change.

By 1988 international concern over the possibility of global climate change leading to global warming was sufficient to cause WMO and UNEP to convene the interdisciplinary Intergovernmental Panel on Climate Change (IPCC), referenced in note 22. The 1990 scientific report of the IPCC stimulated national attention and led to the Framework Convention on Global Climate Change open for signature at Rio in 1992. Yet none of these numerous conferences, studies, and reports have resulted in effective action. They are recorded here as indicative of the slow and often halting movement of national policymakers toward implementing measures that protect the future of humankind and the biosphere but impose restraints on human behavior.

Outer Space

International concern with the space environment became active in October 1957 with the placing in orbit of the Russian artificial satellite Sputnik. Thereafter, international policy development and cooperation in relation to outer space expanded prodigiously. Technologies that would permit human excursion beyond the earth's atmosphere into outer space had already stimulated some political scientists and legalists to consider the need for an international law of outer space. During the cold war of the 1950s and 1960s, Russian-American competition for leadership and presumed advantage in outer space created apprehension that artificial satellites or perhaps the moon itself might be used for military purposes. Consequences of this concern were the United Nations General Assembly's Declaration of the Legal Principles Governing the Activities of States in the Exploration and Use of Outer Space (Resolution 1962 [XVIII], 13 December 1963), and the United Nations Treaty on Principles Governing the Activities of States in the Exploration and Use of Outer Space including the Moon and Other Celestial Bodies (1967). The treaty was ratified by seventy-one states, but not by the two states with space technology capabilities—the United States and the U.S.S.R.[28] In 1988, however, the United States and the U.S.S.R. consummated a bilateral agreement of their own for Exploration and Cooperation in the Use of Outer Space for Peaceful Purposes. The extent of this cooperation was complicated by the U.S. official commitment to Strategic Defense Initiative technology, making outer space a potential war zone.

Will the outer space environment continue to be regarded as an international commons utilized by whatever nations have the technological capability to exploit it? Will national jurisdictions be asserted, effectively as well as formally, upward and outward into space? Or will the uses of space be managed by international agreement? These questions and others relating to outer space have been before the United Nations Committee on the Peaceful Uses of Outer Space, and a body of international space law has been both proposed and developed.[29]

The United Nations has sponsored two international conferences on the exploration and peaceful uses of outer space, both in Vienna—the first, 14–27 August 1968, and the second (UNISPACE '82), 9–21 August 1982. At the 1968 conference, 188 papers were published and abstracted.[30] Among the topics covered were communications, meteorology, navigation, remote sensing by satellites, geodesy, biology and medicine, and indirect benefits of space technology. Papers prepared for UNISPACE '82 have been privately published—many of environmental policy significance.[31] Issues of satellite broadcasting and remote sensing surveillance were among those raised in the general debate with the positions of the delegations following the customary North-South divisions.[32] From the perspective of the delegation of at least one country with a major interest in space, the United States, UNISPACE '82 was only a moderately useful gathering.[33]

On the moon and in orbiting laboratories it would be possible to conduct experiments in astronomy, geophysics, and a number of related sciences that could add significantly to understanding of the environment of the earth. There are also prospects for industrial and medical benefits from weightlessness. Yet because the costs of activities in outer space are very high, because some benefits and risks are almost certain to become common to all mankind, and because of the obligations of signatories to the United Nations treaty on the uses of outer space, this unoccupied region seems a logical candidate for international cooperation. As with exploration and mining of the deep sea bed there are complications and conflicts regarding equitable arrangements among unequal partners. Protection of the space environment against misuse by humans is less difficult than managing the exploitation of space for economic or military purposes. An international institution such as that established for deep sea mining would be a possibility, but it would be vulnerable to the same objections that were leveled against the UN control of deep sea mining enterprise. The very unequal capabilities of nations in space technology raise many questions regarding the forms that an outer space enterprise might take.

As of today, international policy respecting the outer space environment falls into two categories: exploration and experiments. The second of these is in some respects a subset of the first; experimentation may be regarded as a form of exploration. Tests conducted on the surface of the moon by American astronauts and in space capsules by both American astronauts and Russian cosmonauts did not, so far as generally known, include activities that could be potentially harmful to the environment of the earth. However, the first astronauts returning from the moon remained in quarantine for several days as a precautionary measure, a caution informed by scientific studies on the possibilities of inadvertent contamination from outer space, and perhaps inspired by events described in science fiction such as Michael Crichton's *The Andromeda Strain* (1969).[34]

International concern over military activities in space was aroused in the 1960s following experiments by the United States Air Force in the upper atmosphere and outer space environment.[35] Project West Ford (1961) involved the scattering of copper dipoles in the very high upper atmosphere to determine their possible effects upon radio communications. It was discovered that the dipoles had no notable significance, but before this was learned, radio astronomers were distressed by the prospect that if the experiment successfully blocked radio communications, it would cut off radio astronomy on earth for an indefinite period. The Starfish experiment (1962), in which nuclear bombs were exploded in the Van Allen Radiation Belt, was also disturbing to many scientists, who felt that the United States government was inviting unforeseeable effects upon the electromagnetic fields surrounding the earth.

The first concern of people and governments with the outer space environment is that it not be used in a manner deleterious to health, safety, or the advancement of knowledge on earth; the second concern is to increase human understanding of the universe. A third concern, shared by fewer people, involves the possible use of space as a human habitat. The more conservative view of possibilities of human activities in space foresees the use of manned vehicles for merely temporary occupation in connection with medical treatment and materials processing, including manufacture of precision articles such as ball bearings.[36] There is also international scientific interest in direct astronomical observation of the universe beyond the visually distorting atmosphere of the earth. The Hubble telescope has already greatly extended the human view into deep space. The space laboratory and the space shuttle program of the United States National Aeronautics and Space Administration are first steps toward the possible uses of space for these and similar limited purposes.[37]

An extension of these uses would be the establishment of bases on the surface and perhaps under the surface of the moon for carrying on a variety of scientific and commercial activities. Inasmuch as the composition of the moon appears to resemble that of the earth and to contain needed minerals, possibly in substantial quantities and accessible to mining technologies, the extension of resource development activities to the lunar environment (and beyond to the asteroids) might ultimately be feasible.[38] Imaginative architects have already designed moon habitats to permit a relatively permanent occupation of the surface of the moon.[39] If the cost of payloads to the moon could be radically reduced, it has even been speculated that concentrated nuclear wastes from the earth might be stored there in a weather-free environment until future advances in technology would permit their utilization for human purposes.

A more radical proposal regarding the use of outer space is the construction of permanent space habitats—artificial environments simulating terrestrial conditions. The most vigorous proponent of extraterrestrial artificial habitats has been Professor Gerard O'Neill of Princeton University and a group of people known as the L-2 Society.[40] Their proposals envisage construction of very large artificial environments capable of containing soil and atmosphere and providing sustainable ecological conditions.[41] It has been conjectured that the populations would reproduce and continue in existence in these habitats indefinitely.

It has been suggested by some enthusiasts that space colonization could solve the problem of overpopulation on earth. This speculation would appear to have little foundation in either demography, economics, or ecology, and to be unwarranted with respect to the limited foresight and forbearance of human beings. The human ambition to be free from terrestrial restraints has been expressed in myth and fiction, but the bonds are not only those of the earth itself but are built into the human fabric, which is after all derived from the earth. Thus human physiological and psychological dependencies, as they now exist, constitute a major barrier to humanity's adventure in outer space regardless of whatever technology we may devise to take us to distant points in the solar system.[42]

Electromagnetic Environment

Coexistent with the outer space, atmospheric, biospheric, and inorganic terrestrial environments is a fifth environment, which, although usually invisible to humans, has a growing influence on their welfare. It is the electromagnetic environment that surrounds the earth, permeates the atmosphere, and is a phenomenon of the terrestrial planet itself. It might be regarded as a special aspect of the environment of radiant energy, including cosmic radiation, that envelops the earth. As of today, the principal use made by humans of the electromagnetic environment is communications. An international organization, the International Telecommunications Union (ITU), has been created to facilitate international cooperation for this purpose. The ITU is concerned with international telecommunications generally, but it gives particular emphasis to considerations of jurisdiction and equity among nation-states.[43]

Human control over natural electromagnetic phenomena is, however, very limited; use of its various aspects is strictly on nature's terms. Geomagnetic storms, for example, change the electrical characteristics of the ionosphere, disrupt radio transmission, telephone and telegraph cables, electrical power-lines and transformers, and satellite-to-ground communications. In an increasingly electronically connected and automated world, major fluctuations in the electromagnetic environment may have serious international consequences, particularly during times of diplomatic or military crisis.[44]

A study by the National Research Council of Canada observed that the use of radio frequency (RF) transmission has grown almost exponentially. During World War II, transmissions at microwave frequencies became more numerous, adding to the electromagnetic fields in the atmosphere that now may begin to pose problems as a potential hazard.[45]

For the present, the principal international policy questions involving the electromagnetic environment concern the allocation of radio frequencies and the deployment of satellites for communications and remote sensing. Should microwave transmissions of energy from satellites or the moon become feasible, a new international political issue would arise. Health effects of electromagnetic radiation would be cause for apprehension and concern.

Following the launching of Sputnik and American successes with orbiting satellites, it became apparent that a new and internationally significant technology had emerged for which some form of international agreement and control would ultimately be required. In 1989 it was estimated by the National Aeronautics and Space Administration that more than 6,917 satellites were in orbit around the earth, utilizing electromagnetic waves for observation or communication.[46] Unmanned satellites now provide an effective means for observing and monitoring environmental changes in the atmosphere and on the surface of the earth. Weather satellites have markedly improved the accuracy of both long-range and intermediate weather forecasting. The use of satellites for recording changes in the terrestrial environment associated with agriculture, deforestation, desertification, urbanization, plant disease, air and water pollution, and natural catastrophes has not only identified emergent environmental problems but has assisted surveillance and enforcement of law and policy. Remote sensing has now become a significant tool of environmental monitoring and ecological research and of international policy and administration.[47] It has been proposed that a third kind of satellite, for the transmission of energy from the sun to the earth through microwaves might be in part one answer to the need for a continuing source of energy in the future.[48]

International policies in relation to artificially orbiting satellites have related more to functions than to technology. Thus, the International Maritime Organization (IMO) is concerned with satellites as aids to surveillance and communications over the oceans. Meeting in London in September 1976, forty nations adopted a convention and operating agreement to establish the International Maritime Satellite Organization (INMARSAT), with IMO as its international supervisory agency.[49] INMARSAT went into operation in 1982. Meanwhile, cooperative arrangements had been negotiated between the United States Marine Satellite System (MARISAT) and the European Space Agency (ESA), which on 20 December 1981 itself launched MARECS-A, the first Maritime European Communications Satellite.

The principal organization for international telecommunications by satellite is INTELSAT, the International Telecommunications Satellite Organization, officially established 20 August 1971 on the basis of a prior agreement among nineteen countries that set up interim arrangements for a global commercial communications system.[50] The advent of satellites for communication and observation has given rise to a number of international controversies with environmental overtones. A case in point involving INTELSAT is the conflict between proponents of submarine cables and of satellites for international communications. Because of previous invested interest in submarine cables, because of the desire to maintain multiple channels for communication, and because the state of the art is not sufficient to prevent occasional interruptions caused by electromagnetic interference with satellite communications, several communications corporations and national governments favor the continued use and development of cable systems. But influenced by the substantial investment in satellite systems and the obligation of participating nations to pay proportionate shares of the costs and maintenance, opposition to further development of submarine cables has arisen among some members of INTELSAT.

On 3 May of 1974, sixteen European countries holding interest in the TAT-6 cable system meeting in Spoleto, Italy, issued a document entitled "Europe-North America Telecommunications Service—Principles and Policy Development, 1980-1990: The European Views." This declaration known as the Spoleto Document urged continued use of submarine cables for diverse patterns of communication, declaring that cost comparisons between cable and satellite communications were unreliable.[51] The divided interests of nations in communications technology thus extend to uses of the outer space environment and the deep sea bed. As Marvin S. Soroos has pointed out, the radio spectrum and the geosynchronous orbit have moved from essentially technical subjects for decision making among developed nations to issues of limited resources in the international commons.[52] For the present, international concern has focused principally on remote sensing from outer space, notably on photography from satellites.[53]

The United States Earth Resources Sensing System, originally known as the Earth Resources Technology Satellite Program (ERTS), was initiated in 1972. Renamed the LANDSAT program in 1974, it undertook international cooperative projects undertaken or completed in numerous countries and several international organizations on a wide range of applications of remote sensed data.[54] LANDSAT photographs have been useful in developing data on the wide range of environmental and natural resource matters: crop yield forecast, forest management, water resource management, geological surveys and mineral exploration, marine resource assessment, coastal engineering, and identification of potential areas of natural hazards. Although the benefits of the LANDSAT program have been widely diffused with minimal costs to cooperating nations, some nations regard remote sensing as an infringement of their sovereignty. After prolonged debate in the United Nations, no internationally accepted principles covering the use of survey satellites have been adopted.[55]

Ricardo Umali, deputy director general of the Philippines Natural Resource Management Center, has summarized the sovereignty issues as follows: (1) the need for countries operating satellites to obtain prior approval from other nations before surveying them, (2) the right of states to obtain access to all information collected about their country by another country's satellites, and (3) obligation to secure prior consent of a country before transferring information collected regarding it to a third country.[56] But whether agreement regarding satellites and the uses of the electromagnetic environment can be reached remains conjectural in a world in which the technological capabilities of nations vary greatly. The fifty-three-member UN Committee on the Peaceful Uses of Outer Space (COPUOS) has sought a greater role in the uses of outer space, but its Third World members are not well prepared to contend effectively with representatives of the industrial states that have dominated the ITU.

In 1977 an Ad Hoc Committee on Remote Sensing for Development convened by the United States National Research Council recommended that "the United States government should declare soon that remote sensing systems constitute, in effect, an international public utility destined for international governance."[57] But considering the investment by some 120 nations in various aspects of outer space technology, and because of the highly restrictive policies of some nations regarding aerial photographs of restricted areas and certain military and scientific installations or acquisition of other information such as crop estimates that might be politically disadvantageous, establishment of an international coordinative authority is unlikely. An obvious exception is telecommunications, where limitations of accessibility and the obvious necessity for order make regulation essential and hence acceptable.

The controversies over remote sensing and telecommunications were also issues in the larger Third World agitation over the collection and dissemination of information in general. Between 1975 and 1985 an effort was made through UNESCO to establish "The New World Information and Communication Order." The International Commission for the Study of Communication Problems was established and a series of reports issued.[58] The ensuing debate, which included more than the electromagnetic spectrum (e.g., news reports and cinema), did little more than trigger the temporary withdrawal of the United States and the United Kingdom from UNESCO.

The Oceans

The earth has sometimes been described as the watery planet. Nearly two-thirds of its surface lies beneath the sea. Long a realm of fascination, fear, and mystery to humans, the oceans have only recently been penetrated through technologies that have permitted scientists to probe their inner depths and to investigate their origins and understand their complex dynamics.

It was in the warm and shallow margins of ancient seas that life is believed to have originated. The oceans continue to be indispensable elements in the planetary life support systems, being major sources of free oxygen in the atmosphere, sinks for carbon dioxide, and major regulators of the earth's temperature. Climate and weather are powerfully influenced by them and their interaction with land surfaces and the atmosphere. From the marine commons, humans have obtained food, clothing, and organic materials for the making of tools, ornaments, and other artifacts. From the marine commons, people have also obtained salt and, increasingly, a wide spectrum of other chemicals. Although the sea has been a barrier between communities, it has also provided a major means for transportation. Throughout history people have used the energy of the wind for seaborne commerce, but the sea itself may be a major source of energy in the future. Lastly, the sea has been a repository for natural and man-made wastes, acting in some respects as a vast laboratory transforming, but sometimes preserving, the chemical and physical character of materials in its watery depths.

Humanity's efforts to penetrate the mysteries of the oceans, to benefit from their many utilities, and to protect against their dangers have stimulated a vast array of sciences; almost every science has, in some respect, a concern related to the oceans. Collectively these sciences are included under the umbrella of oceanography or the both more and less inclusive designation of marine science and technology.

The investigation and use of the sea and its resources have stimulated many national governments to establish special agencies for the investigation or promotion of marine science and technology. One of the earliest of these was the Institute for Marine Studies (now the International Hydrographic Organization, or IHO), established by the Prince of Monaco. The National Oceanic and Atmospheric Administration of the United States is one of the larger national efforts. There is today a very large and complex international infrastructure for oceanography and marine science and technology, parts of which are described in the following discussion.[59]

Excluded from the discussion here, for instance, have been regional arrangements. These are dealt with elsewhere (notably in Chapter 8), especially in connection with such essentially landlocked international waters as the Mediterranean Sea, the Baltic Sea, the Black Sea, the Persian Gulf, and the North American Great Lakes, which are under the control of the surrounding nation-states.

Agencies and Programs

The institutional infrastructure for marine science and technology is complex, which is explained partly by the multidisciplinary nature of its subject matter, but equally perhaps by the historical circumstances under which international agencies and programs were established. Although an international organization or regime with comprehensive responsibility for or jurisdiction over the oceans has been proposed, none as yet has been created, though one may emerge through implementation of the Law of the Sea Treaty. International action has thus far been largely restricted to scientific investigation, resource conservation, commercial affairs, and navigational safety. From the middle 1960s and into the 1970s a variety of proposals were discussed in the United Nations and in other international gatherings regarding the necessity and possible structuring of an ocean regime. But the growing antagonism between the developed and less developed nations and the extension of national jurisdiction to a two hundred-mile limit delayed agreement for general international jurisdiction over the oceans.

Beginning in 1958 the United Nations held a series of conferences on the Law of the Sea, leading to the signing of a convention at Montego Bay, Jamaica, on 10 December 1982. Although there was general agreement on most of the terms of the treaty, it was nevertheless objectionable in certain details to the principal maritime powers, which declined to ratify it. Over the years the treaty has generated a large mass of commentary, which, however, must be regarded from a policy perspective as speculative because the effects of the treaty as it becomes operative cannot entirely be foreseen.

No analysis in detail will be provided for the United Nations Convention on the Law of the Sea, which is the most comprehensive intergovernmental agreement concerning the international commons. This treaty, comprising 320 articles and 9 annexes, was signed by 119 states, but at the end of the 1980s it still lacked nearly half of the ratifications needed to put it into effect. By December 1993, 158 states had signed on and a sufficient number had ratified for the treaty to enter into force on 16 November 1994.[60]

Exemplifying the necessity for cooperative interaction in studying the problems of the oceans is the Joint Group of Experts on the Scientific Aspects of Marine Pollution (GESAMP), an advisory body nominated by its sponsoring agencies (UNEP, IMO, FAO, UNESCO, WMO, WHO, and IAEA). Its principal task is to provide scientific advice on marine pollution problems to the sponsoring agencies and to the Intergovernmental Oceanographic Commission (IOC). Like many United Nations programs, GESAMP is carried on by specialized working groups, which consider such issues as (1) hazards of harmful substances carried by ships, (2) the interchange of pollutants between the atmosphere and the ocean, (3) implications of sea bed exploitation and coastal area development for marine pollution, (4) the biological effects of thermal discharges in the marine environment, and (5) monitoring biological variables relating to pollution.[61]

Several of the United Nations specialized agencies have functions relating to the oceans, either to facilitate oceanographic research, to assist commercial uses, or to persuade nations to adopt environmental protection measures. Although it is questionable whether these organizations actually make international policies, they provide mechanisms for obtaining the information necessary for collective national action. They also provide forums in which international decisions and cooperative efforts can be undertaken.

The International Maritime Organization (IMO) has responsibilities for protective measures in shipping and navigation and, particularly, for obtaining national cooperation in controlling oceanic pollution, especially by oil. As previously noted, the World Meteorological Organization (WMO) is involved in research that involves the oceanatmosphere interface and has little direct policy relevance. The International Telecommunications Union (ITU) is concerned with submarine cables and with satellites to assist marine navigation; it is obviously an instrument of international policy.

The principal agency in the United Nations system for the scientific investigation of the oceans is the Intergovernmental Oceanographic Commission (IOC) established in 1960 within the general structure of UNESCO.[62] Through its programs at least ninety nations cooperate to promote scientific investigation with a view to learning more about the nature and resources of the oceans. IOC functions are extensive and complex; they include information, direct services, and especially exploration and research. The commission's activities characteristically involve collaboration with member national governments, other UN agencies, and such nongovernmental groups as the ICSU Scientific Committee on Oceanic Research (SCOR).

Under IOC auspices a number of broad-based and long-term research efforts have been organized. The International Decade of Ocean Exploration, 1971–80 (IDOE), was organized within the framework of a larger IOC initiative—the Long-Term and Expanded Programme of Oceanic Exploration and Research (LEPOR).[63] The IDOE was intended to stimulate investigations under LEPOR and especially to improve the capability of member states to participate in oceanographic research.

Continuing services sponsored by IOC include the International Oceanographic Data Exchange (IODE), an information coordination program consisting of Responsible National Oceanographic Data Centres (RNODCs) throughout the world. Major inputs to this system originate in the National Oceanographic Programme (NOP) and the Declared National Programmes of Cooperating States.[64] For example, announcements of oceanic research cruises or of planned marine investigations are made available to all concerned or interested investigators.

Among IOC initiatives and coordinative activities have been the International Indian Ocean Expedition (1959–66), a joint effort of UNESCO, FAO, WMO, and twenty-three countries, and the International Cooperative Investigation of the Tropical Atlantic (ICITA; 1962–65).[65] Directed by an international coordination group, ICITA was carried out in three phases: Equalant I, II, and III. Physical oceanographic data collected during the first two phases were published in 1973 in a comprehensive atlas. The IOC has its own international coordinative body on marine pollution, the Global Investigation of Pollution in the Marine Environment (GIPME), which studies pollution, its causes, and its effects upon the natural ecology of the ocean.[66]

Other environment-related agencies with which the IOC has been associated are the Indian Ocean Biological Centre established by an agreement between India and UNESCO, the Tsunami Warning System in the Pacific (ITSU), and the Integrated Global Ocean Station System (IGOSS). The last is a cooperative undertaking with WMO for monitoring the ocean through automated buoys equipped with detectors for the gathering and transmission of information at the ocean-atmosphere interface. IOC also collaborates with WMO and UNEP in the Global Environmental Monitoring System (GEMS), which maintains surveillance on marine pollution. UNESCO/IOC also sponsors training programs relating to oceanography and fisheries, and since 1973 has published *International Marine Science Newsletter* jointly with FAO.[67] UNESCO, in addition, provides technical assistance to developing countries and encourages educational developments in oceanography.

UNESCO was also the lead agency in the promotion and coordination of the International Hydrological Decade (IHD) (1965–74) and the International Hydrological Programme (IHP) that followed, to advance understanding of the earth's hydrosphere.[68] IHD brought together a number of working groups on global water problems. Among topics studied by working groups were floods and their computation, water balance, techniques of nuclear investigation, hydrological maps, and hydrological forecasting. The marine environment has also been a concern of UNESCO's Man and the Biosphere Programme (MAB)—for example, the ecology of island ecosystems and the effects of human activities on deltas, estuaries, and coastal zones. These UNESCO/IOC activities inform national and international environmental policies but do not directly shape them. Nevertheless, they represent agreement among nations to advance understanding of the oceanic environment cooperatively.

The Food and Agriculture Organization has maintained a major concern with living resources of the sea, its activities relating especially to the sea as a source of food. Its interests, however, extend also to the protection of the marine environment and ecosystems. In 1955 an International Technical Conference on the Conservation of the Living Resources of the Sea was convened by FAO in Rome. Subsequently, in 1958 the Geneva Convention on Fishing and Conservation of the Living Resources of the High Seas was adopted by the first United Nations Conference on the Law of the Sea.[69] FAO's interest in the ocean has included the promotion of a number of regional fisheries councils or commissions. These bodies have generally represented the fisheries industry, and critics have alleged them to be more concerned with dividing the earth than conserving the stock. Continuing concern with threats to fisheries is exemplified by the 1970 FAO Technical Conference on Marine Pollution and Its Effects on Living Resources and Fishing.[70]

In addition to the specialized agencies the International Atomic Energy Agency (IAEA) has been concerned with pollution from radioactivity in the oceans and the dumping of radioactive wastes.[71] The IAEA, however, has not engaged in the monitoring or attempted control of radioactive wastes, confining its efforts largely to maintaining a register of the releases of radioactive substances into surface waters, and establishing guidelines regarding the transportation and dumping of radioactive wastes.

The United Nations Environment Programme is involved with the monitoring of oceanic pollution through its Global Environmental Monitoring System (GEMS), which maintains surveillance of a broader range of environmental phenomena than the oceanic.[72] UNEP is, of course, broadly concerned with the protection of the oceans and associated marine life and ecosystems as essential elements in the planetary biosphere. Its Oceans and Coastal Areas Programme, which includes the formerly separate Regional Seas Programme, is an important instrument for the development of international ocean policy, a role which may be expected to grow, independently and in cooperation with other members of the UN system.

There are, in addition, a large number of intergovernmental and nongovernmental organizations outside the UN system with major oceanographic concerns. Oldest among these is the International Council for the Exploration of the Sea (ICES), established by northwest European maritime countries following meetings in Stockholm in 1899 and in Christiana (Oslo) in 1901.[73] Since its inaugural meeting in 1902, headquarters of the council have been in Copenhagen. Its membership now consists of the major maritime states of Western Europe, but cooperative relations have been established with Canada, the United States, and Japan. The council collects and publishes data supplied by its members, and sponsors cooperative and planned research organized through committees in relation to both particular species of fish and geographic areas.

A somewhat similar arrangement, the International Commission for the Scientific Exploration of the Mediterranean Sea, was established in 1919, and was supplemented in 1949 by the General Fisheries Council for the Mediterranean sponsored by the Food and Agriculture Organization.[74] International organizations for fisheries will presently be considered under the heading "Living Resources."

An obvious focus for international scientific activities relating to the oceans is the International Council of Scientific Unions. Several of the unions, particularly the International Union for Geodesy and Geophysics (IUGG), the Scientific Committee on Oceanic Research (SCOR), and the Scientific Committee on Problems of the Environment (SCOPE) have been involved in various international oceanographic enterprises in cooperation with the several United Nations specialized agencies.

Completing this review of oceanographic organizations are a number of international nongovernmental scientific organizations, among which are the International Association of Biological Oceanography (IABO), the International Association of the Physical Sciences of the Ocean (IAPSO), and the International Hydrographic Organization (the successor to the bureau established by the Prince of Monaco and concerned particularly with the mapping of the ocean). International nongovernmental organizations such as Greenpeace and the Cousteau Society have been actively involved in various aspects of ocean policy, especially regarding living resources of the sea. Of course there are also many other agencies and programs at the national level, both governmental and nongovernmental, with major ocean-related missions that interact internationally.

Living Resources

Since antiquity, marine resources have been objects of exchange, rivalry, and contention among peoples and nations. Competition has heightened as more has been learned about the living resources of the sea—and those derived from the once living (e.g., corals)—and as more uses have been found for them and for their products. But of greater significance for international environmental policy is the role of the oceanic ecosystems in relation to the total balanced functioning of the biosphere. Awareness of these relationships is only beginning to penetrate the consciousness of people and their governments, and scientific understanding of the total biospheric system must be regarded as still rudimentary.

Possibly the least noticed and probably the most important living creatures in the sea are the phytoplankton. These microphotosynthesizers are believed to be the earth's principal emitters of atmospheric oxygen—essential to the continuing vitality of the biosphere. As previously noted, threats to the phytoplankton through chemical pollution or depletion from other causes could pose a serious threat to the continuation of most forms of animal life on earth. These minuscule plants collectively constitute a basic element in a vast and complex food chain, which reaches its apex in human consumption. Ignorance or disregard of the interconnecting character of the oceanic food chain endangers its integrity. Modern technology is capable of seriously depleting various sectors of the food chain and thus affecting all other sectors of the chain with which they are linked.

Overexploitation of the smaller marine animals such as krill in the Antarctic Ocean or anchovies in the Humboldt Current threatens depletion in food-chain linkages that could drastically affect the survival capabilities of other species. Exploitation of krill could have the effect of depleting the major food resource for baleen whales. Overfishing of anchovies has been a factor in greatly reduced numbers of birds in sea bird colonies on the Peruvian coast that were once a major source of agricultural fertilizer.

Although fibrous products such as sponges and pharmaceutical substances such as iodine are derived from living resources of the sea, the principal resource in relation to human needs has clearly been fisheries. A complicated and confusing international institutional structure has developed through which governments have attempted to cooperate in managing both the catch from marine fisheries and the taking of marine animals such as dolphins, seals, and whales. Too often, however, this collaboration has not gone far beyond a division of spoils. In recent years, the concept of maximum yield from the fisheries has begun to yield to the concept of sustainable yield. Nevertheless, not all nations are persuaded that international sustained yield agreements should obstruct opportunities to provide immediate and abundant cheap protein for their populations and profits for the commercial seafood business.

The growing pressure of world population on food supplies and the advancement of fisheries technology seriously stress the stocks of many marine fish and raise crucial questions regarding the future of marine fisheries. During the years 1950 to 1970 the world catch of fish grew nearly 12 percent annually, from twenty-one to seventy million metric tons.[75] Since 1970, however, the world catch has either declined or remained constant. Meanwhile world population has continued to grow, and with it incentives to exploit the available fisheries. Outside of extended national jurisdiction, and except as covered by enforceable international agreements, wild stocks of fish are regarded as a common property resource and hence are open to exploitation by any nation with the requisite technical capabilities. The use of immense drift nets by Japanese trawling for fish in the mid-Pacific has aroused indignation among conservationists in other countries. Governments protest, but at the beginning of the 1990s there was still no comprehensive international law safeguarding the living resources of the sea.

Frederick W. Bell identified three policy areas in which failures of national and international foresight threaten the future of world fisheries: overcapitalization of the fishing fleets and overexploitation of the fisheries; environmental deterioration, especially the degradation and destruction of coastal and estuarial habitats which are the breeding and feeding areas for many fish; and of less general but growing significance, the use of fishery resources for recreational purposes.[76]

The principal coordinating agency for the world fisheries has been the FAO's Department of Fisheries and especially its supervisory council, which consists of senior fishery officers of most of the nations with a significant economic interest in marine fisheries. At least twenty-five bilateral or multilateral organizations have been formed for purposes of international agreements for management of fisheries stocks. These bodies, generally described as "fisheries commissions," undergo organizational changes from time to time, and efforts have been made through the FAO's Department of Fisheries to obtain, where needed, a better coordination among their activities.[77] The organization of world fisheries has evolved into a complex regime that is almost incomprehensible to anyone but experts.[78] Moreover, organizational and jurisdictional problems have been further complicated by extended national jurisdiction to two hundred-mile offshore limits, and there are many regional issues, for example in the western and central tropical Pacific.

Among the more important of these international regional bodies involved with scientific study and management are the Inter-American Tropical Tuna Commission (IATTC), the International Commission for the Conservation of Atlantic Tunas (ICCAT), the Indian Ocean Fisheries Commission (IOFC), and the Indo-Pacific Fisheries Council (IPFC). While the IOFC and the IPFC are concerned with all species of fish within their geographic areas, the IATTC and the ICCAT deal primarily with tuna, tuna-like species, and tuna bait fish. The IATTC is also concerned with several species of dolphin-porpoise found in association with tuna and, frequently, inadvertently captured in the course of fisheries operations.

Observers of the performance of the international fisheries bodies point out that the more effective among them have capabilities for undertaking serious research in marine science and fisheries biology. James Joseph and Joseph Greenough point out that international fisheries commissions with independent research staffs have generally succeeded in carrying out the data collection and research necessary to recommend management measures for conservation of species under their jurisdiction. They concluded that organizations with scientifically qualified personnel such as the Inter-American Tropical Tuna Commission, the International Pacific Halibut Commission, and the International Pacific Salmon Fisheries Commission have made effective use of scientific management and recommendations.[79] The effectiveness of these regional international arrangements is, however, limited by ecological circumstances over which governments have no control. Referring to the tuna and associated fish, Joseph and Greenough observe,

> These species undertake extensive migrations throughout their lives. . . .
> They do not recognize man-made boundaries, nor are their natural geographic distributions comparable with any set of boundaries that might be established by nations in the foreseeable future. . . . If management is to be applied successfully on such stocks, that management must apply over the entire stock range.[80]

For some species, management must be applied over entire oceans; for others, perhaps on a global basis. Measures applying to only a portion of their range will not in the long run be effective. In April 1995, a UN Conference on Straddling Fish Stocks and Highly Migratory Fish Stocks proposed a draft agreement to overcome this difficulty. The treaty was quickly adopted and is expected to provide protection to fish stocks on the high seas.[81]

The difficulty of governing the international exploitation of the ocean commons and of enforcing responsibility for preservation of the living resources of the sea is most dramatically illustrated by the plight of marine mammals, notably the porpoises and the great whales. The history of efforts to protect the whales has been described in other places and will not be repeated here. But beyond the image of the planet Earth itself, the most poignant symbol of the world environment movement has been the whale. It became a representation of the nonhuman living world at the United Nations Conference on the Human Environment. The survival of whales continues to be an object of international anxiety and concern and perhaps better than any other issue illustrates the difficulty of reconciling the multiple conflicting interests and values of nations in the management and protection of the biosphere.[82]

Although less dramatic than the plight of the great whales, the fate of certain species of dolphin has been a matter of domestic and international concern.[83] Notably in the eastern Pacific, some species of tuna are regularly found in association with large numbers of dolphin. The reason for this association is not fully understood, but a consequence has been that large numbers of dolphins have been killed as a result of becoming entangled in the nets of tuna fishermen. In addition, the slaughter of dolphin by Japanese fishermen who regard them as competitors has aroused indignation among conservationists and humanitarians. Inadvertent entrapment of the animals has created difficulties for fishermen who at considerable time and some risk have found it necessary to remove them from the nets.

These and other concerns led the United States Congress in 1972 to adopt the Marine Mammal Protection Act (PL 92-205). Under the terms of this legislation, commercial fishing operations were required to adopt techniques and equipment which would reduce hazards to dolphins. And, of international relevance, the act declared it unlawful to import into the United States fish caught in any manner "proscribed for persons subject to the jurisdiction of the United States." The complex international character of tuna fisheries and markets has made rigorous enforcement of the Marine Mammal Protection Act difficult. Nevertheless, the American market is of great significance to the industry, and hence action by the United States affects the fisheries industries of other countries. As a consequence, the tuna-porpoise problem was placed as early as 1976 on the agenda of the Inter-American Tropical Tuna Commission, and measures have thereafter been explored to see if international agreement can operate to resolve the problem more effectively than unilateral national action.

In 1990 the United States invoked the Marine Mammal Protection Act to ban the import of yellow-fin tuna caught in the eastern tropical Pacific because the incidental catch of dolphins was not in compliance with standard set by the act. Mexico, one of the noncomplying nations, appealed to the General Agreement on Tariffs and Trade on the grounds of unfair treatment. In May and June 1991, a GATT panel found that the U.S. embargo was not justified under the terms of the agreement. This episode led to opposition by environmental and animal protection groups to the reauthorization of the GATT and the adoption of NAFTA in 1993.

At the beginning of the 1980s international policy respecting whales and other marine mammals remained ambiguous, but with evidence of a tilt toward protection. To stave off criticism and to resist pressure for a moratorium on commercial whaling proposed by the United States at the 1972 Stockholm Conference, the International Whaling Commission (IWC) has periodically revised and extended conservation procedures. But many observers have found the commission to be ineffective because of the reluctance of countries with whaling interests to undertake enforcement measures.[84]

The performance of the IWC has been characterized by anomalies. In 1978, as a result of an inquiry into whales and whaling commissioned by the Australian government, that country ceased whaling but retained its membership on the IWC in order to improve the commission's discharge of its responsibilities. At the Thirty-first Annual Meeting of the IWC (London, July 1979), the United States appeared to depart from its hitherto consistent protectionist position to reject the unanimous recommendation of the IWC Scientific Committee that the bowhead whale be totally protected. The objection by the United States, however, was less inconsistent than it appeared. American opposition had been to commercial whaling; the taking of bowheads by Inupiat Eskimos in Alaska was for subsistence, and an integral part of their traditional culture. A subsistence quota finally agreed upon at a special meeting of the IWC in Tokyo, 5–9 December 1977, was small—not exceeding eighteen.[85] Environmental conservationists were divided in opinion, some working with the Eskimo to obtain a subsistence quota. There is obviously a great difference in the taking of a limited number of a depleted species by a small traditional community, and the floating whaling factories of the Russians and Japanese.

At the 1979 London meeting, the Republic of Seychelles proposed the establishment of a whale sanctuary in the Indian Ocean, a three-year moratorium on the commercial taking of all sperm whales, and stiff measures to stop pirate whaling, which had become an international scandal.[86] Investigations by the People's Trust for Endangered Species had revealed open evasions of both IWC regulations and the policies of national governments. Operating under flags of convenience of non-IWC countries, the pirates appeared to be largely Japanese-financed, using crews and ships of multinational origin. The Seychelles proposal would have closed the ports of all IWC members to non-IWC whalers, refused them insurance and registration, and forbidden their own nationals to serve with them.

Efforts of many years climaxed on 23 July 1982. With a twenty-five to seven vote and five abstentions, the IWC voted to end commercial whaling for five years (1986–90) following a three-year phaseout period.[87] But the director general of the International Union for Conservation of Nature and Natural Resources, Lee Talbot, emphasized "the need for continued surveillance to assure that the expressed will of this year's Commission is observed."[88] Major whaling nations voting against the phaseout or zero quota included Japan, the U.S.S.R., Brazil, Peru, Norway, Iceland, and South Korea. In an emergency resolution, the European Parliament called upon Japan, the U.S.S.R., Norway, Chile, and Peru to respect the decision taken by the IWC, and environmental groups in the United States were proposing that the government deny fishing privileges in the oceans within the United States' two hundred-mile economic zone in the event of noncompliance. Brazil, Iceland, and South Korea indicated their intention to abide by the IWC commercial whaling ban.[89]

Marine mammals generally, including the pinnipeds, have been among the creatures most exploited and abused by humans. But these warm-blooded animals have now aroused human interest and compassion, in part because of the high intelligence and learning ability of several of the species.[90] At least one species, Steller's sea cow, localized in the North Pacific was exterminated in supplying ships with fresh meat. The elephant seal and the sea otter (not a pinniped) were nearly extirpated from the west coast of North America. The threat to survival of whales and fur seals and the resulting international response has been noted. But harvesting of harp seal pups off the coast of Newfoundland and the slaughter of porpoises by Japanese fishermen have aroused international protest at nongovernmental levels. Because of their more localized habitat, walruses and manatees have attracted less international concern, although they are endangered through habitat degradation and misapplied technology. Domestic policies have drawn some governments into international controversy—for example, the already discussed implications of the U.S. Marine Mammal Protection Act and the possibility of international sanctions against noncompliance with the IWC whale moratorium. The Canadian government has also been under external pressure to end the commercial taking of harp seal pelts.

Progress toward international protection of marine mammals has been slow. But the change in public attitudes since the 1960s has been dramatic. The

United States government responded to this growth of concern in 1972 with the Marine Mammal Protection Act (PL 92-205), and the Marine Mammal Protection, Research and Sanctuaries Act (PL 92-532). Research on marine mammal behavior, especially of whales and dolphins, has clearly stimulated worldwide public interest. Public awareness has been notably enhanced by the television presentations of Jacques Cousteau and others. A comprehensive four-volume work on *Mammals of the Sea* has been published by FAO, and in cooperation with UNEP a joint plan of action was adopted (1984).[91] Obtaining the cooperation of national governments continues to be difficult—economic and sociopolitical considerations being their paramount considerations. But at least in the United States and Canada, the economic returns from whale watching are now substantially greater than the profits once gained by whaling.

In addition to wild stocks of fish and marine mammals there are other living resources in the sea that are, or should be, subjects of international concern. Marine turtles, for example, are among the more seriously endangered forms of sea life. In response to growing concern, the U.S. Department of State hosted a nongovernmental World Conference on Sea Turtle Conservation, which took place on 26–30 November 1979. More than three hundred participants from forty countries attended and developed a conservation strategy for the most particularly threatened species.[92] Shellfish and crustaceans found in the shallower coastal waters have suffered greatly from land-based pollution. Animals such as sponges and corals have great commercial value and have suffered not only from pollution but also from overexploitation.

An old but expanding area of marine resource development is marine pharmacology. The search for drugs and other chemical products from the sea seems certain to find a place on the agenda of international environmental policy in the very near future.[93] Commenting on prospects for food, drugs, and scientific discoveries from marine research, David H. Attaway of the U.S. National Sea Grant Program observes that "increasingly rapid developments in marine natural products . . . emphasize the importance of the sea's rich storehouse of novel organic molecules," which possess "many interesting and some useful bioactivities and pharmacological properties."[94]

To compensate for the loss of natural resources in the sea and to institute more effective management and control, techniques of aquaculture or mariculture have been developed. Although these technologies hold promise for future food supply, they still present many economic and ecological problems. Farming and conservation of relatively small-sized fish in enclosed areas have been carried on in various parts of the world for a long time (e.g., in Hawaii), but whether the farming of fish in the unbounded ocean can be managed in today's world is dubious.

Legal and institutional arrangements have been sought to ensure that combinations of avarice, ignorance, and necessity do not push the exploitation of marine life to a point of irretrievable depletion and, for many species, extinction. Virtually all international scientific and legal studies of the problem of world marine resources have concluded that some form of international order is necessary to preserve and protect marine life. But short-range national interest and international antagonism have defeated comprehensive efforts in this direction. Political scientist Edward Miles concludes that while major changes are possible in the face of impending or recently experienced "disaster," the most realistic hope for the near future is an incremental growth in the effectiveness of bodies of specific and limited jurisdiction, such as the Food and Agriculture Organisation Committee on Fisheries.[95]

Still, advances in technology may alternatively permit more effective surveillance of the high seas and the policing of extended economic zones. Fisheries adviser to Papua and New Guinea, Peter Wilson, told the 1980 Pacific Telecommunications Conference that satellite surveillance may permit the monitoring of fishing vessels.[96] To do this, he said, would require the cooperation of the Pacific island states. Failure of fishing vessels to cooperate could result in their being barred from territorial waters and economic sanctions taken against their home countries. And the director general of INTELSAT, addressing the same conference, reviewed the entire field of international satellite technology in the Pacific Ocean region, pointing out its significance for new economic and trading relations and its utility in assisting regional cooperation.[97]

Mineral Resources

The extent of mineral wealth within and beneath the sea is largely conjectural because the total area is vast, relative to the area explored. Petroleum, of course, has been discovered and exploited on the outer continental shelves of many of the continents. In certain places on the deep sea floor, mineral concentrations in nodule form provide a potential source of such metals as cobalt, copper, and manganese. The waters of the oceans also contain dissolved minerals, some of which may be significant when, or if, technologies permit their economically feasible concentration and transformation.

Deuterium is an example; it is abundant in seawater and some day may be a significant factor in nuclear fusion technology. For more than four millennia, salt (sodium chloride) has been extracted from seawater largely through evaporation, and bromine and magnesium have been obtained from this source for several decades. There are hopes that boron, aluminum, lithium, fluorine, and uranium might be extracted commercially, although the highly dilute form in which these minerals exist would make their extraction costly.

Deposits of oil and gas, sulfur, coal, tin, and potash are found beneath the outer continental shelf and may be mined by tunneling from shore or by shafts into the seabed. The greater number of known mineral deposits are in areas immediately offshore and are not necessarily subject to international policy. Metal-bearing muds exist in different locations on the sea floor, but difficulties of access and available supplies on land diminish immediate attention as sources for economic development.

A relatively recent problem for the marine environment has been the development of offshore and oceanic structures in the form of deep sea terminals, oil drilling platforms, oil storage tanks, nuclear reactors, solar collectors, airports, and, conjecturally, residential cities.[98] These and other facilities now in use or in contemplation become man-made islands. They are of sufficient environmental concern to have been addressed in Article 60 of the UN Convention of the Law of the Sea. A controversy over a proposal to sink a drilling platform to the bottom of the North Sea is discussed in Chapter 8.

The possible effects of these structures upon the marine environment are not certain, but some apparent risks are involved, especially the threat of pollution. There are also risks to navigation and personal safety, and there are predictable visual and economic effects. In newly independent Belau (Palau), for example, opposition to a mid-Pacific superport proposed by Japanese and American interests caused the issue to be brought before the Security Council of the United Nations.[99] But petroleum reserves are projected to diminish rapidly in the twenty-first century, thus, the supertankers, superports, and massive platforms may be short-lived phenomena—unless they are adapted to other uses. In this case, other problems of marine environmental policy may be anticipated. The offshore incineration of toxic substances presents one such problem, there being fears of spills at sea and uncertainties regarding effects of fallout.

It is, however, the prospect of obtaining mineral wealth from the deep sea bed that has been the primary occasion for considering the ocean as an international commons.[100] The deep sea bed has been beyond the territorial jurisdiction of any nation, and until recently there was no technology adequate to permit its exploration or exploitation. That technology now exists, but is concentrated in a few high-technology societies—these also being the principal consumers of mineral wealth.[101]

A report prepared in 1977 by a group of experts for the United Nations Office for Oceanic Economics and Technology observed that "[o]f the mineral resources of the deep sea-bed, the only major occurrences likely to be exploited in the immediate future are the ferromanganese nodule deposits found in all major oceans and in many different types of sedimentary environments." [102] The report found that "nodules have generated commercial interest not only because of their content of nickel, copper, cobalt and manganese, but also because of a variety of other factors, such as the pressures exerted by political forces." The nodules are unevenly distributed in all of the major oceans and vary over short distances both in quantity and the quality of their metallic values.

In the General Assembly of the United Nations and in the United Nations Conferences on the Law of the Sea, the notion has become endemic that the mineral resources of the deep sea bed are the heritage of all mankind and afford a means of economic assistance to the poorer nations of the world.[103] The proposition has been advanced that development of resources on the deep sea bed should be subject to administrative control and licensure by an international body, with the proceeds being to a large extent distributed among the poorer nations for development purposes.[104]

Under such an arrangement a small number of relatively wealthy, high-technology nations having the technical and financial capability to undertake sea bed mineral developments could proceed, but the conditions under which they operated and the distribution of returns would be determined by an international authority dominated, very likely, by a preponderance of poor, technologically less developed nations.[105] Given the divergent viewpoints, the antagonisms, and the lack of direct responsibility characterizing international bodies of the United Nations, the prospect of high-risk venture in deep sea mining does not at present seem promising.

Not all marine geologists are persuaded that the deep sea bed is in fact a treasure trove of mineral wealth. "Untold wealth," to be sure, but the point that they make is that the actual amount of mineral wealth and the feasibility of obtaining it has not been told because the facts are not known. It does not follow that "untold wealth" is in fact vastly abundant wealth. Indeed, the accessible reserves may turn out to be relatively modest, and returns from their exploitation if divided among the poorer nations of the earth could turn out to be relatively insignificant, and under the circumstances, insufficient to provide financial incentive for deep sea industry to develop.[106]

The Third United Nations Law of the Sea Conference authorized creation of an international authority to control mining of the deep sea bed. Such an authority could hardly be effective without consent of the three major Western economic powers that as of 1989 had declined to sign the treaty: the German Federal Republic, the United Kingdom, and the United States. In 1989 nearly half of the sixty ratifications needed to put the treaty into force were still missing. Deep sea mining continued to be the obstructive issue. On 14 August 1987, however, five countries—Belgium, Canada, Italy, the Netherlands, and the Soviet Union—had signed an Agreement on the Resolution of Practical Problems with Respect to Deep Sea Mining Areas. There was also an exchange of notes between the United States and the parties to the agreement. Then, in June 1994, the UN secretary-general reported to the General Assembly that informal consultations had led to an agreement that appeared to remove obstacles to general adherence to the 1982 Convention on the Law of the Sea. The agreement relating to the implementation of Part XI of the convention (deep sea mining) was adopted on 28 July 1994 by a vote of 121, including the United States and all other industrial states.[107]

Marine Pollution

Because the sea is the ultimate sink for planetary wastes, the vast increase in human activities upon and around it, and the resulting pollution, are a matter of growing international concern.[108]

The effect of these pollutants is particularly damaging in enclosed seas such as the Baltic, the Black, and the Mediterranean. Among the sources of marine pollution are shipping, dredging, mining and drilling, ocean dumping of hazardous materials, and such land-based activities as wetland and estuarial filling and effluent discharges. Some pollutants, such as oil, originate both at sea and from land. Land-based pollutants are transported both by air, as gases and particulate matter in the atmosphere, and by water, as sewage and industrial wastes in rivers. Major sources of land-based wastes are resource development and agricultural activities resulting in the discharge of sediments, pesticides, and fertilizers containing sulfates, nitrates, and chlorinated hydrocarbons. Off the shores of continents and larger islands, domestic sewage has become a serious pollutant. Finally, radioactive wastes offer a present and more serious potential threat to the conditions of the ocean.[109]

Oceanic currents and upwellings make it difficult or impossible to contain toxic chemicals discharged into the sea. For example, pesticides used in Africa have subsequently been identified off the coasts of India and the Caribbean islands. And industrial vinyl chlorides originating in Germany have been found in currents following the coast of Norway and extending to fisheries off the coast of Iceland. Chemical discharges are particularly dangerous because many resist biodegradation and tend to concentrate in living food chains beginning with the oceanic plankton and concentrating in the flesh of edible fish and invertebrates.

Marine pollution from ships has been a matter of growing international concern since the mid-1920s. In 1926 the United States proposed total prohibition of oil discharge from ships. An international conference in Washington during that year declined to go that far, agreeing only on specified zones within which discharge of oil was prohibited. The International Convention for the Prevention of Pollution of the Sea by Oil, signed at London in 1954, applied the zonal concept.[110] As the inadequacies of this approach became apparent, the treaty was amended in 1962, 1969, and 1971. Ratification of the treaty and amendments has been slow; many countries continue to abstain.

In 1973 IMCO (now IMO) convened the International Maritime Pollution Conference in London to draft a treaty dealing comprehensively with all polluting discharges, including oil, from all types of ships.[112] The 1973 International Convention on the Prevention of Pollution from Ships (MARPOL) and its 1978 protocols have superseded the 1954 convention, are comprehensive, and for that reason have been opposed by shipping interests and national governments responsive to their objections.[111] At last on 1 October 1982, with ratification by Italy, the number of adhering states was sufficient to put MARPOL into force a year later (2 October 1983).[112]

Following several disastrous coastal oil spills (notably the wreck of the *Torrey Canyon* off England in 1967), the International Legal Conference on Marine Pollution Damage was convened by IMCO in Brussels in 1969.[113] As an outgrowth of this conference, which related more to coastal damage than to the high seas, the following treaties were signed, the first two of which were ratified with little delay. The first was the International Convention Relating to Intervention on the High Seas in Cases of Oil Pollution Casualties;[114] the second was the International Convention on Civil Liability for Oil Pollution Damage,[115] both becoming effective in 1975. A supplementary document, the International Fund for Compensation of Oil Pollution Damage signed on 10 December 1971, became operative in 1978, having obtained sufficient ratifications to become effective.[116] The conclusion follows that although there is international recognition of the harmfulness of ocean pollution, national economic interests have remained powerful enough to handicap or delay action that would impose cost, inconvenience, or responsibility upon shipping interests.

Although wastes dumped into the oceans constitute a small percentage of the total pollution load of the marine environment, the visibility of the practice and its harmful effects upon localized areas have led to extensive international concern. Two major international treaties deal with ocean dumping. They are the Convention on the Prevention of Marine Pollution by Dumping of Wastes and Other Matter (London Convention of 1972)[117] and the Convention for the Prevention of Marine Pollution by Dumping from Ships and Aircraft (Oslo Convention of 1972).[118] The London Convention is generally applicable to the high seas, whereas the Oslo Convention pertains to the northeast Atlantic. Impetus for these international conventions was provided by a report entitled *Ocean Dumping: A National Policy*, issued October 1970 by the U.S. Council on Environmental Quality. Many of the recommendations contained in this report were written into U.S. statutory law with the passage on 23 October 1972 of the Marine Mammal Protection, Research and Sanctuaries Act.

The greatest practical effect of these treaties is to be found in their annexes, which specify what materials may be dumped, and in the means for their implementation. The 1954 London Convention, for example, permits low-level radioactive waste to be dumped in the northeast Atlantic, a point of contention with environmentalists. Under the Oslo Convention, however, in cooperation with the commission established by the Convention for the Prevention of Marine Pollution from Land-Based Sources (Paris Convention of 1974), a joint monitoring group (JMG) was established to assess the effectiveness of measures taken under these treaties.[119] "If nothing else," a survey of ocean dumping treaties concludes, "these conventions which have come into law since 1970 provide a more precise understanding of the meaning of 'reasonable regard to other state's use of the ocean' as it applies to sea dumping in both a global and regional sense." [120] Thus, these conventions are steps toward responsible management of the international commons of the oceans. As noted in Chapter 8, fifteen European states and the European Community in 1992 signed a Convention for the Protection of the Marine Environment of the North-East Atlantic. When in effect this convention will replace the Oslo and Paris conventions.[121]

Ocean Levels

The level of the seas is known to have changed significantly over geologic time. These changes have occurred primarily during glacial (ice ages) and interglacial or warming periods. When water is frozen into continental ice sheets ocean levels fall and continental shelves now submerged would be exposed. During ice ages, the Bering Strait and the English Channel became dry land valleys. Presently the earth is in a relatively warm or perhaps interglacial period. But massive ice sheets remain over Antarctica and most of Greenland. Were significant proportions of this ice to melt the level of the oceans would be raised, but to what height and consequences has elicited differences of opinion. The worst case would be a catastrophe of global dimensions entailing a multiplicity of policy problems: environmental, social, economic, ethical, and political.[122]

A plausible rise of two hundred feet could inundate much of the most heavily populated areas of the earth, including most of the world's largest cities. Particularly vulnerable are low-lying islands now comprising the Alliance of Small Island States, whose members in the UN General Assembly are numerous.[123] These possibilities are among the reasons for the rapid ratification of the Global Climate Change Treaty, whereas the Montego Convention, or Law of the Sea, Treaty, although in effect since 1994, has many abstainers. How the world would cope with a massive rise in sea levels that would inundate whole countries is, of course, uncertain.

Also uncertain are the causes of ice ages. Interactions between atmosphere and oceans involving oceanic cooling are widely accounted to be factors. Particulates in the atmosphere reflecting solar radiation back into space and reducing the amount of radiation reaching the oceans, if present for sufficient time, might cool the ocean surface sufficiently to deprive the atmosphere of the warming it now receives from the ocean surface. Thus, winter snows would accumulate on land to form glaciers and continental ice sheets. Ocean levels would fall.[124]

Open Spaces: Amazonia and Siberia

Although not common spaces in a legal or political sense, two areas of the earth aside from Antarctica have elicited international concern. They are Amazonia and Siberia. Both are the last large, thinly populated places on Earth, and there are significant ecological values being threatened by inappropriate and often unsustainable development. A comparative study of these areas and the problems they present for international policy has been prepared and edited by Michael Bathe, Thomas Kurzidem, and Christian Schmidt of Johann Wolfgang Goethe University under the title *Amazonia and Siberia: Legal Aspects of the Environment and Development in the Last Open Spaces* (1993). As with the common spaces, the issue of national sovereignty versus international concern is basic.

Politics of Transnational Spaces

The need for an international law for transnational or "common" spaces is a consequence of human preemption of the habitable spaces of the earth, the expansion of advanced technologies, and the globalization of commerce, communication, and transportation. Experience has demonstrated that resources that are held in common without restriction have been vulnerable to destructive exploitation. Garrett Hardin's article "Tragedy of the Commons" (1968) has become its classic description.[125] Awareness of the tendency has led to institutional arrangements designed to correct it, for example, in mutual forbearance written into the 1959 Antarctic treaty and in establishment of the International Whaling Commission to prevent depletion of the common stocks of whales. Yet we have seen that in these instances success is uncertain, owing to the unwillingness of some nations to forgo immediate advantage for long-term universal benefit. Appeals to international or intergenerational equity, to the long-range advantage of humankind, or to the rights of nonhuman organisms have been largely ineffectual in obtaining international consensus on the management of the global commons.

The firmest base for international agreement with respect to protection of the environmental quality of the commons and the management of its resources would appear to be scientific findings derived from mutually agreed-upon methodologies. Yet scientific facts, which exist independent of political ideologies or calculated economic interests, are of themselves insufficient for protective and managerial policies for the commons. And there are philosophical and ideological postures that deny the possibility of factual objectivity. Unless related to a holistic concept of humans-in-biosphere along with a realistic assessment of a sustainable future for mankind, facts alleged to be scientific may as easily be used to support exploitative policies as to protect the integrity of the earth.

No foundation for a political agreement inspired by scientific findings is likely to be firmer than the scientific data supporting it. Hence it is of great importance to the future of the international commons that research into the implications of humanity's relationship to it be pushed forward in an evolutionary, biospheric context. This is especially true of research pertaining to the atmosphere and the oceans, revealing the many interrelationships between these elements and the living world that are important in the consideration of policies conducive to maintenance of life on earth. The international commons will inevitably be a major focus for international environmental policy, but whether, or how, nations may cooperatively manage their activities in the international commons remains to be seen.

To date, the rights and obligations of nations in the international commons have been systematized by legalists as "the law of international spaces." [126] International or supranational regimes have been proposed for the administration of these areas. At present, however, all conventions governing the commons are partial, specialized, or contingent. Only Antarctica has a multipurpose yet restrictive governing regime, but it has no general administrative control. As of 1995, the nations do not appear very close to agreement on any common regime for international spaces. Yet international treaties in effect or proposed indicate an ad hoc evolutionary process toward more inclusive structures of international management. Experience with such arrangements as those for international telecommunications, for the protection of Antarctica, and for management of fisheries and the protection of whales points up the factors that make for success or failure in international cooperation in the commons. Learning from experience and applying its lessons offer the best prospects for a rational management of international spaces in the immediate future. The international commons of the atmosphere, the oceans, Antarctica, and outer space are areas where nations have greater latitude to discover ways of working together, to identify common interests, and to shape institutions of mutual convenience than is possible where their own territorial jurisdictions are involved.

Questions of national and international responsibilities toward the common spaces throw into high profile the assumptions and priorities that determine the policies of governments today. Allowing for inevitable shades of opinion and ulterior purposes, two points of view are identified. The first, represented by U.S. official pronouncements, looks to the commons as a source of minerals and energy and a challenge to technoeconomic development. Monetary wealth and strategic advantage appear to be the primary motivations, with advancement of science being an incidental beneficiary. The second viewpoint, also basically economic, looks to the commons as a potential source of welfare for poor nations. The "heritage of mankind" concept is invoked as a rationale for sharing spoils among the "less developed nations," with investment and operations provided by the developed nations. In neither viewpoint are ecological or environmental values determining factors. The "heritage" concept, no less than the goals of resource exploitation, assumes an earth created for mankind's material benefit. In both perspectives, the three major policy questions are, How may the inevitable environmental damage of development be reduced to "acceptable limits"? By what formula should monetary returns (if any) be distributed? And how, if at all, should costs be shared, and with whom? Ultimately the question of cost must be confronted. Here a fourth viewpoint is relevant, although it is not yet politically dominant in international relations. According to this view, preservation of the ecological or biogeochemical integrity of the common spaces is a priority to which all human uses should conform. Ecological preservation issues could profoundly affect the choice of financial commitments, perhaps calling for a reevaluation of costly expenditures in outer space, Antarctica, and the deep sea bed.

A political battleground for competing priorities has been staked out in the United States, where successive governments have proposed grandiose plans for space science and technology while simultaneously withholding or reducing funds for many aspects of environmental protection. Billions of dollars to build permanent space stations, colonize the moon, and send manned spacecraft to Mars could preempt funds needed to cope with serious environmental problems on earth. Efforts to preserve endangered ecosystems, rectify past environmental errors, and maintain environmental quality are underfunded. Basic science could be deprived of resources if funds were diverted to the instrumental uses of science needed to support manned exploits in space. Relative to environmental expenditures on earth, and by any measure of public expenditure, the costs of the proposed space programs could be difficult to justify.[127]

The international commons are a proving ground for testing humanity's beliefs about its relationships with the earth and the biosphere. If nations cannot cooperate effectively in areas belonging to none of them, if they cannot refrain from predatory extensions of national economic and military ambitions into the common spaces, it is difficult to see how they will fulfill environmental commitments already made in which economic and ideological interests are in conflict. Indications of how these questions will be answered have not appeared during the closing years of the twentieth century. The call by the president of the United States on 11 May 1990 to plant the American flag on Mars by the year 2019 reveals a mind-set that does not augur well for commitments on behalf of planet Earth.[128] In a world in which not all things are possible, some fundamental choices must be made. These choices may shape the future of humanity on earth.

10

Sustainability: Population, Resources, Development

The need for environmental policy in the modern world is in large measure a consequence of a human urge for growth and development and of the
means of obtaining them. In an infinite environment, such ambition might
seek realization indefinitely, but the earth provides no infinite environment
for material growth. Many limitations and hazards of the natural world may
be overcome through science and technology, but the geophysical earth itself
cannot be expanded. Most human needs and wants can be fulfilled from the
biosphere, provided that its capacities for renewal are not impaired. Indeed,
humankind can assist nature to improve upon what evolutionary processes
have produced, and the advances in agriculture, horticulture, manufacturing,
and medicine have afforded numerous examples. Civilization is a composite
result of these advances, but experience shows that we tend to disregard the
requirements necessary to sustain those advances. When humanity's exactions
exceed the capacity of the earth to provide, a breakdown in the life-support
systems follows, the quality of life is diminished, and civilization itself may be
jeopardized. Decline and disintegration of many past civilizations has been
attributed to excessive stress upon their environments.[1]

With the growth of industrialism in the nineteenth and twentieth centuries, the effects of the unbridled exploitation of nature became increasingly evident. The conservation and public health movements were societal
responses — but both were limited by restrictive perceptions of the human-
environment relationship. Until the emergence of the environmental movement during the last third of the twentieth century, problems of resources and
environment were characteristically approached piecemeal, without recognizing their interconnections and interactive effects. While earlier conservation
efforts were not all failures, they fell short of dealing effectively with the problems that they addressed. A basic cause of their inadequacy was the failure to
see nature *whole*. The science of ecology, although anticipated earlier, did not
really emerge until well into the twentieth century. Without ecology, broadly
defined, there was no adequate foundation in science for an environmental

movement. And there was insufficient understanding of the interdependence between the artificial systems of human society and the natural systems of the earth to provide an adequate rationale for an ecologically informed economics.

There is an economy of nature, and there is an ecology of industrial society, both of which are aspects of a total complex living system that Kenneth D. Boulding described as "ecodynamic."[2] The size, complexity, and mobility of this system defies conceptualization. It cannot be dealt with as a whole, but in dealing with its parts its ultimate wholeness must be considered and interactive effects so far as possible ascertained. In pursuit of this necessity the computer has become an indispensable tool. In his book *World Dynamics*, Jay Forrester pointed the way toward enlarging the ability of the human mind to discover hitherto obscured interrelationships and to foresee trends and consequences.[3] The advent and evolution of the computer has permitted an immense advance in environmental science and in the basis for environmental policy.

When we humans undertake to shape policies to guide our behavior in relation to the planetary life support system and the quality of our environment, we are limited by our languages and our mental capabilities. We cannot think of everything simultaneously. Our minds cannot process interactive dynamic complexities as effectively as can powerful computers. We must perforce deal with many of our problems segmentally but without losing sight of their interconnectedness with the rest of life. Specialization has made possible great advances in science-based technology, but it entails the risk of suboptimization—of overdoing one good thing at unnecessary cost to other equally good purposes.

The Sustainability Concept

Initially known as "ecodevelopment" the concept was "rechristened" as sustainable development in the first *World Conservation Strategy* (1980), subtitled *Living Resource Conservation for Sustainable Development.*[4] The concept was adapted as the central theme of the World Commission on Environment and Development (Brundtland Commission) and emphasized in its report, *Our Common Future* (1987).[5] It then became the integrating concept in the 1992 UN Conference on Environment and Development. A voluminous literature has now been published and numerous organized efforts have been initiated to refine and publicize the idea.[6] Sustainable development is defined as the meeting of today's true needs and opportunities without jeopardizing the integrity of the planetary life-support base—the environment—and diminishing its ability to provide for needs, opportunities, and quality of life in the future. As policy, its objective is to prevent developmental or demographic "overshoot" that could result in a socialecological collapse or irreversible impoverishment. Sustainability—or the ability to continue—is then the integrating theme underlying this chapter. The focus will be on the two principal elements in the pursuit of sustainability—population and natural resources (including energy sources). Development is conceived as a process that must be managed with foresight and prudence if the goal of sustainability is to be attained.

Population

The environmental problems of people are direct consequences of their numbers and behaviors; and the impacts of many natural hazards and disasters are attributable to people being in the "wrong place" at the "wrong time." Humans do not generate earthquakes, volcanoes, violent storms, droughts, and floods, although human actions may exacerbate their effects. These events are characteristically localized and transitory; they may be severe, but their impacts are limited, and they seldom diminish the sustainability of society as a whole. Environmental impoverishment and degradation induced by continuous imprudent social behavior are more serious. The environmental issues and problems described in the introduction to this book are chronic and progressive and induced by human activity. Trends in soil erosion, devegetation, toxic contamination, climate change, and resource depletion, among others, are consequences of behaviors inconsistent with requirements for the sustainability or renewability of natural systems. The proclivity of human population to expand is one cause and, in some cases, the foremost cause of environmental impoverishment.

Of ecological misbehaviors, the most obvious, but often vigorously denied, is the overstressing of the environment by sheer numbers of people. When equipped with extractive and environment-modifying technologies, the potential effect of numbers is correspondingly increased. But numbers alone, in excess of the capacity of an environment to sustain their demands upon it, will lead to its degradation and impoverishment. The multiplication of numbers in limited environments has led to the disastrous social and ecological conditions prevailing in many of the world's poor countries. Case studies in man-made human misery are provided (among other countries) by El Salvador, Haiti, Somalia, and Bangladesh. Where the resilience of an environment is exceeded by human demands for space, sustenance, energy, and waste disposal, problems that are simultaneously human and environmental are created. These problems of malnutrition, famine, mental retardation, disease, social unrest, political violence, and exaggerated vulnerability to natural disasters may be removed by either of two means—by a Malthusian decline in human numbers to a level sustainable in a ruined environment or by instituting environment-sustaining policies that will eventually balance the needs and wishes of a population with the environmental capacity to sustain them.

Achieving this balance is the most difficult of environment-related public policies. The beliefs and values of people concerning human numbers are rooted in the distant past when human circumstances differed radically. Population as an issue includes family formation, birth or conception control, and the quality of life itself. The issue is meshed in a cultural context, bound up with considerations of social status (personal, familial, ethnic, and communal), property and legal rights, religion, and concepts of morality and social justice. A denominator common to these issues is social power. In human society, numbers "count"—no less in egalitarian democracies than in expansionist despotisms, no less in market economies than in Marxist socialism. But the relationships among numbers, social power, and environmental impact are complex. Apparent one-to-one cause-effect relationships are apt to be deceptive.

More than any other aspect of public policy, population in its fullest sense is clouded by emotional and intuitive miasmas. Its issues are more often characterized by rationalization than by reality-tested rationality. Although solutions to the problems require technological assistance the "population problem" per se has no technical solution. It is fundamentally a problem of social philosophy, and the nature of its *political* reality, as distinguished from its *ecological* or *environmental* reality, is a construct of the human mind. Political rhetoric may simulate virtual reality obscuring the actual realities of life on earth.

This book is not the place for an in-depth treatment of the population issue—even when restricted to its environmental aspects. There is a large literature bearing upon it, selective references to which are listed in the notes.[7] Acceptance of the reality of the problem of excess numbers has come slowly to modern people generally and to governments—perhaps too slowly to prevent tragic consequences for humanity and disaster to the biosphere. Some countries are now trying to control population growth, but results thus far have not been especially encouraging. One of the more hopeful developments has been the growing acceptance in the Third World of the United Nations population program and consideration of population and family-planning issues at United Nations conferences on population and environment.

The conjectured stabilization of human numbers at perhaps ten billion in the mid-twenty-first century may well be fallacious. Certainly it is hardly optimistic, because in the absence of miracles of science and technology such numbers probably cannot coexist with undomesticated nature. Moreover, excess populations have traditionally sought to alleviate crowding and poverty through mass migration. If many more billions are added to the world's population, social and international violence may be expected to increase beyond the present high levels. The biosphere would inevitably be degraded and impoverished by the pressures to accommodate more people than the natural systems of the world could indefinitely support, even with the aid of technology. With few exceptions, population growth today worsens every problem of natural resources management, energy, biological diversity, environmental contamination, and the quality of life. At its most rational, environmental policy today would seek an immediate end to population growth and in many countries a fallback to lower levels. At present this scenario seems unlikely.

The principal agent for the development of national population policies has been the United Nations Population Fund (UNFPA), established in 1969. During the past two decades, the number of Third World countries with population policies has risen from perhaps six to more than seventy. The mandate of UNFPA is to promote awareness both in developed and developing countries, of the social, economic, and environmental implications of national and international population problems, of the human rights aspects of family planning, and of possible strategies to deal with them in accordance with the plans and priorities of each country. International Conferences on Population met in Bucharest in 1974, Mexico City in 1984, and Cairo in 1994. A World Population Plan of Action was adopted at the Bucharest Conference and amended at Mexico City. In the intervening years the relationship between population and environment attained recognition, reflected in paragraph 8 of the Declaration on Population and Development adopted at the 1984 conference in Mexico:

> In the past decade, population issues have been increasingly recognized as a fundamental element in development planning. To be realistic, development policies, plans and programmes must reflect the inextricable links between population, resources, environment and development. Priority should be given to action programmes integrating all essential population and development factors, taking fully into account the need for rational utilization of natural resources and protection of the physical environment and preventing its further deterioration.

The issue of population has now been clearly joined to policies for environment and development. The rapid and excessive increase in human numbers in relation to resources and environmental resilience defeats the objectives of development. Few people would be satisfied with sustainability at a low level of economy and quality of life. Policies for use and conservation of energy and natural resources cannot be effective unless consistent with policies for population and development, and these policies, to be sustainable, must take account of the capacity of the environment for continuing renewal.

That the population issue has not received more emphasis in this book is in no way indicative of its lack of importance. Excessive population growth in relation to resources, environment, and social stability is the principal obstacle to sustainability and the integrity of the biosphere. The International

Conference on Population, held in Cairo, September 1994 made an incremental advancement in international policy.[8] A United Nations Commission on Population and Development (a subsidiary body of ECOSOC) was established to implement the Programme of Action adopted at Cairo, as reported in the *U.N. Chronicle* 32, no. 2, of June 1995. Control of population growth by whatever means, including voluntary individual choice, is countered by a formidable opposition based on moral, religious, political, and even economic dogma. Nevertheless, and despite the myopia of people and governments, population growth will become a major international policy issue in the twenty-first century.

Resources and Energy

There is a large literature dealing with national and international policies for natural resources and their availability, shortages, flows in international trade, and future prospects. A representative but obviously selective list of references on international resources policy is provided in the notes.[9] Concern here is limited to international environmental implications of resource policy; it does not include resource policy per se, some aspects of which are considered in other chapters.

Before considering the relevance of resource issues for environmental policy, it is necessary to clarify the meaning of resources. Much of the debate on resource policy is confused by semantic differences attributed to the term "natural resources." Policy perspectives may diverge greatly, depending on whether the earth is viewed primarily as a storehouse of resources or as a biosphere. Natural resource managers and economists tend to see nature primarily as a source of raw materials for the production of goods and services and a sink for wastes, whereas environmental protectionists and ecologists look to nature as the complex, interrelating systems whose integrity is essential to the continuity of life on earth. But, as Bruce L. Bandurski has observed, nature is most productive in the long run when sound ecological principles are observed.[10]

In conventional terminology, a resource is a substance or phenomenon amenable to economic "use." Top soil, fresh water, wildlife, and even landscapes may be regarded as natural resources. But the term is more often applied to materials (e.g., fuels and metals) that can be extracted from their environments and treated as commodities. A natural substance or raw material is not a resource unless it has some recognized economic value. This value arises through the demand that created it, and through the technology by which raw material is extracted, processed, and distributed. Thus it has happened that a material that is considered a valuable resource in one society may have a minimal value or no value at all in another. From an environmental perspective, therefore, it is better to consider most natural resources as materials or sources of energy rather than as commodities until they actually enter the economy of human use.

A natural resource may be a property or component of a material (e.g., of a mineral-bearing ore), and the economic value of the material may vary with its quality, utility, and accessibility. Technology and politics influence accessibility and thus affect resource supply. Resources that are known and accessible to present technology at economically feasible costs are called *reserves*. Much of the debate about the abundance or scarcity of resources actually pertains to reserves. Discoveries of new sources of accessible supply and improved technologies for harvesting or extraction increase the size of reserves while adding nothing to the actual volume of resources (e.g., petroleum, timber, silver). Thus, the sheer quantity of these materials may be decreasing simultaneously with increasing reserves and a corresponding temporary decline in their monetary price. Increased market availability and lower prices may induce complacency regarding the possibility of future scarcity and thus encourage excessive or wasteful consumption. Discovery of synthetic substitutes for particular natural resources may reduce the value of reserves and depress their price. Nevertheless, it is more often with reserves that government policies are concerned than with natural resources per se.

This explains why some optimists will inform the public that natural resources, that is, reserves, are increasing and that it is foolish to talk of resource scarcity.[11] Similarly people are told that the world will never "run out" of mineral resources because an infinite amount of them are embedded in rock and sea water and all that is necessary to extract them is an economically feasible technology and the required amount of energy.[12] Listeners usually fail to ask when the requisite technology will be in place and what the required amounts and sources of energy will be. Processed minerals, including metals, are not necessarily destroyed through use, although oxidation (rust) or corrosion may return their elements to the earth. In the absence of a materials management, stockpiling, and recovery system, they may be randomized, dissipated throughout the environment, and rendered unavailable for further human use.

Material resources, like energy, are subject to the laws of thermodynamics and tend in the long run to degrade with use to less accessible states, as has been persuasively argued by Nicholas Georgescu-Roegen. Economists, such as Kenneth Boulding, have broadened the scope of economics to include relevant information from the natural sciences and have provided a more realistic set of assumptions regarding natural resources policies than that offered by conventional abstract economic theory. Yet, national and international policies for natural resources and energy continue to be dominated by concepts that regard ecology and physical science as largely irrelevant to economic policy.[13]

The energy factor frequently determines whether development is economically feasible. Ecological feasibility has seldom been seriously considered. If the monetary costs of obtaining a raw material exceed its commercial price, development may not occur. But under certain circumstances, if the amount of energy required to obtain a raw material exceeds the amount of energy obtainable from that resource, development may nevertheless proceed, as when the form of the energy obtained (e.g., gasoline from oil shale or wheat from energy-intensive mechanized farming) is usable in a way that the energy used to obtain it is not. Energy technology also becomes a determining factor in resource development and usage; commercially feasible atomic fusion or concentrated solar energy equaling the potential of fossil fuels might occasion a reassessment of resources and reserves as presently estimated.

The natural resource policies of any society or nation are therefore determined by the interaction of three factors: the raw materials obtainable within its jurisdiction or otherwise accessible to it, the degree of its technological advancement, and the materials-consuming activities of the society. To the extent that, and as long as, there is complementarity among these factors for any given resource, no policy problem arises directly attributable to resource scarcity. But few countries today enjoy such complementarity. Even large continental jurisdictions such as the United States, Russia, or China depend upon external (and not always reliable) sources for certain critical materials. Moreover, resource development may indirectly give rise to environmental problems (e.g., oil pollution at sea, massive deforestation, or damage from mining).

These considerations suggest that an exclusively economic definition of natural resources will misdirect rather than serve environmental or sustainable economic policy. All economic systems depend for their viability upon certain materials that are difficult to turn into commodities, like uncontaminated air, fresh water, and some nutrients (including trace minerals). A basic error of modern industrial society has been to discount or undervalue material resources—especially those that are renewable if properly managed. Governments are nonetheless gradually coming to recognize that the most beneficial resource development policies will be environmentally conserving.

Natural resources become objects of national and international policy to the extent that the material well-being and environmental conditions of some countries are affected by resources policies of other countries or of international corporate enterprise. Although national governments have often asserted a sovereign right to do as they please with resources under their jurisdiction, there are several philosophical and some legal and technological limitations on their options. There is, for example, the generally accepted principle of international law that a nation should not permit the use of its territory in ways detrimental to its neighbors; a corollary is the less generally accepted proposition that a nation ought not impoverish itself and thus indirectly burden or diminish the world by the mismanagement or destruction of its own natural (or cultural) resources. There is abundant evidence that neither of these limitations on sovereignty is taken seriously by most governments. Nevertheless, the first principle has been invoked in extraordinary cases—for example, in threatened diversion of international rivers and in atmospheric testing of atomic devices. More practical limitations on sovereignty lie in a nation's technical ability to exploit its resources or to control their exploitation by others (e.g., other nations or multinational corporations). The extent of resource development may also be influenced by the world market—a consideration that has induced raw material producers to form international cartels such as the Organization of Petroleum Exporting Countries (OPEC).

Because of the many kinds of resources and the specificity of their respective values and properties, policies regarding resource use are not easily generalized. Attempts by governments to coordinate resources policy assume some principle or value criterion applicable to diverse resources. But principles such as "the greatest good for the greatest number in the long run" fail to define what is "good." Moreover, not all people agree that all resources should be developed, nor what objects should be regarded as resources. In many traditional societies natural sites—forests, mountains, or lakes, for example—are valued for spiritual or aesthetic reasons. Native Americans have often opposed mining and forestry on tribal lands regarded as sacred, and in the United States legislation has been enacted to protect such religious sites from exploitation.[14] Ecological objections have increasingly been advanced against developmental projects of high environmental risk.

The uneven distribution and consumption of natural resources is a major cause of international antagonism over resource development. This circumstance appears to be of greater relevance to international relations and to international economics than to environmental policy. But it affects environmental policy to the extent that it exacerbates international antagonisms, encourages ecologically unwise resource exploitation, and provides national governments with alibis to cover mismanagement of their natural resources and environments. The geophysical fact of uneven resource allocation is a factor, but not the only factor, underlying the demand by the less-developed of the United Nations member countries (the so-called Group of 77) for a "New International Economic Order."

Resource Issues

Few international or environmental policy issues are more confused than those dealing with the status, prospects, implications, and equities of the possession and utilization by nations of natural resources. The issues in public debate are semantically confused and emotionally heated. The facts, such as they are, are often misused; data critical to many important policy questions are missing, and many that exist have a proprietary status—that is, they are the property of certain governments or private resource developers and are not generally made fully available for public information.

The circumstantial differences between the developed and less-developed resource-producing countries inevitably have led to differences in outlook and policy. Each group has invoked sovereignty, economic theories, and national interest in relation to access to markets and to raw materials. International agreements to reconcile differences and stabilize international commerce in raw materials have been initiated, for example the Lomé Conventions between the European Community and less developed African, Caribbean, and Pacific island states. But the task of building environmental considerations into national and international, economically oriented resources policy has only begun—more in intent than actual practice.

The environmental considerations implicit in international resource issues can be made explicit through an examination of three very broad, but basic, resource questions: Is the world resource base of modern society adequate to sustain its future? What are the international implications regarding the use and management of that resource base and its proved reserves in light of the best available knowledge (e.g., is there a limit on the "right" of a nation to mismanage its resources)? Are the benefits of the developed resources of the world distributed equitably, according to appropriate criteria of need or merit? Answers to the second resource question (on a sovereign right to control) imply consideration of the third question (on equitable distribution). And answers to both depend upon assumptions implicit in the first question (on the adequacy of the world resource base to sustain the present industrial order). One may find a variety of answers to these questions among the writings of resource economists and political ethicists. But these questions are not amenable to unequivocal answers. Extensive qualifications and refinement would be necessary in order to obtain clear and constructive analysis of the issues.

Opinions on these resource issues divide generally between two points of view. From the first or proconservation perspective, the finite nature of the earth and its material substances indicates that in the long run natural resources, at least at present utilization rates and population trends, will be insufficient to sustain modern civilization.

In contrast, the pro-utilization viewpoint regards the interchangeability of matter and energy assisted by technological innovation as permitting the indefinite sufficiency of resources. From this latter stance the notion of exhaustible resources is an unscientific delusion: human ingenuity exercised through science and high technology will be able to in some manner supplant or re-constitute whatever nature has created; the world is not about to run out of natural resources. There is, moreover, a presumption that efforts to conserve them may be unnecessary, shortsighted, and probably selfish.[15] The honesty of the environmental conservationists is also questioned—their concern in international resource development being attributed to maintenance of mo-nopoly prices or to the retarding of Third World development.

Pro-utilization advocates observe that prophecies of impending shortages have often proved wrong in the past. They point out that new deposits of min-erals continue to be discovered, that improved technologies allow the extrac-tion and processing of materials formerly unavailable, and that economies and efficiencies in such fields as timber production and agriculture have increased supplies over the years and will most probably continue to do so in the future. This perspective is often associated with a commitment to the vigorous, un-limited growth of populations, material goods, and technologies.[16] In some cases it is plausible to suspect that the optimism regarding the resource base is generated in support of an ideological commitment to unending growth.

But does it follow that while forecasts of shortages have sometimes been demonstrably wrong that the prophecies of plenty are therefore demonstrably right? If it is true that forecasts of resource abundance are sometimes used to support commitments to growth (or at least to counteract attempts to limit growth), it also appears true that advocacy of conservation and the sparing use of scarce resources accompanies a desire to prevent or reduce the marring or diminishing of the natural environment caused by exploitation. Minerals or forests conserved for future generations will not lead to environmental dis-figurement during the present. Thus persons who deeply value natural beauty and ecosystem integrity may oppose uninhibited resource development and hence utilize arguments available to prevent it.

This line of argument is not lost upon those advocates of growth and re-source development, who contend that, if resources are indeed finite, their conservation merely prolongs their future availability at the expense of the present generation. Advocates of development argue that not only is "exces-sive" conservation economically unnecessary, but that it may be politically im-moral because it deprives poor nations and people of opportunities to better their circumstances. To this line of argument advocates of resource conserva-tion respond that environmental quality *is also* a human value and that when resources are regarded as infinitely abundant their exploitation tends to be wasteful and unnecessarily damaging, diminishing not only the quality of the environment but also to the opportunities of future generations. They argue, moreover, that the exploitation of natural materials may work primarily to the advantage of the exploiter and not to society as a whole. Resource conser-vationists point to numerous examples, not only in the Third World but also in developed countries, of resource exploitation that has resulted in the poor people becoming poorer and in society eventually losing not only the material benefits of the resource but the previous quality of the environment as well.

When conservationists accept in theory the proposition that natural resources, being combinations of elementary materials, might one day be put together by humanity more readily than by nature, they also consider the environmental, energy, and opportunity costs of reconstructing nature. Or when they adopt the somewhat different proposition that man-made substitutes for natural products might extend resource availability and possibly diminish damage to the environment (although this is conjectural), they contend that the careful and conservative use of natural resources, so as to prolong their availability, buys time for substitutes to be developed and for materials already fabricated to be recycled for the recovery of needed substances. Laws to encourage resource recovery and recycling have now been enacted in a number of developed countries.[17]

In relatively poor and populous countries such as Egypt and India, almost no material goes to waste. A worn or damaged rubber tire casing that falls off a truck in a crowded Third World street may be promptly cut into sandals. But similar tire casings may be seen along stream banks and in ravines across the American countryside because it does not "pay" anyone to take the trouble of disposing of them in any other way. Analogous examples could be multiplied—discarded refrigerators, automobile bodies, and assorted household castoffs. From Third World perspectives this throwaway economy appears profligate, confirming opinions that the benefits of resource development are unequally distributed—the rich countries having more than they need or can wisely use while the poor countries, as a consequence, remain deprived.

This line of reasoning leads directly to the question whether world resources are managed prudently. The linking of resource conservation and environmental quality is based upon the proposition that resource recovery and recycling reduces environmental dereliction and material waste. Recycling industries may impose some unavoidable burdens on the environment, but these burdens usually are more containable and controllable than those which result from the exploitation of virgin materials and the random discard of resource residuals. For example, a factory to disassemble and reconstitute automobile bodies is less environmentally disfiguring than the automobile "graveyards" scattered across the countryside in the United States and other parts of the world where the automobile has become a widespread public utility.

To the extent that resource usage, recovery, recycling, and substitution have effects transcending national boundaries, resource policy becomes more than a national issue. National positions on international policies regarding natural resources are frequently contradictory. Exclusive national sovereignty over natural resources is universally asserted, yet many countries also demand that other countries refrain from adopting resource or environmental policies which might diminish the export of their natural materials.

This ambivalence regarding international policy was evident in the Action Plan adopted by the 1972 United Nations Conference on the Human Environment. While the nations asserted absolute sovereign control over their resources, they nevertheless adopted as many as 71 out of 106 recommendations variously concerned with international implications of resource management. Some of these dealt with particular natural resource issues at length and in detail. Recommendation 20, for example, consisted of a long list of subrecommendations to implement the proposition "that FAO in cooperation with other international agencies concerned, strengthen the necessary machinery for the international acquisition of knowledge and transfer of experience on soil capabilities, degradation, conservation, and restoration." Recommendation 22 called upon FAO "under its 'war on waste' programme to place increased emphasis on control and recycling of waste in agriculture."

Recommendation 24 dealt at length with various environmental aspects of forest resources and forest management, including the effects of the international trade in forest products on national forest environments and urged that "the forest protection programme recommended by the conference be facilitated through the use of remote sensing techniques employing advanced technologies such as satellites which, using different types of imagery, could constantly survey all forests." But remote sensing technology, which has advanced considerably since the 1972 Stockholm Conference, has raised its own international environmental policy issues, as noted in chapter 9. Not all nations are happy to have their natural resource potential observed by other nations' uninvited satellite eyes.[18]

During the years since Stockholm, the unprecedented cutting of tropical forests, much of it by multinational timber corporations, has led to a new series of natural resource and economic issues. Ecological and quality-of-life aspects of deforestation are considered in the following chapter. Here it should be noted that the economic aspects of international commerce in timber are at last being joined to ecological considerations in agreements such as the Third Lomé Convention between the European Community and African, Caribbean, and Pacific island states, and in the United Nations Conference on Tropical Timber (Geneva, 1983) in which FAO, UNCTAD, and UNEP were involved.[19]

By the mid-1980s destruction of the tropical rainforests, particularly in Amazonia, became the focus of intense international concern. Numerous conferences, exhibits, and resolutions were sponsored by environmental NGOs. Efforts were made, apparently with some success, to change those lending policies of the World Bank that had abetted tropical deforestation. Especially in the United States, environmental conservation groups undertook to purchase some of the foreign debts of developing countries in exchange for agreements to protect natural environments.[20] Possibly encouraging developments involving the Amazonian and Caribbean countries is reported at the conclusion of this chapter.

Recommendations made at the Stockholm and Rio conferences set, at least in principle, an international agenda regarding the protection of wildlife and the world's genetic resources and the conservation of species with medical, aesthetic, or research value. Extensive recommendations were made regarding the protection of the living resources of the sea, the conservation and utilization of water supplies, the development of information on the environmental effects of mining, and on the effects of mineral processing in relation to energy development.

Of particular concern to the developing countries, Recommendation 64 urged that "United Nations' agencies . . . undertake study on the relative costs and benefits of synthetic versus natural products serving identical uses." Addressing alleged inequities of the present economic order, Recommendation 63 called for studies to be conducted "to find out the connection between the distribution of natural resources and people's welfare and the reason for possible discrepancies." Finally, Recommendation 68 requested "the Secretary General, in cooperation with the appropriate agencies of the UN and other international agencies, to promote jointly with interested Governments the development of methods for the integrated planning and management of natural resources, and provide, when requested, advice to governments on such methods in accordance with the particular environmental circumstances of each country."

These latter recommendations lead directly to the third general resource issue—the question of equitable distribution. It is with regard to this last issue that semantic, ethical, and emotional confusion is most pronounced. Whether the uneven distribution of raw natural resources or of the capacity to develop or fabricate substitutes for them is inequitable, is fundamentally a philosophical question; it is, moreover, a question to which well-informed and thoughtful people often reach widely different conclusions.

In their eagerness for resource development many Third World countries find themselves in a quandary. Although politically nationalistic and often socialistic, these countries characteristically lack the technical and economic capabilities necessary to unilaterally develop their resources. To exploit their resources they enter into arrangements with international corporate enterprise and market-economy states. Unless these foreign investors can benefit from such arrangements there is no incentive to invest, and restrictive environmental policies could further discourage foreign developers. Nor does investment from socialist countries hold better promise of consideration for local environmental or social values.

Although neocolonialism, capitalistic exploitation, and other epithets are often invoked to explain why many of the less developed countries fail to benefit more from their mineral riches, the actual reasons appear to be more complex and less contrived. There certainly have been circumstances in which foreign firms have taken advantage of the cupidity or lack of ecological sophistication of public officials in less-developed countries. But James H. Cobbe (1979) in his book on governments and mining companies in developing countries observes that "an explanation can be based on the structures of the industries involved, the location of skills and knowledge embodied in personnel to whom the firms have better access than do the governments, and simple and reasonably plausible assumptions about the objectives of firms and governments and the constraints they face." [21] Many of the less-developed countries face severe personnel skill deficiencies and problems of access to international markets; thus they need and demand technical assistance and training programs.

In the present climate of international relations, Raymond F. Mikesell observes, the prospects of an international code for minerals development are not bright. [22] The World Bank, the United Nations, the European Community, and the OECD have considered establishing guidelines and providing insurance to stabilize relationships between multinational firms and host countries. Yet the conditions advocated by a majority of Third World countries tend to increase the obligations of foreign investors without according them protection for their investments. With the future always uncertain, most governments have been reluctant to consider propositions in which a contract reached at the outset of a mining project should be guaranteed and observed for the life of the project.

Even in developed countries, mining activities frequently raise economic and ecological problems, as occurred with the controversy over mining in a national park on Fraser Island, Australia, by the Dillingham Corporation, or between British Columbia, Canada, and the United States over the Cabin Creek (coal) and Windy-Craggy (gold) mining projects in British Columbia. [23] But it is doubtful that a relationship between a Third World country and a foreign mining firm can ever be wholly satisfactory, no matter what benefits may accrue to the host nation. The presence of foreign mining companies introduces social and economic influences—including consumption patterns, structures of compensation, and lifestyles—that may contrast sharply with those of the local population. The prospect of the foreign mining firm removing resources from a country makes it an almost irresistible target for attack by host country politicians who, harassed by many problems with which they cannot cope, are eager to divert popular discontent.

To consider means to deal more equitably and effectively with these and related economic issues, a meeting on "International Mineral Resources Development: Emerging Legal and Institutional Arrangements" was held in Berlin, 11–15 August 1981, jointly sponsored by the German Federal Republic and the United Nations Department of Technical Cooperation for Development. Case studies were presented on individual mining projects by several of the developing countries represented, notably Botswana, Colombia, Haiti, and Indonesia.[24]

Ways are being found to overcome some impediments to Third World/market-economy cooperation in resource development that is economically, socially, and ecologically acceptable to all parties concerned. Some multinational mineral and forestry development firms have employed ecologists and landscape planners to lessen or avoid adverse side effects. Some have helped developing countries to identify unrecognized opportunities for improving nutrition and living conditions. The president of one firm engaged in international resource transactions (Solvay American Corporation) expresses an attitude that appears to be growing: "Multinational operations and the western democracies must show a greater sensitivity of and flexibility toward the needs of the lesser developed countries. They must not come into negotiations with the LDCs convinced that *they* know what is best for the struggling country and naively believing that *their* solutions to the country's problems will work flawlessly."[25]

While perfect accommodation between socioecological and material-economic values is a utopian ideal, effort must nevertheless be made to approximate it. Unfortunately the almost universal commitment of national governments to continuous economic growth along with the immediate pressing necessities of their growing populations for food and jobs make the outlook doubtful. The future for the world resource base is rendered even more uncertain by the prevailing mind-set in modern society in which natural resources policy is very largely focused on fuels, minerals, and forest products, to the relative neglect of top soil and fresh water. Yet, soil and water have been foci of United Nations conferences, for example, the Water Conference at Mar del Plata, Argentina, and the Conference on Desertification at Nairobi, both held in 1977.

Possibly the most serious natural resource problem now affecting all nations is land degradation. The lack of perception regarding the significance of soil was emphasized by Edouard Saouma, then director of the FAO Land and Water Division, in a report prepared for the 1972 Stockholm Conference:

> Though not generally thought of in such a term, soil is indeed an essential support of human life, not only in relation to our food supply but also for the production of fibre and shelter. Unlike air and water, for which anti-pollution provisions are being established, the use of land is still not guided by any agreed standards. While the demand for land increases at

a very rapid rate through population growth, technological progress and industrial development, soil resources remain fixed. The maintenance of their productivity is therefore of paramount importance.[26]

Many activities in traditional as well as in modern industrial society contribute to the deterioration of soil quality and the loss of agricultural land; losses include soil erosion, loss of fertility, laterization, salinization and waterlogging, dessication, conversion to urban uses, and contamination by toxic wastes. Failure to cope with these destructive processes has had international consequences, yet the actual management of soil resources rests largely with local and private authorities. Governments have been slow to assert a public responsibility for soil resources, being more interested in maximizing current yields from agriculture and forestry than in conserving their resource base. The effects of soil mismanagement are characteristically slow, incremental, and cumulative, so that internationally significant injury may not be evident until irreversible damage has been done. Economistic policies that discount the future and, especially, undervalue renewable natural resources contribute to this process of impoverishment.

A closely related issue of comparable magnitude is the conservation and protection of ground water. Pumping of ground water for irrigation has seriously depleted this resource over large areas. This has led in some cases to desertification and in others to slumping and compacting of the land surface, thus irreversibly impairing soil productivity. As of today, most national leaders would vigorously defend the proposition that their countries have an unequivocal right to treat their soil and water as they please. It is therefore of great importance that stewardship of soil and water become a generally accepted obligation of nations, and that national governments in their own long-range interests effectively administer conserving measures.

It would be premature to be optimistic, yet a more coherent and environmentally sensitive understanding of natural resources issues appears to be developing. Since 1986, annual reports by the World Resources Institute and the Institute for Environment and Development, in collaboration with UNEP, have assessed the state of the resource base that supports the global economy. These and other assessments by the Worldwatch Institute, OECD, and UNEP, and publications such as *Gaia: A Planetary Atlas* (Myers, 1984), are presenting environmental and resources policies in their proper relationships as inseparable aspects of policies for sustainable development and quality of life. One of the more cogent analyses of sustainable resource policy is Robert D. Hamrin's *A Renewable Resource Economy* (Praeger, 1983).

Energy and Environment

The social concern with energy is pragmatic. Energy without purpose may interest the physical scientist, but where considered in relation to human society, it is viewed as a means to an end. This purpose may be no more than physical survival, but a great deal more is involved when the energy needs of modern societies are considered. Yet throughout much of the world today energy appears to be treated as if it were its own purpose—as an end in itself. The uses for which energy is sought are seldom defined more precisely than "maintaining living standards," "providing the fuel for economic growth," or "reinforcing military capability." More explicit statements of purpose for the uses of energy come principally from critics of official and prevailing energy politics. Many of these critics are committed to environmental quality and controlled, limited growth, and their values shape their preferences among energy alternatives. But the controversy over energy policy has been, in fact, a poorly defined and inadequately considered debate on alternative futures. Debate about energy policy ultimately becomes debate about the future.

No modern nation has dealt effectively with the energy issue; none has been conceptually or institutionally prepared to deal with it. The energy issue is not just a matter of supply and demand; it cannot realistically be isolated from its societal context. The problem is not wholly a matter of availability and price, as popularly perceived. If an unlimited supply of very inexpensive energy were suddenly to become available, massive economic, social, ecological, and political problems could ensue. Superabundant energy, without the wisdom or means to control its use, might pose a greater threat to humanity than does the present prospect of scarcity. While many people would probably accept the uncertain risks of an excess to the apparent certainties of deprivation, the misuses of cheap energy have already been documented to a degree suggesting that in the absence of effective social restraints unlimited energy would be a mixed blessing at best and a disaster at worst.

Throughout most of the world, the prospects for energy sufficient to meet continuing demands at present levels of population and technology are problematic. The reasons are: (1) the sources of energy available to meet present needs are very unevenly distributed among and within nations; (2) access to these sources is complicated by political, economic, and environmental factors; (3) all of the presumably available sources upon which advanced technology depends are in some sense limited or finite, and the most versatile of these, petroleum in a relatively accessible state, is one of the more finite; (4) major sources now utilized for energy production have serious long-term side effects that could severely limit or even prohibit their use in the future. The consequences to which this last reason refers include the acid precipitation from coal combustion, chemical and particulate pollutants from fossil fuel emissions generally, gaseous emissions associated with geothermal sources, siltation of hydropower dams, hazards of radioactivity from the spent fuel of nuclear fission reactors, the discharge of heat into the environment from combustion and nuclear reactions, and the effects of CO_2 emissions on global climate change.

Optimism regarding renewable nonpolluting substitutes for these conventional sources should be tempered by the following considerations: (1) although some forms of solar and biologically generated energy have technical feasibility, the prospects for developing them into cost-effective technologies capable of sustaining the energy needs of contemporary industrial societies are uncertain; and (2) as Amory Lovins has observed in his writings on soft energy paths, the kinds of societies sustainable by solar and related sources of energy (e.g., winds and tides) would be distinctly different from existing industrialized political economies.[27] As Lovins points out, the self-renewing, nonpolluting energy sources tend to require use at or near the point of generation—as with solar heating of buildings. From an environmental and social perspective, this decentralization of energy generation and utilization could be advantageous, but it would tend toward quite different configurations of population and of economic and political power than those now prevailing in most developed countries.

Conventional sources of energy, as frequently noted, are very unevenly distributed; even solar energy is nowhere receivable in a continuing flow. Deposits of oil, gas, coal, and uranium as well as areas suitable for obtaining geothermal energy are all concentrated. But some nations high in energy consumption such as Japan, France, Sweden, or the Netherlands are not nations in which conventional energy resources are concentrated. As a consequence there is, for the present, a major and environmentally significant flow of energy resources, particularly of petroleum, from producing to consuming countries.

In a review of the outlook for energy, petroleum, and the economy, researchers for Exxon Corporation conclude that "a plateau in world oil production may be reached some time before the turn of the (20th) century. By that time other sources of energy would be required to supply all of the growth in energy demand. . . . Government policies which respond to the long-term need for transition to new patterns of energy use and supply are essential to minimize the economic costs of this transition."[28]

The MIT Workshop on Alternative Energy Strategies similarly found that the supply of oil will fail to meet the increasing demand before the year 2000, and urged that the change to alternative energy sources must start now. The tenth and last of the conclusions of this workshop was that "the critical interdependence of nations in the energy field requires an unprecedented degree of international collaboration in the future. In addition, it requires the will to mobilize finance, labor, research, and ingenuity with the common purpose never before attained in time of peace; and it requires it now." [29] A study of international cooperation in energy development concluded that national self-sufficiency in all areas of energy production is not possible, and that the transition to renewable energy will be very difficult unless investments are channeled throughout the world in an effort to find additional conventional resources, to conserve them, or to replace them with renewable sources wherever the best opportunity presents itself. [30]

Coal is increasingly being utilized as a "transitional fuel" from present petroleum-dependent economies to newer technologies that might emerge in the twenty-first century. In the Asia-Pacific region alone, its use is expected to double or triple in the remaining decades of this century, with expected increases in production reaching three times present consumption in the United States. But this tremendous increase in the burning of coal comes at a time when the environmental consequences of carbon dioxide (CO_2) and sulfur dioxide (SO_2) emission into the atmosphere (discussed in Chapter 9) are becoming increasing causes of concern. For this reason, the Environment and Policy Institute of the East-West Center in Hawaii has undertaken an international cooperative project on the environmental dimensions of energy policies with particular emphasis on reconciling energy and environmental goals. [31]

Although the World Coal Study was supportive of the proposition that coal would be the principal transitional fuel to renewable energy sources, there are many reasons for uncertainty regarding its future. [32] In their study of the *Geopolitics of Energy*, Conant and Gold identify a number of technical and economic obstacles to a massive conversion to coal as a fuel. And their assessment does not include the ecological consequences emphasized here. They conjecture that "at the end of the century, the increasing use of nuclear power—not coal—may be the key factor in displacing oil in electric power generation." [33]

Nuclear energy also has an uncertain future, although a number of countries, particularly in Western Europe, have chosen to rely heavily upon it. The role of the International Atomic Energy Agency in international energy policy is very restricted, and has little effect upon national decision making. Conant and Gold conclude that there are major obstacles in the way of international agreement on nuclear energy issues. They point out: "It will be difficult to find common ground among: (1) those countries seeking nuclear facilities, including domestic reprocessing and enrichment capacity, in order to achieve some greater degree of energy independence; (2) those countries in which the viability of the domestic nuclear industry depends on the ability to export nuclear technology; and (3) those countries that see in the spread of reprocessing and enrichment facilities an unacceptable risk of nuclear weapons proliferation." [34]

A severe setback for the promotion of atomic energy occurred on 26 April 1986 with the worst accident reported in the history of nuclear energy. A massive release of radioactive material into the atmosphere occurred at the Chernobyl Nuclear Power Plant in the Soviet Union. The catastrophe was not reported until its effects began to be recorded in nearby countries. Antinuclear sentiment was reinforced, and two international treaties were rapidly consummated: the Convention on Early Notification of a Nuclear Accident and the Convention on Assistance in the Case of a Nuclear Accident or Radiological Emergency. Unless there is a breakthrough to commercially feasible atomic fusion and the safe disposal of spent nuclear fuel, the future of nuclear energy is doubtful. Meanwhile the undue complacency of governments regarding attention to alternative energy sources almost guarantees an energy crisis of major proportions by early in the twenty-first century.

Solar energy has been proposed as the only truly permanent and nonpolluting source, and yet serious questions have been raised regarding the safety of various solar energy technologies.[35] Some innovative techniques for the generation of energy have been proposed that might have international repercussions. One of these involves mounting solar cells on satellites placed in outer space in order to beam microwaves to receiving stations on earth. Questions have arisen concerning the safety of this technology for objects or persons that might intercept the microwaves, and there is also the question of the applicability of international law to the use of outer space for this purpose by private as well as public organizations.

Another energy innovation, oceanic thermal energy technology (OTEC), is a means of utilizing the thermal differentials between the various layers of sea water in the equatorial oceans to obtain electrical energy.[36] OTEC is in an early experimental state, and it is too early to know whether it might have international policy implications. For the most part it seems that it would not, although there remains the possibility of dispute if OTEC installations are proposed for closely proximate nations utilizing common bodies of water.

A technoscientific development with significant environmental implications is the effort to produce energy from alcohol obtained from vegetation.[37] This potential source has particular attraction for tropical countries in which conditions of climate and soil permit abundant vegetation. Yet this prospect is not without its problematic aspects—some technical, some political with international dimensions. If food production is diverted to the production of biofuel at the same time that the world faces impending food shortages, obvious policy questions arise. Should the remaining nonagricultural lands be diverted to biofuel production, the loss of wildlife and natural habitat could be severe. It should once again be repeated that the necessity to choose among these alternatives of food, fuel, and wild nature would be less compelling—or even nonexistent—in a world of substantially fewer people or of conservation-oriented societies consuming much less.

Finally, the energy needs for domestic purposes have traditionally been served directly by another biological source—wood. This extensive utilization, however, has created a serious and unforeseen problem over much of the world because of the felling of forests for the burning of firewood and the making of charcoal. A significant part of the deforestation of tropical countries is connected with the need of poor people to obtain cooking fuel. In many of the countries in which this deforestation occurs, no national or local energy alternatives are readily available. A measure of the importance with which fuel wood is regarded in tropical countries is the emphasis accorded it in the 1981 United Nations Conference on New and Renewable Sources of Energy. Preliminary studies urged a special global research and development program on biomass-based energy and new technologies for wood fuels.[38]

One proposal for at least a temporary expedient to retard deforestation has been the planting of rapidly growing tree species in the vicinities of tropical cities. Thus, an immediate source of wood for fuel, charcoal burning, and construction would be available—in theory reducing the impact of human need on mountain forests and forest reserves, which ought to be preserved for watershed as well as genetic, aesthetic, and wildlife purposes, among others.

The resolution of one environmental problem often causes another to appear. Open fires of wood or charcoal carry risks of conflagration to buildings and communities. Exposure to smoke may entail serious impairment of health. For example, the smoke of certain wood has been suspected of possessing carcinogenic properties, making it unsafe to burn for heating or cooking unless adequate ventilation is provided—a feature almost wholly lacking in the huts of numbers of poor people in rural areas of less-developed countries.[39]

Sources of energy may be located in the deep sea bed. Although the present offshore sources of petroleum have been found primarily in shallow seas and outer continental shelves under the territorial jurisdiction of a particular nation, petroleum or natural gas might be discovered in areas not subject to the jurisdiction of any national power. Such resources would, under the concept proposed in the United Nations Conference on the Law of the Sea, be controlled by a United Nations agency, which would then license exploitation, with the benefits going in considerable measure to poor Third World countries. This proposition, however, has been unattractive to First World industrial interests, national and multinational, having the capability for exploiting such oil deposits. A Deep Seabed Agreement was adopted by the UN General Assembly in 1994 alleviating international disagreement.[40]

A somewhat different problem is presented by Antarctica. In addition to the environmentally unacceptable proposition that it might be used as a depository for the storage of atomic waste, there is the possibility for development of its deposits of oil, coal, and natural gas. Under the Antarctic Treaty of 1961 the nations with claims to the region have temporarily set them aside in the interest of its treatment as an international science reserve. If, however, significant energy sources should be discovered, and if the economics and the technology of their exploitation were favorable, there could be a great competitive pressure on the occupying countries to push for the exploitation of those portions of Antarctica over which they have asserted territorial claims. Inasmuch as these claims are not entirely compatible, international conflict could arise.

The seemingly insatiable urge to claim and develop mineral deposits has led to controversy over the future of the Antarctic Continent. Under the treaty of 1959 a moratorium on mining and resource development was accepted. But some parties to the treaty pushed for the right to prospect for minerals, and after eight years of negotiation a new treaty was signed on 2 June 1988 (see Chapter 8). Ratification has been opposed by many environmental and some scientific groups. Australia and France have, thus far, refused to ratify, and because unanimity among the Consultative Parties to the original treaty must prevail, the future of the minerals convention is in doubt. In 1991 the Madrid Protocol to strengthen environmental protection was adopted by the parties, but as of 1995 it had yet to obtain sufficient ratification to put it into effect.

Even when energy sources are congruent with energy demand, a geopolitics of energy is still possible, as the acid precipitation and carbon dioxide issues demonstrate. It appears that further advances in technology will be needed to cope effectively with the problems of energy sources and supplies and to reduce the likelihood of international conflict over energy issues.

Energy Problems and Policies

From the foregoing discussion it is evident that nations have difficulty in cooperating in matters relating to energy policy. Environmental issues complicate already difficult economic and technical problems of energy sources and utilization. International environmental policy with respect to energy is largely concerned with the effects of energy extraction, transportation, and utilization. These concerns are ultimately related to the chosen forms of energy, geographic circumstances, technical problems of utilization, and impacts upon plant and animal health. For example, if the continued burning of fossil fuels threatens to disastrously and irreversibly change the carbon dioxide balance in the atmosphere, a major international conflict could arise between nations that would gain and those that would lose from the resulting climate changes. Alternatively, all nations might have to abandon the use of coal and oil as fuels unless environment protection technologies are developed: acid rain, for example, attributed largely to the sulfur dioxide emissions from coal-fired power plants, has become an international issue.

The deliberate firing of Kuwait's oil wells and discharge of oil into the Persian Gulf during the Gulf War of 1990–91 not only caused severe environmental pollution but stimulated demands for international judicial proceedings and penalties for crimes against the environment. Additional environmental hazards occur with the transportation of petroleum, notably through oil spills at sea and along coastal areas; and the transportation of liquefied natural gas is accompanied by the serious hazards of violent explosions.

International energy programs have been largely confined to data gathering and exchange, research and development, and public information. In each of these areas the greater part of the work is done at the national level. In obtaining access to energy, industrial nations have been competitors. But vulnerability to reduction or cutoff of oil supplies has prompted at least pro forma efforts to coordinate their policies regarding petroleum imports.

On the supply side, industrialized nations have sought institutional arrangements to coordinate their energy policies. Following the Arab oil embargo of 1973 and an Organization for Economic Cooperation and Development (OECD) conference in Washington in February 1974, OECD governments established the International Energy Agency (IEA) to promote cooperative relations with oil-producing countries and to achieve better balance of energy resources within the OECD community. IEA's purpose was to reduce international divisiveness on energy issues rather than to form a consumers' cartel, but it also provides machinery to enable the participating countries to work together rather than at cross-purposes in the event of an oil supply emergency.[41] IEA has not been indifferent to the energy needs of developing countries; a Workshop on Energy Data in Developing Countries was held in Paris from 11–15 December 1978.[42] Monographs on specific energy sources are published, as well as works of worldwide relevance like *World Energy Outlook*, a forecasting and statistical compendium prepared by IEA for OECD.

A more recent and problematic energy concern is the disposition of nuclear wastes. To develop international standardized management practices for nuclear wastes, OECD has established a Nuclear Energy Agency. The International Atomic Energy Agency (IAEA), meanwhile, is responsible for international cooperation on the handling of nuclear materials. Yet a study of current practices concludes that "principal responsibility for the initial formulation of necessary international waste management policies lies within individual nations," although "such initiatives must eventually be replaced by joint international action."[43] So, while IAEA does have a degree of responsibility for nuclear waste disposal, the wastes themselves are in the custody of particular national governments.[44] Nevertheless, there are international and even global hazards involved in the location and custody of the waste material. Propositions to bury the wastes at sea or to place them on the Antarctic continent have aroused environmental opposition in a number of countries.[45]

The discharge of effluents from energy-generating activities has become an international problem in relatively confined seas; thermal discharge from power plants could affect temperature in localized areas of the Baltic Sea, the

Mediterranean, and the Great Lakes. In the Persian Gulf, the heavy traffic of oil tankers and adjacent petroleum installations has caused the surrounding states to negotiate and ratify a treaty to protect the waters of the gulf from contamination, but the treaty was egregiously violated by Iraq in the 1990–91 Gulf War. Extended economic jurisdiction into coastal waters, the two hundred-mile limit, has also created some international problems, particularly where the boundaries of states are convergent, as among the states bordering on the South China Sea. In these areas, a multinational corporation may undertake offshore or open sea drilling, but who has political jurisdiction over energy-producing installations is a question that must always be resolved if projects are to be undertaken.

The ecological conditions of the polar regions pose special problems with respect to hazards associated with the exploitation of minerals and especially of petroleum. Polar environments tend to be ecologically fragile and do not recover easily from prospecting or mining operations. Oil spills on polar seas or ice could have effects more disastrous than those characteristic of more temperate waters. The Canadian government has attempted to forestall some of the hazards of oil pollution in the Arctic through a unilateral act that restricts and controls exploitation in the Arctic territories and territorial waters under its jurisdiction.[46]

International Energy Initiatives

We have noted that nations have been slow to see a common interest in problems related to access to energy sources, but that they have begun to find common concerns in coping with the environmental effects of energy utilization. International efforts to deal with these effects have usually been subsumed under agendas for environmental health and safety and for the prevention and control of environmental pollution rather than under energy policy. Initiatives for containing or correcting adverse effects of energy utilization have largely fallen to the International Maritime Organization (oil pollution at sea), the International Atomic Energy Agency (radioactivity), and UNEP (energy needs and effects generally, including pollution control under the Regional Seas Programme).

International conferences have dealt primarily with two energy-related issues: marine pollution, especially by oil, and nuclear contamination. Pollution by oil has been dealt with in a large number of conferences, which have resulted cumulatively in numerous agreements. A number of these have been formulated as treaties that, although overlapping or supplementary, have often left significant areas of energy-related marine issues uncovered. Meetings convened to deal with the hazards of marine pollution are described largely in Chapters 8 and 9.

Treaties relating to environmental hazards of nuclear energy include the International Convention for the Safety of Life at Sea (SOLAS II, London, 1974) dealing with radioactive contamination by nuclear ships, the International Convention on the Liability of Operators of Nuclear Ships (Brussels, 1962), the Convention Relating to Civil Liability in the Field of Maritime Carriage of Nuclear Materials (Brussels, 1971), the Convention on the Prevention of Marine Pollution by Dumping Wastes and Other Matter (London, 1972), which prohibits the disposal at sea of high-level radioactive waste, and the 1986 IAEA treaties on nuclear accidents (see Appendix D).

In addition, the Treaty on the Prohibition of the Emplacement of Nuclear Weapons and Other Weapons of Mass Destruction on the Sea-Bed and the Ocean Floor and in the Subsoil Thereof relates to the military uses of nuclear energy. Though obviously of great importance to international cooperation regarding energy and environment, the military uses of atomic energy involve a large number of considerations that could not be satisfactorily treated within the scope of this book, although clearly posing global environmental issues.[47] Both civil and military uses of atomic energy have serious environmental consequences, and both uses have been politically divisive. The great cost of nuclear weaponry and power plants is an additional objection, preempting funds and material resources that could be turned to uses of more certain benefit. Atomic energy, moreover, entails opportunity costs extending into an indefinite future. Atomic wastes and decommissioned plants, equipment, and weapons (e.g., nuclear-powered submarines), require perpetual and vigilant care—a legacy that future generations may not welcome. Chapter 22 of *Agenda 21* adopted at the 1992 Rio Conference, which dealt with Safe and Environmentally Sound Management of Radioactive Wastes, contained recommendations to implement the IAEA Code of Practice on the Transboundary Movement of Radioactive Waste.

Reflecting the priorities of national governments, most international initiatives for energy have been for the enlargement of energy sources and supplies. Energy for development has been a continuing concern within the United Nations system and among a large number of scientific investigators.[48] At least four conferences have been held in Geneva on Peaceful Uses of Atomic Energy (1955, 1958, 1964, and 1971).[49] Two major conferences have been convened to consider new energy sources: New Sources of Energy, Rome, 21–31 August 1961, and New and Renewable Sources of Energy (UNERG), Nairobi, 10–22 August 1981. The 1981 conference appears to have had positive, although significantly limited, results.[50] An appraisal in *Science* magazine concluded that

> Reports from the series of technical panels and cross-cutting issues groups generally ranged from good to excellent. A synthesis report put

the various recommendations into a single document . . . providing guidelines for research and projects on the following energy sources: hydropower, fuel wood and charcoal, biomass, solar, geothermal, wind, oil shale and tar sands, ocean, draft animal power, and peat. The delegates . . . were quite responsive to the recommendations by representatives of nongovernmental organizations for further clauses relating energy to environment, fuel wood, or the special needs and roles of women.[51]

Unlike several earlier UN conferences, this one proposed no new UN agency or funds, largely reflecting resistance from the developed countries. An action plan upon which the conference agreed was not provided with a mechanism or means to put it into effect. Nevertheless Charles Weiss, science and technology adviser to the World Bank, observed that "in terms of raising consciousness and focusing attention on renewable energy, it was very successful."[52] Apparently drawing upon previous experience, topics known to lead toward international rancor were excluded from the agenda. The OPEC nations refused to permit global oil policies to be discussed. Nuclear energy was omitted because of concern in the industrial countries regarding nuclear proliferation. Coal was not considered because it is largely concentrated in the industrial countries of the North. With the principal conventional sources of nonrenewable energy largely excluded, conservation was not a major topic.

Preparatory to UNERG, a group of experts meeting in Rome under the chairmanship of Maurice Strong, who had been secretary-general of the United Nations Conference on the Human Environment, produced a long list of recommendations and added the criticism that "the restricted scope of the conference, a product of unfortunate political restraints, is not logical." The group declared that "the energy problem cannot be fragmented; neither analysis of technologies and their role in development, nor any new financial institutions designed to support energy investments, should isolate new and renewable sources from the totality of energy resources."[53]

UNERG illustrates the difficulty of developing global policies in a fractionalized and polarized world. Yet commenting on preparation of the conference, Erik P. Eckholm of the International Institute of Environment and Development observed that "the main problem to date has not been an excess of sterile ideological debate, but a poverty of ideas for action that would do more good than harm in the field of renewable energy. The Conference," he said, "is helping to legitimize renewables, especially in wary developing countries, much as the 1972 Stockholm Conference did for environmental concerns."[54]

In summary, changing patterns of energy supply and generation during the next half-century seem certain to produce changes with implications for the environment and for relations among nations. But it is difficult to forecast with any accuracy just what those changes will be. Their nature will depend much upon the extent to which technology effectively develops substitutes for present energy supplies and, in particular, develops commercially feasible systems for utilizing solar energy.

It should also be noted that within the next fifty to one hundred years and beyond, the progressive siltation of many dams may reduce their efficiency and, possibly, their utility for hydroelectric and irrigation purposes. Inasmuch as suitable dam sites are limited and the best available ones have already been preempted, a resulting reduction of energy or water storage could significantly affect a nation's future in environmental and many other respects. It is, of course, possible that a means will be found to renew the storage capacity of big dams. But the cost of such renewal could be high, and the effort and energy required must be measured against other alternatives.[55]

Development and the Environment

"Development" appears to be a term everyone understands and no one is able satisfactorily to define. The term is characteristically defined by reference to another term that is equally ambiguous—modernization, for example. As a practical matter, "development" is defined by what it is in actual practice: development is as development does.[56] For the United Nations Development Programme, development is defined by the projected and programmed expenditures of the UNDP. Similarly, development activities of the World Bank, the regional development banks, and bilateral aid programs are defined by what they do. Development has been adopted as a designating term, not because it is descriptive or discriminating but because it accentuates the positive. It contains intimations of progress, of growth toward fulfillment, of the unfolding of innate potential. Yet, in fact, conventional development may lead to none of these things: realization may, like the touch of Midas, bring unforeseen and undesired consequences. But it may also bring both tangible and intangible benefits.

"Development and Environment" was one of the topical categories considered at the United Nations Conference on the Human Environment in Stockholm in 1972. It was alleged that the representatives of Third World countries were more concerned about development prospects than about their environmental futures, but the section of the conference report entitled "Development and Environment" was one of the shorter, containing eight recommendations, only one of which (102K) dealt substantively with environmental aspects of development per se. The recommendation went no further than urging enlarged provisions for environmental education, research, and information exchange, "which would help developing countries speed up, without adverse environmental effects, the exploration, exploitation, processing, and marketing of their natural resources."

The other recommendations (103–8) dealt almost wholly with the impact of environmental protection measures in the developed countries upon the export of products from the Third World. Recommendations contained such admonitions as "that all countries present at the Conference agree not to invoke environmental concerns as a pretext for discriminatory trade policies or for reducing access to markets, and recognize further that the burdens of the environmental policies of industrialized countries should not be transferred either directly or indirectly to the developing countries" (103A). This provision was inspired by Third World concern that the ban on pesticides such as DDT in developed countries could be extended to imports from developing countries in which these pesticides were still in use. Accordingly, Recommendation 103B included the provision "that where environmental concerns lead to restrictions on trade, or to stricter environmental standards with negative effects on exports, particularly from developing countries, appropriate measures for compensation should be worked out within the framework of existing contractual and institutional arrangements." [57]

The conference further recommended that the General Agreement on Tariffs and Trade (GATT), the UN Conference on Trade and Development (UNCTAD), and other international bodies should "consider undertakings to monitor, assess, and regularly report the emergence of tariff and non-tariff barriers to trade as a result of environmental policies." [58] Section 106 was directed at the industrialized countries, suggesting that examination be made of the extent "to which the problems of pollution could be ameliorated by a reduction in the current levels of production and the future rate of growth or production of synthetic products and substitutes which, in their natural form, could be produced by developing countries."

Some observers saw a paradox in these recommendations. On the one hand, the developed nations, and the United States in particular, were sharply criticized for an excessive and disproportionate consumption of the world's natural resources. But on the other hand, fear was expressed that developed countries might use substitute materials, synthetics, or environmental regulations to avoid dependence on Third World resources. Moreover, the fact that a large proportion of resources consumed in the United States and other large developed countries came from their own territory was generally overlooked. Throughout the Stockholm Conference, the Third World representatives repeatedly emphasized the absolute sovereign discretion of their governments in disposing of their natural resources in whatever way they chose. Logically, the proposition should also apply to developed countries. But the prospect that developed countries might reduce their imports from developing countries as a consequence of conservation or environmental protection measures and through development of synthetic products and substitutes was viewed with alarm and regarded as an injustice for which "appropriate measures for compensation should be worked out."

Finally, the concluding recommendation of the conference (109) asked that the international environmental considerations arising from these proposals be integrated into the review and appraisal of the international development strategy for the United Nations Second Development Decade in such a way that the flow of international aid to developing countries was not hampered. The recommendation concluded with the admonition that "it should further be ensured that the preoccupation of developed countries with their own environmental problems should not affect the flow of assistance to developing countries, and that this flow should be adequate to meet the additional environmental requirements of such countries." This provision introduced the concept of additionality, which, along with compensation, were principles adopted by the developing countries to protect their interests against the possibility of financial erosion as a consequence of the newly emerged environmental awareness and concern on the part of the developed countries.

The recommendations on "Environment and Development" represented only a small though intensely felt part of the deliberations at Stockholm. Reconciliation of the goals and processes of environmental protection with development was a major task of the Stockholm Conference, and remains a continuing task of UNEP. For many reasons this task cannot easily be accomplished. Activities undertaken for economic or natural resources development have frequently been the direct or indirect causes of environmental damage. For more than three decades, development has been an explicit political objective in many countries whose leaders have often objected to modifying development plans in the interest of environmental quality. The value disparities between developmental and environmental objectives afford abundant occasions for political conflict. Yet in relation to the purposes ultimately sought, much of this conflict is unnecessary; it has often resulted from inadequate formulation of developmental and environmental goals.

Throughout most industrialized countries there is currently a widespread belief that the price of development has been unnecessarily high, and that the payment has been made in the wrong way—through destructive exploitation of the natural and social environment. This so-called price of progress is seldom the least costly or most efficient manner of paying for economic or industrial development if evaluated from a social perspective, and in relation to future consequences. Where the monetary price of a product or service does not cover the costs of environmental protection or social amelioration, it misrepresents its true or ultimate costs. The burden of these costs is thus laid upon the general public (present and future), which must bear them regardless of whether they are commensurate with benefits, or indeed whether there is any benefit at all. Involuntary payment is exacted in the form of resource depletion, environmental pollution, and a lowered quality of life. Payment in this form represents a mismatching of benefits and costs because the ultimate expenses of environmental protection and restoration should logically have been factored into the monetary price of the development project or the product.

This failure to account for all costs of production and to include them in the monetary price has been formulated by economists as the theory of externalities, or costs external to those normally calculated (such as labor, materials, and rent). When environmental or social effects are not included—are externalized in estimating the costs of development—the full costs are not revealed. The public cannot fairly appraise the true cost of development because the monetary price is incomplete; it is not a reliable representation of true cost. People are deceived into believing that the development plans in question are less costly than they may subsequently prove to be. This accounting omission becomes in fact a subsidy for environmental degradation—the amount of subsidy being the difference between the full costs of environmentally sound development and the calculated monetary price that has failed to include so-called external costs.

Where development projects involve irretrievable commitments of resources and maintenance, the range of future opportunities may be forfeited. Opportunities might also be enlarged, but this is not safe to assume in the absence of environmental impact analysis and ecologically sensitive planning. The opportunity costs of development are frequently environmental—for example, in genetic losses when moist tropical forests are cleared for grazing or put into single-species plantation agriculture. The monetary costs of maintaining public works (e.g., irrigation or hydropower) preempt funds that might be allocated to other purposes. Where the advantages of water projects are ultimately lost to siltation or salinization, both the benefits sought and those forfeited are lost. These risks do not argue against all development; they do argue for anticipatory calculation of all ascertainable costs and for careful consideration of environmental impacts.

The technique of cost-benefit analysis, properly applied and with all requisite information available, could provide a rational factor for choosing among development alternatives. But this technique has often been misused to support predetermined plans or preferences. The term "extended cost-benefit analysis" has been used to indicate the inclusion of environmental and social impacts and even unquantifiable data in the analytic process.[59]

Human society is not sustained by environmental quality alone. Civilization draws upon the properties of the earth for much more than biological survival. Even before Stockholm, it was evident that for much of the world there was no way to preserve the natural environment in its pristine state. Humankind's use of the earth being inevitable, environmental policy in large measure implies control over the actions of people. Reconciliation of the goals of development and of environmental quality is thus essentially a reconciliation of values. At least a decade before Stockholm, ways to accomplish this harmonization were being sought. Following several years of preparation, a handbook, *Ecological Principles for Economic Development*, was published by the International Union for Conservation of Nature and Natural Resources

(IUCN) and the Conservation Foundation.[60] A follow-up to this publication was provided by three regional international conferences in Venezuela, Indonesia, and Iran cosponsored by IUCN, UNEP, and other organizations.[61] Although some of these events occurred following Stockholm, the concept of value reconcilement was current at the Biosphere Conference in 1968 and especially at the conference on Ecological Aspects of International Development held in December 1968 at Airlie House, Virginia, under the sponsorship of the Conservation Foundation and Center for the Biology of Natural Systems, Washington University of St. Louis.[62]

At the Stockholm Conference, Third World representatives argued for development as a necessary precursor of environmental protection—as necessary to provide an economic base from which environmental protection measures could be financed. But with growing awareness of the connections between uses of technology and environmental damage, development of conserving or appropriate technologies was gaining interest among many Third World leaders. The work of E. F. Schumacher in Africa was especially influential in promoting an appropriate technology movement that extended to both developing and developed countries.[63] The goal of providing people with ways of working and living that would be environmentally protective rather than environmentally destructive was, however, more easily declared than achieved.

Perhaps as influential as the Stockholm recommendations themselves were the considerations of development and environment issues in the preconference meetings sponsored by the Preparatory Committee for the UN Conference, particularly the meeting of experts at Founex dealing with international economic, developmental aspects of environmental policy. Both before and after the Stockholm Conference, meetings were organized by the ICSU Scientific Committee on Problems of the Environment (SCOPE) in Canberra (1971) and in Nairobi (1974).[64]

The United Nations Environment Programme (1972) and the UNEP-IUCN World Conservation Strategy (1980) have endeavored to keep the Stockholm recommendations on official UN and national agendas. By the end of the decade following Stockholm, it was evident that significant changes had occurred in both the national and international status of environmental aspects of development planning. During the 1980s one could still point to environmental neglect both in the policies of governments and in international cooperation efforts, notably to the inability of the 1980 UN Conference on the Application of Science and Technology to Development to take serious account of environmental issues. Nevertheless, positive accomplishments through governmental and intergovernmental actions in environmental protection exceeded what most objective observers at Stockholm would have thought probable. One consequence of Stockholm and its aftermath has been that, at least in principle, the conflict between environment and development as perceived in 1972 is, to a significant extent, being subsumed by the concept of ecologically sound development, or ecodevelopment, a term now generally displaced by the phrase "sustainable development."[65]

Ecologically sound development remains more an ideal to be realized than a process readily administered. A major obstacle to its attainment has been the unremitting pressure of expanding populations and economies on the natural environment. A second obstacle is the deficiency evident in almost every nation, in comprehensive, integrated planning and management. This want follows from a third obstacle: political and economic policies and structures that have been designed to promote the nonecological type of development that prevailed during earlier decades. As Kenneth A. Dahlberg points out in his study of the Green Revolution, major changes in agriculture or natural resources management or technology require corresponding changes in governmental and economic institutions and infrastructures. Ecodevelopment may not be realizable with prevailing economistic, technocratic policy assumptions and institutional arrangements.[66] The present state of scientific knowledge and of public planning and management are not yet ready to fully realize the ecodevelopment concept. This inadequacy was implied in the Cocoyoc Declaration adopted by the 1974 UNEP/UNCTAD Symposium on Patterns of Resources Use, Environment and Development Strategies. A call was made for new approaches to development that would include "imaginative research in alternative consumption patterns, technological life styles, land-use strategies, as well as institutional frameworks and education requirements to sustain them."[67]

The ecodevelopment concept had not taken hold of governments, perhaps because it was believed to imply a subordination of development activities to ecological consideration. Without losing an environmental content, a new term for a new approach seemed necessary. One was found in the phrase "sustainable development." A new approach to development was initiated in Paris (12–13 September 1979), when representatives of UNEP, UNDP, the World Bank, and several regional organizations with development concerns adopted a draft Declaration of Principles on the Incorporation of Environmental Considerations in Development Policies, Programmes, and Projects. In February 1980, the official declaration was issued and followed up with additional meetings among representatives of the signatories to consider practical means of application. Coordinated by the Committee of International Development Institutions on the Environment (CIDIE), the declaration provided an institutional mechanism for reorienting the lending policies of the multilateral development banks and a forum for the promotion of the sustainable development concept. One effect of the new development approach appears to have been an off-setting of Third World resentment over perceived economic injustices. For more than a decade this resentment corroded international efforts toward environmental protection, and found expression in an effort among developing nations to establish the New International Economic Order (NIEO).[68]

The NIEO was a vehicle for Third World grievances during the 1970s and early 1980s.[69] It found expression primarily through meetings of the United Nations Conferences on Trade and Development (UNCTAD) and the United Nations Industrial Development Organization (UNIDO). Environment was not a consideration in the declarations and manifestos associated with the NIEO. The politics of the Cold War may have been an underlying factor. But with the economic progress of a number of former Third World countries and the collapse of the Soviet Union, the NIEO lost momentum. Its divisiveness was at least partially offset by the concept of sustainable development.

Sustainable Development

An integrative theme for international environmental policy has emerged in the concept of "sustainable development." To be sustainable, development must possess both economic and ecologic sustainability. The concept, which signifies a policy approach or goal rather than a substantive prescription, gained acceptance during the decade of the 1980s. As noted in Chapter 8, the term "sustainable development" was introduced in 1980 by the IUCN-UNEP *World Conservation Strategy*. It began to acquire the status of de facto official policy among governments generally as a consequence of its adoption by the World Commission on Environment and Development (WCED), established in 1983 by the General Assembly of the United Nations, and the publication in 1987 of its report, *Our Common Future*. It was a pervasive theme in Agenda 21 adopted by the 1992 UN Conference on Environment and Development.

"Sustainable development" is self-defining only in a very general sense. It indicates the way in which development planning should be approached. Yet it provides no indication of development goals or priorities nor of the quality of life level of sustainability. Its principal merit is that it modifies the previously unqualified development concept. This modification has an integrative effect because sustainability in any sense requires ecologic or resource life-support base sustainability. A long list has now been accumulated of development projects that failed because of ecological or geophysical oversight or error.[70]

The phrase "sustainable development," integrative in effect, does not in itself provide conceptual integration. As observed in a proposal for a workshop on "Sustainable Development: Principles and Criteria" by the International Institute for Applied Systems Analysis, the concept "is viewed quite differently by industrialists, economists, planners and environmental and ecological scientists, and there is a need to achieve a common understanding between the various groups." The concept has generated a large literature of which the notes provide only a selective sample.[71]

To address this need for definition, a ministerial conference was held in May 1990 in Bergen, Norway, at the invitation of Gro Harlem Brundtland, former prime minister of Norway, who chaired the WCED. Sponsored by the European Economic Commission (EEC), the conference considered the findings of four preparatory workshops. The results of this conference may help to provide more agreement in the substance of sustainable development than previously existed. As of 1996, a large number of symposia, workshops, and conferences on sustainable development had taken place. It is too early at this time to assess their effects. But at least in principle, events and trends previously identified in this book will force governments and international organizations toward the implied ends of sustainable development.

Former UNEP executive director, Mostafa K. Tolba, described the growth of reconciliation between environmental and developmental objectives during the 1970s and 1980s in the introduction to a collection of his addresses on the sustainable development theme:

> Our understanding of the inter-relationships between environment and development has undergone a profound change during the past 15 years. At the end of the 1960s, it was generally believed that it was possible to have either one or the other, but not both simultaneously. In other words, if we wanted development, the price to pay would be a loss in terms of environmental quality, and *vice versa*.
>
> That view has now been completely overturned with the realization that environment and development are interdependent, and are in fact mutually supportive. It is now clear that without environmental protection, it is not possible to have sustained development, and without development, it is not possible to sustain a high quality of our environment and an improved quality of life for all the world's citizens. Thus, what we need is sustainable development, that is, development that can be sustainable over the long term by explicitly considering the various environmental factors on which the very process of development is based.[72]

Another decade may be required before we learn whether sustainable development is a realizable concept and whether it leads to tangible, testable action plans that work toward improving the quality of life and safeguarding the environment. An evaluation of sustainable development is still premature, although it is possible to see the need for clearer definition of its terms. The concept cannot have much value if the words "sustainable development" mean all things to all people. Efforts to realize the concept will probably reveal its essential meaning and validity.

Sustainability was an explicit theme in seven chapters of Agenda 21 and was implicit in many of the others. The chapters in which it played a major role, and their titles, were as follows:

As in all chapters of Agenda 21, research and dissemination of information were emphasized. For an agenda acceptable to all 117 states represented at Rio a considerable blurring of distinctions and indirectness in allusions was necessary. Nevertheless, specific objectives and activities were recommended and important linkages were recognized. Were these objectives and activities to be pursued, they would almost surely lead to policies that the conference could not agree to directly recommend, but which would very likely emerge from objective inquiry at the national level.

Population has been a sensitive issue at all international conferences where it was a relevant factor. Because control of population growth has been the principal point at issue, the subject has been evaded by deferring to national and individual preferences. Sustainability, however, adds a consideration to the population issue that makes evasion difficult. At Rio, "Demographic Dynamics" was accepted as a way to deal with population policy in an oblique manner without using terminology that certain delegations found offensive.

Since the Stockholm Conference in 1972 the population question has slowly and incrementally obtained consideration at international conferences. At Rio, and in the 1994 Population Conference at Cairo, emphasis on women's rights, health care, literacy, and employment opportunities has unacknowledged implications for family size. The test of sustainability applied to the "liberation" and "empowerment" of women clearly has implications for childbearing and family formation. The agenda for sustainable development, applied to women, appears at present to be perhaps the most promising route to control of population growth.

Given the magnitude of the task of establishing sustainable national policies regarding the linkages among population, resources, and development it is difficult to be optimistic. Depending upon how it is defined, the process of development itself could achieve sustainability between the linkages. This balance has been achieved in a number of countries through socioeconomic evolution without governmental intervention. Whether the process can be repeated in many of the developing, high birthrate countries is far from certain.

11
Enhancing the Quality of Life:
Natural and Cultural Environments

Quality of life, or improvement in the conditions for living, is sought by people in efforts to satisfy a broad range of needs and values. But the satisfactions that people seek and the courses of action taken to attain them are often contradictory. The pursuit may lead to consequences that diminish quality. Because individuals are limited in their personal ability to obtain quality in their environments, and because priorities and the requirements for quality differ among people, reconciliation of the differing objectives has increasingly become a task of government and international organizations, urged on by nongovernmental groups. Governments have historically responded to quality of life concerns of at least some of their constituents. In recent years, not only has the range of criteria for an enhanced quality of life enlarged but also the efforts necessary to secure an improvement. Today, collective action to protect the environment is required, a need less urgent when there were fewer people on earth and their technologies had a lesser impact upon the environment.

Indicators and criteria of quality of life would be useful to assist governments in shaping coherent, sustainable objectives that might harmonize the diversity of human needs and preferences. International efforts to discover appropriate measures have already been undertaken, and models of environmental quality trends have been proposed.[1] UNESCO has taken a lead in efforts to obtain international collaboration in clarifying quality of life and environmental concepts, and in developing reliable indicators. Toward this objective a round table organized by the European Coordination Centre for Research and Documentation in Social Sciences was held in Budapest in November 1975. Following this gathering, a meeting of experts on "indicators of the quality of life and environmental quality" was held at UNESCO headquarters in Paris (6–9 December 1976). UNESCO's report concluded that "the meeting and its preparatory activities reflected a deep and far-reaching interest in quality of life research, and unanimously confirmed the desirability of providing an international link for the exchange of experience and results."[2]

This assessment appears to be confirmed by subsequent developments. For example, in December 1980, the International Congress on Applied Systems Research and Cybernetics meeting in Acapulco, Mexico, chose "quality of life" as its organizing theme. The published record of this international gathering indicated the broad scope of interest in quality of life and environment as subject matter for research and for policymaking.[3]

Unfortunately the selection of appropriate indicators has been limited by the present lack of valid and reliable techniques of measurement and by the wide range in perceptions of the nature of quality. Criteria for quality reflect social norms that may or may not be consistent with behavior necessary to maintain a sustainable environment. People may desire an environment of high quality but be unwilling to accept the economic or behavioral constraints required for its attainment. Quality of life is ultimately reducible to personal values, and not all of them relate directly to environmental circumstances. A basic task of environmental policy, therefore, is reconciling social norms and personal values and their expression in social and personal behavior with the requirements for people's continuing existence in the biosphere. Environment is a critical factor in quality of life considerations, because it forms the setting in which all other qualitative values are to whatever degree attained or lost. And environmental quality is especially concerned with protection and preservation of both nature and culture. This conservative position is assumed not because humankind cannot, for some purposes, improve upon nature or the cultural past, but because one cannot improve upon qualities and resources that have been irretrievably degraded or destroyed.

Quality of Life in Environmental Policy

There are today a large number of governmental and nongovernmental international organizations variously concerned with quality of life in the environment. The World Bank and the UN Development Programme (UNDP) have emphasized the economic factors necessary for attainment of quality in life. The World Health Organisation (WHO) and the Food and Agriculture Organisation (FAO) have addressed the conditions of health and nutrition. The International Labour Organisation (ILO) has considered health and safety in the workplace environment. The United Nations Industrial Organisation (UNIDO) has concerns that directly or by implication have quality of life aspects, for example, UNIDO's sponsorship of the International Centre for Genetic Engineering and Biotechnology (ICGEB) in 1983. UNESCO and UNEP administer a broad range of programs bearing directly upon quality of life. The United Nations fund for Population Activities (UNFPA) and the United Nations Centre for Human Settlements (HABITAT) have obvious quality of life relevance. Many quality of life concerns have now been expressed in international programs, treaties, resolutions, and actions by non-governmental organizations intended to protect or enhance particular aspects of the environment.[4]

Characteristic of a majority of these agencies is the attempt to reconcile preservation or enhancement of quality of life with the environmental and social changes attending development efforts, which are themselves intended to improve the quality of life. To understand how this reconciliation may sometimes be achieved, one must examine particular efforts. Following is a selective survey of international efforts to protect the quality of life and environment in relation to natural environments, artificial environments, and natural hazards. These contexts are not mutually exclusive. Human settlements or habitats, for example, occur both in natural and artificial environments; the relationship between culture and nature has always been interactive. Obviously, these do not include all environmental quality of life issues. Efforts to prevent or control various forms of environmental pollution clearly are intended to enhance the quality of life. Improved communication among people via satellites in outer space has qualitative dimensions. But no single logic is sufficient to organize the study of an environment where almost all things are somehow interrelated. Some topics appropriate to this chapter are therefore treated in other places, where considerations other than quality seemed more important.

Protecting Nature: Species and Ecosystems

Diminution in the quality of natural environments may be attributed to any of three general causes, or some combination of them. One of these is the inadvertent and unintended attrition of quality by excessive and continuing human activity—for example, overgrazing of grasslands, heavy and unremitting tourist visitation of natural areas, or overkill of wildlife for food or sport. Deliberate and intentional intervention by humans in nature is another—for example, damming or reversing rivers, draining lakes and marshlands, or disfiguring the landscape through structures, highways, excavations, and military operations. The third encompasses change caused by natural phenomena independent of human activity—for example, climate change or the natural violence of severe storms, volcanoes, or earthquakes. These and other causes of environmental damage are addressed by international agencies or programs, though changes caused by natural phenomena are largely confined to research and to warning as the impact of natural events is usually beyond human control, albeit not always beyond protective adaptation.

Precedents for national and international action on behalf of environmental quality were not perceived initially as either qualitative or environmental. The first international measures for nature protection were primarily economic in purpose, intended for the protection of agriculture or commerce. As noted in Chapter 2, farmers in Central Europe induced governments to enter into international agreements for the protection of migratory birds, and, similarly, commercial interests in fish and furs persuaded governments to enter into international agreements for the allocation of fishing rights and for management of harvest for the North Pacific fur seals. As European governments carved up Africa during the colonial era, it became evident that international cooperation would be necessary to protect the great herds in East Africa that migrated across political boundaries. These practical considerations, reinforced no doubt by ethical and aesthetic values, induced governments to enter into international agreements and institutional arrangements regarding the protection of specific plant or animal species and—although the term was not initially used—ecosystems.

Beyond these economic and aesthetic considerations, peoples and governments became concerned about another group of migrants in the biosphere. These were microorganisms pathogenic to plants, humans, and other animals. Health as an aspect of international environmental protection is considered later in this chapter, but it should be noted here that the well-being of particular species is related in numerous and complex ways to the circumstances of their environments. Conditions of habitat and nutrition have a bearing upon plant and animal health. Epidemic and endemic disease, as well as the morbidity and mortality rates of organisms, may in part be regarded as indicators of the integrity of ecosystems or of the ecological stability of relationships among the organisms present, especially as pertaining to humans.

Individual initiatives have been essential to plant and animal protection, and they continue to influence public and international efforts. There have been instances of the survival of species such as Pere David's deer, the European bison, Przewalski's horse, and the Franklinia tree, resulting primarily from the efforts of persons who cared about the quality of environmental heritage.[5] Among the individual defenders of wildlife in East Africa, Bernard and Michael Grzimek (father and son) accomplished a monumental achievement in bringing public attention to the threat to the animals of the Serengeti Plain.[6] Yet these and similar efforts ultimately required the assistance of many people, and usually the cooperation of governments. Of wider global impact, but often with less public notice, has been the initiating and organizing work of a large number of dedicated persons, of whom Fairfield Osborn, Harold Jefferson Coolidge, and Peter Scott have been outstanding examples. But for the unremitting efforts of such perceptive individuals, governments would not have been moved and international networks of environmental concern would never have been woven or public action would have been longer delayed.

Here a fundamental difference between intuitive and science-based understanding of the biosphere is pertinent. Early efforts at wildlife protection proceeded on the assumption that it was sufficient to prohibit the hunting or collecting of wildlife or, for truly endangered species, to safeguard them in botanical gardens or zoos. It was not readily apparent to many of the early conservationists or governmental protectors of wildlife that the survival of particular plants and animals might be contingent upon the survival of the ecosystem of which they were a part. Thus, while laws and international treaties to protect species multiplied, the destruction of the natural environment and thus the habitat and life-support systems for these species continued. While it is true that hunting accounts for the extinction of many species of animals, it appears to be also true that habitat changes (both man-made and natural) account for a larger number.[7]

Today this close linkage of habitat with survival is widely understood. Governments and international organizations such as FAO, UNEP, and nongovernmental and private groups such as the International Union for Conservation of Nature and Natural Resources (IUCN) and the World Wildlife Fund now act upon the understanding that habitat preservation and species preservation must go hand in hand. It may be true that the survival of some species of plants and animals may be ensured by man in artificial environments such as botanical gardens and zoos for indefinite periods of time. But how long this can continue has yet to be demonstrated. Whatever the case, only a small percentage of living species can be protected by artificial means.

Nature, to be protected, must first be understood. Two major international investigations have been undertaken to advance this understanding. The first, UNESCO's Man and the Biosphere Programme (MAB) is discussed in Chapter 12 of this book. The second is the International Biological Programme (IBP). Of course many scientific investigations, governmental and nongovernmental, continue.

After two years of preparation, the report of the planning committee for the IBP was approved in November 1963 by the Executive Committee of the International Union of Biological Sciences (IUBS) and the Tenth General Assembly of the International Council of Scientific Unions (ICSU).[8] It was officially established in 1964 and continued for ten years. The program was focused on "the biological basis of productivity and human welfare," with the objective of ensuring worldwide study of "(1) organic production on the land, in fresh waters, and in the seas, so that adequate estimates may be made of the potential yield of new as well as existing natural resources, and (2) human adaptability to changing conditions." Coordination was provided through ICSU's Special Committee on the International Biological Programme (SCIBP).

According to its scientific director, the reason for the establishment of IBP was recognition that the rapidly increasing human population called for a better understanding of the environment as a basis for rational management of natural resources.[9] This could be achieved only on the basis of scientific knowledge, which in many fields of biology and in many parts of the world were felt to be inadequate. At the same time it was recognized that human activities were creating rapid and comprehensive changes in the environment. In relation to environmental policy, the value of the IBP lay in its advancement of basic knowledge relevant to the needs of man.

An essential condition for the study of ecosystems is to have some relatively intact ecosystems to study, among other reasons, as "controls." Natural ecosystems—evolved systemic and localized interrelationships among organisms with their environments—are everywhere under stress from human activity. They vary greatly in critical size for survival and in relative toughness and fragility. Protection of nature and of genetic diversity therefore necessitates preservation of as many distinctive types of ecosystems as possible. Many natural areas have been set aside as protected reserves by public and by private organizations. Some of the latter, such as The Nature Conservancy in the United States, have assisted international area protection efforts.

According to the IUCN and World Resources Institute approximately 8,619 nationally and internationally designated natural areas were under protection in 1993.[10] As of the mid-1990s international protection systems include: Biosphere Reserves, 312; World Heritage Sites, 100; and Wetlands of International Importance, 590. National parks (shortly to be considered) are also protected areas, but not all are "natural." Biosphere reserves set aside for ecological research and habitat preservation have been described as "undisturbed natural areas for scientific study as well as areas in which the conditions of disturbance are under careful control."[11] A major objective of the Biosphere Reserve program is to preserve unspoiled examples of the various biogeographic regions of the earth.

Endangered Species

The preservation and protection of species and ecosystems is complicated by disagreements and uncertainties in the scientific community regarding what measures are necessary or feasible. When experts disagree, the preference of governments is often to do nothing, but circumstances often force public action. The greatest threat to the preservation of species and ecosystems today is the inordinate and continuing increase and spread of human population. In spite of protective efforts which are now considerable, the end result of present trends would be to crowd wild nature off the planet. Were human populations stabilized, and geographical distributions somewhat altered (for example, by removing cultivation from agriculturally marginal lands), the prospects for habitat preservation and the survival of wildlife populations could be greatly enhanced. This prospect not being foreseeable, governmental and international efforts at species protection pursue more politically feasible lines of action.

Traditionally, the governmental response to species endangerment is to curtail overexploitation. National legislation and international agreements have focused on hunting, harvesting, and international commercial trade in plants and animals and their products. Within national boundaries, their protection has been and is still regarded as primarily a national responsibility. But the possibility of lucrative international trade in flora and fauna has handicapped enforcement of protective measures. Thus the Endangered Species Treaty and associated national legislation represent a major advance in species protection against one of the more obvious causes of species decimation or extinction.

The Convention on International Trade in Endangered Species of Wild Fauna and Flora (CITES), negotiated in 1973 and effective in 1975, is a major landmark in international cooperation to protect and preserve the natural environment.[12] Following the initiative of the American Endangered Species Conservation Act of 1969, the 1972 United Nations Conference on the Human Environment endorsed an international treaty and recommended that, upon its ratification, secretariat functions be undertaken by the United Nations Environment Programme. As of July 1989, 101 nations had ratified the treaty. Under Article 12 of the treaty, secretariat functions are provided by the executive director of UNEP, who carries on this responsibility through the IUCN and the World Wide Fund For Nature (WWF). Through the IUCN Survival Service Commission, WWF has since 1956 assisted international efforts on behalf of endangered species.[13] Since January 1980, the government parties to CITES have begun to contribute to a special trust fund intended to ensure future implementation of the convention under UNEP/IUCN administration.

Effectiveness of CITES depends upon the identification of plants and animals that are endangered and the continued monitoring of their changing status. These tasks have been provided for many years by IUCN through its *Data Book* series. The *Red Data Book* series classifies species into four categories: extinct, endangered, vulnerable, and rare. Included are descriptions of individual species according to distinguishing characteristics, distribution, estimates of numbers, status, breeding rate in the wild and reasons for decline, protective measures already undertaken or proposed, number held and breeding potential in captivity, and other relevant remarks and references. The *Red Data Books* constitute a basic, reliable, and objective source of information that may readily be consulted by national and other international bodies involved in conservation efforts. Currently listed are more than 1,000 species and subspecies known to be threatened with extinction: 400 birds, 305 mammals, 193 fish, and 138 amphibians and reptiles. Furthermore, the IUCN's Threatened Plants Committee estimates that at least 25,000 species are now endangered.[14]

Supplementing CITES, a treaty to provide greater protection for vulnerable migratory species has been negotiated. The Convention on Conservation of Migratory Species of Wild Animals, signed at Bonn on 23 June 1979, obliges parties to protect endangered migratory species, which includes any threatened "entire population of . . . species or lower taxon of wild animals, a significant proportion of whose members cyclically and predictably cross one or more national jurisdictional boundaries." Further, the treaty makes it incumbent upon signatories to attempt to conclude agreements for the conservation of migratory species whose status is "unfavourable." [15] The parties to the convention acknowledged the need to take action to protect any of these animals from becoming endangered. The extent to which the parties actually take action is, however, another matter.

The convention includes two appendices, the first of which deals with endangered migratory species and includes various provisions for range states (any state that exercises jurisdiction over any part of the range of a migratory species) to conserve the habitats of such species and to prohibit their taking. For species not in this category, the convention is less restrictive. Appendix II lists migratory species "which have an unfavourable conservation status which would significantly benefit from the international cooperation that could be achieved by an international agreement." Provision was made for a conference of the parties, a scientific council, and a secretariat to be provided by the executive director of UNEP.

As a habitat for endangered species, the international commons of the oceans is one of the largest and biologically most important areas of the earth. Marine mammals in particular have been severely stressed by industrial society. Protective measures on their behalf have been considered from differing perspectives elsewhere in this book (notably in Chapter 9). Progress toward effective international protection has been slow, although the North Pacific fur seals have obtained protection at the price of harvested pelts. International efforts toward a broadly inclusive plan of protection have been made by the IUCN, WWF, and the Working Party on Marine Mammals of the Advisory Committee of Experts on Marine Resources Research (ACMRR) of FAO. From these efforts and a joint FAO-UNEP Scientific Consultation in Bergen, Norway, in 1976 there emerged the Draft Global Plan of Action for the Conservation, Management, and Utilization of Marine Mammals.[16] The draft plan was presented to the UNEP Governing Council in May 1981, but the council took no official action toward adopting its provisions. In 1984, however, the Global Plan of Action for Marine Mammals was finally issued.

There are many species, local or regional in their distribution, for which protection, to be effective, must have the cooperative and concerted attention of governments in the area. Efforts to institutionalize this cooperation have been sought through bilateral or regional agreements. For example, although CITES Article 5 prohibits the export, import, and traffic in polar bears or any of their parts and products, few states are directly concerned with the protection of these animals. To provide closer attention to their plight a regional convention on the conservation of polar bears came into force in 1976 with Canada, Denmark, Norway, the Soviet Union, and the United States as parties.[17] With a few exceptions, the agreement prohibits the hunting, capturing, or killing of polar bears. These exceptions include the taking of bears for scientific or conservation purposes, to prevent serious disturbance to the management of other living resources, and by indigenous people using traditional methods in the exercise of their traditional rights and laws. The use of aircraft and large motorized vehicles for capture or killing of the bears is prohibited consistent with local laws. The Soviet government designated Wrangle Island in the Arctic Ocean as a principal preserve for polar bear populations.

Wildlife on the Antarctic continent has been protected by the treaty, signed in 1959. On 7 April 1982 a Convention on the Conservation of Antarctic Marine Living Resources came into force.[18] The growing pressure from commercial fishing in the Antarctic or Southern Ocean, the conflicting interests of nations in the region, and its remoteness from populated areas, will surely complicate its implementation, but monitoring by remote sensing might help (see Chapter 9).

The ability of conservationists to dramatize the threats to endangered species of animals helps greatly in mobilizing public support for protective measures. But people and their governments are not as readily aroused in defense of wild plants. The IUCN in "A Plea for Plants" points out that "conservation action on behalf of world plants is minimal. In the world at large there is precious little awareness of the extent to which our lives depend on plants or of the rate at which they are being wiped out."[19] International protective action thus far has focused on preventing the transmission of diseases or pests harmful to plants of agricultural value — that is, to domesticated plants.

Reinforced by considerations of agricultural economics and public health, the International Plant Convention was signed in Rome, 6 December 1951, becoming effective in April of the following year.[20] The purpose of this treaty has been to obtain international cooperation in preventing the transnational spread of plant pests and diseases, and controlling the importation of plants and plant products. Although the administrative procedures of quarantine, inspection, and issuance of phytosanitary protection certificates may superficially appear simpler than efforts to control the international movement of birds, reptiles, and mammals, appearances are often deceptive. Some of the plant pests such as the Mediterranean fruit fly have a high rate of mobility, and others, notably viruses as in the Dutch elm disease, are difficult to detect until they have established themselves in a new habitat and their destructive effects have become visible. The driving of highways into the equatorial rainforest may have facilitated the spread of viruses from hitherto restricted niches to worldwide epidemics. The HIV virus, which leads to AIDS, is a case in point.

As with many animals, the distribution of many plants, wild and cultivated, is highly localized. For this reason national and regional agreements for plant protection have been negotiated—for example, the Plant Protection Agreement for South East Asia and the Pacific Region signed in Rome in February 1956.[21] Plant protection commissions have also been established for the Near East and Caribbean regions, with their reports published by FAO. The Soviet bloc of nations adopted in 1959 its own version of an agreement concerning Cooperation in the Quarantine of Plants and Their Protection against Pests and Diseases.[22]

Unfortunately, these conventions afford little protection for wild plants, whose species are rapidly being reduced by the spread of human settlements, agriculture, and development, especially in tropical regions. Moreover, the migration or importation of healthy plants (and animals) may lead to disastrous results in environments where their normal ecological controls are not present. And fears have been expressed that genetic diversity among domesticated plants may be reduced by the spread of patented stocks developed through genetic technology and promoted throughout the world by multinational agribusinesses. Whatever the validity of this fear, it does argue the importance of maintaining reserves of diverse plant forms, both wild and domesticated.

Tropical Forests

One of the world's great ecological disasters in the closing decades of the twentieth century has been the pervasive destruction of tropical forests. The forces which have combined to destroy the forests are excessive population growth, poverty, inordinate demand for raw materials in the industrial societies, and technologies that facilitate forest exploitation.

Deforestation occurred much earlier in temperate latitudes, but with consequences very different from those incurred in the tropics. It has been a serious mistake to extrapolate the process of converting forests to farms, as in France or Ohio, to the fundamentally different forest ecosystems of the equatorial regions. In general, the effects of removing forests differ greatly from one region to another depending upon conditions of climate, terrain, and soil; even in temperate regions deforestation has sometimes had disastrous consequences. The ecological impoverishment of Mediterranean and Middle Eastern countries and China through deforestation, among other factors, has been extensively documented.[23]

Unwise deforestation in temperate latitudes has had ill effects locally and regionally, but does not approach the potential for worldwide harm that could follow the loss of tropical forests.[24] These forests are believed to have a significant role in maintenance of the global atmosphere. Further, since their soil may not be convertible to agricultural cultivation, deforestation may worsen present social and economic conditions in many Third World countries. Exponential population growth and a corresponding increase in poverty have not only encouraged governments to sponsor settlement schemes at the expense of the natural environment, but have also forced poor farmers farther and farther into forests and onto marginally productive lands in pursuit of subsistence agriculture. Shortage of energy sources in tropical countries has also led to excessive tree cutting for firewood.

Meanwhile heavy demand for forest products in populous and affluent developed countries has been answered by corporate enterprise equipped with machines for their harvesting in regions that would have been inaccessible as late as a generation ago. As one observer points out, "whole forests can now be utilized—and destroyed—by wood chipping machinery."[25] Propositions like Daniel Ludwig's Jari Project could turn large areas of the Amazonian forest into pulp wood plantations.[26]

Countries possessing tropical forests face a dilemma. On the one hand, there are great short-run pressures to exploit the forest for economic purposes. But the penalties for overexploitation to meet urgent necessities today could be irretrievable ecological, social, and scientific deprivation in the future. Faced with growing populations of poor and hungry people, governments in tropical countries look upon these forests as logical answers to human needs. But the moist forest is not a cornucopia. Much of it is of low fertility, its survival depending upon the integrity of rapid recycling, leak-free nutrient systems. Over large areas, the species richness of the forest has already been greatly diminished because interrelating ecological networks have been cut by roads and settlements. As previously noted, invasion of the tropical rainforest could have the paradoxical effect of spreading hitherto isolated pathogens and simultaneously destroying botanical species of potential medical value.

The nature of the impending ecological disaster is threefold: first, a great loss in genetic and evolutionary diversity; second, a loss in the restorative and reproductive capabilities of those areas that can be reforested; and third, further impoverishment of the people who can no longer derive even the basic necessities from the area. But the losses are not limited to the tropics; they are felt in places far removed. A study published by the Smithsonian Institution reveals a major decrease in the migratory songbirds that have wintered in the mature forests of Central and South America and fly north each spring to breed.[27] Thus, the quality of life and environment diminishes in the northern latitudes with the loss of "one of nature's most colorful vocal and visual exhibits—as woodlands die, so do songbirds."[28]

A description of this syndrome of progressive ecological and social impoverishment is provided in a newsletter by Gary S. Hartshorn, Forest and Man Fellow of the Tropical Science Center in San José, Costa Rica. Describing the tragic confrontation of people and environment in El Salvador, he reports:

Some pine forests occur in the rugged highlands along the Honduran border, but they are being exploited voraciously. Logging activity is usually followed by putting cattle into the more open woods to browse or graze the grass and woody regrowth. Overgrazing inhibits tree regeneration and leads directly to soil degradation through compaction and surface erosion. . . . I was struck by the huge volume of firewood stacked along any and every back road waiting for transport to San Salvador the capital city. Even though El Salvador has virtually no forest left and is rapidly approaching a firewood crisis these are not forest resource problems, but human ecologic problems caused by overpopulation.[29]

In Asia, Africa, and the islands of the tropical seas the relentless felling of the forest proceeds. Under environmentally indifferent dictatorships and in countries suffering severe social disorder, as in Uganda in the 1970s, excessive and wasteful timber cutting has occurred.[30] Alison Jolly describes the poor peasant of Madagascar as "the rainforest's executioner. He is the last frontiersman in a world that has no wilderness to spare. In a defiance of Malagasy law and all principles of conservation he still tackles the virgin forest."[31]

Each contributor to the destruction of the tropical forest benefits in the short run, although no one benefits in the long run. World forestry congresses have repeatedly warned governments about the consequences of the disappearance of the tropical forests. In addition, the Eighth World Forestry Congress, meeting in Jakarta in October 1978, urged action by governments to meet the need for fuel wood and charcoal, which was becoming a "poor man's energy crisis," to emphasize the need to scale industries to the economies of developing countries rather than to encourage economies of scale, and to utilize the available means to deal with watershed management problems.[32]

Strategies to contain the destruction and to preserve the genetic resources of the tropical forests have been proposed. Some have been put into effect. But it is too early to predict the extent to which these efforts may succeed. The Food and Agricultural Organization through its consultative group on international agricultural research and the International Board for Plant Genetic Resources (IBPGR) has provided limited support for research on the genetic potential of certain trees important in agriculture—primarily food-bearing trees.[33] The FAO Panel of Experts on Forest Gene Resources and the FAO staff assist in the coordination of this research, one objective of which is to develop a network of individuals in participating countries so that information regarding opportunities and findings in exploration, collection, and research may be widely disseminated.

The International Council for Research in Agroforestry (ICRAF), which is headquartered in Nairobi, cooperates with UNEP's Ecosystems Task Force in planning and coordinating on a worldwide basis research in combined systems of land management for agriculture and forestry. Areas of study include tropical forests and wetlands. The development of managed tropical ecosystems may be one way to accommodate urgent human needs and so buffer natural genetic forest preserves from destructive exploitation.

The findings of forest ecology research and the efforts of UNEP, FAO, and IUCN to stop the wholesale destruction of tropical forests are beginning to have some effect. For example, a UNEP-sponsored meeting on the world's tropical forests was held from 25 February to 1 March 1980 in Libreville, Gabon. Meetings in themselves do not save forests, but may be a necessary prelude to action. Also in 1980 nine African countries meeting at Yaounde, Cameroon, agreed on articles of a treaty relating to the improved management of tropical forests and ecosystems. Further, with UNEP assistance a regional Centre for Scientific Information and Documentation in Tropical Ecology is to be located in Yaounde.[34] Its main purpose is to provide a bank of ecological data that may be used in socioeconomic development projects. The history of the center illustrates the multi-institutional interaction that characterizes international environmental policymaking. The idea for the institution developed at a UNESCO-sponsored regional meeting of MAB at Kinshasa, Zaire, in 1975. The government of Cameroon offered to host a regional meeting for the center, and UNEP support was approved at the Sixth Session of its Governing Council. In addition to UNESCO, consultation regarding the center and its mission involved WHO, WMO, UNDP, and IUCN.

Tropical agroforestry has been cited as one of the more promising aspects of development; yet there are uncertainties of the extent to which the promise may fulfill the human needs that confront many Third World tropical countries today.[35] Careful site and species selectivity and competent management are essential for success; there have been disastrous failure in agroforestry. In summarizing the dichotomy of management and preservation, two tropical forestry experts, J. P. Lanly and J. Clement, conclude that:

> When looking at forest area statistics or discussing gains and losses in forest areas, two realities have to be kept in mind. One is that any given area of plantation forest is much more productive of industrial wood than an equal area of natural forest. The other is that the loss of natural forest means much more than an economic loss of wood. It means a transformation, damage or loss of whole environmental systems, a disappearance of plant and wildlife species, the destruction of watersheds, the onset of erosion, the possible change of whole water regimes in valleys and low lands. Management of forests is not simply about wood production, but about the state of health and productivity of the environment as a whole, including adjacent and even distant agricultural and urban areas.[36]

The efforts just recounted appear to be leading toward a more general international set of policies for tropical forests. In March 1983, following six years of preparation, the United Nations Conference on Tropical Timber convened in Geneva under the auspices of UNCTAD. The conference, which will reconvene at a later date, adopted all but six of forty-three articles under consideration. Sixty-four countries were represented, and agreement was reached on establishment of a new tropical timber organization. Although the conference primarily represented producers and consumers of timber, the conference chairman, Tatsuro Kunugi of Japan, emphasized that the agreements went beyond traditional commodity arrangements to include a global policy on resource management, taking into account the implications of this policy for other important sectors such as energy, agriculture, food supply, and the preservation of ecosystems. The International Tropical Forest Timber Agreement and Action Plan took effect in 1985 and is being implemented by the International Tropical Timber Organization (ITTO), established in 1987. Ecological considerations have been built into ITTO objectives and activities and will be closely watched by UNEP, IUCN, WWF, and many NGOs.

Nevertheless the destruction of tropical forests continues. Nicholas Guppy believes that an Organization of Timber-Exporting Countries (OTEC) could be the most promising institutional arrangement to control the trade in tropical timber.[37] None of the foregoing efforts seems to have stopped the relentless gnawing away of the tropical forests by poor peasants, or overcome the common interest of international corporations and pliable local officials in the exploitation of timber. Even in developed countries such as the United States and Canada, forests are under constant pressure by corporate interests and their political allies to produce jobs and revenues. A treaty on forest principles was proposed at UNCED in 1992, but failed to obtain sufficient support from countries having tropical forests. The best that could be obtained was a nonbinding Statement of Forest Principles (see Chapter 6).

Critical Wetlands

Another class of threatened ecosystems is wetlands, notably estuaries, marshes, and lakes. The need for international action on behalf of wetlands was recognized as early as 1962 when an International Conference on Wetlands, organized jointly by the IUCN, the International Council for Bird Protection (ICBP), and the International Wild Fowl Research Bureau (IWRB), was held in the south of France at Les Saintes-Maries-de-la-Mer.[38] This conference recommended that a list be compiled of European and North African wetlands of international importance to be considered as a basis for an international convention. At European meetings on waterfowl conservation—in Scotland in 1963 and the Netherlands in 1966—further action was taken toward a convention. A preliminary draft was circulated prior to the International Regional Meeting on Conservation of Wild Fowl Resources convened in Leningrad in September 1968; the meeting resolved "to accelerate the adoption of a convention concerning wetlands conservation." [39] The final text of the Convention on Wetlands of International Importance Especially as Waterfowl Habitat was adopted by the international conference at Ramsar, Iran, on 2 February 1971.[40]

And so, after years of deliberation and preparation, with collateral assistance from the Man and the Biosphere Programme, the wetlands treaty entered into force in December 1975. Its terms required national initiative by each signatory to conserve wetlands as regulators of water regimes and as habitats of distinctive ecosystems. Each adhering nation agreed to designate particular wetlands of international importance for recording on a list to be maintained by the IUCN and the International Waterfowl and Wetlands Research Bureau (IWRB), a nongovernmental body located at Slimbridge, England. In 1994 the World Resources Institute reported that some 590 wetland sites with a total of over thirty-six million hectares had been recorded on the Ramsar List of Wetlands of International Importance. An independent bureau sponsored by IUCN and IWRB administers the treaty to which, as of 1994, seventy-five countries adhered.[41] Although slow to take effect, the Ramsar treaty appears to have gained impetus through the 1980s as a consequence of growing worldwide ecological awareness.

Even so, the pressures of immediate human necessity upon wetlands in many parts of the world are severe. In a report headlined "Ecological 'Disaster' in China," Thomas Land describes the consequences of an effort to transform several million acres of marshland in Heilungkiang Province in northeast China into agricultural land. Falling water tables and soil erosion have diminished expected yields of wheat, and the United Nations University and the Chinese Academy of Sciences have been collaborating to find ways of overcoming these adversities. Nevertheless, the loss of large areas of forest and wetlands to wheat implies an ecological impoverishment that already characterizes vast areas of China. A survey of the environmental constraints on development in China has been prepared by Vaclav Smil for the Office of Environmental Affairs of the World Bank.[42]

For similar reasons relating to food production and economic development, the government of Sudan proposed the draining of the Sudd swamps on the Upper Nile through the Jonglei Canal.[43] In July 1975, the Sudanese government requested UNEP to undertake multisectoral prefeasibility studies on the project. Meanwhile the project has proceeded, although its ultimate ecological, social, and economic effects are uncertain.

With careful management, in some but not all areas, dense population and ecological variety and diversity may coexist. But the practical necessity for governments to assure the feeding of their people and to promote economic growth confirms the proposition that population levels are an inescapable constraint on all efforts to preserve the quality of the biosphere. In the meantime, marshlands are turned into cultivated croplands, habitat for the natural plants and animals disappears, and the ecological impoverishment of the earth continues.

Pastoral Grasslands

A way of life widely disparaged in industrial and agrarian societies is pastoral nomadism. Modern governments dislike mobile populations that are difficult to tax and conscript and often pass illegally across national boundaries. Development experts have generally advised that they be settled. Nevertheless, some ecologists and students of rangeland management have concluded that the rotational grazing, traditional among nomadic pastoralists, may make the best possible use of large areas of semiarid land.[44]

The grasslands issue has been especially significant in North Africa and the Middle East. A report issued by the UN Economic Commission for Western Asia concluded that "[a]nimal husbandry as practiced by the bedouins may be the best adaptation to arid land."[45] Studies of the ecological consequences of the sedentarization of nomads suggest that health and quality of life among nomadic peoples in Central Asia, Iran, and North Africa were generally better before than after settlement. This conjecture is subject to a number of qualifications, including the characteristics of particular ecosystems, the method and quality of settlements, and the accessibility of hygiene and health care to nomadic people. An IUCN study of the ecological consequences of the bedouin settlement in Saudi Arabia suggested that nomad life could be "modernized" without destroying the traditional mobility of its grazing practices, which make optimal, nondestructive use of fragile semiarid lands.[46] The IUCN has urged restoration of the traditional *hema* system of rotational grazing to restore damaged pastoral lands of western Asia, and in cooperation with Oman, Jordan, and Saudi Arabia has been studying the feasibility of combining the advantages of pastoral nomadism with protection of ecosystems and endangered species such as the Arabian oryx.

In central and southern Africa where European and Indian cattle have been imported and intensive grazing practiced, serious ecological harm has ensued. A severe case of conflict between economic pursuits and environmental quality has occurred in Botswana where Z. A. Konezacki found that growth of livestock raising in excess of the land's carrying capacity has led to its deterioration.[47] Private ownership of fenced range land has time and again resulted in overgrazing. But range deterioration occurs as readily on land under common ownership where no one sees a personal advantage in grassland conservation.

In Africa pastoral nomadism, modern animal husbandry, and preservation of the world's most extraordinary assemblage of wildlife are in conflict. Particularly on the higher plains of eastern and southern Africa large herds of wild animals have been threatened by encroachments of domesticated cattle and cultivation. Although the governments of East Africa have tried to protect and preserve this exceptional habitat for wildlife, economic pressures of growing human populations pose a continuing threat to the maintenance of large wild areas. Moreover, poaching has been a continuous hazard. One strategy for enlarging the economic base for East African populations and diminishing the incentives for poaching has been to establish game ranches on which native African grazing mammals can be treated much as domesticated cattle, providing a source of protein for the population as well as hides and trophies. Game ranching has been tried with some success and may yet offer one measure of protection for wildlife habitat on the East and South African plains.

In most of the foregoing examples of the international policy dimensions of nature protection, the counsels of international aid-giving and advisory agencies have often been contradictory. The idea that more be obtained from the land in the long run by taking less today seems paradoxical to governments that are more readily persuaded to push economic growth today and worry (or leave it to others to worry) about the costs tomorrow. Those costs may be heavy; the adverse and ramifying effects of degrading grasslands to deserts are considered in the final section of this chapter.

Preserving Culture: Communities and Artifacts

Culture joins human concern for the natural environment of species and of ecosystems to the artificial environment of cities, rural settlements, managed ecosystems, as well as to technology and art. Culture provides the norms that unite people within and among generations, and this continuity is essential to the cohesiveness and stability of any society. Comparisons among peoples at all stages of development and condition point to the conclusion that people's behavior in relation to their environment is strongly conditioned by their culture. Yet to the extent that cultures are innovative and change (whatever the cause) comes rapidly, tension and even conflict may arise between tradition and conservatism on the one hand and innovation and reformulation of norms on the other.

These processes of cultural resistance and change are local or national to an extent that might appear to preclude international concern. Nevertheless, they have aroused international attention and become the objects of a variety of international environment-related programs and projects on several accounts. First, the conditions of community life, urban or rural, affect the political behavior of a country, including its international relations. Mass migrations from poor or disturbed countries have become a common phenomenon of the late twentieth century. Second, humanitarian motives—or perceived self-interest—persuade nations to help struggling countries cope with problems of urban and rural development that exceed local capabilities. Third, a sense of the universality of human creative experience has spread throughout the world, causing people of different nationalities and cultures to regard all cultural achievements as a common heritage.

Qualities of Urban and Rural Life

In a divided world of sovereign nations there has hitherto been little scope for international environmental policy in relation to localized conditions of living. The principal international concerns have been those expressed through bilateral aid programs and the social and economic programs of the UN specialized agencies. These efforts have included health and safety measures promoted by the European Foundation for the Improvement of Living and Working Conditions and by the ILO and WHO; improvements in rural living assisted by UNDP, WHO, FAO, and more recently by UNEP and the Habitat Centre; and by a variety of programs relating to housing and planning initiated through the Economic and Social Council. Assistance is characteristically offered in the form of education and training, demonstration projects, model codes, and technical advice. Numerous examples could be given of various efforts over the years: for instance, in 1970 the United Nations and the government of Colombia sponsored an interregional seminar on improvements of slums and uncontrolled settlements; and in 1972 the World Bank approved its first loan for self-help housing, including a small, pilot squatter improvement scheme in Senegal. The bank has now assisted squatter-improvement programs in such countries as Botswana, El Salvador, Indonesia, Jamaica, Kenya, Tanzania, and Zambia.[48] The United Nations Centre for Human Settlements/HABITAT has published a selective survey of slum and squatter settlements in Africa, Asia, and Latin America.[49] UNESCO has been concerned with quality of life in many different respects, notably through the Man and the Biosphere Programme, and plays a unique role in the preservation and protection of the cultural heritage of mankind in sites and monuments.

It is unrealistic to separate consideration of the socioecological and environmental problems of Third World cities and rural areas, because their fates are joined together. Nevertheless, a phenomenon of mounting urgency and menace is the rapid and uncontrolled growth of great cities throughout the Third World and the recently industrialized nations. Modern techniques in public health, mechanized mass transit, electricity, and industrialization have permitted populations in developing countries to surge into centers such as Calcutta, Cairo, Djakarta, Lagos, Mexico City, and Sao Paulo. The infrastructures of Third World societies have been almost uniformly unprepared to cope with this development, being especially deficient in the concepts and techniques of demography, urban and regional planning, and public administration. Beyond the façade of tall steel and glass structures of international architecture, these aggregations of people resemble vast, sprawling, disorganized villages. Barbara Ward has described them as "cities that came too soon," and they pose a host of socioecological problems for the societies in which they have arisen.[50] These cities reach out farther and farther into the countryside, depriving the hinterland of the resources needed for agricultural production to feed malnourished urban masses. For great distances surrounding many Third World cities, forests have been totally destroyed in the quest for firewood. The deforestation of the countryside, accompanied very often by overgrazing and improvident cultivation, results in the erosion of soils and the siltation of dams and streams. Gary Hartshorn reports of San Salvador that "ecological disaster is amply evident from the eye searing air pollution, . . . the stench from rivers that are open sewers, the heavy siltation that robbed the '5 de Noviembre Dam' of its hydroelectric capacity in only thirty years, virtually unlimited spraying of DDT and related pesticides on cotton fields. . . . El Salvador is a very real example of Ehrlich's population bomb ticking away."[51]

There is almost no end of prescriptions for alleviating the environmental ills of Third World cities. Many of them are sensible, but few are capable of realization under prevailing conditions. Were populations to stabilize, these reforms might have the desired effect with international assistance reinforcing local efforts. However, in few countries are populations stabilizing. In fact, in spite of declarations and programs, circumstances generally appear to be worsening.

A Symposium on Human Settlements held at Tepoztlan, Mexico, from 30 April to 3 May 1980 resulted in a declaration drafted by twelve Latin American experts on planning and development that appealed to governments to fulfill the principles of the 1976 HABITAT Conference on economic development and the role of human settlements in that process. Unfortunately, the declamatory approach to the problems of settlement does not go far toward solving them. In language suggesting the Cocoyoc Declaration, the Act of Tepoztlan declared that

> poverty will be only overcome through the transformation of the predominant development styles which favor economic growth and the concentration of income, ignoring social equity. Such transformation is not viable if actions are limited to appeals to the conscience of the rich countries and the rich social groups. A necessary precondition is the elimination of those obstacles that oppose the decentralization of power in order to favor basic regional and local organizations.[52]

Whether the numerous acts, declarations, and resolutions regarding quality of life, environment, and human settlements have led to significant accomplishment is uncertain. Obviously, rhetoric alone will accomplish little. At any event, organization and research efforts are necessary precursors to effective action. The United Nations economic commissions have assisted HABITAT and the specialized agencies in attacking problems of human settlements. For example, the Intergovernmental Regional Committee on Human Settlements, established by the Economic Commission for Africa, may be having some effect in stimulating the five African countries receiving technical assistance from HABITAT to cooperate where need be, and to learn from one another's experience. Dealing with very different environmental circumstances, the Economic Commission for Europe in August 1978 at Godthab, Greenland, held a symposium on problems of settlement planning in the Arctic region. Resulting from this seminar is a book that reviews experience from eight Arctic countries over the past two decades in the planning and development of Arctic communities.[53]

HABITAT: Environment of Human Settlements

The environments in which quality of life, or its absence, is most directly experienced by people are those in which they pass their daily existence: residences, immediate neighborhoods, workplaces, and centers of public assembly and institutional custody (e.g., schools, hospitals, and prisons). In these locales are the physical and social aspects of quality of life most clearly integrated into a whole environment.

Rapid population increase, accompanied by rural to urban migration, have created major problems of housing, health, and public services in almost every country. These problems have been most acute in Third World countries where developed resources have been scarce in proportion to needs. Accordingly, international assistance has been sought, with consequences that can most fairly be described as mixed. At the 1972 Conference on the Human Environment, the representatives of the Third World achieved only partial success in obtaining support for the recommendations on "Human Settlements Management" (Recommendations 1–18). But a United Nations Conference on Human Settlements (HABITAT I) did take place in Vancouver, Canada, in 1976 and adopted an action plan that included a voluntary fund for human settlements research and development. The United Nations Department of Economic and Social Affairs had previously maintained a Centre for Housing, Building, and Planning in New York, but after 1977 this center was merged into the Habitat program.

The United Nations Conference on Human Settlements, or HABITAT as it has been more often called, convened in Vancouver from 31 May to 11 June 1976.[54] Its format followed very closely that of the 1972 Conference on the Human Environment. A preparatory committee, assisted by a secretary-general and his staff, guided conference preparation; three regional preparatory conferences were also held during 1975: in Tehran for Asia; in Cairo for Africa; and in Caracas for Latin America. In addition, a regional consultation was held in Geneva under the sponsorship of the Economic Commission for Europe. Reports from these meetings were fed into the deliberations of the preparatory committee. A meeting of international consultants was also held in mid-1975 at Dubrovnik, Croatia, to "seek an international basis for a new interdisciplinary science of human settlements and to assist the HABITAT secretariat in evolving an ideological framework to guide the next stages of the preparatory process for the conference."[55] In a manner similar to preparations for Stockholm, two intergovernmental working groups assisted the HABITAT secretariat to prepare documents regarding (1) a proposed declaration of principles, and (2) consideration of problems of international cooperation for presentation to the preparatory committee.

Another feature of the HABITAT Conference, which has become a routine feature of United Nations conferences, was the Forum of Nongovernmental Organizations. Response to inquiries addressed by the HABITAT secretariat to interested nongovernmental groups exceeded all expectations and elicited several hundred requests for invitations to participate.[56] A Nongovernmental Organization Committee for HABITAT, with offices in the Hague and in Vancouver worked in cooperation with Canadians on arrangements for the HABITAT Forum—a gathering designed to provide for an exchange of views and experience among the nongovernmental groups.

The forum assembled on 27 May at facilities made available by the Canadian government at Jericho Beach, Vancouver. It is not necessary to discuss the activities of the HABITAT Forum here, but it is pertinent to an understanding of the problems of international cooperation relative to quality of environment–quality of life issues that neither the conference nor the forum was able to separate ecological from ideological issues. It was much easier for participants to agree upon minimum quality of life values in human settlements than to agree upon the circumstances necessary to achieve these values.

The preparatory committee and the conference insisted on linking HABITAT deliberations with the Programme of Action on the Establishment of a New International Economic Order. If freed from ideological bias and rhetorical excess, this linkage would not have been unreasonable, because international economic arrangements do have effects upon quality of life and environment. But there are also sociopolitical factors for which the developed world cannot fairly be held accountable. The tendency of many delegates to the conference and the forum to attribute the problems of both urban and rural human settlements in the Third World to the unjust economics of capitalism and colonialism accentuated antagonisms rather than contributed to an atmosphere of mutual problem solving. These allegations of international injustice are not without some basis in historical fact. But they do not stand up well under critical examination because they distort and oversimplify the factors which have diminished the quality of life and economic well-being in an urbanizing, industrializing world.

Following up the Vancouver Conference, UN General Assembly Resolution 32/172 of 19 December 1977 authorized a unified pledging conference to establish the United Nations Habitat and Human Settlements Foundation (UNHHSF). Meanwhile, by Resolution 32/162 of 1977 the General Assembly created a Commission on Human Settlements to develop institutional arrangements for international cooperation. These actions led to the establishment in Nairobi of the United Nations Centre for Human Settlements (HABITAT) through which foundation funds would be administered.

Relations among the foundation (UNHHSF), the commission, the center, and UNEP were initially ambiguous. As with the Governing Council of UNEP, the priorities of the Commission on Human Settlements were developmental rather than environmental. At an African regional conference in 1978, the administrator of the foundation declared that the greatest need, apart from coherent policies, was proper financing and management. Other aspects of fundamental importance were mechanisms for subsidies, access to the housing market in poor neighborhoods, and control of land speculation.[57] Quality of environment—quality of life considerations do not appear to have been of highest priority at HABITAT, but these values may be regarded as implicit in the mission of the center and the foundation; moreover, in the present state of the world and particularly in urban areas and villages, there is little that can be done to improve the quality of the environment without money and management. A HABITAT II UN Conference took place in Istanbul, Turkey, 3–14 June 1996.[58]

World Heritage

Partially offsetting the bleak prospects for the quality of life in many parts of the world is a growing regard for the preservation of notable achievements of humanity's cultural past. Even in countries that now reject the social values and institutions that made those achievements possible, cultural monuments and artifacts are protected with pride as symbolic of the historic skill and creativity of their peoples and the quality of their natural endowments. The heritage concept is based on the assumption that the past has value. Moreover, it is based on the belief that cultural achievements anywhere have meaning for mankind as a whole.

The emergence of the world heritage idea following World War II may be attributed to the convergence of three closely related trends, which led to an extension to international policy of concerns formerly regarded as essentially local or national. The first of these was the international development effort. As a consequence of multinational enterprise and United Nations and bilateral aid programs to promote economic growth, international economic development was homogenizing the world. In the words of a commentator for the International Union for Local Authorities (IULA), "today a kind of 'international' architecture is invading Africa, Asia, and South America, without the slightest attention being paid to the specific character of the countries concerned, their customs, or their habits."[59] A German expert on architectural restoration observed that "people everywhere in the world expect better housing, sanitation, social amenities and opportunities for employment. In the rush of efforts to achieve these goals, traditional buildings, no matter how valuable they may be artistically, are in danger."[60]

A second trend that stimulated concern for the cultural and natural heritage was the rapid expansion of means of transportation and communication. Ideas, styles, and opportunities became more widely disseminated than formerly. This mobility made possible a third trend, which was the enormous increase in international tourism. Nations discovered that their cultural monuments, which had sometimes been viewed as economic burdens, could in fact become significant economic assets. Rapid and relatively cheap air transport made mass tourism feasible. It was the national heritage in distinctive landscapes, wildlife, and historic sites that attracted tourists, and it was these assets that international economic development often appeared to be threatening.

The role of world tourism in the development and implementation of the world heritage concept is very mixed. In 1976 the World Bank and UNESCO jointly sponsored a seminar on tourism for the purpose of analyzing its social and cultural effects. A report on the findings of the seminar described tourism as a "double-edged sword."[61] The ecological impact of tourism has been the subject of several inquiries by the International Union for Conservation of Nature and Natural Resources (IUCN), which has also provided a forum for consideration of the various effects of tourism upon the total environment.[62]

For heavily visited cultural monuments, the sheer number of visitors may be damaging. Prehistoric cave paintings in south central France have been endangered by side effects of heavy visitation. Paradoxically, these monuments suffer from the pressure of tourism that provides resources to maintain and protect a cultural heritage against attrition by the forces of time. Yet in many countries, the dedication of governments to protection of the cultural heritage is dubious, but might be reinforced by international encouragement. Reporting from Italy, one critic deplored the indifference of both government and developers to the degradation of ancient monuments and their surroundings by commercialization.[63]

Prior to the formalities of intergovernmental action, the world heritage idea was promoted by nongovernmental organizations. As early as 1932 an international conference of architects meeting in Athens undertook to formalize proposals for international cooperation in historic preservation. Following the disruptions of World War II, an International Congress of Architects and Technicians of Historical Monuments meeting in Paris (1957) reviewed the idea, and in Venice in May 1964, assisted by UNESCO, delegates from fifty-one countries adopted the so-called Venice Charter, the International Charter of Venice for the Conservation and Restoration of Monuments and Sites.[64]

In the organization and coordination of efforts to implement world heritage conservation, UNESCO has played a leading role.[65] In 1962 the Twelfth Session of the General Conference of UNESCO adopted a Recommendation Concerning the Safeguarding of the Beauty and Character of Landscapes and Sites.[66] This resolution was given practical expression through the International Council of Monuments and Sites (ICOMOS). Founded on 21 June 1965 in Warsaw, Poland, ICOMOS is a follow-up to the Paris and Venice congresses.[67] The purpose of ICOMOS is to "encourage the conservation of the architectural heritage of monuments, sites, historic quarters, and landscapes, to promote research and training in the field, exchange information, and promote the interest in the protection of this cultural heritage by authorities and the general public in all countries."[68] ICOMOS "conducts a broad cooperative international program based on the exchange of knowledge, experience and ideas through a documentation center, symposia, and publication." It has sponsored conferences on the conservation, restoration, and revival of areas and groups of buildings of historic interest, and publishes the periodical *Monumentum*.[69]

Another organization to protect the world heritage is the International Council of Museums (ICOM).[70] This body was founded on 16 November 1946 in Paris. Its purpose is to further cooperation between museums, to protect and promote their interests, widen their influence, and stress the importance of their role in education by the promotion of knowledge and understanding among peoples. The council is a federation of national and international committees representative of various types of museums; it, with the assistance of UNESCO, holds periodic conferences, and maintains a museum documentation center in Paris.

Also aided by UNESCO is the International Centre for the Study of the Preservation and Restoration of Cultural Property (ICCRON).[71] Established in 1958, ICCRON is an intergovernmental body consisting of sixty-one member nations, with headquarters in Rome. Functions of ICCRON include training, research, and exchange of conservation information and specialties among nations. The center, recognized as the foremost international preservation institute, collects and circulates information, initiates and coordinates research, and assists in the training of research workers and technicians. It has contributed significantly to the improvement of standards of preservation and restoration of cultural artifacts. The center's concern is preservation of cultural artifacts from all places and historical periods. Its mission program is perhaps its most publicized activity, center staff having been sent throughout the world on regular and emergency missions—for example, to study such problems as the effects of increased algal growth on prehistoric paintings in the Lascaux Caves in France, the erosion of Byzantine frescos in Göreme, Turkey, and the effects of air pollution on art treasures in Venice.

UNESCO has not only assisted but often taken a lead in mobilizing international action on behalf of cultural monuments and values. In 1955, at the request of the Egyptian government, UNESCO helped to establish a special center for research on the ancient art and civilization of the Nile Valley. In 1964 the lake that would be created by the construction of the Aswan High Dam on the Upper Nile threatened to inundate a number of the most extraordinary monuments of the high period of ancient Egyptian civilization. Among these were four colossal figures of the Pharaoh Rameses II at Abu Simbel, the temples at Philae, and archaeological excavations in Sudanese Nubia. Salvage of the temples and the colossi was beyond the means and inconsistent with the priorities of the United Arab Republic. Through the International Campaign to Save the Monuments of Nubia, financed by donations from fifty-two countries, UNESCO assembled teams of scientists, engineers, and stone cutters to accomplish what must have been the largest task of archaeological reconstruction ever undertaken.[72]

A very different international salvage effort assisted by UNESCO followed the disastrous 1966 floods in Florence and Venice, which damaged many of the finest art treasures of the Italian Middle Ages and Renaissance. An International Campaign for Preservation and Restoration of Cultural Property in Italy was launched on 2 December 1966 to assist the Italian people and authorities and "to receive voluntary contributions from governments, public and private institutions, associations and private persons" for this purpose.[73]

Another instance of UNESCO intervention to protect mankind's cultural heritage has been the restoration of the great Buddhist monument of Chandi Borobudur in central Java.[74] In 1955 the government of Indonesia asked UNESCO to advise on the problem of deteriorating stone in Indonesian monuments and especially in regard to Borobudur. For several years UNESCO-assisted investigations were undertaken, and in 1972 an international consultative committee of archaeological experts was appointed with the advice of the director general of UNESCO. On 29 January 1973 an agreement was signed by UNESCO and the government of Indonesia for a major reconstruction of the monument, and agreements were also consummated between UNESCO and several cooperating governments making voluntary contributions to the effort.

These and other arrangements sponsored by UNESCO in collaboration with national governments and nongovernmental organizations were ad hoc in character. Important to the preservation of the cultural environment in specific instances, they did not provide a systematic, institutional means for extending protection to the wider dimensions of mankind's cultural and natural treasures, and so the idea of a world heritage system began to emerge.

International restoration and preservation projects have also been undertaken by private philanthropy and through bilateral intergovernmental cooperation. For example, restoration of the town of Bhaktapur in Nepal was undertaken cooperatively by the government of Nepal and the Agency for Technical Cooperation of the Federal Republic of Germany.[75] The architecture of the badly decayed city was characteristic of the period regarded as the golden age of Nepal architecture and craftsmanship. Organizations have also been formed on a regional basis, such as the European Committee on Monuments and Sites established as advisory to the Committee of Ministers of the Council of Europe.

Assistance to countries rich in cultural and natural endowments, but poor in the financial, technical, and managerial means to protect them, was imperative if the heritage concept were to be more than a noble idea. But systematic, international means beyond UNESCO's capabilities were not easily conceived; the concept of sovereignty, suspicion of "cultural imperialism," and the inevitable question of finance were complicating factors.

The case for a supranational organization was argued by Dogan Kurban of Turkey (one of the countries richest in cultural sites) in the 1969 Conference on Monuments and Tourism held at Oxford, England. Yet he questioned the feasibility of this approach to preserving the world heritage in a "superheated atmosphere of international politics based on ideological differences." Instead he suggested that "perhaps a body of private, independent individuals from all over the world, with the consent and help of international organizations and national governments and without risk of being labeled by an ideological denomination, might endeavor to initiate such a movement."[76] This approach was, substantially, the one taken.

The idea of a world heritage endowment had been discussed in various forums during the 1960s. It appears that the "World Heritage Trust" was first proposed in 1965 at a White House Conference on International Cooperation. The proposal was to provide a mechanism for "international cooperative efforts to identify, establish, develop, and manage the world's superb natural and scenic areas and historic sites for the present and future benefit of the entire world citizenry." Following up the recommendations of the 1965 conference, Russell E. Train, then chairman of the U.S. Council on Environmental Quality, persuaded President Richard Nixon to propose in his 1971 environmental message the creation of a "World Heritage Trust."[77]

Preparatory to the United Nations Conference on the Human Environment, draft texts of treaties were prepared both by UNESCO and by IUCN, the former emphasizing the cultural heritage and the latter concerned primarily with the natural heritage. One of the preparatory bodies for Stockholm, the Intergovernmental Working Group on Conservation, considered both versions and recommended that a composite finalized draft be prepared. UNESCO convened a special committee of government experts to finalize its own draft convention but subsequently took the working group report and the IUCN proposal into account so that a single draft convention was finally prepared and was endorsed at Stockholm in Recommendations 98 and 99.

At the Seventeenth Session of the General Conference of UNESCO held in Paris, November 1972, the Convention Concerning the Protection of the World Cultural and Natural Heritage was endorsed. The favorable vote in the conference was not unanimous, there being seventy-five member states for, one against, and seventeen abstentions. Moreover, three years passed before a sufficient number of states had ratified to put the treaty into effect. Following Senate approval and signature by President Richard Nixon on 13 November 1973, the United States became the first nation to ratify the convention. On 17 December 1975, after twenty nations had ratified, it became effective.[78] By 1 January 1995, one hundred states had formally joined in the agreement and as of December 1994 some 440 natural and cultural properties had been placed on the World Heritage Reference Map.[79]

There are many problems of definition implicit in the treaty, not the least of which is the criterion of "outstanding universal value" by which sites and areas are judged a part of world heritage. How this phrase is to be interpreted is not entirely clear. Cultural heritage in the treaty is divided into monuments, groups of buildings, and cultural sites. The term "sites" is somewhat ambiguous, as it covers locations of historical, aesthetic, ethnological, or anthropological value. Natural heritage is classified into natural features, special formations and areas, and natural sites. Marine areas were not included, possibly because recommendations considering coastal areas and seascapes were more appropriate to the Law of the Sea Conference. However, in 1988 the International Law Association established a Committee on Cultural Heritage Law, which in 1989 began to prepare a Convention on the Underwater Cultural Heritage. A draft convention was completed in 1993.[80] The IVth World Congress on Natural Parks and Protected Areas meeting in Caracas in February 1994 launched a major program to establish coastal and marine protected areas.[81]

Determination of what monuments, sites, or areas are of outstanding universal value is a function of the UNESCO World Heritage Committee, which is authorized to select from nominations by various governments the ones to be added to the World Heritage List.[82] The treaty establishes nations' responsibilities for identifying and protecting the cultural and natural heritage situated on their territories as well as provisions for international assistance.

Commenting on the treaty, Robert L. Meyer observes that "it remains primarily a convention for international cooperation and assistance. In no way is it a 'world heritage trust' in the sense that the UN or an international body may act as a 'trustee' administering in trust a corpus of internationalized property having extraterritorial status with certain responsibilities assumed by the state where it is situated."[83] To provide a financial basis for international assistance, a World Heritage Fund was established with resources available to it from voluntary contributions, gifts, and bequests from national governments, from UNESCO and the United Nations Development Programme, from proceeds and receipts of events organized for its benefit, and from private sources. Assistance includes making studies, providing experts, technicians and skilled labor, training personnel, supplying equipment, and securing low-interest or interest-free loans, and in exceptional cases, nonrepayable subsidies. Following an agreement between the World Heritage Committee and a given country defining the conditions under which a program of assistance is to be carried out, the recipient government is responsible under Article 26 of the treaty "to continue to protect, conserve, and present a property so safeguarded."[84]

Assistance for National Parks

Separate, but related in spirit to the world heritage idea has been the international movement to encourage and assist the establishment of unique cultural and natural areas as national parks. A Japanese landscape designer and conservationist, Tsuyoshi Tamura, is reputed to have made the first proposal for a World Conference on National Parks at a General Assembly of the IUCN meeting in Athens in 1958. At any event, the First World Conference on National Parks was held in Seattle in 1962 under the joint sponsorships of IUCN, UNESCO, and FAO, and with the cooperation of the United States National Park Service and the Natural Resources Council of America.[85] Under similar sponsorship, the Second World Conference was held in 1972 at the Yellowstone National Park coincident with the one hundredth anniversary of its establishment.[86] A third world conference met in Bali in 1982, under the designation World National Parks Congress and in 1992 the IVth World Congress on National Parks and Protected Areas met in Caracas, Venezuela.[87]

At present, there are no "international parks" administered independently of the governments of particular nations. Functions of the United Nations and its associated agencies with respect to national parks have been confined largely to information and sponsorship. For example, the United Nations List of National Parks and Other Protected Areas has been compiled and periodically updated by the IUCN pursuant to a resolution of the United Nations Economic and Social Council.[88]

In the forefront of efforts to obtain international cooperation in the establishment or protection of national parks has been the IUCN Commission on National Parks and Protected Areas (formerly the International Commission on National Parks). This commission has over the years effectively persuaded national governments and nongovernmental groups throughout the world to set aside appropriate areas for preservation as national parks on the basis of carefully developed criteria.

The extraordinary efforts of individuals, with the assistance of a remarkably effective informational and communication network, have made it possible to bring lapses in the effective protection of national parks to the attention of competent national authorities. Very often through informal nongovernmental intervention, public officials have been persuaded to take remedial action. Such advocacy might have been regarded as an intrusion upon their authority had the suggestions come through formal intergovernmental channels. The cumulative effect of this activity by IUCN was noted by one of its former presidents, Dr. Harold J. Coolidge, when he said that "it will exert a steadily growing influence on each country's conscious appreciation of the extreme importance, from the economic, social, scientific and aesthetic point of view, of this modern method of ensuring the wise and farsighted use of certain areas of its national soil for the benefit of its people now and in the future."[89]

The World National Parks Congress, meeting in Denpasar, Bali, from 11 to 22 October 1982, took a large step toward integrating national park establishment and protection with broader aspects of environment and development. In addition to customary recommendations, a draft action plan was presented and the congress adopted a declaration in which note was taken of the World Conservation Strategy (1980), the UN Charter for Nature (1982), and the contribution of national parks to sustainable development. The IVth World Congress on National Parks and Protected Areas meeting in Caracas, Venezuela, in February 1992 was the largest ever, with four times the attendance of the 1982 World Congress in Bali. The Caracas Declaration set out fourteen principles to promote the role of protected areas and adopted an action plan with twelve priority objectives to strengthen and expand public and private protected area management.[90]

Reducing Natural and Unnatural Environmental Hazards

Although the theme of this book has been the defense of the earth against humankind's misuse of it to its own ultimate detriment, a much older theme in history has been the defense of humans against nature. Science and technology have been the major factors in changing the odds in this struggle—making humans rather than nature the more dangerous aggressor.

There are nevertheless several issues of international relevance where the need is to protect humans from threats in nature, some of which human-kind unwittingly induces. The incidence and effects of natural hazards was for many years the focus of research by Gilbert F. White and his students and associates.[91] Insufficient space precludes dealing with all such issues; some international efforts toward protection from natural hazards are described in other chapters of this book. For example, WMO works toward achiev-ing weather forecasts that might reduce damage from severe storms; regional organizations combat desert locusts; some international river basin com-missions have flood control responsibilities; UNESCO's Intergovernmental Oceanographic Commission has provided for warnings of seismic waves.

Attention here, however, will focus on four areas of environmental policy not treated elsewhere in this book in which man acts defensively—actions sometimes necessitated by previous human deeds. The first topic is environ-mental health—the effort to overcome environmental factors contributing to disease. The second topic is the effort to reverse spread of deserts, particu-larly in Africa and Asia. The third concerns the new threat to health, safety, and environmental quality from man-made substances not found in a natural state on earth. These materials correctly used may improve quality of life, but improperly used or loosed into the environment they can become unnatural threats to human and other forms of life. In each instance the intended ob-jective is a qualitative improvement in human life. A fourth area of policy, preservation of genetic diversity, might, with equal but different logic, be considered in the section "Protecting Nature." But the issue is complex, with major concerns over the uncertain resistance of hybrid plants to environmen-tal hazards and the loss of quality in life if genetic diversity among animals and plants is reduced through displacement of native strains and species by genetically engineered substitutes.

Controlling Disease

The concern of national governments with international conditions relating to health has been expressed concisely in a policy report prepared for the United States Congress:

> Nothing is more international than disease. It recognizes no political boundaries and few natural ones. It moves freely across national fron-tiers and spreads as conditions permit from one area to another. . . . If one extends the problem to include the diseases of plants and animals, there is little doubt today that pathogenic organisms themselves are either already globally distributed or can rather rapidly become so. . . .
>
> Preventive medicine, like disease, is inherently international. Had there not been the problem of preventing the entrance of disease from

one country to another and of controlling the spread of disease within countries, preventive medicine would not have developed as early as it did.[92]

International efforts concerned with the quality of health and its environmental aspects have been undertaken largely through the United Nations system, notably by the World Health Organization (WHO), the Food and Agriculture Organization, and the Pan American Health Organization (an independent associate of WHO).[93] Recently UNDP, UNEP, and HABITAT have acquired responsibilities that impinge upon health. The International Atomic Energy Agency is concerned with that part of health involving nuclear radiation, and the International Labour Organization is involved with conditions affecting health and safety in the working environment.

Policymaking in relation to world health raises two questions for governments, intergovernmental organizations, and private efforts. The first asks how to determine international responsibilities in relation to conditions that are not obviously transnational but may become so. The second wonders how to balance choices between priorities relating to health with those relating to economic development and environmental preservation.

With respect to contagious disease transmitted from person to person, international concerns have been obvious. Nations instituted quarantines to prevent disease from entering their territories. With the greatly increased flow of international traffic in the nineteenth and twentieth centuries, however, and particularly with the advent of air transport, quarantine measures alone have proved to be insufficient. Consequently, it has been in the interest of every nation, so far as it was able, to cooperate in the suppression or eradication of contagious disease everywhere. The campaign against smallpox is a case in point; the only true preventative for this disease being total eradication.[94] Some infectious diseases have extended periods of latency, AIDS being a particular case in point. To test all passengers in international travel for the causal virus would be a practical impossibility. There being no practicable way to quarantine for AIDS, it has become pandemic throughout the world.

The points of origin of many diseases—viral, bacterial, and fungal (e.g., AIDS, Ebola virus, Rocky Mountain spotted fever)—are environmentally localized. One part of a strategy for control of these diseases is to locate the geographical or ecological area from which they have spread or in which they are endemic. A Russian authority on the geography of transmittable diseases, Y. N. Pavlovsky, has described their incidence in *Human Diseases with Natural Foci* (Moscow: Foreign Language Publishing House, 1958) and *Natural Nidality of Transmissable Diseases with Special Reference to the Landscape Epidemiology of Zooanthroponoses* (trans. by Norman D. Levine, Urbana: University of Illinois Press, 1966). International cooperation is essential in the control or eradication of these diseases in which environmental factors are significant.

International cooperation for control of endemic diseases not directly transmitted from person to person has been influenced by humanitarian and economic motives. Debilitating diseases, such as malaria and schistosomiasis, heavily burden the quality of life and economic development throughout much of the tropical Third World. Advances in biomedical science and in statistical reporting and analysis reveal distinct environmental correlates of particular diseases. They sometimes also point toward development activities as contributory to their incidence and spread, schistosomiasis being a case in point.[95] There has also been speculation that the opening of roads in Zaire has contributed to the spread of the deadly Ebola virus.

The environmental context of health in the developing world has been concisely stated by Michael J. Sharpston, a health economist for the World Bank:

> Malnutrition, diseases caused by the contamination of the environment by human wastes, and air borne diseases form the core of the disease pattern of the developing world. This disease pattern is intimately related to large families with children born at short birth intervals, inadequate housing, water supply, sanitation, nutrition, and general hygiene. All these factors work together and reinforce each to emerge as the basic pattern of poverty and disease.[96]

International development efforts to improve living standards and the quality of life may sometimes have adverse results. A report published by the World Bank points out that "the introduction of a work force and its followers into a project area, especially in a previously undeveloped region, may result in the exposure of many non-immune individuals to endemic diseases."[97] It gives numerous examples of environmentally related pathologies, but singles out three as particularly serious: malaria, onchocerciasis, and schistosomiasis.

According to this report, malaria remains probably the single most serious threat to health throughout the tropics and subtropics. Initially a worldwide campaign against malaria coordinated by WHO was highly effective,[98] but it involved the widespread use of DDT, which was harmful to virtually all forms of life when employed in indiscriminate mass spraying. The greater risk of the chemical attack upon malaria, however, was the probability that resistant strains of malaria-bearing anopheles mosquitoes would survive the DDT campaign.

A health risk to workmen imported into tropical areas where a water resource project is under way is onchocerciasis or river blindness. This disease, sometimes leading to loss of vision, occurs chiefly among permanent residents of a hyperendemic area. Vectors of the disease are species of black flies encountered chiefly in East and West Africa and Central America. Large dams constructed for development purposes in Africa have sometimes increased exposure to onchocerciasis, and control of the carrier may add significantly to the cost of these projects.

A third health hazard associated with water development projects in the tropics, and in some temperate areas, is schistosomiasis, also known as bilharzia or snail fever. This parasitic disease is contracted by wading or bathing in water in which the parasite, associated with a certain species of snail, is present. In tropical and subtropical Africa, East Asia, the Caribbean islands, and South America where it is endemic, schistosomiasis appears to be spreading rapidly. Estimates of the number of persons infected range from 150 to more than 200 million. High-level multipurpose dams and year-round irrigation projects have been attractive to development-minded governments and have been assisted by international development programs, but the spread of schistosomiasis has been one of the adverse consequences, offsetting improvements in nutrition and economic well-being.[99] As of the present, schistosomiasis resists efforts toward eradication.

Environmental aspects of health are intertwined with other areas of environmental and economic policy to an extent that precludes any unitary organization inclusive either of all health or of all environmental considerations. How to organize environmental health services thus becomes a policy problem for which no clear answer is apparent. In 1970 a WHO expert committee reported that among national ministries of health, the organizational status of environmental health units varied and that health concerns were inseparable from many other aspects of the human environment. Thus, the committee concluded that "[s]ince this fragmented array of environmental policies and programmes could not be merged into a single organization in the interests of health or any other single social goal, the development of coordination arrangements among sectors and sectoral agencies was the main practical alternative." [100]

Within the United Nations system, health concerns are the collective responsibility of nearly all of the specialized agencies, in particular WHO, FAO, ILO, IMO, and UNESCO, and of programs like UNDP and UNEP. Yet there is need for an institutional focal point for cooperative efforts on behalf of world health, and that has been assumed by the World Health Organisation, which is the principal agency in world health efforts, its mission being one of leadership, mobilization, and coordination.

WHO was established as an outgrowth of the International Health Conference of 1946 and as the culmination of nearly a half-century of international institutional development.[101] The first international health organization was the Pan American Sanitary Bureau (PASB), established in Mexico City in 1902 at the Second Inter-American Conference of the Pan American Union. PASB, now renamed the Pan American Health Organization, is an autonomous agency for the Americas. It functions also as a regional office for the World Health Organisation, in which it nevertheless has an autonomous identity.

The first organized global health effort was undertaken in 1920 when the

League of Nations set up an International Epidemic Commission, followed in 1923 by the establishment of the Health Organisation of the League of Nations. Thereafter, until the outbreak of World War II, the Epidemic Commission or "Paris Office" and the Health Organisation in Geneva operated as parallel, and to some extent, as competing institutions. In retrospect, the Health Organisation of the League appears to have been one of that body's more successful undertakings. It established communications among public health officials in almost all countries of the world, and made large amounts of information available to national health administrators, assembling data and statistics that had not heretofore been available.

The World Health Organisation was established by the international agreement of 22 July 1946 as a specialized agency under Article 57 of the Charter of the United Nations. Its constitution enumerates twenty-two functions of which the common objective is "the attainment by all peoples of the highest possible level of health." Among its advisory panels and expert committees a great many are concerned in one way or another with environmental policy—for example, in relation to air pollution, food additives, cancer, insecticides, nutrition, occupation and health, international quarantine, and, of course, environmental health programs. Certainly the Malaria Eradication and Control Program initiated and coordinated by the World Health Organisation was one of the largest environmental management efforts that has ever been undertaken under international sponsorship.

To provide a structure through which the International Union against Cancer, representatives of governments, and other international organizations can cooperate in various areas of cancer research, WHO has sponsored (1965) the International Agency for Research on Cancer (IARC), headquartered in Lyon, France. Among these areas many relate to environmental influences. Technically IARC is a nongovernmental organization, although participation is by states that are also members of WHO.

The close links among health, environment, and the quality of life were emphasized at the International Conference on Primary Health Care held in Alma Ata, U.S.S.R., in September 1978 under the joint sponsorship of WHO and UNICEF.[102] Primary health care is essentially concerned with disease prevention—the historic emphasis of world health policy. Commenting at Alma Ata on the common basis of primary health care and environmental health management, Dr. O. U. Alozie of UNEP said that "both deal with disease as a consequence of improper management of the environment, requiring appropriate technologies, that is, local participation, low cost materials, bioenvironmental control methods, adequate water supply and proper waste disposal." [103] Thus, in his view, disease prevention and environmental policy cannot realistically be separated.

The quality of water is critical to the quality of life. Water supply is of obvious relevance to arid countries, but it has been a continuing priority on the environmental agenda of most Third World countries, particularly with respect to rural areas. In November 1980, the UN General Assembly declared an International Drinking Water Supply and Sanitation Decade (1981–90) with the slogan "Clean Water and Adequate Sanitation for All by 1990."[104] The movement for the campaign began with recommendations by two environment-related United Nations conferences: Habitat (Vancouver, 1976) and Water (Mar del Plata, 1977). Implementing action occurs at national and local levels, but the widespread concern over present and impending shortages of fresh water has led to the establishment of several international programs of investigation, technical advice, and financial assistance.

A new UNEP program for the environmentally sound management of inland waters (EMINWA) was described in the 1986 Annual Report of UNEP's executive director. The objective is to introduce an integrated approach to the management and development of freshwater resources on a river or lake basin-wide scale.

Among these water-related programs is the WHO surveillance system directed toward problems of water pollution affecting human health. The UNESCO-sponsored International Hydrological Programme studies the ingredients of water and, cooperatively with UNEP, coordinates the World Register of Rivers Discharging into the Oceans (WORRI), which provides information on pollutants carried by rivers into the seas.[105] WMO administers a Hydrology and Water Resources Programme primarily concerned with hydrological networks, that is, water movement. Beginning in 1976, a freshwater quality monitoring program has been carried on jointly by UNEP, WHO, WMO, and UNESCO within the framework of UNEP's Global Environmental Monitoring System (GEMS). The program (GEMS/WATER) involves establishment of nearly four hundred monitoring stations, development of laboratory facilities and analytic methods, and personnel training. A review and assessment of its status after four years of experience concluded that:

> Today, the project GEMS/WATER constitutes an indispensable component in the overall GEMS system. By providing sector-by-sector information on levels of hazardous substances in freshwater, and analysing trends, the project contributes substantially to global health-related monitoring. The collection of data on pollutants carried by major rivers into the oceans serves as a basis for a global assessment of the influx of persistent and hazardous substances and their levels and pathways in marine ecosystems.[106]

International concern with supply and quality of fresh water provides a bridge between the foregoing consideration of environmental aspects of health and the international efforts to combat the spread of deserts.

Combating Desertification

More than one-third of the land surface of the earth is arid, but not all is truly waterless desert. Arid lands may be highly productive of well-adapted life forms. Some of the most dramatic landscapes and distinctive ecosystems on the earth belong to the arid lands or deserts. Once regarded as wasteland, they are today understood as arid land or desert ecosystems that merit protective control against destructive human use, including unwise irrigation. In the United States, several national parks as well as public lands managed under the California Desert Plan are examples of protected desert ecosystems.

Deserts were not considered earlier in this chapter under "Protecting Nature: Species and Ecosystems," although they are appropriate subjects for environmental protection, partly because the only real protection for desert ecosystems, most of which are fragile, is to let them alone. But it is also because some of the most ecologically impoverished deserts of the earth are not "natural," but have been man-made. Large areas of the Indian subcontinent, as well as areas in western and central Asia, in north Africa, and in the Sahelian belt south of the Sahara appear to have been degraded from semiarid, productive lands into true deserts through human mismanagement.

Pressures of growing human populations on arid lands during the present century appear to have contributed to increasing desertification in many Third World countries, notably in the northern subtropical regions of Africa and Asia. The great drought in the broad belt of semiarid land stretching across Africa south of the Sahara, the Sahel, dramatically illustrates the process of desertification.[107]

During an exceptionally favorable phase of climate, the nomadic occupants of the Sahel increased the livestock upon which their livelihood depended. When, in the 1970s, a drier phase in the climatic cycle set in, the continuing and growing pressure of grazing on the semiarid grasslands led to their degradation toward desert conditions. To save the people and their animals governments and international agencies intervened, drilling wells and tapping water stored over millennia in deep but exhaustible aquifers. The nomads were forced to become refugees from their homelands.[108]

The tragedy of the Sahel illustrates the consequences of failure by people, their governments, or international agencies to apply ecological systems thinking to an environmental predicament—or to properly appraise the significance of feedback from their activities. Not only did the well drilling offer no more than a temporary solution to the needs of livestock, but it may have delayed planning for a sustainable future. Yet disaster sometimes stimulates remedial action. International efforts have now been undertaken to find lasting solutions for the Sahel. In this case an innovative effort has been taken through a voluntary group, Le Club des Amis du Sahel, organized by persons associated with OECD, its member governments, and governments of Sahelian states.[109] This group, as intermediary to the aid-giving and aid-receiving countries, assists in expediting and coordinating governmental efforts. In 1988 the IUCN began a three-year program "to bring long-term and sustainable change based on science" to this largely agrarian arid region. The Sahel Programme has become the largest of IUCN's Conservation Development Centre (CDC) programs.

An obvious, but not always correct response to the advancing desert is water. But this flowing resource often moves across national boundaries, and its use for irrigation and other purposes becomes an object for international contention and negotiation. Instances reported elsewhere in this volume involve India and Pakistan over the Indus, Mexico and the United States over the Colorado, and Syria and Israel over the Jordan. It was therefore to be expected that the Governing Council of UNEP, largely representative of developing countries, should promote international efforts to assist with problems of water supply to combat desertification.

Following proposals made at the Third Session of UNEP's Governing Council, the UN General Assembly on 15 December 1975 at its Thirtieth Session (Resolution 3511 [XXX]) called upon UNEP and UNDP to organize a United Nations Conference on Desertification (UNCOD). The conference, held in Nairobi 29 August–9 September 1977, was coordinated through consultation among conference staffs with the United Nations Water Conference which met at Mar del Plata, Argentina, 14–25 March of the same year. Results of both conferences were subsequently reported by the United Nations Institute for Training and Development.[110]

The recommendations reflected ecologically sound and climatologically realistic land-use practices. As with most action plans for environmental protection and restoration, joint and coordinative participation by various agencies was indicated. Among these were the Permanent Inter-State Committee on Drought Control in the Sahel, the UN Sudano-Sahelian Office, UNESCO's Man and the Biosphere Programme, and FAO's program on Ecological Management of Arid and Semi-Arid Rangelands (EMASAR).

To carry out the Plan of Action to Combat Desertification, adopted by the conference, UNEP established a Consultative Group for Desertification Control (DESCON) to help UNEP's Desertification Unit to proceed with specific measures. At its first meeting, proposals for activities in seventeen countries, mostly in Africa, were submitted. Funding for the program, being voluntary, has been uncertain. Moreover the difficulties of applying control measures to traditional ways of life of diverse pastoral and agricultural peoples are great even when funds and technologies are available.

Efforts to control the causes of desertification inevitably involve the management of vulnerable semiarid grasslands. Policies regarding these lands and their often nomadic inhabitants are obviously relevant to efforts to prevent man-made deserts. FAO and UNEP collaborate in a joint effort, the Ecological Management of Arid and Semi-Arid Rangelands Programme (EMASAR) in Africa and the Near and Middle East. Established in 1975 it has explored possibilities for breeding drought resistant forage plants and promoted grassland education and training.

An aspect of desertification arousing growing concern is its effect upon atmospheric dust. Increasing turbidity of dust-laden atmosphere may accentuate drought conditions. Other climatic effects may ensue.[111] Two atmospheric scientists suggest that:

> If the increased dust transport is due principally to poor land use practices and if the dust does modify the radiative properties of the atmosphere to a significant degree, as experiments appear to indicate, then this may be the first relatively clear-cut case study of a possible anthropogenic impact on weather (and perhaps climate) on a macroscopic scale. However, at this time we cannot determine to what degree the observed aerosol increase is attributable to natural physical processes.[112]

These considerations point toward a growth of international concern over national land use policies. When the effects of land degradation spill over national frontiers they become international. An ICSU/SCOPE project on land transformation has great potential significance for international environmental policy. By tradition few things were more exclusively subject to a nation's sovereignty than its territory and control over its land. To the extent that the findings of experimental science and confirmable evidence of the effects of particular land use policies on the biosphere reveal regional or worldwide consequences, the political theory of sovereignty is brought into question. A nation that permits misuse of its land to the jeopardy of other nations may some day be found to be in violation of international law.

There is, however, a controversial aspect to the desertification effort. Some critics allege that the campaign against desertification has been based on erroneous assumptions and insufficiency of reliable data. Thomas and Middleton charge that the approach taken by UNEP has better served the biases of the UNEP bureaucracy than the goal of clarifying the socioecological causes of soil degradation and the spread of desert-like conditions.[113] Climate has been an undeniable factor in the dry-lands environment, but desertification is also a people problem and the push of expanding populations of people and domestic animals into fragile ecosystems is a significant factor apart from climatic change.

Coping with Hazardous Materials

For a wide range of natural threats to health and safety, avoidance or adaptation are the only known remedies. Control measures have been developed for some hazards — for example, cloud seeding to break up severe storms or, with greater success, inoculations against epidemic disease. But many threats cannot wholly be prevented, and public responses are primarily directed toward containing or reducing adverse effects.

In recent decades a new set of hazards has arisen from substances previously nonexistent in nature or present only at low levels or in inert states. These substances have been developed in laboratories through research in chemistry, microbiology, pharmacology, and other sciences as commercial products, and once released into the environment, they have often been found to create hazards previously unknown to human experience.[114] To the extent that they are artificial, not having evolved through natural processes and being alien to the biosphere, they may be called "unnatural." Familiar examples are such substances as DDT (dichlorodiphenyl-trichloroethane) and PCBs (polychlorinated biphenyls). "Organosynthetic pesticides have become a part of the natural and human environment," according to a report attributed to the Central American Institute for Industry.[115] But the presence of these substances in the environment does not mean that either humans or the environment can tolerate their toxic effects. Many of these artificial compounds resist breakdown by natural processes and have a broad range of toxicity. Species-specific pesticides, moreover, encounter resistant strains of the organisms intended for control. The consequence is a new tough strain of bacteria, fungus, or virus for which new or stronger pesticides are sought.

In most developed countries measures of control or prohibition of the more dangerous materials—agricultural and pharmaceutical—have been undertaken. In 1972, the United States adopted the Federal Environmental Pesticide Control Act, and in 1976 the Toxic Substances Control Act was enacted, providing for governmental control over testing, manufacture, use, and sale of chemicals found to be dangerous. Investigations of effects and consideration of control measures were undertaken in various countries and international organizations, for example, WHO, ILO, UNEP, and OECD. In developing countries, however, controls were characteristically weak or nonexistent. Products banned or severely restricted in Europe or North America could be shipped in quantity to the Third World or their manufacturing transferred to receptive countries.[116] Although the International Register of Potentially Toxic Chemicals (IRPTC) in Geneva and the International Programme on Chemical Safety established under the United Nations Environment Programme are sources of information regarding hazardous materials, making this information effective in the policies and practices of national governments is not easily accomplished. Two nongovernmental scientific bodies have been formed to address these hazards—the International Commission for Radiological Protection and the International Commission against Nitrogens and Carcinogens.

During the 1980s, many industrial countries faced domestic crises in disposing of solid, especially hazardous, wastes. Faced with diminishing land disposal sites and restrictions on incineration, a number of countries sought to pay Third World countries to accept these wastes. In some countries, public officials were in effect offered bribes to accept these shipments. These transactions, however, soon aroused indignation in both Third World and developed countries. Some incidents had an element of humor, as in the six-thousand-mile cruise of a garbage-laden barge from New York, seeking in vain a place to dispose of its burden in Central America and the Caribbean and finally forced to return to its home port.

A large body of opinion has now been mobilized to bring international commerce in hazardous materials under international agreement or control. Yet the issue remains controversial in some industrialized countries. In 1980, hearings on export of hazardous substances were held in the U.S. by the Subcommittee on International Economic Policy and Trade of the Committee on Foreign Affairs of the House of Representatives.[117] On 15 January 1981, President Carter by Executive Order 12264 approved a national policy to control, with significant exceptions, the export of hazardous substances by American companies. But shortly after assuming office later in 1981, President Reagan withdrew the Carter order. The argument against the restrictive order as voiced by the National Agricultural Chemical Association is that the United States should not try to write rules for the rest of the world—should not impose United States standards upon other countries.

For the world as a whole, however, a consensus regarding the need for careful surveillance and control appears to be developing. On 17 December 1981, the General Assembly of the United Nations by a vote of 146 to 1 adopted Resolution 37/137, Protection against Products Harmful to Health and Environment. The United States alone failed to support the measure largely on financial grounds.

By 1989, however, the careless shipping and indiscriminate dumping of hazardous wastes were addressed by a 116-nation conference convened at Basel, Switzerland, by UNEP. Reflecting protracted prior negotiations, the conference adopted a Convention on the Control of Transboundary Movements of Hazardous Wastes and Their Disposal, signed on 22 March 1989 by thirty-five nations and the Commission of the European Communities. The United States, the United Kingdom, the Soviet Union, and the German Federal Republic indicated their intention to sign following further examination of the treaty.[118] The treaty entered into force on 5 May 1992, having secured the requisite ratifications.

The magnitude and complexity of the problem can be no more than suggested here. The very large matter of the disposal of radioactive materials, a particular concern of the International Atomic Energy Agency, has a large literature of its own, obviously relevant to environmental policy and presently in a very uncertain state of national control. It is ironic that materials and processes developed to enhance in some respect the quality of life, actually threaten it. Social understanding and institutional means for managing scientific and technological innovation are lagging far behind human efforts to manage the biosphere.

Preserving Genetic Diversity

Awareness of a need to preserve the genetic diversity of the biosphere has been implicit in many of the sections of this book. And as with many other environmental issues, the scope and substance of genetic diversity as an object of international policy can be no more than outlined here.[119] Its emergence as an issue has been recent, and opinions differ as to its urgency and the nature of action to be taken. Although some experts in agriculture and silviculture appear to disagree, substantial numbers of life scientists are alarmed by the rapidly decreasing diversity of the biosphere caused by the extinction of plant and animal species and the disappearance or severe reduction of natural ecosystems.

The crux of the issue is "genocide." This term is customarily applied only to the elimination of genetic types among humans. But humanity has been guilty of genocide against a vast number of life forms. Since prehistoric times, men have systematically—if also inadvertently—eliminated species of plants, animals, and ecosystems numbering in uncalculated thousands. An estimate of the IUCN Threatened Plants Committee that 25,000 species are presently in danger is an indicator of the magnitude of the threat.

The issue of genetic diversity has three aspects: scientific, economic, and moral. The scientific focus confronts the irretrievable nature of extinction with our incomplete and imperfect knowledge of the interrelating species of the earth's biota. The unique properties and interrelationships of myriad life forms are poorly understood. Some species have not been described; others doubtless remain undiscovered. Much of what may be learned from various life forms may contribute significantly to understanding human physiology and behavior. The sciences of genetics, physiology, pharmacology, medicine, and plant and animal sciences generally, benefit from the widest possible range of genetic diversity.

The economic aspect follows from applications based upon scientific knowledge. For example, Chapter 9 points out the importance of marine biota as a source of chemicals and pharmaceuticals. The Green Revolution and genetic selection in plant and animal breeding depend upon diversity among genetic types. Developments in biotechnology, notably in recombinant DNA, are limited, at least in principle, by the availability of genetic material and by understanding of the properties of particular genetic strains. The emergence of biotechnology as a powerful instrument of change in the world makes the preservation of genetic diversity even more important than it would have been without the new capabilities for influencing the course of evolution.

The moral aspect of genetic preservation has gained international recognition. The World Charter for Nature declares that "[e]very form of life is unique, warranting respect regardless of its worth to man, and, to accord other organisms such recognition, man must be guided by a moral code of action." [120] Among the general principles enunciated by the charter is the imperative that "[t]he genetic viability on earth shall not be compromised; the population levels of all life forms, wild and domesticated, must be at least sufficient for their survival, and to this end necessary habitats shall be safeguarded."

The idea that humanity shares rights with other living things and is not morally free to pursue its advantage to the destruction of other occupants of the earth has a long history. Although largely ignored or rejected by the technocratic economistic societies of the nineteenth and twentieth centuries, the concept always had some followers. Obviously, genocide also occurred in preindustrial societies, as often perhaps from ignorance regarding the requirements for the survival of a species as from indifference.

Today, hazards to genetic diversity come from two main and interrelated causes. The first and common factor is the ever growing demand of expanding human populations for food, fuel, and fiber from the environment. Specific threats to genetic diversity are the conversion of natural areas to human uses for agriculture, animal husbandry, scientific forestry, urbanization, and mass recreation, with attendant use of biocides, dispersion of toxic substances, and pollution from by-products of energy production (e.g., emissions from the burning of fossil fuels).

A second, relatively recent and less certain hazard to genetic diversity comes from the replacement of varieties of plants, notably grains, by patented hybrid germplasm. The spread of monocrop agriculture and standardization of agricultural products associated with agribusiness and the Green Revolution is perceived by some observers as threatening genetic diversity and increasing vulnerability of food supplies to unforeseeable disruptions. Kenneth A. Dahlberg concludes that extensive adoption of Green Revolution hybrids has involved the large-scale displacement of traditional cultivars, making them more scarce and in many cases threatening their extinction.[121] Plant breeding has obvious benefits for human society, and has been an international scientific enterprise. Whether the patenting of plant and animal forms under the laws of any one country would confer international legal rights prejudicial to experimental work in genetics remains to be seen.

The question has become an international issue with North–South overtones. In 1961 a group of developed countries formed the International Union for the Protection of New Plant Varieties (UPOV). Although UPOV is a public intergovernmental organization, its primary purpose is to protect private rights to patented plant materials. But the developing countries tend to regard genetic resources as part of the common heritage of mankind. At the Twenty-first Conference of FAO in November 1981, Mexico led a movement for an international convention that would ensure the conservation of genetic resources for agriculture and guarantee their free availability and exchange among all nations. The move was successfully opposed by the developed countries, but the issue remains a subject of controversy.

The Conference of FAO adopted a voluntary International Undertaking on Plant Genetic Resources, an issue that had been debated for several years, to be implemented by an intergovernmental commission. The undertaking is an agreement less binding than a treaty and has been accepted by at least seventy-five FAO member states.

Applications of the new microbial biotechnology, especially to agriculture and medicine, were assisted by UNESCO and UNEP as early as 1975 with establishment of the International Network of Microbiological Resources Centres (MIRCEN). Through this network of eight centers, exchange programs in training, information, and the distribution of useful strains of microorganisms are carried on by research institutions (of which there were at least seventeen in 1988 and doubtless more since).[122] In 1983 the United Nations Industrial Development Organization (UNIDO) established the International Centre for Genetic Engineering and Biotechnology (ICGEB). The center is intended to assist developing countries through dissemination of information of special relevance to energy, food, and health. In 1986 dual headquarters were opened in New Delhi, India, and Trieste, Italy.[123]

Preservation of genetic diversity has been one of the reasons for the Man and the Biosphere (MAB) program of biosphere reserves and for the network of biogenetic reserves recommended by the European Committee for the Conservation of Nature and Natural Resources. In addition to MAB, several UN agencies have been involved in the preservation of genetic diversity. Experimental research in agriculture and seed storage centers are coordinated by the Consultative Group on Agricultural Research (CGIAR) housed at the World Bank. In 1974 FAO, UNEP, and CGIAR established the International Board for Plant Genetic Resources (IBPGR), which has emphasized genetic preservation through seed banks, although there is considerable risk of loss of viability involved in this method. Preservation of genetic diversity for animal species as well as plants is a high-priority concern of UNEP and of the UNEP/IUCN World Conservation Strategy and in the world movement for national parks.

At the national level, the United States government, for example, held an interagency conference in November 1981 on strategy for maintenance of biological diversity.[124] Among specific measures for preservation identified for United States bilateral aid programs were: (1) introduction of appropriate small farmer cropping systems to replace slash and burn agriculture; (2) establishment of natural (ecological) reserve systems; (3) training programs in tropical ecology and conservation; and (4) increased U.S. financial aid for university research relating to biogenetic diversity.

Not all species threatened with extinction are wild. Many domesticated species of plants and animals are endangered—some with unique and potentially valuable genetic characteristics. The Rare Breeds Survival Trust (now international), founded in England by Joe Henson, established the first of a growing number of rare breeds survival centers.[125] Some breeds of cattle, pigs, and sheep are of ancient lineage, descendants of species wild in prehistoric times. A unique case in Turkmenistan is the recovery of the Akhal-Teke horse, condemned as economically useless by the former Soviet Union. The International Association of Akhal-Teke Breeders promotes this allegedly first pure-blood horse in history.[126] The survival centers are thus gene banks for genetic preservation.

As with efforts to abate environmental pollution, preservation of genetic diversity impinges upon a wide variety of human activities, economic interests, and scientific shortcomings—as the ecological requirements for the survival of many species are poorly understood or unknown. The territorial requirements for certain species are of a magnitude or specialized character that would preclude human use of many areas of land and sea now being encroached upon for economic purposes. Reserve areas are often too small to provide the desired protection.

If the present international declarations, charters, action plans, and treaties were substantially observed, the genetic richness of the biosphere would be preserved—at least from human hazards. But whether these commitments will be backed by action is uncertain. A new kind of ecological morality appears to be spreading around the world. Governments in theory are committing their policies and resources to enhancing the quality of life. What is theory today may be more fully realized in practice tomorrow.

Practical measures have been undertaken. In 1989 the IUCN organized a workshop in Washington, D.C., of economists and conservationists to demonstrate that economics can contribute to conservation and that there are significant economic advantages to be gained from biological diversity.[127] In 1990, the possibility of a draft international convention was under consideration that would supplement and integrate existing international agreements to conserve biological variety and diversity. The *IUCN Bulletin* for July-September 1989 reported that a draft Convention on the Conservation of Biological Diversity was ready for review. "Fifty draft articles which address conservation measures and provide for the creation of a funding mechanism" were prepared by the IUCN Environmental Law Centre in Bonn. The draft treaty was a major item on the 1990 agenda of the Governing Council of UNEP. Another step toward protecting genetic and ecological diversity was taken at the July 1990 Houston economic summit of the world's seven leading industrial nations. A proposal was adopted to ask the World Bank to initiate a pilot program for easing financial burdens (e.g., international debt) that have led to destructive exploitation of the Amazonian rainforest.

The Convention on Biological Diversity was signed by 158 governmental representatives at the UN Conference on Environment and Development in June 1992. The United States conspicuously declined to sign, presumably to protect U.S. economic interests abroad. In 1993, however, the Clinton administration agreed to sign. Ratification by the signatories will doubtless require several years.

12

Strategies for Global Environmental Protection

The record of both organized and individual efforts to safeguard the biosphere and the quality of the human environment has been impressive, and the commitment of governments within the last two decades to protection of the biosphere is without precedent. While specific international agreements had addressed particular environmental problems prior to both the Biosphere Conference of 1968 and the United Nations Conference on the Human Environment of 1972, no worldwide concerted or comprehensive approach toward international responsibility for the safeguarding of the biosphere had occurred. Today, however, an extensive and complex network of intergovernmental, nongovernmental, and scientific organizations addresses a broad range of international environmental problems. The twenty years between the Stockholm Conference in 1972 and the UN Conference on Environment and Development at Rio de Janeiro in 1992 culminated in a most extensive and detailed statement of international goals and national responsibilities—Agenda 21—a "roadmap" for international policy and action into the twenty-first century. Treaties and other international agreements have been negotiated to such an extent that environmental protection is now recognized as a significant aspect of international law.[1]

What outcome may we anticipate from these worldwide efforts in defense of earth? Popular attitudes toward environmental protection are ambivalent. In technologically and scientifically advanced countries, opinion polls show strong public support in principle for environmental protection.[2] On specific issues and in competition with other priorities public support is often much less certain. People in large numbers seem to prefer positive, reassuring assessments of environmental trends to objective reports based on factual and holistic evidence. Critics of environmentalism welcomed Gregg Easterbrook's *A Moment on the Earth* (1995), which they regarded as refuting the so-called "doom and gloom" warnings of future environmental disaster. But the scientific community of the world did not agree, as witness *The World Scientists Warning to Humanity* (1995).[3]

So recent is this comprehensive international effort that it would be unrealistic to expect more than a beginning to have been made. While tangible accomplishments have been reported in the preceding chapters, it must be conceded that, as of the mid-1990s, the principle results of international cooperation were investigations of the causes of environmental problems and identification of needs for action. Action programs have been initiated, but most of the action has yet to be undertaken. The unprecedented environmental efforts of the last quarter of the twentieth century are no match for the magnitude of the problems.

An international structure for environmental policy is now in place, and some experience with intergovernmental cooperative environmental programs has been acquired. But governments have found it easier to sign declarations and to collaborate in joint scientific investigations such as the International Geophysical Year, the International Indian Ocean Expedition, and the Global Atmospheric Research Programme than to fulfill environmental agreements through regulatory measures of their own, or through conformity to international policies or standards. International environmental cooperation even in scientific endeavors has not been an easy achievement.

One cannot be sure that these hard-won beginnings will succeed. As the implementation of a particular treaty moves toward the point of action, those problems and practices that made the action necessary are confronted. At this point environmentally concerned people and their governments face a task as difficult as humans ever face—the changing of human behavior. It is not yet clear that a sufficient number of people will become sufficiently convinced that their well-being depends upon the preservation of the integrity of the biosphere to cause their governments to act and to act together. And whether such consensus may be achieved in time to prevent serious impoverishment of the biosphere and severe diminution of environmental quality cannot be foreseen. Destructive and irreversible environmental effects such as significant change in the composition of the earth's atmosphere could occur before people and nations are prepared to transcend the barriers that limit their cooperation in mutual self-interest.

The Uncertain Human Dimension

The question of mankind's ability to protect the biosphere is a multiple question of understandings, values, priorities, and their behavioral consequences—not primarily one of technical possibilities. Technology may be a powerful instrument of environmental protection and improvement, but its uses may also be destructive. Where technologies are incompatible with environmental protection (e.g., certain persistent pesticides), restraint in their use is the rational though often impolitic alternative. Looking only at technical possibilities employed with informed discretion, the prospects appear good; and viewed alone, technological advancement might justify an optimistic assessment of mankind's ability to manage the biosphere. But technologies are not employed in isolation from other things; their effects transcend their immediate applications. As their power and applicability increase, it is increasingly important to know how and where they may be safely used. Much of what people need to know about the biosphere and about themselves in order to be responsible custodians probably remains to be learned. Scientific understanding of the environment and of human behavior as it affects the earth is growing—and with adequate investment might advance more rapidly. The need for more rapid advancement is evident from the record of the environmental impacts of science-based technology: know-how has outrun the guidance afforded by fundamental knowledge.

Conceding the shortfalls between aim and achievement among international agencies and national governments, experience in the organizational and technical aspects of environmental protection and management has been increasing. Environmental sensitivity is beginning to characterize public policies in many countries, and public opinion surveys in several major industrial nations consistently show widespread and growing commitment in principle to environmental protection. Social demand for environmental protection, like technological innovation, threatens to outrun the scientific knowledge needed for soundly based environment development policies.

The principle hazards to positive prospects for environmental policy are social. Political instability, ignorance, avarice, rapid population growth, disorder, and violence are deterrents to all aspects of environmental protection. The contention that the environment must wait until social justice is achieved could well lead to the failure of both objectives. Inertia in dealing effectively with the basic social problems of the world's peoples—poverty, overpopulation, environmental degradation, and violation of human rights, among many others—could also weaken the basis for international cooperation necessary for environmental protection. The social theories which underlie the policies of contemporary political systems have shown little success in coping with the enormous problems of social condition and behavior that threaten the integrity of most nations of the world today. Hence the pessimistic probability that real progress toward responsible international custody of the biosphere may be slowed, halted, or even reversed by international and intranational social-political conflict.

The depth of informed environmental concern among the world's peoples and governments, however real, is difficult to ascertain. Environmental awareness, although spreading nearly everywhere, is still concentrated in a few advanced countries. The declared intention of the Stockholm Conference and of UNEP to combine socioeconomic reform and environmental protection received more tangible operational expression in the Rio Conference and Agenda 21. Still, peoples and governments almost everywhere have been harassed by problems of economic and political instability, ethnic and sectarian violence, crime and corruption that distracts them from the fundamental problem of achieving a sustainable relationship with the Earth.

The enormously increased impact of humankind upon the physical and living systems of the earth makes ever more important an understanding of human behavior. The role of humans in changing the face of the earth has been described at length and in detail.[4] But implicit in much of this literature is an assumption that if people understood the harmful consequences of environmental mismanagement, their behaviors would change. However, the ways in which people relate to their environments and to other living species appear to be influenced by their entire cultural matrix. If this is true, and if the tendencies of modern cultures have caused the continuing attrition and destruction of the biosphere, then fundamental cultural change will be required if the many efforts now focused in various aspects of environmental and biospheric protection are to succeed. There may be psychogenetic factors that affect human environmental behavior (i.e., territoriality and urge to dominate), but as yet we know too little to speculate upon this possibility.

The human dimension of environmental protection is the complex range of attitudes and behaviors, embedded in culture, that account for the ways in which humanity impacts the environment. Foreknowledge of environmental consequences is only one of many (and often stronger) considerations influencing human behavior. Immediate advantages, personal and social, have often outweighed long-range considerations. For as long as it pays for people to destroy living systems unnecessarily, to deplete nonrenewable resources, and to degrade the quality of the biosphere, they may be expected to do so except as influenced by social or political restraints. To attempt to protect the environment against the preferences of people for its exploitation would entail certain costs and uncertain effectiveness. More effective and less costly would be a pervasive popular attitude of respect for the environment and regard for its self-renewal that would reduce the need for regulatory or coercive protection. If the future of world cooperation depends upon the fulfillment of unrealizable goals and purposes, the prospects for international protection of the biosphere are dubious. But it is possible that the peoples of the world have learned more about working together than might be inferred from the literature of international relations or the rhetoric of nationalism and class warfare. The evolution of international environmental policy described in this book makes this possibility plausible.

For informed persons who hope for a better world, the prospect is discouraging, but there remains the outside possibility of a sufficient reversal of destructive trends to redirect the world toward a more sustainable future. To realize this possibility, it will probably be necessary to clear away some of the misconceptions and wishful premises that have led modern industrial society to its present predicament.

World Change in Retrospect

To describe an international order as changing is to imply that it is moving from a circumstance that *was* to a circumstance that is *becoming*. The condition of all human societies has been transitional, but the duration of specific conditions, along with rates of change, have varied greatly over historical time. For a variety of reasons—ecological, economic, technological, and psychological—it seems clear that the present time is a period of massive, swift, fundamental change. But it is change with cross-currents and contradictions.

What may be called the old international order developed during the last five hundred years. It began with the expansion overseas of the principal European nations following their formation at the end of the fifteenth and the beginning of the sixteenth centuries. The political consequences of this expansion were the conquest and settlement of the American continents and Australia, the partition and colonization of Africa, and the European domination of large parts of Asia.

By the beginning of the twentieth century this European-dominated international order had reached its maximum extent, and beginning in 1914 with World War I the order began an uneven but continuous political decline throughout the balance of the century.[5] Reasons for the decline of this international order are directly related to the consequences of its expansion, to the consequences of the changes that it wrought—ecological, cultural, ideological, demographic, and technological—throughout the world.[6]

During the five centuries following 1492, European-initiated science and technology transformed the earth. This enormous extension of European economic power and technology was aided by the Industrial Revolution. Industrialization enjoyed an immense bonus from the exploitation of forests, soils, minerals, and fossil fuels that were accessible to development and seemingly undiminished by human use. By the twentieth century much of the cream had been skimmed off the earth's resource base, but demands upon resources and the environment had grown. Although in monetary terms the cost of most resources continued to decline until the latter third of the century, the increased uses to which these resources were put, together with demands upon them by growing numbers of people everywhere, created economic environments of impending scarcity rather than of abundance.[7] In the long run, environmental degradation and impoverishment will be followed by economic impoverishment. Short-term affluence may result from excessive exploitation of the material resource base. But technology alone cannot sustain an economy if erosion of the natural resource base deprives it of the materials required for meeting human needs. The argument that resources may be created and substitute materials may be found may be true, in a limited, technical way. But the ultimate end of economic activity is not merely survival; it is rather the improved quality of life. Economic affluence might yield little satisfaction in an environmentally impoverished world.

A major and lasting effect of the old international order of European political dominance was massive acculturation: traditional ways of life were changed almost everywhere. The process was often tragic. The movement of Europeans to the Americas, the Pacific islands, southern Africa, and northern Asia resulted in the degradation or extinction of numerous traditional cultures and civilizations. African Negro slavery and Asian contract labor accompanied European colonization and led to the transplantation of African and Asian populations and cultures to tropical and subtropical America. At the end of the European colonial period, reverse colonization brought large numbers of Asian and African peoples as migrant workers, refugees, or students into the former colonizing nations of Europe and to the United States and Canada. Distinctive national and cultural traits and values were visibly disintegrating in the industrialized or so-called First World during the last third of the twentieth century. The ethnic and economic bases that had supported nationalism as a political ideology for five centuries were eroding in the nations that had heretofore been the great powers. But in many of the Third World and former colonized states, nationalistic feeling was intense. Perspectives on international cooperation have differed among the older and newer states; legacies of colonization continue to complicate international and environmental policy.

A second important effect of the old international order was development of a complex network of economic interconnections and interdependencies. Interpretations of the significance of interdependence vary widely: some have read into it social and political implications that find little support in the actual behavior of nations. There is a large contemporary literature devoted to the celebration of interdependence among peoples and nations. International interdependence has been described in a generalized and quasi-ethical sense as both necessary and desirable, yet much of this literature tends toward economic rationalization or moral exhortation. It often confuses interdependency with interconnectedness, suggesting by inference that interrelationships among nations imply unavoidable interdependencies. Quasi-interdependencies that are voluntary, and could be terminated readily, are too often not differentiated from dependencies of fundamental and lasting character. There sometimes appears to be an assumption that legal equality should be considered wholly apart from economic equality in matters of international interdependence. One need only compare the economic strength of the great continental powers with that of small island states to see the implausibility of this assumption. Nations, often for economic reasons, are not equally able to fulfill international obligations. Interdependent nations are not necessarily economic or political equals.

Viewed globally, interconnectedness is a characteristic of the modern world and is one of the changes brought about over the centuries by the old international order. This network of relationships is continued by multinational corporations, which, of course, had antecedents in the Dutch and English trading companies of the seventeenth, eighteenth, and nineteenth centuries. But examined at closer range, it appears that interconnectedness involving interdependency becomes in the case of certain nations, outright dependency with relatively little reciprocity. Thus, to picture international relations in the modern world as a matrix of mutual and reciprocal interdependencies is to fabricate an idealized representation of the present order that may obstruct consideration of its real problems.

Bilateral and multilateral interdependencies among particular nations exist chiefly among those most highly developed, as for example between the United States and Japan; explicitly among Canada, the United States, and Mexico under the North American Free Trade Agreement; or among the countries of the European Union. If, however, one examines relationships between developed and less developed nations, reciprocal dependencies are less easily generalized. For those nations whose economies are heavily dependent upon the production of raw materials from natural resources, there is dependency upon consumers in the developed world. But the developed countries often have options with respect to where they obtain their raw materials. The nature of resource flows and their effects upon environmental policy will be considered presently. It is sufficient here to observe that the dependencies that have been created by modern economics and technology do not necessarily bring a competitive world together in a relationship of mutual self-interest.

Thus, present-day efforts to preserve and protect the biosphere take place among multiple contingencies in which perceived national interests and the factors determining the survival of peoples are interpreted in radically different ways. The difficulty in dealing with so large and complex a subject as international environmental policy is increased by lack of common understanding regarding the fundamental causes of peoples' discontents. Not only have people seldom developed a common perception of the forces and trends with which they are contending, but there has also been a lack of agreement upon the fundamental meanings of words and concepts. And so before dealing specifically with some of the more important environment-related concepts of the changing international order, it is necessary to understand how differing understandings regarding the meaning of words and concepts complicate the ability of nations to cooperate in shaping mutually acceptable policies.

Conceptual Obstacles to International Order

Differences of opinion and debate over the political economy of the environment are complicated by disagreement, often unperceived, over the meaning and implications of fundamental terms. Vagueness and ambiguity in word meaning lead to confusion over concepts, which exacerbates political and ideological differences to the point that they may be, or appear to be, irreconcilable. Following are some examples of conceptual confusion that have proved to be obstructive of environmental policy and international cooperation.[8]

Ecology and economics may appear to be opposed, which they may be, in the sense that opposite sides of a coin are—they are different aspects of the same thing. Energy, for example, is drawn from resources in the planetary environment and then returned as residual particulates, radioactivity, gases, water vapor, and heat. The applications of energy impact upon the environment through agriculture, mining, manufacturing, construction, transportation, and other activities. It therefore follows that public policies for energy, natural resources, the environment, the economy, and health cannot realistically be separated.

Nor can they be separated from policies concerning so-called growth and development. The qualifier "so-called" is attached to these terms because their use, while widespread and often official, is ambiguous until clarified by agreed understanding or by practical example. Politicians, economists, and planners act mainly as if the meaning of growth and development is so well understood that no explanation is needed. Unfortunately what everybody *knows* is often what nobody really *understands*. And if one asks the meaning of these terms, the reply is usually a synonym, not an explanation. Qualified definitions may sometimes be more explicit—or at least their implications may be deduced from the context of their usage.

Thus "economic growth" implies in the vernacular a rising standard of living, more economic activity, and a larger gross national product—but with the particular style or quality of life seldom specified. The assumption that economic growth always correlates with more food, clothing, shelter, health, education, and security is contrary to the experience of many people in countries where economic growth has been substantial. It may be argued that in poor, less developed countries, people in the aggregate are better off than they would have been without "growth"; some people at least appear to be better off. This argument, however, is questionable when one considers the extent of rural poverty and urban squalor in the Third World (the same argument is hardly less questionable in the First World). Even with commitments to planning for human betterment assisted by United Nations agencies, development banks, and bilateral programs, economic growth has fallen far short of its alleged results. And because of the way in which it has been promoted, and its benefits distributed, such growth may have increased social disorder, material and psychological deprivation, and political discontent.

The term "development" is equally misleading. It is literally no more than a synonym for a certain kind of growth and progress. Some conventional definitions of development equate it with terms hardly less ambiguous—for example, with modernization, democratization, or (less commonly today) industrialization. Taken literally, growth and development are *processes;* they are terms without content, implying nothing per se about the substance of goals or outcomes. Yet there must be some content in a term about which hundreds of articles, books, and official reports have been written, and in the name of which governments and international agencies have employed experts, administered programs, and expended funds.

Assuming that the terms "growth" and "development" do represent real concerns that many people believe to be (or should be) the major focus of relations among nations today, these concerns are not well served by the language used to identify them. "Growth" and "development" are terms that everyone ostensibly understands but no one can define to general satisfaction in simple unequivocal language. To discover what these terms mean to those who use them one must look to practice—to what is done in the name of growth and development. Clarification of the development concept and reconciliation of ecological and economic values has been sought through the qualifying concept of "sustainable development" (see Chapter 10). As previously noted, "sustainability" tells us nothing about what the development process will sustain and at what level of environmental quality. Still, the term has won widespread acceptance and to this extent lowers a barrier to international environmental cooperation.

Beyond these ambiguous concepts the jurisprudential doctrine of "national sovereignty" is a politically resistant barrier to international environmental cooperation (as it is also in other areas of policy). Rhetorical assertions of national sovereignty continue to be heard, but the imperatives of geophysical hazards to all nations are pushing their governments toward modification of their asserted freedom to act as they please in relation to their natural resources, industrial practices, and the environment. A rhetorical strategy to evade the sovereignty roadblock is the concept of "merged sovereignty," which in fact occurs when nations seek to realize their own national objectives through treaties with other nations having interests in common.

Salient features of the international environmental movement have been openness, diversity, complexity, dynamism, and purposiveness. No other international movement—world peace included—engages so large a number of individual participants, so many and such diverse forms of cooperative effort, so broad a range of human skills and interests, and such complex interrelationships among governmental and nongovernmental organizations. For persons not accustomed to organizational complexity, the matrix of interactive relationships that shape international environmental policy could indeed be incomprehensible. It defies description in detail, and the chapters of this book have, of necessity, examined its aspects selectively, and largely at a high level of generality.

Although some aspects of the environmental movement are aesthetic or atavistic, its greater and more compelling function has been an intermediary role between the environmental findings of the sciences and the policies and actions of governments and intergovernmental organizations. Credible forecasts from the sciences in matters such as global climate change, ozone layer depletion, acid precipitation, toxics in food chains, effects of deforestation and loss of biodiversity—among many other issues—have provided the substance for environmental policies and a geopolitics of the planet Earth.

Implications for the sciences are far-reaching. Every science, from agronomy to zoology, is somehow drawn into environmental investigation; and beyond the continuing mandatory agency missions of a specialized and narrow focus, interactive interdisciplinary approaches are developing, notably through the large-scale environmental investigations described in Chapters 9 through 11 of this volume. The outcomes of these investigations frequently contain important implications for the nonscience disciplines and the humanistic and the science-based professions.

The foundations of sound environmental policy require a substratum of understanding of the earth's atmosphere, hydrosphere, biogeochemical cycles, ecological interrelationships, and basic human needs and behaviors. But these fields of investigation are collectively insufficient to provide a holistic approach to environmental policymaking. In order to build their findings into public and international policies, an even wider range of disciplines must be involved: law, medicine, economics, ethics, aesthetics, engineering, and the design arts, among many others. An integrated interdisciplinary use of scientific and technical knowledge is required to assist the formulation of policy. There are, of course, precedents for the collaboration of science and policy, but what is different today is the greatly increased scope of the relationships and their international character. This increasing recourse to science for guidance in policymaking remains in addition to policies predetermined in the absence of full knowledge of their impacts, costs, and benefits, as in the issue of global climate change. Indeed, the criteria for judgment in science and politic are very different. Science draws conclusions on weight of evidence or probabilities. It seldom expresses its conclusions in absolutes. Politics, in contrast, often alleges self-evident truths and unequivocal conclusions based on opinion.

Science may not be counted upon to transcend introverted nationalism, but it does have that capability. Scientific knowledge is not immune to political manipulation, but there are remedies for the resultant misrepresentation. The growth of science in the less-developed countries is gradually enabling their leaders to assess national capabilities more realistically and to participate with greater confidence in international cooperative efforts. This growing scientific component of international environmental cooperation is a factor that can help to offset political antagonisms. To the extent that representatives of governments can agree upon the implications of demonstrable facts, they may to that extent also agree to cooperation in the policies implied, even though they continue their disagreements on other matters.

Science in principle is a system of thought and action open to all who can fulfill its requirements of preparation, accuracy, and candor. Throughout a large part of the world, and even under ostensibly despotic regimes, science and scientists may have a measure of freedom from political control that enhances their ability to collaborate in international environmental research and planning. Scientific associations are prominent among the many nongovernmental organizations (NGOs) that play important roles in international environmental affairs. The NGOs, moreover, are not wholly dependent upon governments for their information or agendas, and the relative autonomy of science thus reinforces the relative independence of environmental NGOs from political and ideological divisions.

The importance of the NGOs in international environmental policymaking can hardly be overemphasized. NGOs have been influential, both within and among nations, and they have been the instigators of numerous treaties and international cooperative arrangements. Because NGOs form extragovernmental networks among as well as within nations, they are less constrained by the characteristic inhibitions of diplomatic protocol and bureaucratic procedure. In both the forming and execution of international policy they may act more rapidly and directly, and with less risk to national sensitivities, than can the official intergovernmental agencies. Some of the more prominent among these are listed in Appendix B.

With environmental policy increasingly built into the normal bureaucratic structures and agendas of governments and intergovernmental agencies, one cannot safely predict the future role of NGOs in international environmental politics. The role of the multinational corporation as a special class of NGO is ambiguous. Its position in international environmental politics is widely believed to be neutral or negative. But multinationals are increasingly drawn into environmental policymaking, and their influence has been felt in the Law of the Sea Conferences, in UNEP's Regional Seas Programme, in the deliberations of IMO, in debates on orbiting satellite communication, in energy and natural resources policies, and in government policies for the transportation and management of hazardous materials. Support of international environmental protective measures might well be consistent with long-term corporate interests. But it is uncertain whether many of these firms presently see it in that context. Meanwhile, the International Chamber of Commerce has established an International Environmental Bureau, headquartered in Geneva, which has been actively involved in a broad range of environmental issues in which industrial research and development are needed. In addition, the New York City-based World Environment Center is a clearinghouse for environmental information published largely by business firms. Annually it awards a Gold Medal for International Corporate Environmental Achievement pursuant to its mission "to contribute to sustainable development by strengthening the management of industry-related health and safety practices worldwide." At the time of the Rio Conference a Business Council on Sustainable Development was formed under the leadership of Swiss industrialist Stephan Schmiedheiny and published a case book of thirty-eight studies of business initiatives under the title *Changing Course: A Global Business Perspective on Development and Environment* (MIT Press, 1992). Corporate enterprise today needs reliable environmental information, a need documented in 1984 for the U.S. Council on Environmental Quality by Russell E. Train, president of the World Wide Fund for Nations (WWF).[9]

The most significant augury for the future importance of NGOs in world environmental affairs has been their widespread increase in numbers and members during the post-Stockholm decade. In 1980 the Environment Liaison Centre (ELC) in Nairobi reported that of the more than one thousand NGOs represented "half are from less developed countries."[10] More significantly the ELC reported in 1982 at the tenth anniversary of the Stockholm Conference that "there are 2,230 nongovernmental environmental organizations in developing countries of which 60 percent were formed in the last ten years, and 13,000 in developed countries of which 30 percent have formed in the last ten years." Of particular political significance is the fact that these NGOs could provide a constituency for the official agencies for environmental protection now established in most countries and could function as receptors and disseminators of the concepts embodied in the World Conservation Strategy. The number of environmental NGOs continues to grow. The *1987 Annual Report* of the executive director of UNEP stated that ELC "coordinates a network of over 6,000 NGOs dedicated to the protection and improvement of the environment."[11] The *BNA International Environment Daily* (8 June 1992) estimated that two thousand NGOs were in some respects represented at the 1992 UN Conference in Rio.

In summation, a dynamic, composite, and flexible structure, partly official, partly unofficial, has emerged during the two decades following Stockholm. This interactive structure has formed and grown so rapidly that many experts in international affairs are unaware of its extent or significance. Its purpose—international environmental protection—is novel in substance and scope, while including elements that existed before environment became a focus of public policy. Obstacles to its effective functioning lie in inappropriate human responses to impending environmental disasters, and in the antagonistic postures of nations divided within and against one another. Hope for international cooperation toward safeguarding the human environment and the biosphere lies in the demonstrated capacity of human beings, and their governments, to transcend their differences where mutual interests and mutual survival should be evident.

Protecting the Biosphere: Methods and Strategies

Although the cumulative record of declared international intent to protect nature and the human environment is impressive, the continuing and growing impact of human activities upon the biosphere leaves the future in doubt. Almost invariably performances fall short of promises. Today the attitude of many people who influence and determine the policies of the world's governments increasingly appears to be divided between the desire to protect the natural world and the quality of the environment and the desire to promote economic growth and ideological political objectives. If international conservation measures are to be realized, the different and often conflicting purposes must be reconciled.

The necessity for reconciliation grows out of the dangers inherent in the policy options available today. Some choices in themselves could be severely destructive to the environment, but a risk to environmental conservation is that in an effort toward a reconciliation or balancing of values, the integrity of the biosphere as a whole—its species and ecosystems—may be expendable. For example, values cannot always be balanced by shared or multiple uses of the same area or resource. Some values, such as those inherent in preservation of the great herds of African wildlife, cannot be balanced with the value of agriculture or unlimited tourism in the same place. Some ecological conditions cannot be compromised by incompatible uses and still be preserved. Nations committed by treaty to protection of the environment can be equally committed to policies that make this protection difficult. Even when government experts agree upon the desirability, even necessity, of environmental protection, they may differ fundamentally upon the method or extent of such protection.

Before Stockholm the international environmental movement was a spectrum of many different interests and efforts with little organized interrelationship. The Stockholm Conference set the United Nations Environment Programme (UNEP) in motion, providing an official intergovernmental focus for interaction among organizations concerned with international environmental issues. The UNEP headquarters and secretariat provided a point of convergence for representation of nongovernmental environmental organizations through the Environment Liaison Centre. Tendencies toward exclusiveness that once characterized the older NGOs, such as the International Council of Scientific Unions (ICSU) and the International Union for Conservation of Nature and Natural Resources (IUCN), diminished with the growth of opportunities for constructive collaboration. Thus by the 1980s an international network of environmental concern had developed outside the international intergovernmental agencies such as UNESCO and UNEP, and this network played a vital role in the implementation of these agencies' programs.

It is difficult to assess the strength of the nongovernmental network for environmental policy. Its functioning can be observed in relation to particular issues such as the Convention on International Trade in Endangered Species or protection of whales by the International Whaling Commission. Its influence upon policy is greatest and most critical at national levels in persuading governments to negotiate, ratify, and abide by treaty commitments. The active role of NGOs in promoting the several international environmental treaties endorsed at Stockholm and Rio offers at least inferential support for belief in their importance as factors in the development of international environmental policy through governments.

More certain is their importance in inducing an attitude of support for international environmental policy among the people of the world. Environmental protection literally begins at home. Unless environmental quality is valued at local and national levels of society, it is not likely to be a high priority in governmental and international affairs. The scientific, technical, and legal components of international policy are essential to its effectiveness, but without the added human dimension of value commitment, policy remains no more than possibility. Such components afford means to action, but the action depends on human purposiveness. There may already be sufficient information to overcome a great number of the world's pressing environmental problems. But it is not adequately used because environmental quality falls behind other priorities within the power structure of most countries.

National policies and intergovernmental agreements are essential but insufficient to achieve international environmental protection. The basic strategy for global environmental protection must also be directed toward popular understanding and evaluation. The need for such effort has been recognized by leaders in the international environmental movement in both governmental and nongovernmental organizations. More by convergence than by conscious design a de facto four-phase strategy has emerged. Its logical components are fact finding, programming, education, and activation. Fact finding and programming provided by the large number of international scientific investigations and the action plans adopted by the UN Conference on the Human Environment, the United Nations Conference on Environment and Development, and the UNEP Governing Council have been referred to extensively throughout this book. Three international programs in particular illustrate methods and strategies now being employed to reshape the human dimension of environmental attitudes and values. They are Man and the Biosphere, the UNESCO/UNEP Environmental Education Programme, and the World Conservation Strategy. Their common outcome should be an elevation of environmental priorities among the world's nations.

Man and the Biosphere

UNESCO's Man and the Biosphere (MAB) has been described by one of its principal architects, Michel Batisse, as "an international programme of concerted scientific cooperation among countries, directed towards the quest for practical solutions to the concrete problem of management of land resources and human community systems." [12] It addresses the fact-finding component of an environmental policy strategy but does so explicitly in relation to people.

Initiated officially in 1971, MAB was a direct consequence of the Biosphere Conference of 1968 and a corollary to the International Biological Programme of 1964 to 1975 sponsored by the International Council of Scientific Unions.[13] MAB is an international interdisciplinary program of research that emphasizes an integrated ecological approach to the study of relationships between man and the environment for the purpose of developing within the natural and social sciences a basis for the rational use and conservation of the resources of the biosphere. Organized around fourteen major research themes, it represents an extraordinary effort toward international interdisciplinary planning and research.

After passing through phases of defining objectives and priorities and developing global and regional work plans, MAB became operational in 1976. Detailed planning for MAB occurs largely through nearly one hundred national committees under the oversight of an International Coordinating Council, and functioning through more than nine hundred field research projects in most of the participating countries. A major function of MAB and its collaborating nations is the establishment of protected areas of ecological significance (biosphere reserves). As of 1994 the World Resources Institute reported that 312 biosphere reserves (noted in Chapter 11) have been established in sixty-eight countries. In addition to scientific values, establishment of the reserves has an important symbolic significance. It is an explicit recognition of mankind's involvement in the biosphere and is a moral commitment to responsible custody of the living world, the future of which now lies heavily in human hands.

It might appear superficially that MAB research activities, being localized on a national basis, would have relatively little international policy relevance, but Michel Batisse points out that

> it is the diversity of situations and conditions, and the complementarity of national approaches and efforts, that provide the very basis for meaningful international cooperation. For if this diversity did not exist, it would be sufficient to have one research project somewhere, and to disseminate its results all over the world. From this aspect the proliferation of national MAB activities is in itself a healthy feature of the international programme, and should remain so as long as the necessary linkages and exchanges of information between projects are maintained.[14]

MAB, therefore, presents an appropriate model for an international agency in environmental policy and administration. It provides an arrangement for the development and dissemination of program-relevant knowledge, and its efforts are also directed toward assisting the application of this knowledge to the realities of environmental management. Integration and synthesis of scientific learning are needed for its effective application to public policy and management. The availability of such knowledge, and the ability of governments to use it, are conditions essential to transnational cooperation in coping effectively with environmental problems, either on a bilateral, a regional, or a global scale. Such integrated, policy-adaptable information is not often available and, when it is, may be overlooked or perhaps rejected for political reasons. Nevertheless, without relevant knowledge, international environmental cooperation can hardly occur. MAB's purpose is to enlarge that knowledge base and extend its accessibility. Its activities confirm a new and constructive form of geopolitics.

It is significant that one of the research teams of MAB has dealt with the perception of environmental quality.[15] This topic reaches the very heart of popular attitudes toward the relationships between people and their environments and is obviously critical to the roles that governments play in shaping and protecting the human environment. Thus, the success of MAB's efforts, and of others like them, to develop and extend environmental understanding at national and international levels is highly relevant to prospects for the success of UNEP and the World Conservation Strategy. They both require public perception of qualitative aspects of the environment to achieve their goals.

An interesting innovation in environmental education was developed by the secretariat of the Man and the Biosphere Programme on the occasion of MAB's tenth anniversary. In connection with an international conference/exhibition entitled "Ecology in Practice: Establishing a Scientific Basis for Land Management," held in Paris from 22 to 29 September 1981, a set of thirty-six illustrated panels was prepared demonstrating various aspects of applied ecology. The exhibit was designed to be displayed anywhere in the world with text in local languages.[16]

Depending upon one's expectations, the accomplishments of MAB might be regarded as having a relatively modest impact proportionate to its ambitious objectives and the formidable nature of the challenge. But if one considers the novelty of the efforts, the organizational constraints and financial limitations of UNESCO, and the generally unbiospheric priorities of national governments, MAB deserves a more positive assessment. It has broken ground in international cooperative environmental investigation upon which its own future efforts may build. MAB ought not to be expected to produce initial results spectacularly greater than independent scientific research in the participating states has produced. But it is easy for sceptics to underestimate the subtle, quiet changes in public attitude that discovery and information make over the years. And the effect of MAB is to disseminate policy-relevant scientific information among all participating countries. For this, time is required, and it is too early today to fully assess the significance of MAB. Several decades may be necessary before it will be possible to do so. Under UNESCO sponsorship an international conference to review the bioreserve program was held in Minsk (Byelorussia S.S.R.) 26 September–2 October 1983.

Environmental Education

Recommendation 96 of the 1972 United Nations Conference on the Human Environment called upon "the organizations of the UN system, especially UNESCO . . . [to] take the necessary steps to establish an international programme in international education, interdisciplinary in approach, in-school and out-of-school, encompassing all levels of education and directed toward the general public." In January 1975, UNESCO and UNEP launched jointly an International Programme for Environmental Education. Following an extensive survey of the state of environmental education in the world's countries, an International Environmental Education Workshop was held at Belgrade, Yugoslavia, 13–28 October 1975. An outcome of this workshop was *The Belgrade Charter: A Global Framework for Environmental Education.*[17] The charter, which began with a call for a new global ethic and a tacit endorsement of the United Nations Declaration for a New International Economic Order, identified two goals of environmental action:

1. For each nation, according to its culture, to clarify for itself the meaning of such basic concepts as "quality of life" and "human happiness" in the context of the total environment, with an extension of the clarification and appreciation to other cultures, beyond one's own national boundaries.
2. To identify which actions will ensure the preservation and improvement of humanity's potentials and develop social and individual well-being in harmony with the biophysical and man-made environment.

The Belgrade workshop reviewed and evaluated plans for regional seminars, and made preparations for an international conference on environmental education held in 1977. Following the workshop the UNESCO/UNEP Environmental Education Programme initiated a newsletter, *Connect*, in which international activities concerning environmental education were reported. *Connect*'s first number, January 1976, in addition to the Belgrade Charter, included a survey of previous international conferences relating to environmental education, and subsequent numbers have summarized and evaluated trends and accomplishments since inception of the International Environmental Education Programme in 1975.

The culmination of the first phase of the International Environmental Education Programme (IEEP) was the Intergovernmental Conference on Environmental Education held at Tbilisi, Georgia, 14–26 October 1977. The conference issued a declaration and forty-one recommendations, which were intended to constitute a framework and guidelines for the development of environmental education (EE) at national, regional, and international levels.[18] The Tbilisi Declaration set forth eight guiding principles for national action:

—Environmental education should consider the environment in its totality—natural and man-made, ecological, political, economic, technological, social, legislative, cultural, and esthetic.

—Environmental education should be a continuous life-long process, both in-school and out-of-school.

—Environmental education should be interdisciplinary in its approach.

—Environmental education should emphasize active participation in preventing and solving environmental problems.

—Environmental education should examine major environmental issues from a world point of view, while paying due regard to regional differences.

—Environmental education should focus on current and future environmental situations.

—Environmental education should examine all development and growth from an environmental perspective.

—Environmental education should promote the value and necessity of local, national, and international cooperation in the solution of environmental problems.

Because the recommendations included a broad spectrum of related activities on the part of numerous international organizations, UNEP convened in June 1978 a joint programming exercise among ten principal United Nations agencies and other concerned international organizations. The objective was to review related work planned or in progress in the several agencies and, if possible, to work out cooperative arrangements. In its March 1982 issue, *Con-*

nect reviewed progress in international environmental education since Stockholm and Tbilisi.[19] This assessment, and the conclusions of the UNEP Session of a Special Character, lead to the inference that the UNESCO/UNEP program has been generally successful in cultivating throughout the world an awareness of the importance of environmental education and has begun to create a network of communication among national and international environmental education programs. But the task of actual education in many countries is just beginning. Meanwhile, in early 1984, an International Society for Environmental Education was in the process of formation.

A second major UNESCO/UNEP conference on environmental education, the International Congress on Environmental Education and Training, was held 17–21 August 1987 in Moscow. The agenda, as reported in *Connect*, included "a review of progress and trends in EE since the Tbilisi Conference; the state of the environment and its educational and training implications; relations between intergovernmental environmental-scientific programmes and EE and training; presentation of a draft international strategy of EE and training through the 1990s." One of five symposia, coincident with the congress, considered the role of biosphere reserves in the "dissemination of ecological knowledge and the training of ecological specialists."[20]

Three international conferences on environmental education (1981, 1985, 1989) have been organized by the Indian Environmental Society and the Department of the Environment of India. Emphasis in all of these efforts has been on environmental education in developing countries. It is conceivable that in a few decades the quality of environmental education in these countries could be better than that of the presently developed countries, in which curricula are under the control of long-established educational bureaucracies.

UNESCO has assisted the incorporation of environmental education into the school curricula of many Third World nations. This innovation may contribute over time to the changing perception of the importance of environmental issues in countries where little environmental awareness formerly existed. Popular environmental education in the field may influence attitudes. In many countries national parks, biosphere reserves, demonstration plots, and protected natural areas are visited by their citizenry, and especially by children. To the extent that teaching about and experience of natural processes and systems are included in these visitations, public attitudes may be influenced. In time, these developments may remove the perception gap between generations regarding relationships among environment, human behavior, and human welfare; children may teach their elders.

Regional and national workshops and demonstration projects on environmental education have been reported in *Connect*. But the effectiveness of efforts in education and toward substantive action is contingent on the availability of reliable knowledge concerning human-environment relationships.

Therefore scientific inquiry of a broadly multidisciplinary and interdisciplinary character is basic to both education and action. The formidable task of overcoming barriers of illiteracy, poverty, and political indifference cannot readily be accomplished. But without the foundation being laid by the UNESCO/UNEP program, national action over much of the world could not be taken. Emphasis could thereafter turn increasingly to activation.

This was the thrust of the 1992 report of the Rio Conference, Agenda 21, Chapter 36, "Promoting Education, Public Awareness and Training." The recommendations of the 1977 Tbilisi Intergovernmental Conference were endorsed and a program of action proposed. The recommendations of the 1990 World Conference on Education for All: Meeting Basic Learning Needs (Jomtien, Thailand) and its Framework for Action were also endorsed at Rio.

World Conservation Strategy

Development of scientific information and understanding, as well as education to disseminate findings and cultivate attitudes, provides a basis for activating people and governments toward actual and specific environmental protection measures. In order to bring a sharper focus to the tasks of national and international environmental protection and to provide policy guidance on how objectives can be realized, the World Conservation Strategy was launched in March 1980 after three years of intensive effort organized through the IUCN with the sponsorship of UNEP and the financial assistance of the World Wildlife Fund. The strategy has been described as "designed principally to draw the attention of both decisionmakers and the general public to the urgent need for the conservation of the world's land and marine ecosystems as an integral part of economic and social development."[21] Its goal is activation.

The three main objectives of the World Conservation Strategy as stated in its executive summary are "(a) to maintain essential ecological processes and life support systems, (b) to preserve genetic diversity and (c) to ensure the sustainable utilization of species and ecosystems."[22] The summary identifies six main obstacles to these objectives. In large measure they reflect the consequences of the dualistic view of man and nature associated with the paradigm dominant in modern technoeconomic society. These obstacles were identified as: (1) the belief that living resource conservation is a limited sector rather than a process that cuts across, and thus must be considered by, all sectors; (2) consequent failure to integrate conservation with development; (3) a development process that is often inflexible and needlessly destructive due to inadequacies in environmental planning [and management]; (4) a lack of capacity to conserve, due to inadequate legislation and lack of enforcement; (5) a lack of support for conservation due to lack of awareness (other than at the most superficial level) of the benefit of conservation; and (6) the inability to deliver conservation-based development where it is most needed, that is, in the rural areas of developing countries.

The World Conservation Strategy as a published document is essentially a statement of goals and targets. Divided into three categories, the first states the objectives of conservation, and the requirements for their achievement; the second lists the priorities for national action; and the third denotes the priorities for international action. Under the international category, the document specifies a need for developments in law and international assistance, for programs specifically directed toward tropical forests and arid lands, for a global program for the protection of genetic resource areas (such as biosphere reserves), and, finally, for priorities for the international commons (the atmosphere, the open oceans, and Antarctica). Strategies at the regional level were advocated for international river basins and seas (considered in Chapter 8), and regional arrangements and priorities for sustainable development (e.g., in forestry), were recommended. The emphasis on sustainable development provides a way to integrate the objectives of the World Conservation Strategy with those of the Third United Nations Development Decade.

The language of the World Conservation Strategy is free from the heated rhetoric which characterizes New International Economic Order documents. Nevertheless, the strategy recognizes the fundamental anomalies and irrationalities that have led to serious social, economic, and ecological deterioration in many countries, and calls attention to the destructive consequences of human poverty for natural resources and the quality of the environments. It seeks national and international action that would put the world's nations on the road to sustainable economies. While IUCN intends to monitor and periodically assess the effectiveness of the World Conservation Strategy, financial support and governmental cooperation are essential to this effort and are not easily obtained. Nevertheless, the logic of its purpose and the scientific validity of its substance assure the World Conservation Strategy a prominent place among developments of long-range significance in international environmental affairs.

How significant the strategy will prove to be depends upon its practical use. It is the culmination of more than a decade of thought and experience relating to environmental conservation. Its purpose and challenge have been summarized by Lee M. Talbot, environmental scientist and former director general of IUCN, in an address (19 March 1980) before the Royal Society of Arts in London:

The Strategy now exists, it has been introduced to the world, and it has already achieved significant results. In one sense, then, this represents the culmination of a major effort. In a broader sense, however, it represents the start of a new phase in conservation. For while the Strategy is the most ambitious effort yet undertaken in international conservation, in a historical perspective it is simply a part of the ongoing process. The challenge now is to make the Strategy work—to see that its recommendations are implemented—and, most important, to see that it really does serve as a focus for cooperation of all segments of the world society to achieve common goals to maintain a world in which human welfare, and survival, is possible.[23]

During the first decade of the strategy, forty-five countries developed national plans. IUCN (the World Conservation Union) has been the principal coordinator of the project. By 1989 IUCN, UNEP, and WWF agreed that a new intensified phase of the strategy was needed with special focus on sustainable development and environmental conservation. A new document was scheduled for 1991, timed so that it could be an input to preparation for the United Nations Conference on Environment and Development in 1992.

As numerous treaties confirm, and as this book has illustrated, a world structure for environmental policy exists. It is a heterarchial network of cooperative relationships, however, and does not command. In a world divided, the only feasible way to make the strategy work is through a persuasive interpretation of the implications of environmental issues in which the well-being of nations and all humankind is involved. The identification of areas of mutual concern and the formulation, so far as possible, of mutually advantageous lines of action would seem to be the most promising route toward realization of World Conservation Strategy objectives.

The IUCN and associated international and nongovernmental organizations identified throughout this book provide a structure in which leadership develops outside of the confines of intergovernmental politics and diplomacy. Thus a constructive and cooperative politics of world conservation might achieve lasting results apart from the conventional politics of nation-states and, in time, might help to improve the character of conventional international political behavior.

This book has surveyed the emergence and dimensions of efforts to protect the quality of the world environment, but it cannot lead to firm conclusions regarding the future. Still, the cumulative record of national and international effort toward environmental protection justifies modest hope. Were populations declining, and international and social tensions abating, a cautious optimism for the successful achievement of World Conservation Strategy goals would be justified. Even were governments persuaded, and many are not, that environmental conservation is in their long-range interest, their ability to adopt and enforce effective environmental protection measures is limited.

Barring unforeseen events, the ecological quality of the environment for all living things seems almost certain to suffer a net decline in the decades ahead. The twenty-first century seems almost certain to be ecologically poorer than any of the centuries preceding it. Were it not for the international programs just described, and the conservation efforts in many countries, the losses could be much greater. It is difficult to believe that a change in human perceptions and values could occur on a scale and within a period of time to reverse this pessimistic conclusion. Yet human history has recorded abrupt and unpredicted changes in social behavior. Events since the year 1972 suggest that peoples and their governments are, with relative rapidity, developing an appreciation of the consequences of continuing along the path of ecologically heedless exploitation of the earth. This consideration reinforces the significance of the World Conservation Strategy which, as Thomas E. Lovejoy of the World Wildlife Fund has declared, is also a strategy for survival.[24]

Defense of Earth in a Divided World

How to translate the World Conservation Strategy, the World Charter for Nature, Agenda 21, and the environment-related objectives proposed by various scientific and professional bodies into practical political action is now the critical task of international environmental policy. The legacies of past practice and political indifference are deeply entrenched, and fundamental change in human behavior, in individuals or institutions, is never easy. The conference resolutions, treaties, protocols, programs, and agencies that provide the structure of official international environmental policy also provide the goals of action and often the means. But the energy to realize these intentions derives from personal human commitment intensified through organized, purposive social effort and programmed through government and governmental initiatives.

To defend the environmental future of the earth in a divided world, effective action must be based on a realistic assessment of possibilities. Expectation of an imminent upsurge of worldwide ecological morality would hardly be realistic, but belief in the possibility of a gradual progression toward a universal environmental ethic would find support in actual experience. Such an eventuality is implicit in the many and diverse efforts toward international environmental protection described in this book. During the late 1980s, however, an international upsurge in environmental awareness did occur. In the United States the news magazine *Time* dedicated its first 1989 issue to "The Planet of the Year" (the earth); in Toronto, Canada, the *Sunday Sun* (19 November 1989) brought out a special sixteen-page supplement asking "Can We Bring Our Environment Back from the Brink?" There were many other declarations and calls for action. At their July 1989 summit meeting in

Paris, the political heads of the seven principal industrial democracies placed environmental concerns among the top items of their agendas, but doubtfully high on their priorities.

To achieve practical results, however, the world must be taken as it is. Its divisions and antagonisms are givens—circumstances that cannot readily be changed. Divisiveness among nations is expressed primarily through their political parties and governments—which are often primary agents of international antagonisms—and not through the people governed. Yet governments must be moved if environmental policy is to be activated. Governments must simultaneously be influenced from within, and induced from without, to work with other governments and international organizations.

NGOs representing transnational environmental interests and values, of which the International Union for Conservation of Nature and Natural Resources (IUCN) is a notable example, are indispensable means toward energizing the world environment movement. The constituent national member organizations of IUCN represent international environmental concerns within their own countries (but not necessarily the priorities of their governments), avoiding the rejection reflex that frequently characterizes governmental reaction to perceived importunities from without—especially from intergovernmental organizations.

Whatever strengthens and assists nongovernmental international organizations such as ICSU, WWF, and the affiliates of MAB, among many others, contributes to the prospects for realizing a rational international order of policy and practice for the world environment. In relation to what is needed for effective environmental protection, even at present limits of international receptivity, all programmatic efforts—in research, in education, and in popular activation—are chronically underfunded. How to increase their support is a problem deserving early attention and ingenuity by governments, international organizations, and international philanthropy.

Further, existing institutional arrangements may not be adequate to provide for all aspects of international biospheric protection. At least three areas of deficiency may be identified which do not appear to be covered by existing institutions or agreements: environmental impacts of international commerce and investment, transnational means to implement environmental protection agreements in the international commons (i.e., monitoring compliance), and means to rehabilitate socioecologically bankrupt nations.

No present means adequately protects against environmental damage incident to international investment and resource development. National laws and the policies of some international investment agencies, such as the World Bank, ostensibly provide safeguards against environmentally damaging development. The 1980 "Declaration of Environmental Policies and Procedures Relating to Development" adopted by banking and development institutions, and the implementing Committee of International Development Institutions on the Environment (CIDIE), provide reinforcement for good intentions.[25] The committee meets periodically for coordination of effort and exchange of information and experience. Meanwhile the World Bank has adopted a post-Rio Strategy to follow through on Agenda 21. The strategy includes implementation of a comprehensive environmental assessment procedure "to ensure that all development options under consideration are sound and sustainable."[26]

But economic incentives characteristically override environmental considerations. Private investment, especially by multinational corporations, is not easily influenced nor its ramifications easily contained. International resource developers need national collaborators and usually find them. In 1995 the government of Surinam was approached by several Southeast Asian timber companies to log 40 percent of the country's largely intact rainforest. The president of Surinam denounced the opposition of foreign environmental organizations as "eco-colonialism." The combination of a relatively poor country possessing natural resources coveted by acquisitive corporate enterprise and with pliable political officials is hard to resist.[27] The consequences of this collaboration are not necessarily environmentally damaging, but they have often been so—Daniel Ludwig's Jari Project in the Amazon forest being a case in point. Perhaps nations need some process of mutually agreed upon international review, independent of the inclinations of funding and investment agencies, to assess the national, international, and environmental costs, broadly defined, of international investment proposals and transfers of industrial technologies.

The 1982 UNEP Session of a Special Character in its concluding declaration addressed this issue, stating that "[a]ll enterprises including multinational corporations, should take account of their environmental responsibilities when adopting industrial production methods or technologies, or when exporting them to other countries. Timely and adequate legislative action is important in this regard." In 1991 (26 February) twenty-five countries and the European Community, meeting at Espoo, Finland, signed the Convention on Environmental Impact Assessment in a Transboundary Context. But by 1996 only six governments had ratified the treaty. As previously noted (Chapter 8) the Madrid Protocol to the Antarctic Treaty provides for environmental assessment in that regime, which as of 1995 was still not yet in effect.

As to the second deficiency, no solution is presently in sight for an adequate international or transnational arrangement to monitor or police international environmental protection agreements in the international commons. The history of the United Nations Conference on the Law of the Sea gives no reason for optimism that an institutional solution to replace and extend present multilateral arrangements can be found in the foreseeable future. And international treaties, such as those relating to endangered species or to whaling, make insufficient provision for collective enforcement. There is growing recognition of the need to monitor and report the observance of treaty obligations. Institutions of governance, short of actual government, may be necessary.

The third deficiency in the present international structure for environmental policy is the absence of any institutional arrangement to assist the administration of countries faced with socioecological collapse. If there are no countries actually in this state today, some are nearly so. Socioecological insolvency means that a state has exhausted its material means of self-support and no longer provides to its people the elementary services of government. Some countries may have already reached this condition, which, however, is masked but scarcely concealed by infusions of foreign monetary aid.

More than two decades ago at an international conference on environment and development I suggested the need for an arrangement under which a severely handicapped nation might voluntarily place itself under some form of international receivership but with a more acceptable name.[28] There would obviously be many difficulties in implementing such an arrangement. The operational weakness of the United Nations was glaringly apparent in its ineffectual peace-keeping missions in Somalia and Bosnia in the 1990s. The proposition is probably utopian for the 1990s, but might become more realistic in the twenty-first century. The concept ought not be relegated to the limbo of unthinkable thoughts in a world in which there can be no assurance that a real need for such an arrangement will not arise.

Similar in objective was a proposal by the Soviet Union in 1989 that the UN General Assembly consider establishing a Centre for Emergency Environmental Assistance to send international groups of experts to ecological disaster areas. To a limited extent, this function is currently under taken by the Office of the United Nations Disaster Relief Coordinator (UNDRO). It is intended, however, to respond to relatively short-term emergencies, for example, volcanoes, earthquakes, and floods rather than long-term chronic socioecological disintegration.[29]

In its concluding declaration the UNEP 1982 Session of a Special Character (Chapter 5) took cautious cognizance of the threat of environmental disruption in developing countries. But consistent with prevailing United Nations rhetoric, it phrased the issue as one of conventional technical and economic assistance. Point 7 of the declaration stated that "[d]eveloped countries, and other countries in position to do so, should assist developing countries affected by environmental disruption in their domestic efforts to deal with their most serious environmental problems." The declaration assumes an obligation to assist, but does not address the capacity to receive, which is often the most difficult aspect of international assistance.

The disastrous effects of war upon all aspects of the human environment have been extensively documented and will not be recounted here.[30] Modern warfare is the utter antithesis of an ecologically conserving relationship with the earth. Were the energies and resources now devoted to military activities made available for defense of the earth, the prospects for the human and environmental future would be infinitely brighter. But this book has been intended to record and describe the emergence of the environment as an aspect of international policy, and only incidentally attempts to deal with threats to the environment that international environmental policy has not primarily addressed. War, the exponential growth of human populations, and the profligate consumption of natural resources are the principal threats to the human environment and the biosphere, and their great importance to the environmental future bears no relationship to the attention given them in this book. But the environmental movement (as distinguished from policy) focuses attention on each of these menaces. Chapter 7 identifies war crimes against the environment committed by the government of Iraq in the 1990–91 Gulf War, and ecological warfare by the United States in Vietnam was severely criticized.

To the extent that nations learn to work together in resolving common problems in the environment, they may discover patterns of interaction and elements of mutual understanding that will assist efforts to reduce the extent of violence and improvidence in the world. Environmental destruction as a strategy of warfare has historically been treated as an inevitable consequence of combat. Today a distinction is being drawn between destruction incident to military necessity and unnecessary damage to the environment as an expression of hatred or of total disregard for the consequences of the action.

13
A Changing World Order:
Into the Twenty-First Century

That we live in a world of rapid change is axiomatic. More accurately, we live in a world undergoing multiple changes, some of which contradict one another. The phenomenon of change is worldwide, but the impacts and implications of the changes differ among different sectors and circumstances of human societies. Generalizations about the effects of rapid multiple change should therefore be taken for what they are—estimates of probabilities not predictions. No generalization is universally true; but the more reliable of them may identify tendencies having implications for the future—even for the cultural or physiological survival of humanity and perhaps even for the sustainability of the complex living systems of the earth.

Just as our present was shaped in the past, so the future is being shaped in the present. However, the complex multiplicity of evolving forces and effects makes forecasting an exercise in identifying probabilities—hardly certainties. Change is an interactive process and can lead to unforeseeable consequences, and perhaps our most realistic (or cautionary) model for assessing the general direction of change is chaos theory. When we do forecast our collective future based on present and foreseeable trends, we are attempting to estimate the consequences of present human tendencies. In making such forecasts, we should employ the most reliable data and make appropriate use of statistical and mathematical analyses with particular attention to probability. We should recognize with due humility that we may not know all that we need to know to ascertain all aspects of the future toward which we are tending. But the fact of uncertainty should not derogate from the use of impact assessment and trend analysis in forecasting probabilities of which policy should take account: more is known today than ever before about cause-consequence relationships, and policy-informing knowledge should not be dismissed because it may be incomplete.

Among the things that we know, the behavior of the physical world may be the most accurate. For the biological world, our ability to predict behavior is less accurate. It is least accurate for the prognosis of human behavior, especially of large, diverse, and complex societies (e.g., modern society). While opinion polling and behavioral studies reveal certain human tendencies, they often raise more questions than they answer. In attempting to ascertain social perceptions and preferences, we rely instead to a large extent upon their representation in institutions, especially those of government.

For this reason, the emphasis in this book has been heavily on institutions of governance. These institutions comprise not only formal governments, but also international organizations, regimes, agreements, and customary behaviors that in effect govern without being governments. Collectively these institutions are creating a new aspect of international policy. Because the numerous factors that comprise the planetary environment are now seen to be interactive and are not bounded by political jurisdictions, the new international policies of nations are unavoidably geopolitical. The new geopolitics goes far beyond the old geopolitics of competition for territorial, military, or economic advantage. It is now the use of politics in managing human behavior in relation to the opportunities and limitations of life on earth, which means a politics that takes, in the broad sense, geographical (i.e., planetary environmental) factors into account.

The Environmental Movement and the New Geopolicy

The emergent politics on the planet Earth differs in an important respect from earlier geopolitical strategies of nations. At least in regard to environmental policy, most nations appear willing to cooperate in principle, but often with reservations on particulars. Uncontrolled international economic or industrial competition could have, and has had, environmentally damaging consequences. We have seen how the move toward economic cooperation and merger in the European Union necessitated common or comparable environmental policies. Similarly, the North American Free Trade Agreement was accompanied by a "side agreement" for environmental cooperation. Today the same convergence is occurring on a global scale, and national economic and foreign policies generally will need to be examined in relation to their broader social and environmental effects.[1]

The behavior of governments today, and especially of democracies, tends to be reactive. Environmental policy is a conspicuous example of concerns and responsibilities forced upon often reluctant governments by organized groups of citizens (NGOs), backed by findings from the environmental sciences. The various roles of international and national environmental NGOs have been described in many places throughout this book. Their dual function of transnational collaboration on environmental issues (especially in relation to the UN agencies) and pressure on governments in their respective countries has been a driving force behind the development of international environmental policy.

It is not possible to foresee what the world order will become in the course of the twenty-first century. A series of global models have been developed to forecast possibilities, but their predictive reliability is as yet undemonstrated.[2] Still, from all of them, the wisdom of initiating environmental and resource conservation measures may be inferred. There is no way to know how long the international order will remain in its present unstable state. Equally uncertain is the time required for a critical mass of peoples and nations to reach a level of ecological understanding and political maturity sufficient to implement an international environmental policy that could promise a sustainable future for man and the biosphere.

In assessing the state of international environmental policy in 1982 at the end of the post-Stockholm decade, Robert D. Munro found a paradox:

> At the national level, public responses and opinion polls indicate that public awareness and support remains high. However, at the international level, there has been a noticeable decline in governmental, political and financial support—a trend which is in stark contradiction to the increase in the extent and seriousness of the worldwide environmental problems and challenges which governments must confront together in the 1980s and later.[3]

By 1988 and the fortieth anniversary of IUCN (now the World Conservation Union) the political circumstances had changed. The World Conservation Strategy launched in 1980 was surely a factor contributing significantly to the report of the World Commission on Environment and Development (Brundtland Commission and UNEP's *Environmental Perspective to the Year 2000 and Beyond*). Organized public opinion had moved governments, and Dr. David Munro (a former director general of IUCN) was leading the development of a Second World Conservation Strategy Project.

From the cumulative record of environmental policy development reported in this volume, it may be inferred that the relative decline in governmental support for international environmental cooperation (if in fact there was a decline) has been temporary. Governments have not reverted to pre-Stockholm positions, but worldwide economic difficulties preempted political priorities almost everywhere. Moreover, the great majority of national political leaders in the 1980s came to maturity before the environment was a matter of public awareness or concern. The programs of research, public information, and education now under way, and the institutional machinery and operational strategies now in place, are likely to be more enduring than governments of the day and their leaders. If these efforts are even moderately effective, pronounced changes in the international environmental policies of nations should be evident by the end of the 1990s.

A survey of global environmental concern undertaken by the World Environment Center in 1982 revealed a 500 percent increase in governmental environment and natural resource management agencies during the post-Stockholm decade.[4] At least 144 countries were found to have such agencies; 105 of these were in the Third World, up from 11 in 1972. Increases also occurred in the industrialized countries, only fifteen of which had environmental agencies at the time of the Stockholm Conference. These figures, together with the growth in numbers and members of environmentally concerned NGOs, especially in Third World countries, suggest that worldwide environmental concern has been increasing. Decline in United States financial support for international environmental programs during the 1980s and 90s was a serious negative development, but is explainable by circumstances peculiar to United States politics. It does not reflect a world trend and is not necessarily indicative of future policy.

In a world of nations, developments at national levels affect the course of international cooperation. During the decade of the 1970s, the United States government played an active and generally positive role in international environmental protection. Beginning in 1981, however, a new national administration took a series of actions suggesting a retreat from international cooperation in environmental affairs. The United States withdrew support for the draft treaty negotiated at the United Nations Conference on the Law of the Sea. It was the only nation voting against the UN Charter for Nature and the General Assembly Resolution for Protection against World Trade in Materials Harmful to Health and Environment, easing its own regulations on export of hazardous materials as well. Further, the United States reduced its financial contribution to UNEP and attempted to phase out its participation in MAB.

The paradox of the nation that was in the leadership of the environment movement at Stockholm now officially standing against the rest of the world on international environmental measures was more apparent than real. The majority of negative votes by the United States delegation to the General Assembly were not always in opposition to the principles at issue, but rather reflected opposition to any increase in United States contribution to the budget of the United Nations.[5]

Environmental protection in the United States throughout the 1970s and 80s was bipartisan, and opinion surveys indicated that popular "environmentalism" remained strong, cutting across lines of social class, occupation, and economic status. However, a polarization of the environmental issue has to some extent occurred in a number of developed countries. In Germany and the United States polarization occurred between parties for and against the environmental movement. Opposition to environmentalism emerged when some people with chiefly economic interests discovered the scope and direction of the environmental movement, which they had not foreseen.

New ecology-oriented ideological parties or action groups appeared in

France (Ecology Party), in New Zealand (Values Party), in Australia (the Green Bans), and in Germany (Green Lists).[6] Some of these parties or action groups were ephemeral (e.g., in Australia and New Zealand) partly because the major parties adopted their environmental agenda, partly because their agendas were too narrow for broad political appeal. In the United States, disaffection with the negative environmental policies of the Reagan administration catalyzed environmental activism in the 1988 national elections. Environmental policies were major factors in 1988-89 elections in Canada, Norway, and the Netherlands. Nor were socialist countries immune to clashes between ecology and politics. In 1981 the government of Poland acted to suppress an environmental organization of Polish students,[7] but following the peaceful overthrow of the Communist regime in 1989, Green politics became legitimate.

The prospect of ecological political movements in the developed countries is enhanced by the trend, now worldwide, to link the environmental issue with efforts to create sustainable economies, extending the concept to include ecological sustainability. This trend, while not necessarily incompatible with economic development, is clearly incompatible with many prevalent development practices in agriculture, energy generation, foresty, mining, transportation, and urbanization. Conflicts are thus arising between advocates of maximum production now and advocates of ecologically optimal, sustained yield. Nations divided internally on basic economic-environment issues are not likely to cooperate effectively in international environmental policies and programs. In Brazil, for example, the shifting balance of political power between prodevelopment and proprotection forces make its future policy unpredictable.

In its tenth anniversary report in 1982, UNEP pointed to an evolution of environmental issues growing out of changing perceptions of the environment. Evidence of worldwide change and growth in environmental perception was cited from three basic sources: decisions by United Nations bodies revealing governmental acceptance of new principles, growing numbers of scientific findings concerning the environment that have been gaining public acceptance, and practical experience in dealing with environmental issues. A body of precedent-setting law and practice has begun to take on the character of an international constitution for the world environment. Among the basic documents of this constitution are the Final Report of the Biosphere Conference of 1968, the Declaration and Action Plan of the United Nations Conference on the Human Environment, the World Conservation Strategy, the World Charter for Nature, the Brundtland Commission report, and Agenda 21. Normative goals and standards have been set by voluntary international action against which the future conduct of nations in relation to the environment can be measured.

A growing public awareness of environmental problems and national commitments in principle to environmental protection—and the probability that environmental conditions will worsen—suggest that national and international policymaking will almost certainly be concerned with environmental problems for the foreseeable future. The costs to governments of belated attention to environmental hazards may be high, but the consequences of failure to address them may become socially, politically, and morally unacceptable. The social and political divisions of the world burden and delay the implementation of environmental policies to which nations have in principle agreed. The record of experience shows that under the continuing pressure of events and organized public opinion, action is eventually forthcoming. The question for many environmental issues is whether the action will be timely enough to avoid the consequences of delay.

In achieving an effective response to global environmental problems there is a wide but slowly narrowing gap between what is needed and what is presently possible. International environmental agreements today are negotiated and implemented (where they are implemented) in the political context of antagonistic cooperation. A worldwide consensus among informed persons regarding responsible custody of the earth, its soils, waters, wildlife, and built environments is visibly growing but is yet insufficient to fundamentally influence the behavior of most governments and multinational organizations. Yet the history of the emergence of the international movement recounted in the preceding chapters reveals an extraordinary growth of environment-related concepts, institutions, and influence within less than half a century.

The need for institutions for environmental governance better fitted to the task has been voiced repeatedly by students of environmental policy and world affairs.[8] The building of consensus is proceeding in many ways, in many places, exemplified in its initiating stages by the environmental education movement, the MAB programs, and the World Conservation Strategy. Networks of NGOs are extending and consolidating, and multinational corporate bodies, economic and philanthropic, are being drawn into closer relationships with the evolving structure of international cooperation.

The emergence of this loosely coordinated structure for cooperation offers a very great challenge to the analytic capabilities of organization theorists. The biosphere is too large, too diverse, and too complex to be "managed" by any centralized coordinating authority. Decentralization of responsibility and action is a practical necessity, but of equal necessity is cooperative action across national boundaries and within the unpreempted regions of the air, oceans, Antarctic, and outer space. Science assists the process of defining the tasks of international environmental action, but building a coherent and effective structure for global environmental collaboration is a collective work of the art of governance—no science provides a blueprint. Humanity was never given an organization manual for spaceship Earth.[9]

The international environmental movement since Stockholm has been a large-scale experience in social learning. If the peoples of the world can learn and then act together in time in relation to those matters essential to their mutual survival, they may be able to maintain their diversities and differences and simultaneously to improve the quality of their lives. Defense of the earth should be possible in a world divided in many ways, yet united in cooperative endeavors to safeguard the viability of the planet upon which the survival of the human race depends.

The year 1992 marked five centuries since the European "discovery" of America. The succeeding half-millennium witnessed the preemption of the entire earth by the human species and the rise of national governments, followed by European colonization throughout the world. During these centuries both war and commerce took on worldwide dimensions. The natural environment of the earth underwent the most devastating assault since the beginning of history until, near the end of the twentieth century, the reaction of the environmental movement emerged.

The Big Change

The decades at the end of the twentieth century and the beginning of the twenty-first century mark a major discontinuity in the history of civilization. From a retrospect in the future, the environmental movement will be seen as an integral part of this fundamental change in the relationship of human society to the earth—and of the peoples of the earth to one another. More than conjecture, this transformation of the world is being forced by changing circumstances. The assumptions and conditions that made the modern world are nearing their end. One cannot be certain what the new world political economy will be like, but the old exclusively nation-centered politics and the exuberant exploitation of people, resources, and environment can no longer be the general rule. Intransigent governments will doubtless persist, but the relative and absolute shrinking of the real resource base (as distinguished from temporarily marketable reserves) and the impact of contamination on all aspects of the human environment will force changes that no amount of self-serving political power can indefinitely resist, but may resist long enough to impoverish the biosphere.

The capacity of the human mind to simultaneously entertain conflicting values, thoughts, and priorities has long been recognized. People may avoid a sense of contradiction through techniques of denial. They tend to reject proposals perceived to be inimical to their self-interests. This tendency presents a difficulty for environmental issues that are complex, extended over time, and characterized by uncertainties in detail although not necessarily in ultimate consequences. A broadening of perceptual horizons has made possible the rise of an environmental movement during the last half of the twentieth century. Public opinion has changed to a degree that has moved governments. And yet the degree of commitment is uncertain. Max Nicholson sums up the circumstance in his book *The Big Change: After the Environmental Revolution:*

> Vigorous currents are converging toward a far more integrated and whole view of human life and its meaning, and of human affairs and their conduct, than ever before. Perhaps the Big Change will assume a more positive and constructive shape than has so far seemed likely, and the healing forces which are so desperately needed by mankind will begin to work faster. Perhaps, if the worst comes to worst, catastrophe may be less than total, and those who survive it may be in a frame of mind to make a fresh start. To attempt prophecy is pointless, but to do everything possible to avert what current prophecy often predicts is still worthwhile.[10]

National policies in the future to succeed must take account of their geopolitical implications, not out of altruism perhaps, but certainly out of prudence and aversion to failure. Commerce, technology, war, mass human migration, and concepts derived from the sciences have transformed the world of sovereign nations and separate cultures into a complex, dynamic multiethnic world system that is struggling to find appropriate and feasible means for its governance. World government is not necessarily implied, but more coherent and effective instruments for collective action have become necessary.

It is important to understand that environmental policy is absolutely basic to these changes and that the linkage with sustainable development is essential to bringing into harmonious relationships these erstwhile aspects of human behavior that humans have heretofore separated. Without intending to do so, modern man has made geopolicy inevitable. This inevitability holds far-reaching implications for international relations and traditional diplomacy.[11] The doctrine and dogmas of inalienable national sovereignty are being modified de facto to accommodate the imperatives of international environmental cooperation. For example, Latin American states in which exclusive sovereign rights were most vigorously asserted at Stockholm now accept the necessity to join with other nations in environmental protection measures in their own, and the broader, international interest. On 31 March 1989, a "summit" conference of Latin and Caribbean states adopted the Declaration of Brazilia and, although reasserting the sovereign right of each state "to administer freely its natural resources," declared their willingness to cooperate on environmental issues while appealing to the industrial nations to augment contributions to developing countries and to UNEP for environmental protection purposes.[12] Later, on 6 May 1989, the eight Amazonian basin states meeting in Manaus, Brazil, although reasserting their sovereign rights over their resources, declared support for an Amazonian Special Environmental Commission and a Special Commission on Indigenous Affairs. These states had withheld approval of the UN World Charter for Nature in 1982, but now on their own initiative appeared ready to address the problems of environmental protection within the context of sustainable development as they interpreted it.

In the practical realm of national priorities and budgets, the impact of environmental imperatives may be most visible on armaments and military commitments. Warfare is increasingly seen as incompatible with environmental integrity—especially when waged with chemical and nuclear weaponry, the effects of which could persist far beyond the immediate impact on human life and health. Moreover, there is also increasing recognition that the principal threats to human safety and security are less likely to come from the armed forces of nations than from the degradation of the biosphere. All humanity is put at risk by disintegration of the ozone layer and by climate change of uncertain incidence. All of the environmental issues enumerated in Chapter 1 of this book pose threats to security in all its aspects, as well as to the quality of life. Vast and costly outlays for military defense will neither protect a nation from internal decay nor from the transboundary effects of environmental misconduct elsewhere. A rethinking of national defense and security strategies is needed. Faced with environmental problems and hazards of truly planetary proportions, the present allocation of financial resources to military purposes has become indefensible. In a world still rampant with violence and aggression, total disarmament is unrealistic—but a very large universal diminution of military outlays is not only possible, it is a practical necessity if the resources essential for the world's present environmental commitments are to be mobilized.

Without intending to do so, humanity has brought upon itself the necessity to manage its impact on the whole earth. The earth was encompassed by technologies of communications, manufacturing, and commerce long before the environmental implications of these developments were perceived. All of these developments were advanced and accelerated by applications of innovative science. Freed, it seemed, from the Malthusian restraints of the past, human populations increased rapidly, and through unprecedented migrations have changed the demographic and ethnic patterns over large areas of the world. As distinctive national cultures begin to fade, centralizing nationalism as a political unifier has diminished in developed countries, but increased in many Third World nations. Transboundary and international communication among people and organizations concerned with environmental issues has been greatly facilitated. A consequence has been the rise to prominence and influence of the NGOs, documented throughout this book. The changes in the condition of the world occurring during these last three decades of the twentieth century have been too great to be fully grasped by most people who have lived through them. The realities of the world have changed, and so therefore have the realities of politics and national policies. National policies recently thought to be realistic have become impracticable—and policies formerly regarded as impractical and utopian are becoming acceptable but not everywhere yet accepted.

What the full effects of the new geopolicy—or policy in a planetary context—will be cannot be foreseen. Global cooperation may coexist with a trend toward local, regional, and provincial autonomy. Yet the need for international collaborative measures, especially on economic and environmental matters, has been answered by the formation of the European Union and the North American Free Trade Agreement and by an unprecedented number of environment-related treaties, only the more comprehensive of which are listed in Appendix E. This apparently contradictory trend toward local and international responsibility may be viewed as a logical manifestation of the admonition "to think globally and act locally." Nations need not lose their cultural identity and integrity by cooperating with other nations in matters of common necessity. Indeed, international environmental policy has been directed to protecting and restoring the cultural and ecological distinctiveness of nations—as in the World Heritage Programme and UNESCO's Man and the Biosphere. The security and future of humankind on earth depends greatly upon the reformation of human behavior in relation to the biosphere. It thus seems plausible that in the twenty-first century the environment may become the primary subject of international policy, but the way in which this subject will be defined is uncertain.

If not a blueprint for environmental policy in the twenty-first century, Agenda 21 provides a checklist of goals and guiding principles that could preserve the quality of life on earth and prevent avoidable environmental catastrophe. Yet as Norman Meyers has pointed out, we do not know what unforeseen dangers are now in the making.[13] It may already be too late to avoid some environmental adversities. Depletion of the stratospheric ozone layer and global warming may be too far advanced to avoid consequences harmful to humanity and to many forms of life on earth. The prospect of resource scarcity (e.g., water) is troublesome. Resource deprivation can lead to civil disorder, political instability, and pressure to migrate en masse to countries more favorably situated. Resource wars, civil and international, may characterize the twenty-first century.[14]

A political danger that might have been foreseen was a counterrevolution in the 1990s against the environmental movement, notably in the United States and Canada. The reaction sought to replace regulatory measures with market incentives and common law remedies and was thus chiefly applicable to legal systems based on English common law as contrasted with civil law systems prevailing in Europe and most of the rest of the world. While appropriate market incentives were also favored by many environmental advocates, the antienvironmentalists, including those in government, went much further.

With polls of public opinion continuing to show strong worldwide support for environmental protection, the United States government appeared to be out-of-step with the rest of the world. This was not because of a shift of public opinion; it was a consequence of certain characteristics of American politics—notably the inordinate role of economic interests and money in political campaigns, and the peculiarities in the congressional committee system.

This book records the history of the environmental movement and policy response in the twentieth century. It places the environmental movement on a trajectory in time, but cannot project the future time dimension of the curve although its direction seems probable. Were the changes between the Stockholm and Rio conferences to be strengthened and extended, one might be hopeful for the future. But projections of population growth into the twenty-first century obstruct any reliable forecast of the environmental future.[15] Retention of any unequivocal right to reproduction could be the point of vulnerability that negates most of the goals and achievements of the environmental movement. In a world of nations numbers "count," and unless the environmental movement can effectively encompass the population issue, the international conflicts of the twenty-first century may be over who inherits an ecologically impoverished world.

Appendix A
Abbreviations

ABEDA	Arab Bank for Economic Development in Africa
ACASTD	Advisory Committee on the Application of Science and Technology to Development (UN)
ACC	Administrative Committee on Coordination (UN)
ACMRR	Advisory Committee of Experts on Marine Resources Research (FAO)
ADB	African Development Bank
AGRIS	International Information System for the Agricultural Sciences and Technology (FAO)
ALECSO	Arab League Educational, Cultural, and Scientific Organisation
AOSIS	Association of Small Island States
ASDB	Asian Development Bank
ASEAN	Association of Southeast Asian Nations
ASOC	Antarctic and Southern Ocean Coalition
ASPEI	Association of South-Pacific Environmental Institutions
AT&T	American Telephone and Telegraph Corporation
ATCPs	Antarctic Treaty Consultative Parties
ATS	Antarctic Treaty System
BAPMoN	Background Air Pollution Monitoring Network
BIOMASS	Biological Investigations of Marine and Antarctic Systems and Stocks
BMB	Baltic Marine Biologists
CAB	Caribbean Development Bank
CARICOM	Caribbean Community
CCAMLR	Convention on the Conservation of Antarctic Marine Living Resources
CCAS	Convention on the Conservation of Antarctic Seals
CCITT	International Telephone and Telegraph Consultative Committee (ITU)

CCMS	Committee on Challenges of Modern Society (NATO)
CEC	Commission of the European Communities
CEPT	European Conference of Postal and Telecommunications Administrations
CEQ	Council on Environmental Quality (U.S.)
CGIAR	Consultative Group on Agricultural Research
CI	Conservation International
CIDIE	Committee of International Development Institutions for the Environment
CIOMS	Council for International Organisation of Medical Sciences
CITEL	Committee for the Inter-American Telecommunications
CITES	Convention on International Trade of Endangered Species of Wild Fauna and Flora
CLASP	Centre for Law and Social Policy
CMEA	Council for Mutual Economic Assistance
COMSAT	Consortium of Private Telecommunications Organizations (U.S.)
CONABIO	National Biodiversity Use and Documentation Commission
COP	Convention of the Parties
COSPAR	Scientific Committee on Space Research (ICSU)
CPPS	Permanent Commission for the South Pacific (Chile, Ecuador, Peru)
CRAMRA	Convention for the Regulation of Antarctic Mineral Resource Activities
CSAGI	Comité Special de L'Année Geophysique Internationale (ICSU)
CSCE	Conference on Security and Cooperation in Europe
CSD	Commission on Sustainable Development
CPPS	Permanent Commission for the South Pacific
DESCON	Consultative Group for Desertification Control (UNEP)
DOEM	Designated Official for Environmental Matters
DPCSP	Department for Policy Coordination and Sustainable Development
EBRD	European Bank for Reconstruction and Development
EC	European Community
ECA	Economic Commission for Africa (UN)
ECAFE	Economic Commission for Asia and the Far East (UN)
ECB	Environment Coordination Board (UN)
ECDIN	Environmental Chemical Data and Information Network
ECE	Economic Commission for Europe (UN)
ECJ	European Court of Justice

ECLAC	Economic Commission for Latin America and the Caribbean (UN)
ECOSOC	Economic and Social Affairs Council (UN)
ECR	European Court Reports
ECWA	Economic Commission for Western Asia (UN)
EDB	European Development Bank
EEA	European Environment Agency
EEB	European Environmental Bureau (Brussels)
EEC	European Economic Community
EEZ	Exclusive Economic Zone
EFTA	European Free Trade Association
EFZ	Exclusive Fishing Zone
EIA	Environmental impact assessment
EIB	European Investment Bank
ELC	Environment Law Centre (IUCN)
ELC	Environment Liaison Centre (Nairobi)
EMASAR	Ecological Management of Arid and Semi-Arid Rangelands (FAO)
EMINWA	Environmentally Sound Management of Inland Waters
ENMOD	Convention on Prohibition of Military or Any Other Hostile Use of Environmental Modification Techniques
EPA	Environmental Protection Agency (U.S.)
ERDF	European Regional Development Fund
ERTS	Earth Resources Technology Satellite Program (U.S.)
ESA	European Space Agency
ESCAP	Economic and Social Commission for Asia and the Pacific
EU	European Union
EYE	European Environment Year (1987)
FAO	Food and Agriculture Organization
FCC	Federal Communications Commission (U.S.)
FOE	Friends of the Earth
FOEI	Friends of the Earth International
GAO	General Accounting Office (U.S.)
GARP	Global Atmospheric Research Programme
GATT	General Agreement on Tariffs and Trade
GDP	Gross Domestic Product
GEF	Global Environmental Facility
GEMS	Global Environmental Monitoring System
GESAMP	Group of Experts on Problems of Marine Pollution
GFCM	General Fisheries Council for the Mediterranean
GIPME	Global Investigation of Pollution in the Marine Environment (IOC)
GNP	Gross National Product

GRID	Global Resource Information Database
G7	Group of Seven (Advanced Industrial States)
GWP	Gross World Product
HELCOM	Baltic Marine Environment Protection, or Helsinki, Commission
IABO	International Association of Biological Oceanography
IADB	Inter-American Development Bank (also listed as IDB)
IAEA	International Atomic Energy Agency
IAPSO	International Association of the Physical Sciences of the Ocean
IARC	International Agency for Research on Cancer (WHO)
IATTC	Inter-American Tropical Tuna Commission
IBP	International Biological Programme (ICSU)
IBRD	International Bank for Reconstruction and Development (World Bank)
IBRGR	International Board for Plant Genetic Resources (FAO)
IBSFC	International Baltic Sea Fishery Commission
IBWC	International Boundary and Water Commission
IC	Interim Baltic Marine Environment Protection Commission
ICAO	International Civil Aviation Organisation
ICBP	International Council for Bird Protection
ICC	International Coordinating Council (MAB)
ICCAT	International Commission for the Conservation of Atlantic Tunas
ICCRON	International Centre for the Study of the Preservation and Restoration of Cultural Property
ICEL	International Council for Environmental Law
ICES	International Council for the Exploration of the Sea
ICGEB	International Centre for Genetic Engineering and Biotechnology (UNIDO)
ICITA	International Cooperative Investigation of the Tropical Atlantic
ICJ	International Court of Justice (World Court)
ICOM	International Council of Museums
ICOMOS	International Council of Monuments and Sites
ICRAF	International Council for Research in Agroforestry
ICRP	International Commission on Radiological Protection
ICRW	International Convention for the Regulation of Whaling
ICS	International Chamber of Shipping
ICSEAF	International Commission for the Southeast Atlantic Fisheries
ICSU	International Council of Scientific Unions

IDA	International Development Association
IDB	Inter-American Development Bank (also listed as IADB)
IDOE	International Decade of Ocean Exploration
IEA	International Energy Agency (OECD)
IEB	International Environmental Bureau (Geneva)
IEEP	International Environmental Education Programme
IEEP	Institute for European Environmental Policy
IFAD	International Fund for Agricultural Development
IFC	International Finance Corporation
IFHP	International Federation for Housing and Planning
IFLA	International Federation of Landscape Architects
IFRB	International Frequency Registration Board
IGADD	Intergovernmental Authority on Drought and Desertification
IGBP	International Geosphere-Biosphere Programme
IGOSS	Integrated Global Ocean Station System
IGY	International Geophysical Year
IHD	International Hydrological Decade
IHO	International Hydrographic Organization (Monaco)
IHP	International Hydrological Program
IIASA	International Institute for Applied Systems Analysis
IIED	International Institute for Environment and Development
IJC	International Joint Commission (U.S./Canada)
ILO	International Labour Organisation (UN)
IMCO	Intergovernmental Maritime Consultative Organization (UN)
IMF	International Monetary Fund
IMO	International Maritime Organization (UN)
INC	International Negotiations on Climate
INFOTERRA	International Referral Service (UNEP)
INIS	International Nuclear Information System (IAEA)
INMARSAT	International Maritime Satellite Organization
INPFC	International North Pacific Fisheries Commission
INTECOL	International Association for Ecology
INTELSAT	International Telecommunications Satellite Organization
IOC	International Oceanographic Commission (UNESCO)
IODE	International Oceanographic Data Exchange
IOFC	Indian Ocean Fisheries Commission
IOI	International Ocean Institute
IPCC	Intergovernmental Panel on Climate Change (also listed as IPGCC)
IPCS	International Programme on Chemical Safety
IPFC	Indo-Pacific Fisheries Council

IPGCC	Intergovernmental Panel on Global Climate Change (also listed as IPCC)
IPHC	International Pacific Halibut Commission
IPSFC	International Pacific Salmon Fisheries Commission
IRPTC	International Register of Potentially Toxic Chemicals
IRS	International Referral Service (INFOTERRA)
ISSC	International Social Science Council
ITF	International Transport Federation
ITSU	Tsunami Warning System in the Pacific
ITTA	International Tropical Timber Agreement
ITTO	International Tropical Timber Organization
ITU	International Telecommunications Union (UN)
IUBS	International Union of Biological Sciences
IUCN	International Union for Conservation of Nature and Natural Resources (World Conservation Union)
IUFRO	International Union of Forest Research Organizations
IUGG	International Union of Geodesy and Geophysics
IULA	International Union of Local Authorities
IWRB	International Waterfowl and Wetlands Research Bureau
IWC	International Whaling Commission
JMC	Joint Monitoring Group
JOC	Joint Organizing Committee (GARP)
KAP	Kuwait Action Plan
LANDSAT	Earth Resources Sensing System (U.S.)
LD72	London Convention 1972 (previously known as LDC)
LDC	London Dumping Convention
LEPOR	Long-Term and Expanded Programme of Oceanic Exploration and Research
LRTAP	Convention on Long-Range Transboundary Air Pollution
MAB	Man and the Biosphere Programme (UNESCO)
MARC	Monitoring and Assessment Research Centre (SCOPE)
MARECS	Maritime European Communications Satellite
MARISAT	Marine Satellite System (U.S.)
MARPOL	International Convention for the Prevention of Marine Pollution from Ships
MEDI	Marine Environmental Data Information Referral System (UNESCO)
MED POL	Coordinated Mediterranean Pollution Monitoring and Research Programme
MEMAC	Marine Emergency Mutual Aid Centre
MIGA	Multilateral Investment Guarantee Agency
MIRCEN	Microbiological Resources Centre

MTO	Multilateral Trade Organisation
NACEC	North American Commission on Environmental Cooperation
NAFO	Northwest Atlantic Fisheries Organisation
NAFTA	North American Free Trade Agreement
NAMMCO	North Atlantic Marine Mammals Conservation Organization
NASA	National Aeronautics and Space Administration (U.S.)
NATO	North Atlantic Treaty Organization
NC	Nordic Council
NCAR	National Center for Atmospheric Research (U.S.)
NCCP	National Climate Change Plan
NEAFC	North-East Atlantic Fisheries Commission
NGOs	Nongovernmental organizations
NIEO	New International Economic Order
NOAA	National Oceanic and Atmospheric Administration (U.S.)
NOP	National Oceanographic Programme (IOC)
NOSS	National Oceanic Satellite System
NPFSC	North Pacific Fur Seal Commission
NPT	Non-Proliferation Treaty
NWF	National Wildlife Federation
OAS	Organization of American States
OAU	Organization of African Unity
ODA	Official Development Assistance
OECD	Organization for Economic Cooperation and Development
OICMA	International African Migratory Locust Organization
OMVS	Organization for Development of the Senegal River Basin
OPEC	Organization of Petroleum Exporting Countries
OTEC	Oceanic Thermal Energy Technology
OTEC	Organization of Timber Exporting Countries
PACD	Plan of Action to Combat Desertification
PACOM	Paris Convention on Pollution
PACs	Programme Activity Centres (UNEP)
PAHO	Pan American Health Organization
PCIJ	Permanent Court of International Justice (predecessor to the ICJ)
PCSP	Permanent Commission of the Conference on the Use and Conservation of the Marine Resources of the South Pacific
PECC	Pacific Economic Cooperation Conference
PLUARG	Pollution from Land Use Activities Reference Group (IJC)

PPP	Polluter-Pays Principle
PREPCOM	Preparatory Committee (for UN Conference on the Human Environment)
RNOCDS	Responsible National Oceanographic Data Centres (IOC)
ROCC	Regional Oil Combating Centre
SAARC	South Asian Association for Regional Cooperation
SACEP	South Asia Cooperative Environment Programme
SADCC	Southern Africa Development Coordination Conference
SCAR	Scientific Committee on Antarctic Research
SCEP	Study of Critical Environmental Problems
SCIBP	Special Committee on the International Biological Programme (ICSU)
SCOPE	Scientific Committee on Problems of the Environment (ICSU)
SCOR	Scientific Committee on Oceanic Research (ICSU)
SEDESOL	Social Development Secretariat
SIAP	Inter-American Planning Society
SMIC	Study of Man's Impact on Climate
SOLAS	Convention for the Safety of Life at Sea
SPC	South Pacific Commission
SPEC	South Pacific Bureau for Economic Cooperation
SPREP	South Pacific Regional Environmental Programme
STAP	Scientific and Technical Advisory Panel
STWG	Scientific-Technological Working Group (HELCOM)
TFAP	Tropical Forests Action Plan
UN	United Nations
UNCED	United Nations Conference on Environment and Development, Brazil, 1992
UNCHE	United Nations Conference on the Human Environment, Stockholm, 1972
UNCHS	United Nations Centre for Human Settlements (HABITAT)
UNCLOS	Conference on the Law of the Sea (UN)
UNCTAD	United Nations Conference on Trade and Development
UNDP	United Nations Development Programme
UNDRO	Office of United Nations Disaster Relief Coordinator
UNEP	United Nations Environment Programme
UNERG	Conference on New and Renewable Sources of Energy (UN)
UNESCO	United Nations Educational, Scientific, and Cultural Organization
UNFPA	United Nations Fund for Population Activities

UNGA	United Nations General Assembly
UNHHSF	United Nations Habitat and Human Settlements Foundation
UNIDO	United Nations Industrial Development Organization
UNISIST	Intergovernmental Programme for Cooperation in the Field of Scientific and Technical Information (UNESCO)
UNISPACE	United Nations Conference on the Exploration and Peaceful Uses of Outer Space
UNITAR	United Nations Institute for Training and Research
UNSCEAR	United Nations Scientific Committee on Effects of Atomic Radiation
UNSO	United Nations Statistical Office
UPOV	International Union for the Protection of New Plant Varieties
USAID	United States Agency for International Development
WARCs	World Administrative Radio Conferences
WCED	World Commission on Environment and Development (the Brundtland Commission)
WCMC	World Conservation Monitoring Centre (New York)
WCP	World Climate Programme
WCS	World Conservation Strategy
WEC	World Environment Center (New York)
WECAFC	Western Central Atlantic Fishery Commission
WEDF	World Environment and Development Forum
WFC	World Food Council
WFEO	World Federation of Engineering Organizations
WHO	World Health Organization
WICEM	World Industry Conference on Environmental Management
WMO	World Meterological Organization
WORRI	World Register of Rivers Discharging into the Oceans
WPL	World Peace through Law
WRI	World Resources Institute
WTO	World Trade Organization
WWF	World Wide Fund for Nature (formerly World Wildlife Fund)
WWW	World Weather Watch
YIEL	Yearbook of International Environmental Law

Appendix B
Representative Listing of International
Organizations and Programs

United Nations System

Administrative Committee on Coordination	ACC
Economic and Social Affairs Council	ECOSOC
International Court of Justice	ICJ
Office of United Nations Disaster Relief Coordinator	UNDRO
United Nations Centre for Human Settlements	HABITAT
United Nations Conference on Trade and Development	UNCTAD
United Nations Development Programme	UNDP
United Nations Environment Programme	UNEP
United Nations Fund for Population Activities	UNFPA
United Nations Habitat and Human Settlements Foundation	UNHHSF
United Nations Institute for Training and Research	UNITAR
United Nations Scientific Committee on Effects of Atomic Radiation	UNSCEAR

UN Regional Commissions

Economic and Social Commission for Asia and the Pacific	ESCAP
Economic Commission for Africa	ECA
Economic Commission for Europe	ECE
Economic Commission for Latin America and the Caribbean	ECLAC
Economic Commission for Western Asia	ECWA

Specialized and Affiliated Agencies

Food and Agriculture Organization	FAO
International Oceanographic Commission (UNESCO)	IOC

International Atomic Energy Agency	IAEA
International Civil Aviation Organization	ICAO
International Maritime Organization (formerly Inter-governmental Maritime Consultative Organization, IMCO)	IMO
International Telecommunications Union	ITU
United Nations Educational, Scientific, and Cultural Organization	UNESCO
United Nations Industrial Development Organization	UNIDO
World Health Organization	WHO
World Meteorological Organization	WMO

International Funding Agencies

African Development Bank	ADB
Arab Bank for Economic Development in Africa	ABEDA
Asian Development Bank	ASDB
Caribbean Development Bank	CAB
European Development Bank	EDB
European Investment Bank	EIB
European Regional Development Fund	ERDF
Global Environment Facility	GEF
Inter-American Development Bank	IDB
International Bank for Reconstruction and Development (World Bank)	IBRD
International Development Association	IDA
International Finance Corporation	IFC
International Monetary Fund	IMF

Non-UN Intergovernmental

Arab League Educational, Cultural, and Scientific Organization	ALESCO
Association of Southeast Asian Nations	ASEAN
Baltic Marine Environment Protection, or Helsinki, Commission	HELCOM
Committee on Challenges of Modern Society (NATO)	CCMS
Council for Mutual Economic Assistance	CMEA
European Environment Agency	EEA
European Community	EC
European Free Trade Association	EFTA
European Union	EU

General Agreement on Tariffs and Trade	GATT
International Boundary and Water Commission	IBWC
International Council for the Exploration of the Sea	ICES
International Energy Agency (OECD)	IEA
International Hydrographic Organization (Monaco)	IHO
International Joint Commission (U.S./Canada)	IJC
International Telecommunications Satellite Organization	INTELSAT
Nordic Council	NC
North American Commission for Environmental Cooperation	NACEC
North American Free Trade Agreement	NAFTA
Organization for Economic Cooperation and Development	OECD
Organization of African Unity	OAU
Organization of American States	OAS
South Asia Cooperative Environment Programme	SACEP
South Pacific Commission	SPC
World Trade Organization	WTO

Nongovernmental Scientific

International Association for Ecology	INTECOL
International Council of Scientific Unions	ICSU
Scientific Committees:	
on Antarctic Research	SCAR
on Oceanic Research	SCOR
on Problems of the Environment	SCOPE
on Space Research	COSPAR
International Geosphere-Biosphere Programme	IGBP
International Institute for Applied Systems Analysis	IIASA
International Social Science Council	ISSC
International Union for Conservation of Nature and Natural Resources (World Conservation Union)	IUCN

Nongovernmental Quasi-Scientific, Technical, and Professional

Centre for Law and Social Policy	CLASP
Council for International Organization of Medical Sciences	CIOMS
Environment Law Centre (IUCN at Bonn)	ELC
Environment Liaison Centre (Nairobi)	ELC
European Environmental Bureau (Brussels)	EEB
Institute for European Environmental Policy	IEEP

International Council for Bird Protection	ICBP
International Council for Environmental Law	ICEL
International Council for Research in Agroforestry	ICRAF
International Council of Monuments and Sites	ICOMOS
International Council of Museums	ICOM
International Environmental Bureau (Geneva)	IEB
International Federation for Housing and Planning	IFHP
International Federation of Landscape Architects	IFLA
International Institute for Environment and Development	IIED
International Ocean Institute	IOI
International Tropical Timber Organization	ITTO
International Union of Forest Research Organizations	IUFRO
International Union for Local Authorities	IULA
World Environment Center (New York)	WEC
World Federation of Engineering Organizations	WFEO
World Peace through Law	WPL
World Resources Institute	WRI
World Wide Fund for Nature (formerly World Wildlife Fund)	WWF

International Programs and Services

Committee of International Development Institutions for the Environment	CIDIE
Consultative Group for Desertification Control (UNEP)	DESCON
Environmentally Sound Management of Inland Waters	EMINWA
Global Environmental Monitoring System	GEMS
Global Investigation of Pollution in the Marine Environment (IOC)	GIPME
Global Resource Information Database	GRID
Intergovernmental Panel on Climate Change	IPCC
Intergovernmental Programme for Cooperation in the Field of Scientific and Technical Information (UNESCO)	UNISIST
International Agency for Research on Cancer (WHO)	IARC
International Biological Programme (ICSU)	IBP
International Board for Plant Genetic Resources (FAO)	IBPGR
International Centre for Genetic Engineering and Biotechnology (UNIDO)	ICGEB
International Decade of Ocean Exploration	IDOE
International Environmental Education Programme	IEEP
Integrated Global Ocean Station System	IGOSS
International Hydrological Program	IHP

International Information System for the Agricultural
  Sciences and Technology (FAO)   AGRIS
International Maritime Satellite Organization   INMARSAT
International Nuclear Information System (IAEA)   INIS
International Programme on Chemical Safety   IPCS
International Referral Service (UNEP)   INFOTERRA
International Register of Potentially Toxic Chemicals   IRPTC
International Telecommunications Satellite Organization INTELSAT
Group of Experts on Problems of Marine Pollution   GESAMP
Long-Term and Expanded Programme of Oceanic
  Exploration and Research   LEPOR
Man and the Biosphere Programme (UNESCO)   MAB
Marine Environmental Data Information Referral System
  (UNESCO)   MEDI
Conference on New and Renewable Sources of Energy
  (UN)   UNERG
Study of Critical Environmental Problems   SCEP
Study of Man's Impact on Climate   SMIC
World Climate Programme   WCP
World Conservation Monitoring Centre (New York)   WCMC
World Weather Watch (WMO)   WWW

Marine Fisheries Management

General Fisheries Council for the Mediterranean   GFCM
Indian Ocean Fisheries Commission   IOFC
Indo-Pacific Fisheries Council   IPFC
Inter-American Tropical Tuna Commission   IATTC
International Commission for the Conservation of
  Atlantic Tunas   ICCAT
International Commission for the Southeast Atlantic
  Fisheries   ICSEAF
International North Pacific Fisheries Commission   INPFC
International Pacific Halibut Commission   IPHC
International Pacific Salmon Fisheries Commission   IPSFC
International Whaling Commission   IWC
North-East Atlantic Fisheries Commission   NEAFC
Northwest Atlantic Fisheries Organization   NAFO
Permanent Commission for the South Pacific (Chile,
  Ecuador, Peru)   CPPS
Western Central Atlantic Fishery Commission   WECAFC

Appendix C
Events of Significance for
Protection of the Biosphere, 1945–1995

1945	Establishment of the United Nations Organization, UNESCO, and FAO
1946	Establishment of the World Health Organization
1948	Establishment of the International Union for Conservation of Nature and Natural Resources, following an international conference sponsored by UNESCO and the Government of France
1949	United Nations Scientific Conference on the Conservation and Utilization of Resources, 17 August–6 September, Lake Success, New York
1951	International Meteorological Organization, established in 1878, reconstituted as the World Meteorological Organization
1954	World Conference on Population, sponsored by the Economic and Social Affairs Department of the United Nations, 31 August–10 September, Rome
1955	International Technical Conference on the Conservation of the Living Resources of the Sea, 18 April–10 May, Rome
1955	United Nations Scientific Committee on Effects of Atomic Radiation established
1955–71	Geneva Conferences on the Peaceful Uses of Atomic Energy
1956	Establishment of the International Atomic Energy Agency
1957–58	International Geophysical Year
1958–82	United Nations Conferences on the Law of the Sea
1959	Intergovernmental Maritime Consultative Organization activated after delay of ten years (International Maritime Organization after 22 May 1982)
1959–66	International Indian Ocean Expedition
1961	United Nations Conference on New Sources of Energy, 21–31 August, Rome
1962	First World Conference on National Parks, 30 June–7 July, Seattle, U.S.A.

1962–65   World Magnetic Survey

1963   Conference on Application of Science and Technology for the Benefit of the Less Developed Areas, 4–20 February, Geneva

1963–64   International Cooperative Investigations of the Tropical Atlantic

1964–75   International Biological Programme

1965–74   International Hydrological Decade

1967   Implementation of World Weather Watch, under sponsorship of the World Meteorological Organization

1968   First United Nations Conference on the Exploration and Peaceful Uses of Outer Space, 14–27 August, Vienna

1968   UNESCO Intergovernmental Conference of Experts on the Scientific Basis for Rational Use and Conservation of the Resources of the Biosphere, 4–13 September, Paris

1968   United Nations General Assembly Resolution 2398 (3 December) on the Problems of Human Environment, initiating Stockholm Conference

1969   Conference on Monuments and Tourism, 7–11 July, Oxford, England, UNESCO/ICOMOS/OECD

1969   First landing on moon, U.S. Apollo XI, 20 July

1970   European Conservation Year, sponsored by the Council of Europe

1970   FAO Technical Conference on Marine Pollution and Its Effects on Living Resources and Fishing, 9–18 December, Rome

1970   Scientific Committee on Problems of the Environment (SCOPE) established by the International Council of Scientific Unions

1971   Symposium on Problems Relating to Environment, Economic Commission for Europe, 2–5 May, Prague

1971   International Conference on the Environmental Future, 27 June–3 July, Helsinki and Jyvaskyla, Finland

1971   Founex Panel of Experts on Development and Environment

1971–80   International Decade of Ocean Exploration

1972   Limits to Growth report to Club of Rome

1972   United Nations Conference on the Human Environment, 5–16 June, Stockholm

1972   Second World Conference on National Parks, 18–27 September, Yellowstone National Park, U.S.A.

1972   Establishment of the United Nations Environment Programme (UNEP) by the General Assembly, 15 December

1973   First European Ministerial Conference on the Environment, Council of Europe, 28–30 March, Vienna

1974   ICSU/SCOPE Symposium on Environmental Sciences in Developing Countries, 11–13 February, Nairobi

1974	Regional Seas Programme initiated by UNEP
1974	Declaration on the Establishment of a New International Economic Order
1974	World Food Conference, 5–16 November, Rome
1975	Belgrade Charter: Global Framework for Environmental Education, 28 October, Belgrade
1976	United Nations Conference on Human Settlements (HABITAT), 31 May–11 June, Vancouver, Canada
1977	United Nations Water Conference, 14–25 March, Mar del Plata, Argentina
1977	Second International Conference on the Environmental Future, 5–11 June, Reykjavik, Iceland
1977	United Nations Conference on Desertification, 29 August–9 September, Nairobi
1977	UNESCO/UNEP Intergovernmental Conference on Environmental Education, 14–26 October, Tbilisi, Georgian Republic, U.S.S.R.
1978	International Conference on Primary Health Care, 6–12 September, Alma Ata, Kirghiz Republic, U.S.S.R.
1979	World Climate Conference, 12–23 February, Geneva
1979	United Nations Symposium on the Interrelations between Resources, Environment, Population, and Development, 6–10 August, Stockholm
1979	United Nations Conference on Science and Technology for Development, 21 August–11 September, Vienna
1980	Publication of the UNEP-IUCN *World Conservation Strategy*
1980	Declaration of Environmental Policies and Procedures Relating to Economic Development by Committee of International Development Institutions on the Environment (CIDIE)
1981	International Conference on Plant Genetic Resources, 6–10 April, Rome
1981	United Nations Conference on New and Renewable Sources of Energy, 10–21 August, Nairobi
1981–90	International Drinking Water Supply and Sanitation Decade
1982	UNEP Governing Council, Session of a Special Character, 10–18 May, Nairobi, commemorating Tenth Anniversary of the United Nations Conference on the Human Environment, June 1972
1982	International Whaling Commission votes phaseout of commercial whaling over a three-year period, 23 July
1982	Second United Nations Conference on the Exploration and Peaceful Uses of Outer Space, UNISPACE '82, 9–21 August, Vienna

1982   World National Parks Congress, 11–22 October, Denpasar, Bali, Indonesia

1982   World Charter for Nature adopted by the UN General Assembly, 28 October

1982   Conference on Environmental Research and Management Priorities for the 1980s, 23–26 November, Rättvik, Sweden

1983   United Nations Conference on Tropical Timber, 14–31 March, Geneva

1984   International Society for Environmental Education and World Council for the Biosphere established, 23–29 February, New Delhi

1984   Declaration of Luxembourg, 9 April, EFTA and EC

1986   International Geosphere-Biosphere Programme initiated by ICSU

1986   New Alliance Ecumenical Conference at Assisi, Italy, sponsored by WWF

1987   Declaration of Noordwijk, 5 October, EFTA and EC

1987   World Commission on Environment and Development Report: *Our Common Future*

1988   Intergovernmental Panel on Climate Change established by WMO and UNEP

1988   Declaration of Rhodes, 3 December, by EC Council of Ministers

1988   UN General Assembly, Resolution on Protection of Global Climate for Present and Future Generations of Mankind, 6 December

1989   International Meeting of Legal and Policy Experts on the Protection of the Atmosphere, Ottawa, February

1989   Hague Declaration on the Environment, 31 March; Latin American-Caribbean environmental "summit"

1989   Helsinki Declaration on Protection of the Ozone Layer, 2 May

1989   The Amazon Declaration, Manaus, (Brazil), 6 May, presidents of Amazonian treaty states

1989   UNEP Decision on Global Climate Change, Nairobi, 25 May

1989   Communique of Seven Heads of Industrial Democracies Concerning the World Environment, Paris, July 1989

1989   Noordwijk Declaration on Atmospheric Pollution and Climate Change, 6–7 November; Ministerial conference representing 67 countries and 11 international organizations

1989   UN General Assembly Resolution (44/228) on a United Nations Conference on Environment and Development (UNCED) to be held in Brazil in 1992, 22 December

1991   Declaration of Rovaniemi, Espoo, Finland, 14 June

1991	The Dobris Assessment: Report on the State of the Pan-European Environment, Conference of European environmental ministers at Dobris Castle, Czechoslovakia
1992	Agenda 21, UNCED, Rio de Janeiro
1992	Rio Declaration on Environment and Development, UNCED, Rio de Janeiro
1992	Statement of Principles on Forests, UNCED, Rio de Janeiro
1992	Signing of Conventions on Climate Change and Biodiversity, UNCED, Rio de Janeiro

Appendix D
Selected Treaties of Environmental Significance, 1946–95

Note: For a comprehensive compilation of environmental treaties see B. Rüster, B. Simma, and M. Bock, eds., *International Protection of the Environment: Treaties and Related Documents*, 30 vols. (Dobbs Ferry, N.Y.: Oceana). Supplements have been published as Second Series 1990–. The principal standard compilation of treaties sponsored by the United Nations is the *United Nations Treaty Series (Treaties and International Agreements Registered or Filed and Recorded with the Secretariat of the United Nations)*. For recent treaties to which the United States is a party, see *United States Treaties and Other International Agreements*. Comprehensive lists of international environmental treaties may be found in *Environmental Quality: The Twentieth Annual Report of the [U.S.] Council on Environmental Quality* (Washington, D.C.: Council on Environmental Quality, 1990), Appendix C; and in *Breakthrough*, Global Education Associates (Summer–Fall 1989): 18–19. In addition, see Alexandre Charles Kiss, ed., *Multilateral Treaties in the Field of the Environment* (Nairobi: UNEP, 1982); and Peter H. Sand, ed., *The Effectiveness of International Agreements: A Survey of Existing Legal Agreements* (Cambridge: Grotius, 1992).

1946   Convention for the Regulation of Whaling signed

1948   Convention of the Inter-Governmental Maritime Consultative Organization

1949   FAO Agreement for the Establishment of a General Fisheries Council for the Mediterranean signed 24 September, Rome, in force 3 December 1963

1951   FAO International Plant Protection Convention signed 6 December, Rome, in force 3 April 1952

1952   International Convention for the High Seas Fisheries of the North Pacific signed 9 May, Tokyo

1954   Convention for the Prevention of Pollution of the Sea by Oil; see 1969 Amendments to the 1954 Convention

1958   Convention on the High Seas

1958 Convention on the Continental Shelf

1958 Convention of the Territorial Sea and Contiguous Zone

1958 Convention of Fishing and Conservation of the Living Resources of the High Seas

1958 Antarctic Treaty signed 1 December, establishing the south polar region as an international scientific reserve

1963 Treaty Banning Nuclear Weapons Tests in the Atmosphere, in Outer Space, and Under Water (Partial Nuclear Test Ban Treaty) signed 5 August, Moscow

1964 International Charter of Venice for the Conservation and Restoration of Monuments and Sites

1967 Treaty on Principles Governing the Activities of States in the Exploration and Use of Outer Space, Including the Moon and Other Celestial Bodies (London, Moscow, Washington)

1968 OAU African Convention on the Conservation of Nature and Natural Resources signed 15 September, Algiers

1969 Agreement for Co-operation in Dealing with Pollution of the North Sea by Oil (Bonn)

1969 Convention Relating to Intervention on the High Seas in Cases of Oil Pollution Casualties

1969 Convention on Civil Liability for Oil Pollution Damage

1969 Vienna Convention on the Law of Treaties

1971 Convention on the Establishment of an International Fund for Compensation for Oil Pollution Damage

1972 Agreement on Great Lakes Water Quality, Canada and the United States (extended in 1978 and 1987)

1972 Convention Concerning the Preservation of World Cultural and Natural Heritage adopted

1972 Convention on the Prevention of Marine Pollution by Dumping of Wastes and other Matter (London Convention)

1973 Convention on International Trade in Endangered Species of Wild Flora and Fauna (CITES)

1973 Convention for the Prevention of Pollution by Ships (MARPOL); 1978 Protocol to the 1973 Convention for the Prevention of Pollution by Ships

1974 Convention for the Prevention of Marine Pollution from Land-Based Sources (Paris Convention)

1974 Convention on the Protection of the Marine Environment of the Baltic Sea Area (Helsinki Convention)

1974 Nordic Convention on the Protection of the Environment

1975 Convention Concerning the Protection of the World Cultural and Natural Heritage (in effect following required number of ratifications)

1975   Convention on Wetlands of International Importance Especially as Waterfowl Habitat in force

1976   Convention on Protection of the Rhine against Chemical Pollution

1976   Convention for the Protection of the Mediterranean Sea against Pollution (Barcelona Convention), Protocol for the Prevention of Pollution of the Mediterranean Sea by Dumping from Ships and Aircraft, and Protocol Concerning Cooperation in Combating Pollution of the Mediterranean Sea by Oil and Other Harmful Substances in Cases of Emergency

1978   Action Plan for the Protection and Development of the Marine Environment and the Coastal Areas of Bahrain, Iran, Iraq, Kuwait, Oman, Qatar, Saudi Arabia, and the United Arab Emirates

1978   Kuwait Regional Convention for Cooperation on the Protection of the Marine Environment from Pollution; and Protocol Concerning Regional Cooperation in Combating Pollution by Oil and Other Harmful Substances in Cases of Emergency

1978   Amazon Pact Treaty

1978   Convention of Prohibition of Military or Any Other Hostile Use of Environmental Modification Techniques (ENMOD) in force

1979   Convention on Conservation of Migratory Species of Wild Animals signed 23 June, Bonn

1979   Convention on Long-Range Transboundary Air Pollution; 1985 Protocol on the Reduction of Sulphur Emissions or Their Transboundary Fluxes by at Least 30 percent; 1988 Protocol Concerning the Control of Emissions of Nitrogen Oxides or Their Transboundary Fluxes; 1991 Protocol on the Reduction of Volatile Organic Compounds

1980   Convention on the Conservation of Antarctic Marine Living Resources signed 20 May 1980, in force 7 April 1982

1981   Convention for Cooperation in the Protection and Development of the Marine and Coastal Environment of the West and Central African Region (Abidjan Convention) and Protocol Concerning Co-operation in Combating Pollution in Cases of Emergency

1982   United Nations Convention on the Law of the Sea, in force 16 November 1994

1983   Convention for Protection and Development of the Marine Environment of the Wider Caribbean Region; and Protocol on Co-operation in Combating Oil Spills in the Wider Caribbean Region

1985   Vienna Convention for the Protection of the Ozone Layer; 1987 Montreal Protocol on Substances that Deplete the Ozone Layer; 1990 London Revisions to the Montreal Protocol

1986   Convention on Assistance in the Case of a Nuclear Accident or Radiological Emergency

1986   Single European Act

1988   Convention on the Regulation of Antarctic Mineral Resources
Activities

1989   Convention on Control of Transboundary Movements of Hazardous
Waste and Their Disposal, Basel, in force 24 May 1992

1990   Convention on Oil Pollution Preparedness, Response, and Cooperation (OPRC), sponsored by the International Maritime Organization, signed in London 30 November 1990, in force 13 May 1995

1991   Convention on Environmental Impact Assessment in a
Transboundary Context, signed 25 February, Espoo, Finland (not in
force)

1991   Convention on the Ban of Exports into Africa and the Control of
Transboundary Movement and Management of Hazardous Wastes
within Africa signed 29 January, Bamako, Mali (not in force)

1991   Protocol on Environmental Protection to the Antarctic Treaty
signed 4 October, Madrid

1991   Canada-United States Agreement on Air Quality

1992   Treaty on European Union signed 17 February, Maastricht

1992   Convention on the Protection and Use of Transboundary
Watercourses and International Lakes signed 17 March, Helsinki
(not in force)

1992   UN/ECE Convention on the Transboundary Effects of Industrial
Accidents signed 17 March, Helsinki

1992   Convention for the Protection of the Marine Environment of the
Baltic Sea Area signed 9 April, Helsinki

1992   United Nations Framework Convention on Climate Change signed
9 May, New York

1992   Convention on Biological Diversity signed 5 June, Rio de Janeiro

1992   Convention for the Protection of the Marine Environment of the
North-East Atlantic signed 22 September, Paris (not in force)

1993   North American Free Trade Agreement, the Environmental Side
Agreement

Appendix E
Environmental Soft Law: Declarations,
Resolutions, Recommendations, Principles

16 June 1972	Stockholm Declaration and Principles—UN Conference on the Human Environment
26 November 1972	Protection, National Level, of the Cultural and National Heritage—Recommendation (UNESCO)
14 November 1974	Implementation of the Polluter-Pays Principle—Recommendation (OECD)
11 May 1976	Equal Right of Access in Relation to Transfrontier Pollution—Recommendation (OECD)
17 May 1977	Implementation of a Regime of Equal Right of Access and Non-Discrimination in Relation to Transfrontier Pollution—Recommendation (OECD)
19 May 1978	Conduct in the Field of the Environment for the Guidance of States in the Conservation and Harmonious Utilization of Natural Resources Shared by Two or More States—Principles (UNEP)
1 February 1980	Environmental Policies and Procedures Relating to Economic Development adopted by Major Development Banks
29 April 1980	South Pacific Declaration from the Conference on the Human Environment in the South Pacific
1 May 1982	World Soils Policy (UNEP)
31 May 1982	Legal Aspects Concerning the Environment Related to Offshore Mining and Drilling within the Limits of National Jurisdiction (UNEP)
29 October 1982	World Charter for Nature (UNGA)

The author acknowledges with appreciation permission to select these citations from *International Soft Law: Collection of Relevant Instruments*, edited by W. E. Burhenne, selected and compiled by Marlene Jahnke (Dordrecht, Netherlands: Martinus Nijhoff, published under the auspices of the International Council of Environmental Law, 1995).

23 November 1983	International Undertaking on Plant Genetic Resources—Resolution (FAO)
24 November 1983	International Undertaking on Plant Genetic Resources—Report of the Director-General (FAO)
19 April 1985	Montreal Guidelines for the Protection of the Marine Environment against Pollution from Land-Based Sources (UNEP)
26 June 1985	Draft Principles Relating to Remote Sensing of the Earth from Space
28 November 1985	International Code of Conduct on the Distribution and Use of Pesticides (FAO)
1 December 1985	General Policies and Principles of Environmental Protection (Gulf Cooperation Council)
14 October 1986	Environment and Development—Declaration (Arab League)
10 February 1987	Chemicals in International Trade: Guidelines for the Exchange of Information (UNEP)
17 June 1987	Goals and Principles of Environmental Impact Assessment (UNEP)
25 November 1987	Second International Conference on the Protection of the North Sea
11 December 1987	Environmental Perspective to the Year 2000 and Beyond (UNGA)
11 December 1987	Environmental Perspective to the Year 2000 and Beyond Resolution (UNGA)
5 March 1988	Regional Strategy on Environmental Protection and Rational Use of Natural Resources Covering the Period up to the Year 2000 and Beyond (ECE)
22 April 1988	Decision on Co-operation in the Field of Environmental Protection and Water Resources (ECE)
31 March 1989	The Hague Declaration on the Environment
11 April 1989	The Hague Declaration on Tourism (Inter-Parliamentary Conference)
21 April 1989	Statement of the International Congress Concerning a More Efficient International Law on the Environment and Establishment of an International Court for the Environment within the United Nations System
2 May 1989	Helsinki Declaration on the Protection of the Ozone Layer
1 June 1989	Charter on Ground-Water Management (ECE)
16 June 1989	Kampala Declaration on Environment and Sustainable Development

7 July 1989	Application of the Polluter-Pays Principle to Accidental Pollution—Recommendation (OECD)
11 July 1989	Tarawa Declaration on Drift-Net Fishing
21 October 1989	The Langkawi Declaration (Commonwealth Countries)
24 November 1989	The Castries Declaration (Organisation of Eastern Caribbean States)
3 December 1989	Declaration of Rhodes (by EC Council of Ministers)
7 December 1989	European Charter on Environment and Health
21 December 1989	The Cairo Compact: Toward a Concerted World-Wide Response to the Climate Crisis
24 January 1990	Moscow Declaration of the Global Forum on Environment and Development for Human Survival
6 April 1990	Statement of Ministers Responsible for the Environment (EFTA)
2 May 1990	Declaration of Environmental Interdependence and Decisions—International Interparliamentary Conference on the Global Environment
4 May 1990	International Conference on Global Warming and Climate Change: African Perspectives
5 May 1990	Bergen Conference: Joint Agenda for Action (ECE)
5 May 1990	Bergen Conference: Ministerial Declaration (ECE)
1 June 1990	Code of Conduct on Accidental Pollution of Transboundary Inland Waters (ECE)
3 June 1990	G15 Leaders Joint Communique (Developing Countries)
7 November 1990	Declaration of the Second World Climate Conference (SWCC)
4 December 1990	Prohibition of the Dumping of Radioactive Wastes (UNGA)
30 January 1991	Pan-African Co-ordination Policy and Strategy for Co-operation for Environment and Sustainable Development
30 January 1991	Bamako Commitment on Environment and Development
31 January 1991	Integrated Pollution Prevention and Control—Recommendation (OECD)
31 January 1991	Reduction of Transfrontier Movements of Wastes—Recommendation (OECD)
31 January 1991	Use of Economic Instruments in Environmental Policy—Recommendation (OECD)
7 March 1991	Tlatelolco Platform on Environment and Development

12 April 1991	Final Declaration of the Second World Industry Conference on Environmental Management (WICEM II)
29 April 1991	The Law of the Non-Navigational Uses of International Watercourses—International Law Commission
14 June 1991	Arctic Environmental Protection Strategy
19 June 1991	Beijing Ministerial Declaration on Environment and Development—Declaration (UNGA)
3 December 1991	Policy Statement from Meeting of Ministers on Environment and Development (OECD)
17 December 1991	The European Energy Charter
11 April 1992	Environment and Development: The Views of Parliamentarians on the Main Directions of the United Nations Conference on Environment and Development and Its Prospects
14 June 1992	Rio Declaration on Environment and Development
14 June 1992	Forest Principles—Principles for a Global Consensus on the Management, Conservation, and Sustainable Development of all Types of Forests
14 June 1992	Agenda 21—United Nations Conference on Environment and Development
30 April 1992	Declaration by the Ministers of the Environment of the Region of the United Nations Economic Commission for Europe (ECE) and the Member of the Commission of the European Communities Responsible for the Environment
17 June 1993	General Guidelines for the Sustainable Management of Forests in Europe
18 September 1993	Building Global Human Security—The Bonn Declaration

Notes and References

Because of the wide range of subject matter dealt with in this book and the escalating volume of relevant literature, no separate set of bibliographical references is provided. For leading topics, representative listings of background and supporting works have been included. For some topics multiple sources are provided for follow-up on the assumption that an interested reader may lack ready access to some of the source and supplementary materials. Wherever possible the topics are covered in historical depth, and for this reason earlier citations are provided as well as those more recent. In examining the historical development of an agency or a program, early accounts of its activities are useful parts of the institutional record. Moreover, the latest published work is not always the most informative. Most publications listed in the second edition of this work (1990) have been retained and reprinted here. Some journals have ceased publication, but remain useful sources of information and are cited in the text.

It is the record of historical progression, growth, and development of international environmental policy that the author regards as the principal contribution of this work. Books treating environmental law and policy are increasing in number. A representative listing is provided under note 4 of the introduction. Note 5 lists thirty-one periodical publications that will keep readers updated on all but the more technical aspects of environmental policy. New journals, however, keep appearing. A more comprehensive listing of publications may be found in *Ulrich's International Periodicals Directory*, which includes entries in languages other than English and journals of a more specifically scientific or technical character. Given the broad scope of this book and the voluminous, but also specialized, character of much of the literature, it is inevitable that some items that should have been included will be omitted from the citations. The author hopes that any significant omissions will be brought to his attention.

Readers interested in environmental policy in general and international or global environmental issues in particular should be aware of four encyclopedias issued in 1994–95. Each volume is different and has particular advantages: *Environmental Encyclopedia*, edited by William P. Cunningham et al. (Detroit, Mich.: Gale Research International, 1994) xxi + 981 pp.; *The Encyclopedia of the Environment*, edited by Ruth A. Eblen and William Eblen (Boston: Houghton Mifflin, 1994) xvii + 864 pp.; *The Environment Encyclopedia and Directory*, no editor given (London: Europa Publications, 1994) xvii + 381 pp.; *Conservation and Environmentalism: An Encyclopedia*, edited by Robert Paehlke (New York: Garland, 1995) xxxiii + 771 pp.

Introduction

1. Jacques Ellul, *La Technique* (1954). English translation, *The Technological Society*, by John Wilkerson (New York: A. A. Knopf, 1964).
2. Robert Nisbet, "Environmentalism," *Prejudices: A Philosophical Dictionary* (Cambridge, Mass.: Harvard University Press, 1982), p. 101.

3. Clarence J. Glacken, *Traces on the Rhodian Shore: Nature and Culture in Western Thought from Ancient Times to the End of the Eighteenth Century* (Berkeley: University of California Press, 1967); George Perkins Marsh, *Man and Nature, or, Physical Geography as Modified by Human Action* (New York: Scribner, 1864; reprinted by Harvard University Press, 1965); William L. Thomas et al., eds., *Man's Role in Changing the Face of the Earth* (Chicago: University of Chicago Press, 1956); B. L. Turner et al., eds., *The Earth as Transformed by Human Action: Global and Regional Changes in the Biosphere over the Past 300 Years* (Cambridge: Cambridge University Press, 1990). Clive Ponting, *A Green History of the World: The Environment and the Collapse of Great Civilizations* (New York: St. Martin's Press, 1992).

4. Among alternative approaches to the broader dimensions of international environmental policy during the last two decades are the following books: Richard N. Barrett, ed., *International Dimensions of the Environmental Crisis* (Boulder, Colo.: Westview Press, 1982); Lynton K. Caldwell, *In Defense of Earth: International Protection of the Biosphere* (Bloomington: Indiana University Press, 1972); Kenneth A. Dahlberg, Anne T. Feraru, and Marvin S. Soroos, eds., *Environment and the Global Arena: Actors, Values, Policies, Futures* (New York: Holt, Rinehart and Winston, 1983); Eric Eckholm, *Down to Earth: Environment and Human Needs* (New York: W. W. Norton, 1982); Richard Falk, *This Endangered Planet: Prospects and Proposals for Human Survival* (New York: Random House, 1971), and *A Global Approach to National Policy* (Cambridge, Mass.: Harvard University Press, 1975); David A. Kay and Harold K. Jacobson, eds., *Environmental Protection: The International Dimension* (Totowa, N.J.: Allanheld, Osmun, 1983); David A. Kay and Eugene B. Skolnikoff, eds., *World Eco-Crisis: International Organizations in Response* (Madison: University of Wisconsin Press, 1972); Alexander King, *The State of the Planet: A Report Prepared by the International Federation of Institutes for Advanced Study (IFIAS), Stockholm* (Oxford: Pergamon Press, 1980); David W. Orr and Marvin S. Soroos, eds., *The Global Predicament and World Order* (Chapel Hill, N.C.: University of North Carolina Press, 1979); Dennis Pirages, *The New Context for International Relations: Global Ecopolitics* (North Scituate, Mass.: Duxbury, 1978); Jan Schneider, *World Public Order of the Environment: Towards an International Ecological Law and Organization* (Toronto: University of Toronto Press, 1979); and Harold Sprout and Margaret Sprout, *Toward a Politics of the Planet Earth* (New York: Van Nostrand Reinhold, 1971). In addition, see Martin W. Holdgate, Mohammed Kassas, and Gilbert F. White, eds., *World Environment 1972–1982* (Dublin: Tycooly International for UNEP, 1982); Annual Report of the Executive Director of UNEP, *The State of the World Environment 1988* (Nairobi: UNEP, April 1989); *The State of the Environment 1985* (Paris: OECD, 1985); *State of the World Report(s)* of the Worldwatch Institute; and the *World Resources* reports of the World Resources Institute.

Among more recent books, see Lamont C. Hempel, *Environmental Governance: The Global Challenge* (Washington, D.C.: Island Press, 1996); Gareth Porter and Janet Welsh Brown, *Global Environmental Politics*, 2d ed. (Boulder, Colo.: Westview Press, 1995); Sheldon Kamieniecki, ed., *Environmental Politics in the International Arena: Movements, Parties, Organizations, and Policy* (Albany: State University of New York Press, 1993); Nicholas Polunin and Sir John H. Burnett, eds., *Surviving with the Biosphere: Proceedings of the Fourth International Conference on Environmental Future* (4th ICEF) (Edinburgh, U.K.: Edinburgh University Press, 1993); Patricia W. Birnie and Alan E. Boyle, *International Law and the Environment* (Oxford: Clarendon Press, 1992); Edith Brown-Weiss, ed., *Environmental Change and International Law* (Tokyo: United Nations Press, 1992); Caroline Thomas, *The Environment in International Relations* (London: Royal Institute of International Affairs, 1992); Ian H. Rowlands and Malory Greens, *Global Environmental Change and International Relations* (Basingstoke, U.K.: Macmillan, 1992); Andrew Hurrell and Benedict Kingsbury, eds., *The International Politics of the Environment* (Oxford: Oxford University Press, 1992); Lynton K. Caldwell, *Between Two Worlds: Science, the Environmental Movement, and Policy Choice* (Cambridge: Cambridge University Press, 1990); John McCormick, *Reclaiming Paradise: The Global Environmental Movement* (Bloomington: Indiana University Press, 1989).

Among periodicals widely available and useful for regular reporting of developments in international environmental policy are the following: *Ambio,* Royal Swedish Academy of Sciences/IUCN, Stockholm, 1972–; *Ceres,* Food and Agriculture Organization (FAO), Rome, 1968-1988, 1990–; *Connect,* United Nations Educational, Scientific, and Cultural Organization (UNESCO), Paris, 1976–; *Desertification Control,* United Nations Environment Programme (UNEP), Nairobi, 1976–; *Environment,* Heldref, Washington, D.C., 1958–; *The Environment in Europe,* Institute for European Environmental Policy, Bonn, 1976–; *Environmental Conservation,* Elsevier Sequoia (for the Foundation for Environmental Conservation), Lausanne, Switzerland, 1974–; *Environmental Law and Policy,* (Kluwer since 1992) 1975–; *The Environmentalist,* Elsevier Sequoia, Lausanne, 1981–; *European Environment Review,* Graham and Trotman, London, 1987–; *Habitat News,* UN Centre on Human Settlements, Nairobi, 1979–; *Impact of Science on Society,* UNESCO, Paris, 1950–; *International Environmental Affairs,* University Press of New England, Hanover, N.H., 1989–; *International Environment Reporter,* Bureau of National Affairs, Washington, D.C., 1978–; *IMS Newsletter,* UNESCO, Intergovernmental Oceanographic Commission, 1973–; *IMO News,* International Maritime Organization (formerly Intergovernmental Maritime Consultative Organization), 1977–; *International Environmental Yearbook* (A Directory), Madrid, 1993–; *IUCN Bulletin,* International Union for Conservation of Nature and Natural Resources, Gland, Switzerland, 1964–; *Mazingira,* Tycooly International Publications (with support of UNEP, 1977-1985), Dublin, 1977–; *Nature and Resources,* UNESCO, Paris, now published by Parthenon, 1965–; *Naturopa,* European Information Centre for Nature Conservation, Strasbourg, 1974–; *OECD Observer,* Organization for Economic Cooperation and Development (OECD), Paris, 1962–; *Our Planet,* UNEP, 1989–; *The Siren,* UNEP, Regional Seas Programme (now OCA), Nairobi, 1978–; *Unasylva,* Food and Agriculture Organization, Rome, 1947–; *Unesco Courier,* UNESCO, Paris, 1948–; *Uniterra,* UNEP, Nairobi, 1974–; *World Environment Report,* Center for International Environment Information (since 1980, World Environment Center), New York, 1974–; *World Health,* World Health Organization (WHO), Geneva, 1948–; *WHO Bulletin,* World Health Organization, Geneva, 1947–; *WMO Bulletin,* World Meteorological Organization (WMO), Geneva, 1952–. There are, in addition, numerous periodicals largely confined to specialized, technical, or national aspects of environmental policy. Many of these are referenced in the notes.

5. Eric Ashby, *Reconciling Man with the Environment* (Stanford, Calif.: Stanford University Press, 1978).

6. Ernest Becker, *The Revolution in Psychiatry—A New Understanding of Man* (New York: Free Press of America, 1964), p. 222.

7. *Environmental Sabbath Newsletter* (Earth Rest Day) 2, no. 3 (June 1990) (New York: United Nations Environment Programme, DC2-803).

1. Comprehending the Environment

1. Social learning is basic to the emergence of international environmental concern. For further discussion of the process, see Edgar S. Dunn, *Economic and Social Development: A Process of Social Learning* (Baltimore, Md.: Johns Hopkins University Press, 1971); and Donald N. Michael, *On Learning to Plan and Planning to Learn: The Social Psychology of Changing towards Future Responsive Societal Learning* (San Francisco, Calif.: Jossey Bass, 1973). Also Lester W. Milbrath, *Envisioning a Sustainable Society: Learning Our Way Out* (Albany: State University of New York Press, 1989).

2. *Funk and Wagnalls New Practical Standard Dictionary I* (New York, 1946): 43.

3. Claude Bernard, *An Introduction to the Study of Experimental Medicine,* translated by Henry Copley Green with new foreword by I. Bernard Cohen and introduction by Lawrence J. Henderson (New York: Dover, 1957).

4. Roderick Seidenberg believes that all historic time is transitional. His concept of a post-historic steady-state society challenges the argument that human society can retain its vitality indefinitely after having established a condition of homeostasis equilibrium. See

Post-historic Man: An Inquiry (Chapel Hill: University of North Carolina Press, 1950); and a review by Bentley Glass, "A Biologic View of Human History," *Scientific Monthly* 73 (December 1951): 363–68.

5. W. (V) I. Vernadsky (Volodymyr Vernad'ski, 1863–1945), "The Biosphere and the Noosphere," *American Scientist* 30 (January 1945): 8.

6. Leo Bagrow, *History of Cartography,* translated from the German by D. L. Paisey, revised and augmented by R. A. Skelton (Cambridge, Mass.: Harvard University Press, 1966 edition).

7. Armando Cortesao, "Nautical Science and the Geographical Revolution," *Impact of Science on Society* 4 (Summer 1953): 111–18.

8. Ibid., 118.

9. See G. R. Crane, *Maps and Their Makers: An Introduction to the History of Cartography* (London: Hutchinson's University Library, 1964) (on the International Map of the World on the scale of 1/1 million, see 163–65). On the technology of mapping, see George D. Whitmore, Morris M. Thompson, and Julius L. Speert, "Modern Instruments for Surveying and Mapping," *Science* 130 (October 23, 1959): 1059–66. For examples of specialized types of mapping, see V. A. Kovda, "The Need for International Cooperation in Soil Science," *Nature and Resources* 1 (September 1965): 10–16; and O. Franzle, "Geomorphological Mapping," *Nature and Resources* 2 (December 1966): 14–16. See also Dean S. Rugg, "The International Map of the World," *Scientific Monthly* 72 (April 1951): 233–40; and Edward L. Stevenson, *Terrestrial and Celestial Globes: Their History and Construction,* vols. 1–2 (New Haven, Conn.: Yale University Press for the Hispanic Society of America, 1921).

10. Thomas F. Gaskell, *Under the Deep Oceans: Twentieth-Century Voyages of Discovery* (London: Eyre and Spottiswood, 1960); and George S. Ritchie, *Challenger: The Life of a Survey Ship* (New York: Abelard-Schuman, 1958).

11. Hugh Odishaw, "The International Geophysical Year and World Politics," *Journal of International Affairs* 13 (Winter 1959): 47–56; and Walter Sullivan, *Assault on the Unknown: The International Geophysical Year* (New York: McGraw-Hill, 1966).

12. David A. Davies, "Geophysics and Its Impact on International Affairs" in *Modern Science and the Tasks of Diplomacy,* Karl Braunias and Peter Meraviglia, eds. (Graz, Austria: Verlag Styria, 1965), pp. 103–16.

13. Cited by Jack Major in "Historical Development of the Ecosystem Concept" in *The Ecosystem Concept in Natural Resource Management,* George M. Van Dyne, ed. (New York: Academic Press, 1969), p. 11.

14. Richard Brewer, *A Brief History of Ecology: Part 1—Pre-Nineteenth Century to 1919,* Kalamazoo: Western Michigan University, Occasional Papers of the E. C. Adams Center for Ecological Studies (November 22, 1960), 18 pp. On the growth of ecological concepts, see W. Frank Blair, "Ecology and Evolution," *Antioch Review* 14 (Spring 1959): 47–55; and Joel Hagen, *An Entangled Bank: The Origins of Ecosystem Ecology* (New Brunswick, N.J.: Rutgers University Press, 1992); and Frank B. Golley, *A History of the Ecosystem Concept in Ecology* (New Haven: Yale University Press, 1993).

15. "The Use and Abuse of Vegetational Concepts in Terms," *Ecology* 6 (1935): 284–307.

16. LaMont Cole, "The Ecosphere," *Scientific American* 198 (April 8): 83–96.

17. On the "where" of animal populations, see Marston Bates, *The Forest and the Sea: A Look at the Economy of Nature and the Ecology of Man* (New York: Random House, 1960); V. C. Wynne-Edwards, *Animal Dispersion in Relation to Social Behavior* (London: Oliver and Boyd, 1962); G. Evelyn Hutchinson, "Homage to Santa Rosalia or Why Are There So Many Kinds of Animals?" *American Naturalist* 93 (May–June 1959): 145–59; and publications on biogeography and zoogeography.

18. L. K. Frank, "Time Perspectives," *Journal of Social Philosophy* 4 (July 1939): 293–312; J. N. Mills, "Human Circadian Rhythms," *Physiological Review* 46 (January 1966): 128–71; Stanley R. Mohler, Robert Dille, and H. L. Gibbons, "The Time Zone and Circadian Rhythms in Relation to Aircraft Occupants Taking Long-Distance Flights," *American Journal of Public*

Health 58 (August 1968): 1404-9; and Marc Richelle, "Biological Clocks," *Psychology Today* 3 (May 1970): 33-35, 58-60.

19. "Not Peace, but Ecology," in *Diversity and Stability in Ecological Systems—Brookhaven Symposia in Biology*, no. 22 (Upton, N.Y.: Brookhaven National Laboratory, 1969): 151-58.

20. For a historical treatment of theories of environmental influences, see the following: Franklin Thomas, *The Environmental Basis of Society: A Study in the History of Sociological Thought* (New York: Century, 1925). H. B. van Loon, "Population, Space and Human Culture," *Law and Contemporary Problems* 25 (Summer 1960): 397-405; and Paul Ward English, "Landscape, Ecosystem, and Cultural Perception: Concepts in Cultural Geography," *Journal of Geography* 68 (April 1968): 198-205. A standard source is Harold Sprout and Margaret Sprout, *The Ecological Perspective on Human Affairs with Special Reference to International Politics* (Princeton, N.J.: Princeton University Press, 1965).

21. Robert S. Platt, "Environmentalism vs. Geography," *American Journal of Sociology* 52 (March 1948): 351-58; and Gordon R. Lathwaite, "Environmentalism and Determinism: A Search for Clarification," *Annals of the Association of American Geographers* 56 (March 1966): 1-23.

22. See, for example, John B. Calhoun, "Population Density and Social Pathology," *Scientific American* 206 (February 1962): 139-48; and "Space and the Strategy of Life," *Ekistics* 29 (June 1970): 425-37. Note also Frederick Sargent II and Demitri B. Shimkin, "Biology, Society, and Culture in Human Ecology," *BioScience* 15 (August 1965): 512-15.

23. *Future Shock* (New York: Random House, 1970).

24. (New York: Macmillan, 1913). Reprinted in 1958 with an introduction by George Wald (Boston, Mass.: Beacon Press) (Paperback no. 68).

25. Ibid. (Macmillan, 1924 printing, pp. 274-75.

26. Translated from the French by Bernard Wall with an introduction by Julian Huxley (London: William Collins, 1959; Harper Torchbook Edition, 1961).

27. *Fitness of the Environment*, p. 312.

28. Note for instance, A. D. Voute, "Ecology as a Teleological Science," *Acta Biotheoretica* 18 (Series A, 1968): 143-64.

29. Note especially chapter 6, "The Fitness of the Environment," 2d ed. (Princeton, N.J.: Princeton University Press, 1955).

30. Ibid. (New York: Harper Torchbook Edition, 1962).

31. Ibid., p. 212.

32. (Glencoe, Ill.: Free Press, 1960).

33. "The Meddlers" in *Voices from the Sky* (New York: Pyramid Books, 1967), 162.

34. See "The Biosphere," *Scientific American* 223 (September 1970), whole issue. Note especially the introductory article, "The Biosphere," by G. Evelyn Hutchinson, 45-53; and also by Hutchinson, "The Biosphere or Volume in Which Organisms Actually Live," in *The Ecological Theater and the Evolutionary Play* (New Haven, Conn.: Yale University Press, 1926), 1-26.

35. Librairie Felix Alcan, 232 pp. The volume consists of two essays, "La Biosphère dans le cosmos," and "Le Domaine de la vie." The appendix is a letter to the Society of Naturalists of Leningrad, Feb. 5, 1928, "L'Evolution des espèces et la matière vivante."

36. 30 (January 1945), 1-12.

37. Ibid., p. 4.

38. Ibid., p. 9.

39. Ibid., pp. 9-10.

40. Ibid., p. 9.

41. *Phenomenon of Man*, pp. 94-95.

42. J. E. Lovelock, *Gaia: A New Look at Life on Earth* (New York: Oxford University Press, 1979). For an earlier and similar hypothesis regarding the biosphere, see Alfred Redfield, "The Biological Control of Chemical Factors in the Environment," *American Scientist* 46 (September 1958): 205-21.

43. Peter Bunyard and Edward Goldsmith, eds., *Gaia: The Thesis, the Mechanisms, and the Im-*

plications (Camelford, U.K.: Wadebridge Ecological Centre, 1988); Richard A. Kerr, "No Longer Willful, Gaia Becomes Respectable," *Science* 240 (22 April 1988): 393-95; Stephen Schneider, "A Goddess of the Earth? The Debate on the Gaia Hypothesis," editorial in *Climate Change* 8, no. 1 (February 1986): 1-4.

44. *Operating Manual for Spaceship Earth* (Carbondale: Southern Illinois University Press, 1969). Also included in William R. Ewald, Jr., ed., *Environment and Change* (Bloomington: Indiana University Press, 1968).

45. "The Energy Cycle of the Biosphere," *Scientific American* 222 (September 1970): 74.

2. Growth of International Concern

1. Peter H. Haas, Robert O. Keohane, and Marc A. Levy, *Institutions for the Earth: Sources of Effective International Environmental Protection* (Cambridge, Mass.: MIT Press, 1993).

2. E.g., *Proceedings of the United Nations Scientific Conference on the Conservation and Utilization of Resources: 17 August–6 September 1949*, A/Conf. 10/7 (New York: United Nations, 1956); and *Application of Science and Technology for the Benefit of Less Developed Areas*, 4–10 February (Geneva: United Nations, 1963).

3. UNESCO, "Final Report of the Intergovernmental Conference of Experts on the Scientific Basis for Rational Use and Conservation of the Resources of the Biosphere, held at UNESCO House, Paris, 4–13 September 1968," SC/MD/9 (Paris, 9 January 1969), p. 5. See also p. 9.

4. Public Law 91-190, *United States Code*, vol. 42, secs. 4321–47 (1970).

5. René Dubos, "The Biosphere: A Delicate Balance between Man and Nature," *UNESCO Courier* 20 (January 1969): 14.

6. See *International Union for the Protection of Nature* (Brussels: Imprimeri M. Hayez, for the Union, 1948). For an extensive account of the efforts of the International Union for Conservation of Nature and Natural Resources (IUCN) and its associated organizations, see Robert Boardman, *International Organization and the Conservation of Nature* (Bloomington: Indiana University Press, 1981). For a valuable firsthand account of voluntary citizen efforts in the United States, see John C. Phillips and Harold J. Coolidge, *The First Five Years: The American Committee for International Wildlife Protection* (Cambridge, Mass.: American Committee for International Wildlife Protection, 1935).

7. Sherman Strong Hayden, *The International Protection of Wildlife* (New York: AMS Press, 1970), pp. 92–93; initially printed as Columbia University Studies in the Social Sciences, no. 491 (New York: Columbia University Press, 1942).

8. Ibid., pp. 92–94. Note references to documentary sources.

9. "Convention on the Conservation of European Wildlife and Natural Habitats," Berne, 18 September 1979. For terms of the draft convention, see *Environmental Policy and Law* 5 (February 1979): 52–58; and *European Treaty Series*, vol. 104 (September 1979). See also *Explanatory Report Concerning the Convention* (Strasbourg: Council of Europe, 1979); and "Hope Was Born in Bern," *Naturopa* 34–35 (1980): 30–31, 34–35. On ratification, see *Environmental Conservation* 9 (Summer 1982): 140. See also "Council Directive of 2 April 1979 on the Conservation of Wild Birds," *Official Journal of the European Communities* 22 (25 April 1979): L103/1–L103/18.

10. Hayden, *International Protection of Wildlife*, p. 77. For text of the treaty, see *United States Statutes at Large*, vol. 39, pt. 2, pp. 1702–5.

11. *United States Statutes at Large*, vol. 50, pt. 2, pp. 1311, 1316.

12. Charles I. Bevans, comp., *Treaties and Other International Agreements of the United States of America, 1776–1949*, 12 vols. (Washington, D.C.: U.S. Department of State, 1968–74), vol. 3, pp. 639–60 (hereinafter cited as Bevans, *Treaties*). For a contemporary report, see Harold J. Coolidge, "A New Pan American Treaty," *Science* 92 (15 November 1940): 458–60. For a detailed account of the early implementation of the treaty and the conservation problems

of Latin American countries, see ibid., "Notes on Conservation in the Americas," *Chronica Botanica* 7 (July 1942): 155–63.

13. Harold J. Coolidge, "A World Approach to Nature Protection," *Proceedings of the Inter-American Conference on Conservation of Renewable Natural Resources, Denver, Colorado, September 7–20, 1948*, publ. no. 3382 (Washington, D.C.: U.S. Department of State, n.d.). Dr. Coolidge urged the establishment of a World Convention for Nature Protection, an objective not realized until the negotiation of the endangered species and world heritage treaties following the 1972 United Nations Conference on the Human Environment.

14. The following communication from Harold J. Coolidge shows why statements regarding policy based only on official documents may be misleading: "Under the Convention the Pan American Union was supposed to carry out its implementation, and this job was assigned to their agricultural division headed by Dr. [José] Colom who employed a full-time officer with money raised by our Committee [American Committee for International Wildlife Protection] to follow-up with the recommendations of the Convention.

"Bill Vogt was appointed to do this job and was actively engaged in it until his book entitled *Road to Survival* [1948] was published about the same time as Fairfield Osborn's book entitled *Our Plundered Planet*. Unfortunately, Vogt lost his job because he had offended some of the Catholic officials on the governing Board of the Pan American Union [over the issue of population growth and birth control]. Before this happened he played a significant role in preparing the Denver Conference that discussed in detail the Pan American Convention and gave it a true shot in the arm.

"Unfortunately, Dr. Colom had no real interest in conservation, and after Bill Vogt was fired his office was continued for a while by Annette Fluger until the money ran out." Letter from Coolidge to Caldwell, 8 January 1980.

15. Victor H. Martínez, "Hacia la creación del Sistema Interamericano para la Conservación de la Naturaleza," *Ambiente y Recursos Naturales* 4, no. 2 (April–June 1987): 12–34.

16. *League of Nations Treaty Series*, vol. 172 (Geneva, 1936), no. 3995, pp. 241–42.

17. "African Convention," *IUCN Bulletin*, n.s., 2 (October–December 1968): 68.

18. See "A Convention Is Born," *IUCN Bulletin*, n.s., 10 (June 1979): 41, and "The Bonn Convention Concluded," *Environmental Policy and Law* 5 (June 1979): 135.

19. Bevans, *Treaties*, 1: 804–13; 8, pt. 2: 2283–341. The revised treaty was signed 9 February 1957 and proclaimed 15 November 1957. In 1976 the treaty was amended and extended for twenty-two years.

20. U.S. Department of Commerce, National Oceanographic and Atmospheric Administration, *Report of the National Marine Fisheries Service for the Calendar Year 1978* (Washington, D.C., 1978). For statistical data, see ibid., p. 36.

21. Bevans, *Treaties*, 3: 26–33.

22. Bevans, *Treaties*, 4: 248–58.

23. Noel Simon, "Of Whales and Whaling," *Science* 149 (25 August 1965): 942–46. See also International Whaling Commission, *Twentieth Report of the Commission* (London, 1970), and preceding issues. Also Scott McVay, "Can Leviathan Long Endure So Wide a Chase?" *Natural History* 80 (January 1971): 36–40. For post-Stockholm comments, notes, and references see Chapter 9.

24. *The International Council for the Exploration of the Sea* (Charlottenlund, Denmark: International Council for the Exploration of the Sea [ICES], 1971). For this and other marine organizations, see Kamil A. Bekiashev and Vitali V. Serebriakov, *International Marine Organizations: Essays on Structure and Activities*, trans. Vitali V. Serebriakov (The Hague: Martinus Nijhoff, 1981).

25. See M. J. Girard, "Note on the General Fisheries Council for the Mediterranean," in "Papers Presented at the International Technical Conference on the Conservation of the Living Resources of the Sea," held at UN, FAO, Rome, 8 April–10 May 1955, (A/Conf. 10/7, 1976), pp. 262–65; and Bekiashev and Serebriakov, *International Marine Organizations*, pp. 233–51.

26. For background, see Wilbert McLeod Chapman, "The Theory and Practice of International Fisheries Commissions and Bodies," in *Gulf and Caribbean Fisheries Institute 20th Annual Session* (Miami: University of Miami, November 1967), pp. 77–105; Brian J. Rothschild, ed., *World Fisheries Policy: Multidisciplinary Views* (Seattle: University of Washington Press, 1972); and J. A. Gulland, *The Management of Marine Fisheries* (Bristol, U.K.: Scientechnica, 1974).

27. See Chapter 7 or Chapter 9 for references.

28. Hannes Jonsson, *Friends in Conflict: The Anglo-Icelandic Cod Wars and the Law of the Sea* (London: Hurst, 1982).

29. See *American Journal of International Law* 20 (special supplement, July 1926): 230–41; 20 (October 1926): 752–53.

3. The Road to Stockholm

1. On the concept of a changing social paradigm as used in this volume, see Peter Schwartz and James Ogilvy, *The Emergent Paradigm: Changing Patterns of Thought and Belief*, International VALS Report no. 7 (Menlo Park, Calif.: Stanford Research Institute, April, 1979); William Catton and Riley Dunlap, "A New Ecological Paradigm for Post-Exuberant Society," *American Behavioral Scientist* 24 (September–October 1980): 15–47; Ronald Inglehart, *Culture Shift in Advanced Industrial Society* (Princeton, N.J.: Princeton University Press, 1990) and "Public Support for Environmental Protection," *P.S. (Political Science and Politics)* 28, no. 1 (March 1995): 57–72; Paul R. Abramson and Ronald Englehart, *Value Change in Global Perspective* (Ann Arbor: University of Michigan Press, 1995).

2. The history of these organizational efforts is detailed by Robert Boardman, *International Organization and the Conservation of Nature* (Bloomington: Indiana University Press, 1981); and Sherman Strong Hayden, *The International Protection of Wildlife* (New York: AMS Press, 1970). See also Harold J. Coolidge, "New Horizons in International Conservation," in *Transactions of the Thirteenth North American Wildlife Conference, March 8, 9 and 10, 1948* (Washington, D.C.: Wildlife Management Institute, 1948), pp. 142–50; and Coolidge, "International Activities in the Protection of Nature," in *Exploring Our National Parks and Monuments*, ed. Devereux Butcher (Boston: Houghton Mifflin, 1949), pp. 273–78. For a personal perspective on the growth of ecological awareness, see E. Barton Worthington, "The Ecological Century," *Environmental Conservation* 9 (Spring 1982): 65–70.

3. Coolidge, "New Horizons in International Conservation," p. 149.

4. "Background and Objectives of the Conference," of UN Scientific Conference on the Conservation and Utilization of Resources (Lake Success, 17 August–6 September 1949), in *Proceedings of the United Nations Scientific Conference on the Conservation and Utilization of Resources*, A/Conf. 10.7, 1956, pp. 145–66 (hereinafter, *Proceedings of the Conference on Resources*). For reports following this conference, see Francis W. Carpenter, "Conservation of World Resources: A Report on U.N. Scientific Conference," *Bulletin of the Atomic Scientists* 5 (November 1949): 313–14; and Carl N. Gibboney, "The United Nations Conference for the Conservation and Utilization of Resources," *Science* 110 (23 December 1949): 675–78; and Kamil A. Bekiashev and Vitali V. Serebriakov, *International Marine Organizations: Essays on Structure and Activities*, trans. Vitali V. Serebriakov (The Hague: Martinus Nijhoff, 1981), pp. 252–406.

5. *Proceedings of the Conference on Resources*, p. 13.

6. Ibid., p. 13.

7. Ibid., p. 15. The abstract of Dr. Clark's paper added his personal opinion that the "real origin [of concern with unlimited population growth] appears to lie in a feeling of race superiority on the part of Europeans and Americans which the rest of the world bitterly resents." (Dr. Clark was the father of seven children.)

8. UNESCO and IUCN, International Technical Conference on the Protection of Nature (Lake Success, 22–29 August 1949), *Proceedings and Papers of the International Technical Conference on the Protection of Nature*, Brussels and Paris, 1950.

9. UN Conference on the Application of Science and Technology for the Benefit of the Less Developed Areas (Geneva, 1963), *Science and Technology for Development: Report on the United Nations Conference*, 8 vols. (New York: United Nations, 1963).

10. For the Geneva "Peaceful Uses" conferences, see International Conference on the Peaceful Uses of Atomic Energy, *Proceedings, Geneva, 8 August through 20 August 1955*, 16 vols. (New York: United Nations, 1956); *Proceedings, Second Geneva, 1 September through 13 September, 1958*, 33 vols. (Geneva: United Nations, 1958); *Proceedings, Third Geneva, 31 August through 9 September 1964*, 16 vols. (New York: United Nations, 1965); *Proceedings, Fourth Geneva: Jointly Sponsored by the United Nations and the International Atomic Energy Agency 6 through 16 September 1971*, 15 vols. (New York and Vienna: United Nations and International Atomic Energy Agency [IAEA], 1972). For the Rome "New Sources" conference, see United Nations Conference on New Sources of Energy (Rome, 21-31 August 1961), *Report of the United Nations Conference on New Sources of Energy*, E/3577/Rev. 1, 1962, and *Proceedings: General Sessions*, 7 vols., 1963-64, E/CONF.35/2.

11. On the Biosphere Conference, see Michel Batisse, "Can We Keep Our Planet Habitable?" *UNESCO Courier* 22 (January 1969): 4-5; Harold J. Coolidge, "World Biosphere Conference: A Challenge to Mankind," *IUCN Bulletin*, n.s., 2 (October-December 1968): 65-66; Raymond F. Dasmann, "Conservation and Rational Uses of the Environment," *Nature and Resources* 4 (June 1968): 2-5; UNESCO, "Final Report of the Intergovernmental Conference of Experts on the Scientific Basis for Rational Use and Conservation of the Resources of the Biosphere, held at UNESCO House, Paris, 4-13 September 1968," SC/MD/9 (Paris, 9 January 1969); and "International Conference on the Biosphere," *UNESCO Chronicle* 14 (November 1968): 414-18. For the ecological thinking underlying the conference, see the following report (initial draft by Raymond F. Dasmann) submitted by UNESCO and FAO to the Economic and Social Council (ECOSOC): *Conservation and Rational Use of the Environment Report of March 12, 1968*, E/4458, 1968. For a discussion of the report, see *Nature and Resources* 4 (June 1968): 2-5.

12. UNESCO, "Final Report on the Biosphere."

13. Pierre Auger, "Plan for an Institute of Studies for the Better Utilization of the Globe," in *The Population Crisis and the Use of World Resources*, ed. Stuart Mudd (Bloomington: Indiana University Press, 1964), pp. 384.

14. See *Beauty for America: Proceedings of the White House Conference on Natural Beauty, May 24-25, 1965* (Washington, D.C.: U.S. Government Printing Office, 1965).

15. For a full report of recommendations, see Richard N. Gardner, ed., *Blueprint for Peace: Being Proposals of Prominent Americans to the White House Conference on International Cooperation* (New York: McGraw Hill, 1966), which presents several committee reports that describe international environmental programs and proposals in detail in, e.g., chaps. 7 and 11-16.

16. Public Law 91-190, United States Code, vol. 42, secs. 4321-47 (1970). See also Lynton K. Caldwell, *Science and the National Environmental Policy Act: Redirecting Policy through Procedural Reform* (University: University of Alabama Press, 1982).

17. *Weekly Compilation of Presidential Documents* 4, no. 23 (June 1968): 906.

18. UN Economic Commission for Europe (ECE), *ECE Symposium on Problems Relating to Environment: Proceedings and Documents of a Symposium Organized by the Economic Commission for Europe (With a Study Tour of Ostrava, Czechoslovakia and Katowice, Poland, held at Prague, Czechoslovakia, May 2-5, 1971)* (New York: UN Economic Commission for Europe, 1971). See also Jon McLin, "European Organizations and the Environment," West Europe Series, no. 7, *American Universities Field Staff Reports*, no. 2 (1972), pp. 1-11.

19. For papers presented, see Nicholas Polunin, ed., *The Environmental Future: Proceedings of the First International Conference on the Environmental Future, held in Finland from 27 June to 3 July 1971* (New York: Barnes and Noble, 1972).

20. UN, ECOSOC, E/446/Add. 1, 1968, in ECOSOC, *Official Records*, Forty-fifth Session, Annexes, Agenda Item 12.

21. UN General Assembly, *Official Records*, Twenty-seventh Session, 1629th meeting, 13 December 1967.
22. UN, "Report of the Secretary General," E/4553, 11 July 1968.
23. UN, ECOSOC, Resolution 1346 (XLV), 30 July 1968, in ECOSOC, *Official Records*, Forty-fifth Session, Supplement no. 1, p. 8.
24. UN General Assembly, Official Records, Twenty-third Session, Plenary Meetings, Verbatim Records, 1733d meeting, 3 December 1968. See also *UN Monthly Chronicle* 6 (January 1969): 35–41. For the procedural steps leading to Resolution 2398 (XXIII) of the General Assembly, see "Problems of the Human Environment," chap. 14, in *Yearbook of the United Nations 1966* (New York: United Nations, 1968), pp. 473–77.
25. UN Secretary-General, "Problems of the Human Environment: Report of the Secretary General," E/4667, 26 May 1969, summarizes activities and programs of UN bodies relative to the human environment.
26. UN General Assembly, "Reports and Other Documents of the Preparatory Committee for the United Nations Conference on the Human Environment," A/Conf.48/PCI-17. These are mimeographed and unbound, the "Official Reports" of the committee sessions being: First Session, New York, 10–20 March 1970, A/Conf.48/PC/6; Second Session, Geneva, 8–19 February 1971, A/Conf.48/PC/9; Third Session, New York, 13–24 September 1971, A/Conf.48/PC/13; Fourth Session, New York, 6–10 March 1972, A/Conf.48/PC/17.
27. UN Conference on the Human Environment, *Information Letter*, G.E. 71-12626, 30 June 1971, pp. 3–4; G.E. 71-15629, 31 July 1971, pp. 2–7.
28. See Robert J. Bazell, "Human Environment Conference: The Rush for Influence," *Science* 174 (22 October 1971): 390–91. *The United Nations and the Human Environment, Twenty-Second Report* (New York: Commission to Study the Organization of Peace, April 1972); *Institutional Arrangements for International Environmental Cooperation: A Report to the Department of State by the Committee on International Environmental Programs, Environmental Studies Board, National Academy of Sciences — National Academy of Engineering, June 1972* (Washington, D.C.: National Academy of Sciences, 1972); and Philip A. Douglas, ed., *Unifying Nations for Biosurvival: An International Symposium* (Washington, D.C.: National Wildlife Federation, 1973).
29. *Development and Environment: Report and Working Papers of a Panel of Experts Convened by the Secretary-General of the United Nations Conference on the Human Environment, held at Founex, Switzerland, June 4–12, 1971* (Paris: Mouton, 1972). See also International Conciliation 586 (January 1972), and *In Defense of the Earth*, UNEP Executive Series, vol. 1 (Nairobi: UNEP, 1980).
30. See the following UN documents on particular regional economic commission environment and development seminars: Economic Commission for Africa, "Report of the First All-African Seminar on the Human Environment, held at Addis Ababa, August 23–28, 1971," D/CN.14/532; Economic Commission for Asia and the Far East, (after 1974, "Pacific" replaced "Far East"), "Report of the Seminar on Development and Environment, held at Bangkok, August 17–23, 1971," E/CN.11/999; Economic Commission for Latin America, "Latin American Regional Seminar on Problems of the Human Environment and Development, held at Mexico City, Mexico, September 6–11, 1971 — Under the Co-Sponsorship of the Economic Commission for Latin America and the Governments of Latin America," ST/ECLA/Conf.40/Lf; Economic and Social Office at Beirut, "Report of Regional Seminar on Development and Environment held at Beirut, Lebanon, September 27–October 1, 1971, in Cooperation with the Secretariat of the United Nations Conference on the Human Environment," ESOB/DE/i; General Assembly, Preparatory Committee for the UN Conference on the Human Environment, Third Session, 1324 September 1971, "Regional Seminars on Development and Environment: Note by the Secretary-General Addendum," A/Conf.48/PC.11/Add.2. Also relevant to Stockholm preparations are the following economic commission documents: United Nations, "Report of the Regional Seminar on the Ecological Impli-

cations of Rural and Urban Population Growth, held at Bangkok, August 25–September 3, 1971," E/CN.11/L.312; Economic Commission for Latin America, *The Human Environment in Latin America: Proceedings of the 14th Session of the United Nations Economic Commission for Latin America, held at Santiago, Chile, April 27–May 8, 1971, With a Note by the United Nations Secretariat* (Santiago, 1971).

31. See the following Stockholm Conference documents for reports on Founex, the Canberra meeting, and the regional seminars: United Nations, *Development and Environment* (Subject Area 5) A/Conf./10; *Development and Environment: Report and Working Papers of a Panel of Experts Convened by the Secretary-General of the United Nations Conference on the Human Environment, held at Founex, Switzerland, June 4–12, 1971,* Annex 1; *Environmental Problems in the Developing Countries. Basic Issues: Summary of the Report Prepared by SCOPE in Cooperation with the Secretariat of the Conference, held at Canberra, Australia, August 24–September 3, 1971,* Annex 2; and *Regional Seminars on Development and Environment,* Annex 3.

32. Study of Man's Impact on Climate (SMIC), *Inadvertent Climate Modification* (Cambridge, Mass.: MIT Press, 1971). For a related American study, see William H. Matthews, William W. Kellogg, and Gershon DuVall Robinson, eds., *Man's Impact on the Climate* (Cambridge, Mass.: MIT Press, 1971).

33. UN Conference on the Human Environment, *Information Letter,* no. 1 (30 June 1971).

4. The Stockholm Conference

1. For accounts of the conference and its outputs, see: R. Stephen Berry et al., "What Happened at Stockholm: A Special Report," *Science and Public Affairs: Bulletin of the Atomic Scientists* 28 (September 1972): 16–56; Robert Gillette, "Human Environment Conference: Citizen Advisers Muddle Through," *Science* 174 (29 October 1971): 479–81; Nigel Hawkes, "Human Environment Conference: Search for a Modus Vivendi," *Science* 175 (18 February 1972): 736–38; Hawkes, "Stockholm: Politicking, Confusion, but Some Agreements Reached," *Science* 176 (23 June 1972): 1308–10; Brian Johnson, "The Settlement of Stockholm," *Ecologist* 3 (March 1973): 87–88; Jon McLin, "Stockholm: The Politics of Only One Earth," West Europe Series, no. 7, *American Universities Field Staff Reports,* no. 4 (1972): 5–12; McLin, "The United Nations System and the Stockholm Conference," West Europe Series, no. 7, *American Universities Field Staff Reports,* no. 3 (1972): 1–7; Louis B. Sohn, "The Stockholm Declaration on the Human Environment," *Harvard International Law Journal* 14 (Summer 1973): 423–515; E. Thomas Sullivan, "The Stockholm Conference: A Step toward Global Environmental Cooperation and Involvement," *Indiana Law Review* 6, no. 2 (1972): 267–82; UN, *Report of the United Nations Conference on the Human Environment, Stockholm, 5–16 June 1972,* A/CONF.48/14/REV.1 (New York, 1973), p. 77 (Sales no. E.73.11.A.14); UN, Centre for Economic and Social Information, *Environment Stockholm: Declaration, Plan of Action, Recommendations, Resolutions, Papers Relating to the United Nations Conference on the Human Environment, held at Stockholm, Sweden, June 5–16, 1972* (Geneva, 1972); and David E. Luchins, *The United Nations Conference on the Human Environment: A Case Study of Emerging Political Alignments, 1968–1972* (Ann Arbor, Mich.: University Microfilms International, 1980). Conference documents (A/CONF.48/1–12) were issued individually. A collected set was issued in three volumes by the U.S. Department of State and reproduced by the National Technical Information Service, Springfield, VA (PB-206 618-1, 2, and 3) March 1972, variously paged.

2. *Environment Stockholm* describes the proceedings.

3. Hans H. Landsberg, "Can Stockholm Succeed?" *Science* 176 (19 May 1972): 749.

4. "What Happened at Stockholm—A Special Report," *Science and Public Affairs: Bulletin of the Atomic Scientists* 28 (September 1972): 44.

5. Ibid., p. 36.

6. Ibid., p. 56.

7. See Johan Kaufmann, "Size and Membership of the Conference," in *Conference Diplomacy: An Introductory Analysis* (Dobbs Ferry, N.Y.: Oceana Publications, 1968), pp. 57–59. See also Hawkes, "Human Environment Conference," pp. 736–38; and "Vienna Convention on Diplomatic Relations, Vienna, 18 April 1961," in *Treaties and International Agreements Registered or Filed and Recorded with the Secretariat of the United Nations*, vol. 500 (1964), no. 7310, p. 124.

8. Frances Gendlin, "Voices from the Gallery," in *Science and Public Affairs: Bulletin of the Atomic Scientists* 28 (September 1972): 26–29.

9. "What Happened at Stockholm," pp. 46–48.

10. Harold H. Leich, "The Environment Conference at Stockholm," *Appalachia* 38 (15 December 1972): 118–24.

11. *Time*, 19 June 1972, p. 55, comparing the miscellaneous groups, individuals, and causes represented at Stockholm to the huge rock music festival occurring at Woodstock, New York, 15–17 August 1969.

12. Tim E. J. Campbell, "The Political Meaning of Stockholm: Third World Participation in the Environment Conference Process," *Stanford Journal on International Studies* 8 (Spring 1973): 138–53.

13. Henry J. Kellerman, "Stockholm and the Role of Science," *BioScience* 23 (August 8): 485–87.

14. Robert J. Bazell, "Human Environment Conference: The Rush for Influence," *Science* 174 (22 October 1971): 390–91.

15. Reported by Professor di Castri at the Joint Senate-House Colloquium on International Environmental Science, Old Supreme Court Chamber, Washington, D.C., 25–26 May 1971. See *International Environmental Science Proceedings of the Joint Colloquium before the Committee on Commerce, U.S. Senate, and the Committee on Science and Astronautics, House of Representatives*, 92d Cong., 1st sess., Serial no. 92-13, pp. 31–43.

16. A convenient source of the Declaration and Action Plan is *Report of the United Nations Conference on the Human Environment*. For effect of the declaration on international law, see Sohn, "The Stockholm Declaration," pp. 513–15.

17. On these principles, see the following publications by IUCN: Shadia Schneider-Sawiris, *The Concept of Compensation in the Field of Trade and Environment*, IUCN Environmental Policy and Law Paper, no. 4 (Morges, Switzerland, 1973); Yvonne I. Nicholls, comp., *Source Book: Emergence of Proposals for Recompensing Developing Countries for Maintaining Environmental Quality*, IUCN Environmental Policy and Law Paper, no. 5 (Morges, Switzerland, 1973); and Scott MacLeod, *Financing Environmental Measures in Developing Countries: The Principle of Additionality*, Environmental Policy and Law Paper, no. 6 (Morges, Switzerland, 1974).

18. C. C. Bailey, *The Aftermath of Chernobyl* (Dubuque, Iowa: Kendall/Hunt, 1989).

19. Maurice F. Strong, "One Year after Stockholm: An Ecological Approach to Management," *Foreign Affairs* 51 (July 1973): 703; quoted passages immediately following are from this article. See also William Hadley Kincade, "International Environmental Action: The Road after Stockholm," *Earth Law Journal* 2 (1976): 311–15.

20. "Resolution on Institutional and Financial Arrangements, Governing Council for Environmental Programmes," in *Report of the United Nations Conference on the Human Environment*, Recommendations 2a and 2b.

21. "United Nations Conference on the Human Environment, Report of the Second Committee," A/8901, in UN General Assembly, *Official Records*, Twenty-Seventh Session, 2112th Plenary Meeting, 15 December 1972, Agenda Item 47; and see "Resolutions Adopted by the General Assembly during its Twenty-Seventh Session, 19 September–19 December 1972," in UN General Assembly, *Official Records*, Twenty-seventh Session, Supplement no. 30, A/8730, pp. 42–48 (Resolutions 2994 [XXVII] to 3604 [XXVII]).

22. For an early discussion of the significance of the headquarters of international organizations, see C. Wilfred Jenks, *The Headquarters of International Organizations: A Study of Their Location and Status* (London: Royal Institute of International Affairs, 1945).

5. Post-Stockholm Assessment

1. See "Twenty Years Since Stockholm," *United Nations Environment Programme: Annual Report of the Executive Director* (Nairobi: UNEP, 1992).
2. O. P. Diwendi and Dhirendra K. Vajpeyi, eds., *Environmental Policy in the Third World: A Comparative Analysis* (London: Marsell, 1995).
3. For official descriptions of UNEP, see *UNEP: What It Is, What It Does, How It Works* (undated leaflet); and *UNEP, The United Nations Environment Programme*, Na.78/5921-5000 (Nairobi, 1979). Proceedings and reports of the UNEP Governing Council and secretariat are available in most libraries covering general UN documentation. Note especially UNEP, *An Environmental Bibliography: Publications Issued by UNEP or under Its Auspices 1973–1980* (Nairobi, 1981). See also Bernice Heilbrunn, "U.N.E.P. Mandates: Stockholm Recommendations and Governing Council Decisions," *Earth Law Journal* 1 (1975): 161–67; William Hadley Kincade, "International Environmental Protection," (editorial), pp. 311–17; Fukashi Utsunomiya, "The United Nations Environment Program and the United Nations Environment System," in *Transnational Environmental Policy: Aspects and Prospects*, Open Grants Paper, no. 5 (Honolulu: East-West Center, February 1977); and especially Norman N. Miller, "The United Nations Environment Programme Africa/General, no. 1, *American Universities Field Staff Reports*, no. 17 (1979).
4. See Resolution 2997 (XXVII), as recommended by the Second Committee, A/8901, adopted by Assembly on 15 December 1972, Meeting 2112, by 116 votes in favor with 10 abstentions, in *Yearbook of the United Nations, 1972* (New York: United Nations, 1975), pp. 331–33. For background, see "International Organizational Implications for Action by the Conference including Financial Implications," in Preparatory Committee for the United Nations Conference on the Human Environment, Fourth Session, "Report of the Secretary General" (New York, 6–10 March 1972), A.Conf.48/PC/15, pp. 99–103.
5. In addition to the reports of the Governing Council, see *UN Monthly Chronicle* 10 (July 1973): 83–84. For an account of the politics and deliberations of the First Session of the Governing Council, see Vanya Walker-Leigh, "On from Stockholm," *The World Today* 29 (December 1973): 543–46. Detailed accounts of meetings of the Governing Council have been reported in the journal *Environmental Policy and Law*.
6. See UNEP Governing Council, "Introductory Statement by the Executive Director, April 15, 1975," UNEP/GC/L.27, p. 10.
7. Note opening statement of the "Introductory Report of the Executive Director to the Third Session of the Governing Council," UNEP/GC/28, 5 February 1975: "The Executive Director's introductory report to each session of the Governing Council gives him an opportunity to put before the Governing Council those broad policy issues which, in his view, would help produce a fruitful general debate and serve to guide more detailed discussions subsequently. Because the policy issues will differ from year to year, it is not possible to adopt a standardized format for the reports." For annual state of the environment reports, see *United Nations Environment Programme: Annual Review*.
8. UNEP Governing Council, *Earthwatch—Related Research, Evaluation and Review: A Background Paper* (Nairobi, 2 April 1979), Item 7 of provisional agenda; and *Earthwatch—An In-Depth Review*, UNEP Report, no. 1, and Supplement nos. 1–3 (1981).
9. For a condensed version by the editors, see *Environmental Conservation* 9 (Spring 1982): 11–29.
10. See R. E. Munn, *Global Environmental Monitoring System (GEMS) Action Plan for Phase One*, SCOPE Report 3 (Toronto: International Council of Scientific Unions, 1973). Also see *Earthwatch—An In-Depth Review;* and Michael D. Gynne, "The Global Environment Monitoring System (GEMS) of UNEP," *Environmental Conservation* 9 (Spring 1982): 35–42.
11. See *Earthwatch—An In-Depth Review*, UNEP Report, no. 1, Supplement no. 1; and *Environmental Data*, Supplement no. 2, (Nairobi: UNEP). The latter supplement describes

collaboration between the UN Statistical Office (UNSO) and UNEP in the development of environmental statistics. See also *The Environmental Data Book: A Guide to Statistics on the Environment and Development* (Washington, D.C.: World Bank, 1993). On the International Register of Potentially Toxic Chemicals (IRPTC), see *Earthwatch—An In-Depth Review*, UNEP Report, no. 1, pp. 10–11; and UNEP Governing Council, "The Environment Programme: Report of the Executive Director, 30 January 1980," UNEP/GC.8/s, pp. 18–21. Also see Alexander I. Kuckerenko and Jan W. Huismans, "The International Register of Potentially Toxic Chemicals (IRPTC) of UNEP," *Environmental Conservation* 9 (Spring 1982): 59–63. For current information, see the *IRPTC Bulletin: Journal of the International Register of Potentially Toxic Chemicals* devoted to information on hazardous chemicals.

12. See *International Organization* 29 (Summer 1975): 893–901. For text, see *In Defence of the Earth: The Basic Texts on Environment: Founex, Stockholm, Cocoyoc* (Nairobi: UNEP, 1981), pp. 110–19.

13. On the functions of international conferences, see Norman L. Hill, *The Public International Conference: Its Functions, Organization, and Procedure* (Stanford, Calif.: Stanford University Press, 1929).

14. See the following UN reports: "Report of Habitat: United Nations Conference on Human Settlements, Vancouver, 31 May–11 June 1976," A/Conf.70/15, 1976; "Report of the United Nations Water Conference, Mar del Plata, 14–25 March 1977," E/Conf.70/29, 1977; and "Report of the United Nations Conference on Desertification, Nairobi, 29 August–9 September 1977," A/Conf.74/35, Nairobi, n.d.

15. See leaflet *Environment Liaison Centre* (Nairobi: Environment Liaison Centre, n.d.); and "UNEP to Expand Work with Global NGOs," *Uniterra* 5 (November–December 1980): 1, 12. For successive years, see the publication *The State of the Environment: Selected Topics* (Nairobi: Pergamon Press, for UNEP). This publication should be distinguished from the executive director's annual report to the Governing Council.

16. Robert E. Stein and Brian Johnson, *Banking on the Biosphere: Environmental Procedures and Practices of Nine Multilateral Development Agencies* (Lexington, Mass.: Lexington Books, 1979). On the Paris meeting (12–13 September 1979) addressing development concerns, see UNEP, "Meeting with Multilateral Development Financing Agencies to Adopt a Draft Declaration of Principles on the Incorporation of Environmental Considerations in Development Policies, Programmes and Projects," UNEP/GC.8/INF, Nairobi, 1979. For final text, see "Declaration of Environmental Policies and Procedures Relating to Economic Development, adopted at New York, 1 February 1980," *International Legal Materials* 19 (March 1980): 524–25. Signatories included UNEP, UNDP, World Bank, African Development Bank, Arab Bank for Economic Development in Africa, Asian Development Bank, Caribbean Development Bank, Inter-American Development Bank, Commission of the European Communities, and the Organization of American States. The European Investment Bank subsequently (1983) adhered to the declaration. For brief reports on CIDIE-UNEP relationships, see the annual reports of UNEP's executive director for the years 1987 (p. 20) and 1988 (pp. 16–17).

17. Michael J. Brenner, "The Intergovernmental Oceanographic Commission and the Stockholm Conference: A Case of Institutional Non-Adaptation," *International Organization* 29 (Summer 1975): 771–804; and Brenner, *The Science Advisory Function: The Case of International Marine Environment Policy*, Sio Reference Series (La Jolla: University of California, Scripps Institution of Oceanography, August 1973).

18. *Foreign Affairs* 51 (July 1973): 690.

19. Barbara Maslam, "Global Marine Pollution Treaty Has Been Ratified," *World Environment Report* 8 (30 November 1982): 1–2.

20. Robert E. Stein, "Environmental Mediation of Transboundary Disputes," paper prepared for the annual meeting of the Air Pollution Control Association, Montreal, 22–27 June 1980 (Washington, D.C.: Environmental Mediation International, 1980); and Stein, "The Uses of Mediation to Settle Canadian-U.S. Environmental Issues," *Quarterly Newsletter of the*

Standing Committee on Environmental Law of the American Bar Association (February 1982): 5–7.

21. UN General Assembly, "Development and International Co-operation: Report of the 2nd Committee, Part II," A/35/592/Add.1, 27 November 1980, pp. 11–12.

22. For a retrospective on the 1972 conference, see Michel Batisse, "Stockholm Ten Years Later," *Nature and Resources* 18 (January–March 1982): 3–4; "Stockholm Revisited," *Uniterra,* no. 1 (1982): entire issue; William Dampier, "Ten Years after Stockholm: A Decade of Environmental Debate," *Ambio* 11, no. 4 (1982): 215–28; Jeannie Peterson, "Ten Years After Stockholm UNEP Looks Back—and Forth," *Ambio* 11, no. 4 (1982): 71; Patricia Scharlin, "The United Nations and the Environment: After Three Decades of Concern, Progress Is Still Slow," *Ambio* 11, no. 1 (1982): 26–29; Raisa Scriabine, interview of Strong, "Strong Assesses the Decade since Stockholm," *Ambio* 11, no. 4 (1982): 229–31; and for trends since Stockholm, Eckholm, *Down to Earth;* UNEP, "Review of Major Achievements in the Implementation of the Action Plan for the Human Environment: Report of the Executive Director," UNEP/GC.10/INF.1, Nairobi, 26 January 1982; "The Environment—Ten Years On, 1972–1982," *Uniterra,* no. 2 (1982): entire issue; and *Review of the Global Environment: 10 Years after Stockholm,* Hearings before the Subcommittee on Human Rights and International Organizations of the Committee on Foreign Affairs, House of Representatives, 97th Cong., 2d sess., 30 March; 1 and 20 April 1982 (Washington, D.C.: Government Printing Office, 1982). For an evaluation of the Stockholm strategy and related environmental programs, see David A. Kay and Harold K. Jacobson, eds., *Environmental Protection: The International Dimension,* (Totowa, N.J.: Allenheld, Osmun, 1983), chap. 13, pp. 310–32.

23. UNEP, *The Environment in 1982: Retrospect and Prospect,* UNEP/GC/SSC/2, 29 January 1982, p. 22. For a report on the session, see "UNEP: Session of a Special Character," *Environmental Policy and Law* 9 (September 1982): 2–28.

24. David E. Bloom, "International Public Opinion on the Environment," *Science* 269 (21 July 1995): 354–358.

25. UN General Assembly, "Draft World Charter for Nature. Report of the Secretary General," Thirty-sixth Session, Agenda Item 23, A/36/539, 13 October 1981, Annex I, Appendix II, pp. 14–15.

26. Ibid., Annex II, p. 1.

27. UN General Assembly, "Consideration and Adoption of the Revised World Charter for Nature," Agenda Item 21, Provisional Verbatim Record of the Forty-eighth Meeting . . . , New York, 28 October 1982, A/37/PV.48.5ee. UN General Assembly, World Charter for Nature, A/Res/37/7, 9 November 1982. Text reprinted in *Environmental Policy and Law* 10 (January 1983): 30–31. See also *Environmental Policy and Law* 10 (April 1983): 37, 70; and Wolfgang E. Burhenne and Will Irwin, *The World Charter for Nature: A Background Paper* (Berlin: Erich Schmidt Verlag, 1983).

28. For brief accounts of these resolutions, see Libby Bassett, "UN Votes to Adopt World Nature Charter," *World Environment Report* 8 (30 November 1982): 2–3; and "UN Votes to Protect People from Harmful Imports," *World Environment Report* 9 (15 January 1983): 1–2.

29. Provisional Verbatim Record of the Forty-eighth Meeting . . . , New York, 28 October 1982, pp. 87–90.

30. UN Resolutions and Decisions Adopted by the General Assembly Forty-fourth Session, 22 December 1989, Resolution 228.

31. *International Legal Materials* 28, no. 5 (September 1989): 1311–13. On the Declaration of Manaus, see *Kessing's Record of World Events: News Digest for May 1989,* p. 36654, and *Facts on File,* March 10, 1989, pp. 156–57.

32. United Nations Environment Programme. *1989 Annual Report of the Executive Director* (Nairobi: UNEP, 1990).

6. Rio de Janeiro and Agenda 21

1. See *Report of the United Nations Conference on Environment and Development, Rio de Janeiro, 3–14 June 1993* vols. 1–3 (New York: United Nations, 1993). Among the readily comprehensive and accessible assessments of UNCED are "Earth Summit—Judging Its Success," *Environment* 34, no. 8 (October 1992); *Environmental Policy and Law* 22, no. 4 (August 1992); and *IUCN–World Conservation Union Bulletin* 23, no. 3 (September 1992). Note also *The Global Partnerships for Environment and Development: A Guide to Agenda 21*, United Nations Publication 92-1-100481-0 (1992), and *Draft Agenda 21, Rio Declaration, Forest Principles*, United Nations Publication 92-1-1-482-2. See also Nicholas Robinson, Parvez Hassan, and Françoise Burhenne-Guilmen, eds., *Agenda 21 and UNCED Proceedings*, 5 vols. (Dobbs Ferry, N.Y.: Oceana Publications, 1992); and Nicholas C. Yost, "Rio and the Road Beyond," *Environmental Law* (Quarterly Newsletter of Standing Committee on Environmental American Bar Association) 11, no. 4 (Summer 1992): 1–6; *Colorado Journal of International Law and Policy* 4, no. 1 (Winter 1993): entire issue; Michael Grubb et al., *The Earth Summit Agreements: A Guide and Assessment* (London: Royal Institute of Public Affairs, 1993); and Stanley Johnson, ed., *Earth Summit* (Dordrecht, Netherlands: Kluwer Academic, 1992).

2. United Nations, *Earth Summit: Rio Declaration and Forest Principles* (Final Text), DPI/1299-October 1992-3M (New York: UN Department of Public Information, 1992).

3. United Nations, *Earth Summit: Press Summary of Agenda 21* (Final Text), DPI/128-2nd printing, November 1992-5M (New York: UN Department of Public Information, 1992); and *Report of the United Nations Conference on Environment and Development Rio de Janeiro, 3–14 June 1992* vols. 1–3 (New York: United Nations, 1993).

4. United Nations, *United Nations Framework Convention on Climate Change* (Final Text for Information Media—Not an Official Record), DPI/1300-October 1992-3M (New York: UN Department of Public Information, 1992).

5. United Nations, *Earth Summit: Convention on Biological Diversity* (Final Text for Information Media—Not an Official Record), DPI/1307-October 1992 (New York: UN Department of Public Information, 1992).

6. Notably by Jack Manno in *Advocacy and Diplomacy in the Great Lakes: A Case History of NGO Participation in World Environmental Politics* (Syracuse, N.Y.: Great Lakes Research Consortium, SUNY College of Environmental Science and Policy, 1992).

7. For "environmentalist" misgivings on NAFTA see "Trading away the Environment," *Amicus Journal* 14, no. 4 (Winter 1993): 9–10. But an analysis of the environmental impact of NAFTA in a paper by Peter M. Emerson and Alan C. Nessman (Environmental Defense Fund), in *Review of Marketing and Agricultural Economics* 63, no. 2 (August 1995): 243–55, finds many of the fears regarding NAFTA to be unfounded.

8. See Stewart A. Hudson and Rodrigo J. Prudencio, *The North American Commission on Environmental and Other Supplemental Environmental Agreements* (Washington, D.C.: National Wildlife Federation, 4 February 1993); and Janine Ferretti, *Elements of an Effective North American Commission on the Environment* (Toronto: Pollution Probe, 4 March 1993). The name finally selected for the agency was the North American Commission on Environmental Cooperation. A secretariat was located in Montreal.

9. Lawrence T. Woods, "Nongovernmental Organizations and the United Nations System: Reflecting upon the Earth Summit Experience," *International Studies Notes* 18, no. 1 (Winter 1993): 9–15. See also "The Role of NGOs in the Post Rio Era," *Network*, no. 21 (Geneva: Centre for our Common Future, November 1992), pp. 6–7; and Thomas Princen and Matthias Finger, *Environmental NGOs in World Politics: Linking the Local and the Global* (New York: Routledge, 1994).

10. Committee on Science, Engineering, and Public Policy, *Policy Implications of Greenhouse Warming* (Washington, D.C.: National Academy Press, 1991). See also note 4, above.

11. This necessity was recognized at the 1972 UN Conference at Stockholm. See Lynton K.

Caldwell, "Organizational and Administrative Aspects of Environmental Problems at the Local, National, and International Levels," in *Organization and Administration of Environmental Programs*, ST/ESA/16 (New York: UN Department of Economic and Social Affairs, 1974), pp. 12–36.

12. Stephen Schmidheiny, ed., *Changing Course: A Global Business Perspective on Development and the Environment* (Cambridge, Mass.: MIT Press, 1992). See also Barry Sadler and Brian Hull, *In Business for Tomorrow: The Transition to Sustainable Development* (Ottawa: The Conference Board of Canada, 1990).

13. *Brundtland Bulletin*, no. 17, October 1992, p. 3.

14. David W. Cook, "The State in Nature-Society Relations: Explaining Patterns in the Ratification of Global Environmental Treaties," M.A. thesis, University of Colorado, 1990.

15. "General Assembly Creates CSD," *Network*, no. 21 (Geneva: Centre for Our Common Future, November 1992).

16. Nicholas C. Yost, "Rio and the Road Beyond," *Environmental Law*, 11, no. 4 (Summer 1992): 1.

17. Albert Gore, *Earth in the Balance: Ecology and the Human Spirit* (Boston: Houghton Mifflin, 1992).

18. Jay W. Forrester, "Counterintuitive Behavior of Social Systems," *Technology Review* 73 (January 1971): 53–68.

19. See full-page appeal *New York Times*, 14 April 1992, p. A12. See also *GATT: Implications on Environmental Laws*, Hearings before the Subcommittee on Health and the Environment of the Committee on Energy and Commerce, House of Representatives, 102 Cong., 1st sess. (27 September 1991); and *Trade and Environment*, Hearing before the Subcommittee on International Trade, Committee on Finance, Senate, 102 Cong., 1st sess. (25 October 1992). Also see *The Environmental Effects of Trade* (Paris: OECD, 1994). A case can be made, however, that appropriately controlled global trade may raise national environmental standards as often as depress them. See David Vogel, *Trading Up: Consumer and Environmental Regulations in a Global Economy* (Cambridge, Mass.: Harvard University Press, 1995).

20. Kiichi Miyazawa, "Creating a Policy Framework for the Rio Process," *Earth Summit Times* (14 September 1992), pt. 2, p. x.

21. World Commission on Environment and Development, *Our Common Future* (Oxford: Oxford University Press, 1987).

22. Cited by Peter M. Haas, Marc A. Levy, and Edward A. Parson in "Appraising the Earth Summit: How Should We Judge UNCED's Success?" *Environment* 34, no. 8 (October 1992): 7.

23. Ibid.

24. David Runnals, "Triumph or Disaster?" *IUNC—World Conservation Union—Bulletin* 23, no. 3 (September 1992): 19–20.

25. See Hilary E. French, *After the Summit: The Future of World Governance* (Washington, D.C.: Worldwatch Institute, March 1992); and French, *Partnership for the Planet: An Environmental Agenda for the United Nations* (Washington, D.C.: Worldwatch Institute, 1995).

7. International Structures for Environmental Policy

1. For more comprehensive or detailed descriptions of the structure of international environmental cooperation as it has evolved, see Bernd Rüster, Bruno Simma, and Michael Bock, eds., *International Protection of the Environment: Treaties and Related Documents*, 30 vols. and index (Dobbs Ferry, N.Y.: Oceana Publications, 1975–82); continued after vol. 30 in looseleaf under the title *International Protection of the Environment: Conservation and Sustainable Development* (1995); D. A. Kay and H. K. Jacobson, eds., *Environmental Protection: The International Dimension* (Totowa, N.J.: Allanheld, Osman, 1983); A. LeRoy Bennett, *International Organizations: Principles and Issues*, 2d ed. (Englewood Cliffs, N.J.: Prentice-Hall, 1979);

Harold K. Jacobson, *Networks of Interdependence: International Organizations and the Global Political System* (New York: Knopf, 1979); P. L. DeReeder, *Environmental Programmes of Intergovernmental Organizations—With Special Reference to the Sphere of Interest of the Chemical Industry* (The Hague: Martinus Nijhoff, May 1977); and Amos J. Peaslee, *International Governmental Organizations: Constitutional Documents*, 3d ed., rev., 5 parts (The Hague: Martinus Nijhoff, 1974).

By way of background, many of the UN specialized agencies as well as national governments published descriptions of their environmental activities in connection with the UN Conference on the Human Environment in 1972, and country statements accompanied the report of the 1992 UN Conference on Environment and Development (vol. 3). For what was possibly the most comprehensive but concise description of international environmental activities at that time, see U.S. Congress, Senate Committee on Commerce, 1972 *Survey of Environmental Activities of International Organizations*, 92d Cong., 2d sess., February 1972. Obviously these functions, at least in specifics, change over time. But their record provides a set of mileposts in the development of international environmental policies and points for comparison with later policies and programs. Additional references describing environment-related programs of particular agencies will be cited under notes to the chapters in which they receive detailed consideration.

For a comprehensive listing of environmental organizations see T. C. Trzyna and E. V. Coan, eds., *World Directory of Environmental Organizations* (Claremont: Center for California Public Affairs, Sequoia Institute [Sierra Club], 1976); Bruno Dente, ed., *Environmental Policy: In Search of New Instruments* (Dordrecht, Netherlands: Kluwer Academic Publishers, 1995).

2. Livingston Hartley, "Challenge to the Environment: Some International Implications," *Orbis* 14 (Summer 1970): 490–99; J. R. Huntley, *Man's Environment and the Atlantic Alliance*, 2d ed. (Brussels: NATO Information Service, 1972); Gunnar Randers, "NATO's International Governmental Cooperation on Environmental Management," in *Managing the Environment: International Economic Cooperation for Pollution Control*, ed. Allen V. Kneese et al. (New York: Praeger, 1971), pp. 343–48; and Maria Rita Saulle, *NATO and Its Activities: A Political and Judicial Approach on Consultation* (Dobbs Ferry, N.Y.: Oceana Publications, 1979). For accounts of the Committee on Challenges of Modern Society (CCMS), see *NATO Letter* 18 (January, February, May, July–August, and December 1970).

3. Cited for the historical record only: Council for Mutual Economic Assistance (CMEA) Secretariat, *Information on Co-operation of the CMEA Member Countries in the Field of Environmental Protection and Improvement and the Related Rational Use of Natural Resources* (Moscow: CMEA Secretariat, February 1977). See also Linda Ervin, comp., *The Council for Mutual Economic Assistance: A Selective Bibliography* (Ottawa, Canada: Carleton College, Norman Patterson School of International Affairs, 1975). For agreement between CMEA and UNEP, see *Uniterra* 4 (November–December 1979): 8.

4. See Organization of African Unity (OAU), *Report of the Second Meeting of the Committee of the African Convention for the Conservation of Nature and Natural Resources: The Council of Ministers Eleventh Ordinary Session, held in Algiers, September 1968* (Addis Ababa: OAU Secretariat, 1968), pt. 2, CM/232; "African Convention," *IUCN Bulletin*, n.s., 2 (October–December 1968): 68; Wolfgang E. Burhenne, "The African Convention for the Conservation of Nature and Natural Resources," *Biological Conservation* 2 (January 1970): 105–14; and Kai Curry-Lindahl, "The New African Conservation Convention," *Oryx* 10 (September 1969): 116–29. On the Mweka College, see Eric Robbins, *The Ebony Ark: Black Africa's Battle to Save Its Wildlife* (New York: Taplinger, 1970), pp. 1–2.

5. See *Uniterra* (OAU special supplement, May–June 1981): entire issue. See also Paul Richards, "African Environment and the IAI," *IAI Bulletin* 46, no. 1 (1976): 3. International African Institute (IAI) publications frequently relate to environmental issues, e.g., Paul Richards, ed., *African Environment: Problems and Perspectives* (London: IAI, 1975).

6. See U.S. Council on Environmental Quality, "International Events," in *Environmental Quality, 1976: The Seventh Annual Report of the Council on Environmental Quality, September 1976* (Washington, D.C.: Government Printing Office, 1976), p. 144. And note the following reports published by OAS: *Final Report of the Inter-American Specialized Conference to Deal with Problems Relating to the Conservation of Renewable Natural Resources in the Western Hemisphere, held at Mar del Plata, Argentina, 18–22 October 1965*, OEA Series C/VI.9.2 (Washington, D.C.: Pan American Union/Organization of American States, 1966); and *Report of the Technical Meeting on Education and Training for the Administration of National Parks, Wildlife Reserves and Other Protected Areas, held at Merida, Venezuela, September 25–29, 1978*, SG/Series P/III.1. The latter report [in Spanish] was jointly financed by OAS and UNESCO (1978). See also OAS, *Report of the Meeting of Experts on Conservation of Marine Mammals and Their Ecosystems, held at Puerto Madryn, Argentina, 12–16 September 1977*, OEA Series J/XI (Washington, D.C.: Inter-American Council for Education, Science, and Culture, OAS, 1978), and *Seminar on Environment and Dams*, vols. 1–2 (Montevideo, Uruguay: University of the Republic and OAS, 1977 [report in Spanish]). For papers and resolutions on an early international conference to promote implementation of the 1940 Convention on Nature Protection and Wild Life Preservation in the Western Hemisphere, see U.S. Department of State, *Proceedings of the Inter-American Conference on Conservation of Renewable Natural Resources: Denver, Colorado, September 7–20, 1948*, publication no. 3382 (Washington, D.C.: Department of State, Division of Publications, Office of Public Affairs, 1949).

7. See OECD, *OECD History—Aims—Structure* (Paris: OECD, 1973); *OECD at Work for the Environment* (Paris: OECD, 1973); and *OECD and the Environment* (Paris: OECD, 1976). See also "The State of the Environment," *OECD Observer* 98 (May 1979); *OECD in Brief* (1990); and *Annual Reports*.

8. *Environment Policies for the 1980s* (Paris: OECD, 1980), pp. 60–61, and *OECD and the Environment* (Paris: OECD, 1986) (a valuable report, 220 pages in length).

9. *Activities of the OECD in 1977* (Paris: OECD Information Service, 1973).

10. *OECD at Work for the Environment*, p. 7.

11. *Activities of the OECD in 1977* (Paris: OECD Information Service, 1978), pp. 43–48.

12. Nancy K. Hetzel, *Environmental Cooperation among Industrialized Countries: The Role of Regional Organizations* (Washington, D.C.: University Press of America, 1980), pp. 61–71.

13. *The State of the Environment in OECD Member Countries* (Paris: OECD, 1979), pp. 9–14.

14. For historical details concerning OECD environment programs (and others in the Western Europe–North Atlantic areas), see, in addition to OECD reports, the following: Hetzel, *Environmental Cooperation among Industrialized Countries*; Marshall E. Wilcher, *Environmental Cooperation in the North Atlantic Area* (Washington, D.C.: University Press of America, 1980). OECD, *Environmental Problems in Frontier Regions* (Paris: OECD, 1979), Annex 1, pp. 55–90, summarizes agreements and institutional structures among the twenty OECD member states with common terrestrial frontiers (Japan and the United Kingdom, e.g., are not included).

15. See *The Environment: Challenges for the '80s: Proceedings of the Special Session of the OECD Environmental Committee held on 1st April, 1981, on "OECD and Policies for the '80s to Address Long-Term Environment Issues"* (Paris: OECD, 1981).

16. *Economic and Ecological Interdependence: A Report on Selected Environment and Resource Issues* (Paris: OECD, 1982). For environment-relevant publications of OECD see *OECD Publications Catalogue* (issued periodically) and *Environment: A List of Monographs*, updated on 10 February 1994.

17. For general background on UN environmental activities, see the annual *Yearbook of the United Nations* (New York: United Nations Office of Public Information); the annual *United Nations Handbook* (Wellington, New Zealand: Ministry of External Relations and Trade); Brian Johnson, *The United Nations System and the Human Environment* (Brighton, U.K.: University of Sussex, Institute for the Study of International Organization, 1971); Claude M. Stanley,

Environmental Management by the United Nations (Muscatine, Iowa: The Stanley Foundation, 1972); and Harry Winton, comp., *Man and Environment: A Bibliography of Selected Publications of the U.N. System* (New York: Unipub, 1972). For more detail on the structure of the UN system, see L. M. Goodrich, E. Hambro, and A. Simons, *Charter of the United Nations* (New York: Crowell, 1959); H. Nicholas, *The United Nations as a Political Institution*, 4th ed. (New York: Oxford University Press, 1971); Louis B. Sohn, *United Nations in Action, University Casebook Series* (Mineola, N.Y.: Foundation Press, 1968); United Nations, *United Nations System of Organizations: Members of the United Nations, the Specialized Agencies and the International Atomic Energy Agency, and the Contracting Parties of the General Agreement on Tariffs and Trade . . .* (New York: United Nations Office of Inter-agency Affairs and Co-ordination, 1976); *Editor's Guide to the United Nations* (New York: United Nations Association of the United States of America, 1978); Amos Yoder, *The Evolution of the U.N. System* (Bristol, Pa.: Crane Russak, 1989).

18. UN Office of Public Information, *Charter of the United Nations and Statute of the International Court of Justice.*

19. Ruth B. Russell, *The General Assembly: Patterns, Problems, Prospects* (New York: Carnegie Endowment for International Peace, 1970). For a personal and highly critical view of the politics of the General Assembly during the 1970s, see Daniel Patrick Moynihan, *A Dangerous Place* (Boston: Little, Brown, 1978).

20. UN Security Council, *Official Records*, Eighth Year, 629th Meeting, S/3108, 27 October 1953; and *Official Records*, Eighth Year, Supplement for October, November, and December 1959, S/3106, 12 October 1953; S/3116, 20 October 1953; S/3122, 23 October 1953.

21. UN Trusteeship Council, "Petition for the People of Palau Islands Concerning the Trust Territory of the Pacific Islands," T/Pet.10/121, 5 July 1977. For background, see Ken Brower, "Turning Paradise into Petrodollars," in *Earthworks: Ten Years on the Environmental Front*, ed. Mary Lou Van Deventer (San Francisco: Friends of the Earth, 1980), pp. 79–86.

22. Gerard J. Mangone, ed., *UN Administration of Economic and Social Programs* (New York: Columbia University Press, 1966), pp. 1–36.

23. E.g., United Nations, "Problems of the Human Environment: Report of the Secretary-General," E/4667, 26 May 1969.

24. Walter W. Sharp, "Program Coordination through the Economic and Social Council," in *UN Administration of Economic and Social Programs*, pp. 102–15; and *Conference on United Nations Procedures, held at Mohonk Mountain House, New Paltz, NY, 21–24 May 1976* (Muscatine, Iowa: Stanley Foundation, 1976), pp. 19–27.

25. Robert W. Gregg, "Program Decentralization through the Regional Economic Commissions," in *UN Administration of Economic and Social Programs*, pp. 231–48. For example, note the inclusion of major environmental regulations in the Economic Commission for Africa (ECA)/OAU Economic Summit in April 1980, which produced the Lagos Plan of Action and the ECA/UNEP "Seminar for Lawyers in the Development of Environmental Protection Legislation in the ECA Region," which took place in Addis Ababa in September-October of 1980. See *Environmental Policy and Law* 6 (November 1980): 171.

26. UN ECE, *ECE Symposium on Problems Relating to Environment, held at Prague, Czechoslovakia, May 2–15, 1971*. See also Jon McLin, "European Organizations and the Environment," pp. 1–11; and D. N. Leff, "A Meeting in Prague," *Environment* 13 (November 1971): 29–33.

27. See "Type of Organizations Included," *Yearbook of International Organizations, 1978*, 17th ed. (Brussels: Union of International Organizations and International Chamber of Commerce, 1978), pp. 5–9.

28. Peter I. Hajnal, *Guide to United Nations Organization, Documentation, and Publishing* (Dobbs Ferry, N.Y.: Oceana Publications, 1977), pp. 57–92.

29. *UN Chronicle* 16 (July–October 1979): 45–47; and Anne C. Roark, "UN Technology Meeting Lacked Clear Direction," *Science* 202 (21 September 1979): 1236–38.

30. *United Nations Handbook 1992* (Wellington, New Zealand: Ministry of External Relations and Trade), pp. 104–105.

31. "The Global Environment Facility," *Environmental Data Report 1993–94*. Prepared for UNEP by the GEMS Monitoring and Assessment Research Centre in cooperation with the World Resources Institute, Washington, D.C., and the U.K. Department of the Environment, London, pp. 384–94; "Instrument for the Establishment of the Restructured Global Environment Facility," *International Legal Material* 33 (March/September 1994): 1273–1310; and "Global Environment Facility (GEF)," 1991 *Annual Report of the Executive Director of UNEP*, pp. 4, 21.

32. Durwood Zaelke, Paul Orbuck, and Robert F. Hausman, eds., *Trade and the Environment: Law, Economics, and Policy* (Washington, D.C.: Island Press, 1993); Patrick Low, ed., *International Trade and the Environment*, World Bank Discussion Paper 159, (Washington, D.C.: World Bank, 1992); *The Environmental Effects of Trade* (Paris: OECD, 1992); and "The WTO is Born," *Focus: GATT Newsletter*, no. 107, special issue, May 1994. See also David Vogel, *Trading Up: Consumer and Environmental Regulation in a Global Economy* (Cambridge, Mass.: Harvard University Press, 1995).

33. Gheorghe Elian, *The International Court of Justice* (Leiden, Netherlands: A. W. Sijthoff, 1971); and John King Gamble and Dave D. Fischer, *The International Court of Justice: An Analysis of a Failure* (Lexington, Mass.: Lexington Books, 1976), bibliographical references; and Robert Y. Jennings, "The International Court of Justice after Fifty Years," *American Journal of International Law* 89, no. 3 (July 1995): 493–505.

34. Stuart F. Clayton, Jr., "The International Atomic Energy Agency: An Expanding Role in the Post-Chernobyl World," *North Carolina Journal of International Law and Commercial Regulation* 12 (Spring 1987): 269–75; Diana K. Brown, "Chernobyl: Its Implications for International Atomic Energy Regulation," *Michigan Yearbook of International Legal Studies* 9 (1988): 367–82; Richard E. Levy, "International Law and the Chernobyl Accident: Reflections on an Important but Imperfect System," *University of Kansas Law Review* (Fall 1987): 81–131; Michael A. Heller, "Chernobyl Fallout: Recent IAEA Conventions Expand Transboundary Nuclear Pollution Law," *Stanford Journal of International Law* (Summer 1987): 651–64. See also A. O. Adede, *The IAEA Notification and Assistance Conventions in Case of a Nuclear Accident: Landmarks in the History of the Multilateral Treaty-Making Process* (London: Graham and Trotman, 1987); also Christopher Flavin, *Reassessing Nuclear Power: The Fallout from Chernobyl*, Worldwatch Paper 75 (Washington, D.C.: Worldwatch Institute, 1987).

35. See descriptions of agencies in Peaslee, *International Government Organizations;* and De-Reeder, *Environmental Programmes of Intergovernmental Organizations.* See also Hebe Spaull, *The Agencies of the U.N.: A Survey of Economic and Social Achievements* (London: Ampersand, 1967), which is brief, popular, and dated but provides useful background information. Many agencies publish separate reports on their environmental activities that are cited where pertinent. For a more detailed account of the requirements for specialized agency status, see Bennett, *International Organizations*, pp. 244–46.

36. UN, FAO, International Technical Conference (Rome, 18 April–10 May 1955), *Papers Presented at the International Technical Conference on the Conservation of the Living Resources of the Sea*, A/Conf.10/7, 1956.

37. Bertil Bolin, "The ILO [International Labour Organisation] and the Working Environment," and Ernesto Mastromattea, "Safety and Health in Industrial Society," *Earth Law Journal* 2 (1976): 7–19; and Lawrence David Levien, "A Structural Model for a World Environmental Organization: The ILO Experience," *George Washington Law Review* 40 (March 1972): 464–95.

38. *IMCO: What It Is, What It Does, How It Works* (London: IMCO [Intergovernmental Maritime Consultative Organization], June 1968). The most comprehensive treatment is R. Michael M'Gonigle and Mark W. Zacher, *Pollution, Politics, and International Law: Tankers at Sea* (Berkeley: University of California Press, 1979). For current activities, see *IMO [International Maritime Organization] News.* See also "International Maritime Organization," *IMS [International Marine Science] Newsletter*, no. 32 (Summer 1982): 6, 8.

39. On the establishment of UNESCO and its beginnings, see articles by Reinhold Niebuhr

and Charles S. Ascher in *International Organization* 4 (1950): 3–26. For a general account as of its date of publication, see Walter H. C. Laves and Charles A. Thomson, *UNESCO: Purpose, Progress, Prospects* (Bloomington: Indiana University Press, 1957); *Man and His Environment: An Overview of UNESCO's Involvement* (Paris: UNESCO, 1979), previously published as chap. 7 of *Thinking Ahead: UNESCO and the Challenge of Today and Tomorrow* (Paris: UNESCO, 1977); and "Unesco's Role in Marine Science," *IMS Newsletter* 1, no. 32 (Summer 1982): 7–8. Current activities of UNESCO are reported in a number of periodicals, e.g., *Unesco Courier, Unesco Chronicle, Nature and Resources, IMS Newsletter, Impact of Science on Society,* and *Connect.*

40. DeReeder, *Environmental Programmes of Intergovernmental Organizations.*

41. See Hector R. Acuna, "The Pan American Health Organization's 75 Years of International Cooperation in Public Health," *WHO Chronicle* 31 (December 1977): 479–85. For background, see Neville M. Goodman, "The Pan American Sanitary Organization and Bureau," in *International Health Organizations and Their Work* (Philadelphia, Pa.: Blakiston, 1952), pp. 242–51.

42. Acuna, "75 Years of International Cooperation in Public Health"; and "Trends and Activities in the WHO Regions, 1976–77," *WHO Chronicle* 32 (May 1978): 175–92.

43. *A Brief Survey of the Activities of the World Meterological Organization Relating to Human Environment* (Geneva: WMO, 1970); David Arthur Davies, "The Role of WMO in Environmental Issues," in David A. Kay and Eugene B. Skolnikoff, eds., *World Eco-Crisis: International Organizations in Response* (Madison: University of Wisconsin Press, 1972), pp. 161–70, current activities are reported in the *WMO Bulletin.* See also "World Meteorological Organization," *IMS Newsletter* no. 32 (Summer 1982): 5–6.

44. *World Weather Watch,* WMO-183.TP.92 (Geneva: WMO, 1965). See also *WMO Bulletin* (October 1968): 172–81; Thomas F. Malone, "World Weather Watch," *Science* 154 (4 November 1966): 678–79; and "New Dimensions of International Cooperation in Weather Analysis and Prediction," *Bulletin of the American Meteorological Society* 49 (December 1968): 1134–40.

45. See Robert E. Stein and Brian Johnson, *Banking on the Biosphere: Environmental Procedures and Practices of Nine Multilateral Development Agencies* (Lexington, Mass.: Lexington Books, 1979); and James E. Lee, "Environmental Considerations: Development Finance," *International Organization* 26 (Spring 1972): 337–47. Publications by the International Bank for Reconstruction and Development (IBRD) regarding its environmental policies are referenced in notes to Chapter 6.

46. *International Legal Materials* 19 (July 1980): 837–38. For an account of CIDIE and the Declaration, see "CIDIE to Address Looming Crisis," *Our Planet* 1, no. 2/3 (1989): 11.

47. Philippe G. Le Prestre, "The Ecology of the World Bank: Uncertainty Management and Environmental Policy," Ph.D. diss., Indiana University, 1982; and *The World Bank and the Environmental Challenge* (Selinsgrove, Pa.: Susquehanna University Press, 1989); and Mohamed T. El-Ashry, "The World Bank's Post-Rio Strategy," *EPA Journal* (April–June 1993): 22–25.

48. For distinction between intergovernmental and nongovernmental organizations, see "Types of Organizations Included," *Yearbook of International Organizations;* United Nations Economic and Social Council, "Review of Consultative Arrangements with Non-governmental Organizations," (Resolution 288 [X], 27 February 1950), *Official Records, Fifth Year, Tenth Session, Resolutions, Supplement No. 1* (Lake Success, N.Y.: United Nations, 1950). There is a large literature on voluntary international associations. See the periodical *International Associations;* and Kjell Skjelsbark, "The Growth of International Nongovernmental Organizations in the Twentieth Century," in *Transnational Relations and World Politics,* ed. Robert O. Keohane and Joseph S. Nye, Jr. (Cambridge, Mass.: Harvard University Press, 1971), pp. 70–92; Kai Curry-Lindahl, "Background and Development of International Conservation Organizations and Their Role in the Future," *Environmental Conservation* 5 (Autumn 1978): 163; and *Directory of Non-Governmental and Development Organizations in OECD Member*

Countries (Paris: OECD, 1992). See also Russell J. Dalton, *The Green Revolution: Environmental Groups in Western Europe* (New Haven, Conn.: Yale University Press, 1994).

49. United Nations Institute for Training and Research (UNITAR), *Non-Governmental Organizations in Economic and Social Development: A UNITAR Conference at Schloss Hernstein, Austria*, n.d. [1975?]. Includes colloquium at New York, 6 October 1975.

50. UNEP, "Relationships with Non-Governmental Organizations: Note by the Executive Director," UNEP/GC/77, 26 January 1976.

51. See the leaflet by Environment Liaison Centre, *Environmental Activities of Non-Government Organizations (NGOs) Related to UNEP Programmes: Report and Directory* (Nairobi, 1980); and Tryzyna and Coan, *World Directory of Environmental Organizations.*

52. International Council of Scientific Unions (ICSU), *Year Book 1995* (Paris: ICSU, 1995); Wallace W. Atwood, "International Council of Scientific Unions," *Science* 128 (19 December 1958): 1558–61; Harold S. Jones, "The International Council of Scientific Unions," *Endeavor* 18 (April 1959): 88–92. See also U.S. Congress, House Committee on Foreign Affairs, *International Council of Scientific Unions and Certain Associated Unions: Hearings before the Subcommittee on International Organizations and Movements of the Committee on Foreign Affairs*, 89th Cong., 1st sess., 1956, p. 40. For a more general treatment, see Joseph Needham, *Science and International Relations: 50th Boyle Lecture at Oxford* (London: Blackwell Scientific Publications, 1949), note "International Unions," pp. 6–12.

53. International Council of Scientific Unions, *SCOPE: Scientific Committee on Problems of the Environment* (Paris: Scientific Committee on Problems of the Environment, SCOPE Secretariat, January 1980), p. 1. Also *Encyclopedia of Associations*, 16th ed. (Detroit, Mich.: Gale Research, 1981), vol. 1, no. 4887, and the annual *International Council of Scientific Unions Yearbook* (Miami, Fla.: ICSU Press).

54. *International Council of Scientific Unions Yearbook 1989*, p. 123. See Thomas F. Malone, "Mission to Planet Earth: Integrating Studies of Global Change," *Environment* 28, no. 8 (October 1986): 6–11, 39–41; also T. F. Malone and J. C. Roederer, eds., *Global Change*, the proceedings of a symposium sponsored by the International Council of Scientific Unions (ICSU) during its twentieth General Assembly in Ottawa, Canada, on 25 September 1984 (Cambridge: Cambridge University Press for ICSU, 1985); for the book that seemingly catalyzed this effort, see J. E. Lovelock, *Gaia: A New Look at Life on Earth* (New York: Oxford University Press, 1979).

55. *IUCN Bulletin* 19, 7–12, Special Issue (1988): 21.

56. For a firsthand report on the founding of IUCN, see Harold J. Coolidge Jr., "The Birth of a Union," *National Parks Magazine* 23 (April–June 1949): 35–38. Accounts of the work and history of IUCN have been published in the following issues of the journal *Biological Conservation:* E. J. H. Berwick, "The International Union for Conservation of Nature and Natural Resources: Current Activities and Situation," 1 (April 1969): 191–99; J. P. Harroy, "L'Union Internationale pour la Conservation de la Nature et de ses Resources: Origine et constitution," 1 (January 1969): 106–10. See also a descriptive leaflet, *This Is IUCN*, published by the Union; the *IUCN Bulletin*; and the *Yearbook.* See R. Boardman, *International Organization and the Conservation of Nature* (Bloomington: Indiana University Press, 1981); and articles in *Environmental Conservation* 10, no. 1 (Spring 1983). For what is probably the initial publication of the Union and a summary of the Fountainebleau conference, see *International Union for the Protection of Nature* (Brussels: Imprimeri M. Hayez, for the Union, 1948). See also *IUCN Bulletin* 20 1–3, 7–12, 1989, 40th anniversary issue. Also John McCormick, *Reclaiming Paradise: The Global Environmental Movement* (Bloomington: Indiana University Press, 1989).

57. For the Arusha Conference, see *IUCN Bulletin*, n.s., 2 (August 1961): 1, 7; and ibid. (December 1961): 1, 7–8. For subsequent conferences, see the following: on the Conference on Conservation of Nature and Natural Resources in Tropical South East Asia (Bangkok, Thailand, 1965) see Lee M. Talbot, "The Bangkok Conference," *IUCN Bulletin*, n.s., 19 (April–June 1966): 1–2; on the Conference on Renewable Natural Resources (San Carlos de Bariloche,

Argentina, 1968) see E. H. H. Berwick, "The Bariloche Conference," *IUCN Bulletin*, n.s., 2 (April–June 1968): 49–50, plus special supplement giving text of resolution adopted; on the Conference on Productivity and Conservation in Northern Circumpolar Lands (Edmonton, Canada, 1970) see W. A. L. Fuller, "Conference on Productivity and Conservation in Northern Circumpolar Lands," *IUCN Bulletin*, n.s., 2 (April–June 1970): 125–26; on the Conference on the Rational Utilization and Conservation of Nature on the Island of Madagascar (Tananarive, Malagasy Republic, 1970) see *IUCN Bulletin*, n.s., 2 (July–September 1970): 133–34; and on the South Pacific Conference on National Parks and Reserves (Wellington, New Zealand, 1975) see *IUCN Bulletin*, n.s., 6 (April 1975): special supplement.

58. See *World Wildlife Yearbook, 1961–67* (Morges, Switzerland: World Wildlife Fund [WWF], 1968); and *WWF Yearbook 1982* (Gland, Switzerland: WWF, 1982), a comprehensive survey of all aspects of wildlife protection.

59. Max Nicholson, *The New Environmental Age* (Cambridge: Cambridge University Press, 1987), p. 44. Also *Yearbook of International Organizations 1978*, A1722. For an account of the establishment of the International Council for Bird Preservation (ICBP) and the negotiations preceding the Paris Treaty of 1902, see *Bulletin of the International Committee for Bird Protection* (New York: ICBP, 1927). The *XIII Bulletin of the International Council for Bird Protection* lists sixty-four national sections, several of which, however, were under reorganization. The International Waterfowl Research Bureau located in England is sponsored by the ICBP. The official name of the ICBP has been changed twice; see Boardman, *International Organization*, p. 30.

60. Principles of international law do not change rapidly. New applications arise, but many treaties, chronologically dated, continue to be relevant, informative, and often pertinent to recent events.

For general treatments of international environmental law or for discussion of its concepts and principles, see the following: David Hunter, Julia Sommer, and Scott Vaughan, *Concepts and Principles of International Environmental Law: An Introduction* (Nairobi: UNEP, 1994); Phillipe Sands et al., *Principles of International Environmental Law*, 3 vols. (Manchester, U.K.: Manchester University Press, 1994–1995); Phillipe Sands, ed., *Greening of International Law* (New York: New Press, 1994); Lakshaman D. Guruswamy et al., *International Environmental Law and World Order: A Problem-Centered Coursebook* (St. Paul, Minn.: West Publishers, 1994); Harold Hohmann, ed., *Basic Documents of International Environmental Law* (London: Graham and Trotman, 1992); Alexandre Charles Kiss and Dinah Shelton, *International Environmental Law* (London: Graham and Trotman, 1991); Frances Cairncross, "Agora," *American Journal of International Law* 84, no. 1 (January 1990): 190–212; Edith Brown Weiss, *In Fairness to Future Generations: International Law, Common Patrimony, and Intergenerational Equity* (Tokyo and Dobbs Ferry, N.Y.: United Nations University and Transnational Publishers, 1989); Frances Cairncross, "The Environment: The Politics of Posterity," *Economist* (2 September 1989): 3–18; Alexandre Charles Kiss, ed., *Selected Multilateral Treaties in the Field of the Environment*, 2 vols. (Nairobi: UNEP, 1983); *Trends in International Environmental Policy and Law*, Michael Bothe, project coordinator (Berlin: Erich Schmidt Verlag, 1980), a project in collaboration with the IUCN Environmental Law Centre; Jan Schneider, *World Public Order of the Environment: Toward an International Ecological Law and Organization* (Toronto: University of Toronto Press, 1979); Alfred Rest, *International Protection of the Environment and Liability: The Legal Responsibility of States and Individuals in Cases of Transfrontier Pollution* (Berlin: Erich Schmidt Verlag, 1978); Aida Luisa Levin, *Protecting the Human Environment: Procedures and Principles for Preventing and Resolving International Controversies* (New York: United Nations Institute for Training and Research, 1977); Bo Johnson, *International Environmental Law* (Stockholm: Liberforlag, 1976); Alexandre Charles Kiss, *Survey of Current Developments in International Environmental Law*, Environmental Policy and Law Paper, no. 10 (Morges, Switzerland: IUCN, 1976); *Selected Documents on International Environmental Law: Selected and Arranged by the British Institute of International Law and Comparative Law*

(Dobbs Ferry, N.Y.: Oceana Publications, 1975); Alexandre Charles Kiss, ed., *The Promotion of the Environment and International Law* (Leiden, Netherlands: Sijthoff, Hague Academy of International Law, 1975); Ludwik A. Teclaff and Albert E. Utton, eds., *International Environmental Law* (New York: Praeger, 1974); and John Lawrence Hargrove, ed., *Law, Institutions and the Global Environment* (Dobbs Ferry, N.Y.: Oceana Publications, 1972).

See also UNEP, *Compendium of Legislative Authority* (Oxford: Pergamon Press, 1978) and UNEP, "International Conventions and Protocols in the Field of the Environment: Report of the Executive Director," UNEP/GC/61/Add.2, 5 February 1976. These status reports are periodically submitted to the Governing Council of UNEP pursuant to its decision 24 (III) of 30 April 1975. These reports were subsequently collated in the *Register of International Conventions and Protocols in the Field of the Environment*.

For a continuing listing of conventions and protocols, see *Register of International Conventions and Protocols in the Field of Environment*, UNEP/GC/Inf.5, 7 February 1977; Supplement 1 (15 December 1977); Supplement 2 (7 March 1979); Supplement 3 (15 January 1980); Supplement 4 (27 January 1981); Supplement 5 (26 November 1981). A sixth supplement was published in 1983, but because the series had become voluminous and difficult to obtain, a consolidated compilation entitled *Register of Treaties and Other Agreements in the Field of the Environment* (May 1985) was issued (UNEP/GC/11 Rev. 1). Supplements have followed. Compilations published in the United States include *Environmental Laws and Treaties: Reference Guide* (Congressional Research Service, Library of Congress) and the *International Environment Reporter* (Washington, D.C.: Bureau of National Affairs).

61. A clear and concise account of international lawmaking through the United Nations system may be found in Hannes Jonssan, *Friends in Conflict: The Anglo-Icelandic Cod Wars and the Law of the Sea* (London: C. Hurst, 1982), pp. 23–30.

62. *ICEL References*, an international bibliographical service, has been published since 1969, 4 issues per year.

63. "Ad Hoc Meeting of Senior Government Officials Expert in Environmental Law: Conclusions and Recommendations of Montevideo," *Environmental Policy and Law* 8 (January 1982): 31–35. The Montevideo meeting has generated several ad hoc working groups of experts for various aspects of environmental law. See UNEP, *1984 Annual Report of the Executive Director*, pp. 30–32. More recent general works include Réné-Jean Dupuy, ed., *The Future of the International Law of the Environment: Workshop, The Hague, 12–14 November 1984* (Dordrecht, Netherlands: Martinus Nijhoff, 1985).

64. For international law in relation to environmental disasters, see Ved P. Nanda and Bruce Bailey, *Challenges for International Environmental Law: Seveso, Bhopal, Chernobyl, the Rhine and Beyond* (Washington, D.C.: World Peace through Law Center, 1987); also in *Law/Technology* 21 (Summer 1989): 1–49.

65. Linda A. Malone, "The Chernobyl Accident: A Case Study in International Law Regulating State Responsibility for Transboundary Nuclear Pollution," *Columbia Journal of Environmental Law* 12, no. 2 (1987): 203–41; Kenneth Jost, "Chernobyl Mishap: Payments Unlikely in Wake of Fallout," *Chicago Daily Law Bulletin* 132 (13 May 1986): 1; "The Chernobyl Nuclear Plant Accident and International Law," *Australian Law Journal* (October 1986): 586–87.

66. See *American Journal of International Law* 33 (January 1939): 182; ibid. 35 (October 1941): 684; and A. C. Kiss, "Trail Smelter Case" in *Survey of Current Developments* (Morges, Switzerland: International Union for Conservation of Nature and Natural Resources, 1975), pp. 4346. A. P. Rubin, "Pollution by Analogy: The Trail Smelter Arbitration," *Oregon Law Review*, no. 3, pt. 1 (1971): 259–82; D. H. Dinwoode, "The Politics of International Pollution Control: The Trail Smelter Case," *International Journal* 27 (Spring 1972): 219–35; and Charlotte K. Goldberg, "The Garrison Diversion Project: New Solutions for Transboundary Disputes," *Manitoba Law Journal* 11, no. 2 (1981): 177–89.

67. International Court of Justice, *Reports of Judgments, Advisory Opinions and Orders* (1973): "Nuclear Test Cases" (*Australia v. France*), 22 June 1973, pp. 99–133; (*New Zealand v. France*),

23 June 1973, pp. 135–64. Also Alastair Mattheson, "French Nuclear Testing and International Law," *Rutgers Law Review* 24, no. 1 (1969): 144–70. See also L. F. E. Goldie, "The Nuclear Test Cases: Restraints on Environmental Harm," *Journal of Maritime Law and Commerce* 5 (1973–74): 491.

68. Louis B. Sohn, "The Stockholm Declaration on the Human Environment."

69. Arthur H. Westing, *Warfare in a Fragile World: The Military Impact on the Human Environment* (London: Taylor and Francis, 1980); "Symposium on Environmental Crimes: Crimes against the Environment, Current Policies and Future Trends in Environmental Criminal Enforcement," *Columbia Journal of Environmental Law* 16, no. 2 (1991): 201–34; UN General Assembly, Ad Hoc Committee on the Establishment of an International Criminal Court, 3–13 April 1995. *Summary of Proceedings* A/AC.244/2.21 April 1995; and James Crawford, "The ILC Adopts a Statute for an International Criminal Court," *American Journal of International Law* 89, no. 2 (April 1995): 404–16.

70. Law Reform Commission of Canada, *Protection of Life: Crimes against the Environment*, Working Paper 44 (Ottawa, Canada: Law Reform Commission, 1985).

On the environmental effects of the Iraqi war widely condemned as criminal and violative of numerous international legal provisions see: *Kuwait: Report to the Secretary-General on the Scope and Nature of Damage Inflicted on the Kuwaiti Infrastructure during the Iraqi Occupation* (United Nations Public Information Department, 1991); *The Environmental Role of the National Oceanic and Atmospheric Administration and the U.S. Coast Guard in the Persian Gulf Conflict*, Hearing in the U.S. House of Representatives, 102d Congress (Oversight Hearing on the Ecoterrorism Inflicted as a Result of the Persian Gulf War) 17 October 1991; *The Environmental Aftermath of the Gulf War*, A Report for the U.S. Senate Gulf Pollution Task Force, Committee on Environment and Public Works, 2 March 1992; Margaret T. Okorodudu-Fubara, "Oil in the Persian Gulf: Legal Appraisal of an Environmental Warfare," *St. Mary's Law Journal* 23, no. 15 (1991): 125–218; Anthony Leibler, "Deliberate Wartime Damage: New Challenges for International Law," *California Western International Law Journal* 23 (Fall 1992): 67–137; Paul C. Szasz, "The Gulf War: Environment as a Weapon," *American Society of International Law: Proceedings of the 85th Annual Meeting, Washington, D.C., April 17–20, 1991* (Washington, D.C.: ASIL, 1991).

71. See N. Papadakis, *International Law of the Sea: A Bibliography* (The Hague: Martinus Nijhoff, 1980). A selective list of references on the public order of the oceans and the law of the sea would include: R. P. Anand, *Origin and Development of the Law of the Sea: History of International Law Revisited* (The Hague: Martinus Nijhoff, 1983); Ross D. Eckert, *The Enclosure of Ocean Resources: Economics and the Law of the Sea* (Stanford, Calif.: Hoover Institution Press, 1979); John Lawrence Hargrove, ed., *Who Protects the Ocean?* (St. Paul, Minn.: West Publishing Co., 1976); Ann L. Hollick, *U.S. Foreign Policy and the Law of the Sea* (Princeton, N.J.: Princeton University Press, 1981); Douglas M. Johnston, ed., *The Environmental Law of the Sea*, Environmental Policy and Law Paper, no. 18 (Gland, Switzerland: IUCN, 1982); Myres S. McDougal and William T. Burke, *The Public Order of the Oceans* (New Haven, Conn.: Yale University Press, 1962); Gerard J. Mangone, *Law for the World Ocean* (London: Stevens and Sons, 1981); K. R. Simmonds, ed., *Cases on Law of the Sea* (Dobbs Ferry, N.Y.: Oceana Publications, 1976); Rao P. Sreenivasa, *The Public Order of Ocean Resources: A Critique of the Contemporary Law of the Sea* (Cambridge, Mass.: MIT Press, 1975); Don Walsh, ed., *The Law of the Sea: Issues in Ocean Resource Management* (New York: Praeger, 1979).

72. "Convention on the High Seas," 29 April 1958, *United Nations Treaty Series*, vol. 450 (1963), no. 6465, p. 82; or U.S. Department of State, *U.S. Treaties and Other International Agreements*, vol. 13, pt. 2, 2319, TIAS No. 6465 (Article 25).

73. Rudolph Preston Arnold, "The Common Heritage of Mankind as a Legal Concept," *International Lawyer* 9, no. 1 (1975): 153–58; and Zlatibor R. Milovanovic, "National Sovereignty, the Common Heritage of Mankind and the Role of the International Seabed Authority: Towards a New Law of the Sea," Ph.D. diss., Temple University, 1978.

74. On UNCLOS III, see Bernard H. Oxman, "The Third United Nations Conference on the

Law of the Sea: The Ninth Session (1980)," *American Journal of International Law* 75 (April 1981): 211–56; *San Diego Law Review* 19, no. 3 (April 1982): entire issue; ibid. 20, no. 3 (1983): entire issue; Philip Allott, "Power Sharing in the Law of the Sea," *American Journal of International Law* 77 (January 1983): 1–30; and Lawrence A. Howard, "The Third United Nations Conference on the Law of the Sea/Custom Dichotomy," *Texas International Law Journal* 16, no. 3 (1981): 321–45.

75. Bernard D. Nossiter, "Sea Law Signed by 117 Nations; U.S. Opposes It; 46 Other Countries Also Refuse to Back Treaty," *New York Times*, 11 December 1982, late ed., pp. 1, 6. See also editions of 9 December, p. A10, and 11 December, p. A11. For the text of the treaty, see Third United Nations Conference on the Law of the Sea, *United Nations Convention on the Law of the Sea*, A/CONF.62/122 (New York: United Nations, 7 October 1982); and *Final Draft Act of the Third United Nations Conference on the Law of the Sea*, A/CONF.62/121 (New York: United Nations, 21 October 1982). For the position taken by the United States, see U.S. Congress, House Committee on Merchant Marine and Fisheries, *Law of the Sea: Hearings before the Subcommittee on Oceanography and the Committee on Merchant Marine and Fisheries, House of Representatives*, 97th Cong., 22 October 1981; 23 February, 20, 27 July 1982 (Washington, D.C.: Government Printing Office, 1982).

76. Jan Schneider, *World Public Order of the Environment Towards an International Ecological Law and Organization* (Toronto, Canada: University of Toronto Press, 1979), p. 199.

77. R. Van Ermen, "NGOs—A Force to Be Reckoned With," *Naturopa*, no. 64 (1990): 18.

78. Richard B. Bilder, *The Settlement of International Disputes*, University of Wisconsin Sea Grant College Program, Technical Report 231 (Madison: University of Wisconsin Press, 1976), p. 60.

79. Cyril Black and Richard Falk, "Introduction: The Structure of the International Environment," *The Future of the International Legal Order*, 4 vols. (Princeton, N.J.: Princeton University Press, 1972).

80. Peter H. Sand, ed., *The Effectiveness of International Environment Agreements: A Survey of Existing Legal Instruments* (Cambridge: Grotius, 1992) and *Lessons Learned in Global Environmental Governance* (Washington, D.C.: World Resources Institute, 1990).

8. Transnational Regimes and Regional Agreements

1. The concept is familiar chiefly to students of international organization, among whom there are differences in definition. See Thomas Gehring, *Dynamic International Regimes: Institutions for International Environmental Governance* (Frankfurt am Main: Peter Lang, 1994); Volker Rittberger, ed., *Regime Theory and International Relations* (Oxford: Clarendon Press, 1993); Oran Young, *International Cooperation: Building Regimes for Natural Resources and the Environment* (Ithaca, N.Y.: Cornell University Press, 1989); Stephen Krasner, ed., *International Regimes* (Ithaca, N.Y.: Cornell University Press, 1983).

2. See *IUCN Bulletin*, n.s., 10 (August–September 1979): 69, 77; and *IUCN Bulletin*, n.s., 12 (March–April 1981): 12. Peru and Bolivia were subsequently joined by Argentina, Chile, and Ecuador in a treaty known as the La Paz Convention. See note 14 for Canada–United States agreements, and *World Environment Report* 9 (15 May 1983).

3. B. P. Uvarov, "Efforts to Control Locusts in Africa Described," *Science* 130 (4 December 1959): 1564–65; and Stanley Baron, "No Frontier in the Fight against the Desert Locust," *Ceres* 1 (September–October 1968): 32–42. For treaties establishing international organizations for locust control, see the following references in Amos J. Peaslee, *International Governmental Organizations: Constitutional Documents*, 3d ed., rev. (The Hague: Martinus Nijhoff, 1974), pt. 2: "Desert Locust Control Organization for Eastern Africa," pp. 58–59; "International African Migratory Locust Control Organization," pp. 159–66; "International Migratory Locust Control Organization for Central and Southern Africa," pp. 292–93; and "Joint Anti-Locust and Anti-Aviarian Organization," pp. 536–37.

4. "Commission for Controlling the Desert Locust in the Eastern Region of Its Distribution

in South-West Asia," in Amos J. Peaslee, *International Governmental Organizations: Constitutional Documents*, 3d ed., rev., pt. 2: 36–46.

5. See United Nations, "Legal Problems Relating to the Utilization and Use of International Rivers," in *Yearbook of the International Law Commission*, 1974, A/5409 (New York, 1976) vol. 2, pt. 2: 33–264, and UN General Assembly, "Report of the International Law Commission of the Work of Its Twenty-Ninth Session, 9 May–29 July 1977," *Official Records*, Thirty-Second Session, Supplement no. 10, A/32/10, p. 304; and "Report of the International Law Commission on its Thirty-Fifth Session, 3 May–22 July 1983," Official Records, Thirty-eighth Session, Supplement no. 10, A/38/10, pp. 140–83. See also Ralph Zacklin and Lucius Caflisch, eds., *The Legal Regime of International Rivers and Lakes* (The Hague: Martinus Nijhoff, 1981). The definitive document on the law of international waters is the *Helsinki Rules on the Uses of International Waters*, adopted by the International Law Association at its fifty-second conference, held in Finland, August 1966, published by the International Law Association (London, 1967).

6. See J. D. Chapman, ed., *The International River Basin: Proceedings of a Seminar on the Development and Administration of the International River Basin* (Vancouver, Canada: University of British Columbia, 1963); David C. LeMarquand, *International Rivers: The Politics of Cooperation* (Vancouver, Canada: University of British Columbia, Westwater Research Centre, 1977); Albert Lepawsky, "International Development of River Resources," *International Affairs* 39 (October 1963): 533–50; A. H. Carretson, C. J. Olmstead, and R. D. Hayton, *The Law of International Drainage Basins* (Dobbs Ferry, N.Y.: Oceana Publications, 1967); B. R. Chauhan, *Settlement of International Water Law Disputes in International Drainage Basins* (Berlin: Erich Schmidt Verlag, 1981); Henry C. Hart, *Administrative Aspects of River Valley Development* (New York: Asia Publishing House, 1961); M. Kassas, "An Environmental Science Programme for an International River Basin: A Case Study," in *Water Management for Arid Lands in Developing Countries*, ed. Asit K. Biswas et al. (New York: Pergamon Press, 1980), pp. 79–89; and Gilbert F. White, *Environmental Effects of Complex River Development* (Boulder, Colo. Westview Press, 1977). An extensive series of studies of multipurpose river basin development was undertaken and published by the Economic Commission for Asia and the Far East, e.g., *Multi-Purpose River Basin Development*, Parts 2A, 2B, and 2C (New York and Bangkok, 1955–59). These publications were a part of the United Nations Flood Control Series. For specific rivers, see *International Rivers–Some Case Studies: Congo, Danube, Indus, Jordan, Niger, Nile, Mekong, Rio de la Plata*, Occasional Paper no. 1 (Bloomington: Indiana University Department of Geography, 1965). See also Peter Beaumont, "The Euphrates River—An International Problem of Water Resources Development," *Environmental Conservation* 5 (Spring 1978): 35, and "International Water Resources Law: Report of the Committee," *Report of the Sixtieth Conference of the International Law Association*, Montreal 1982, pp. 531–52. This conference also included reports on legal aspects of the conservation of the environment, pp. 157–82; space law, 479–530; and air law 558–593.

7. J. P. Chamberlain, *The Regime of International Rivers: Danube and Rhine* (New York: Columbia University Press, 1923).

8. See Alexandre Charles Kiss and M. Prieur, "Institutional and Administrative Practice Relating to the Management of the Environment and Land-Use Planning in the Frontier Regions of the Upper Rhine Basin," in *Environmental Protection in Frontier Regions*, pp. 23–60.

9. See *Yearbook of International Organizations, 1978*, 17th ed., A1538. For a more complete account of the Saar-Moselle-Mid-Rhine region, see Uta Hulshoff, "Protecting the Environment in the Saar-Lorraine Region," in *Environmental Protection in Frontier Regions*, pp. 266–81; "International Commission for the Protection of the Moselle against Pollution"; Axel Gosseries, "The Scheldt and Meuse Rivers," *European International Law Review* 4, no. 1 (January 1995): 9–14; *International Legal Materials* 34 (1995): 851–53; and "International Commission for the Protection of the Rhine," in Peaslee, *International Governmental Organizations*, pt. 5: 423–25, 428–29.

10. "Niger River Commission—Summary," "Organization and Development of the Senegal River," and "Convention Establishing the Organization of the Senegal River," in Peaslee, *International Governmental Organizations*, pt. 5: 558–59, 580–82. See also *Yearbook of International Organizations*, 1978, A3036; and "Senegal River Scheme Faces Strong Criticism: Environmental Impact Cited," *Ceres* 15 (January–February 1982): 8–10.

11. See *Africa South of the Sahara 1981–82*, 11th ed. (London: Europa Publications, 1982), p. 124; "Lake Chad Basin Commission," *Yearbook of International Organizations*, 1978, A4496. Peter H. Sand, "Development of International Water Law in the Lake Chad Basin," in *Zeitschrift für Auslandischer Öffentliches Recht and Volkerrecht* 34, no. 1 (April 1974): 52–82. Note detailed referencing regarding riparian regimes in general and those of Africa in particular. See also B. O. Tonwe, "The Lake Chad Basin Commission," *Journal of African Law* 16 (Autumn 1972): 343–44.

12. Gilbert F. White, "The Mekong Plan," *Scientific American* 208 (April 1963): 49–59; and White, "Vietnam: The Fourth Course," *Bulletin of the Atomic Scientists* 20 (December 1964): 6–10. See also Henri Lorgere, "The Mekong Delta Model," *Science Journal* 4 (July 1968): 70–75. For historical background, see A. D. Burnett, "The Mekong and the Rivers of Southeast Asia," in *International Rivers—Some Case Studies*, pp. 68–77. For more recent activities, see [Interim] Committee for Coordination of Investigations of the Lower Mekong Basin, *Annual Report, 1982* (Bangkok: Economic and Social Commission for Asia and the Pacific [ESCAP], 1982). The interim committee has also published *Environmental Impact Assessment: Guidelines for Application to Tropical River Basin Development* (Bangkok: ESCAP, 1982), and a periodical, *Mekong News*. See also Agreement on the Cooperation for Sustainable Development in the Mekong Basin, April 1995, *International Legal Materials* 34 (July 1995): 864–80.

13. For text of treaty, see James Barros and Douglas M. Johnson, *The International Law of Pollution* (New York: Free Press, 1974), pp. 165–70. See also *Yearbook of International Organizations*, 1981, 19th ed., FO968-fg. For background, see Robert Dean Sawvell, "The Indus River System," in *International Rivers—Some Case Studies*, pp. 54–67 (note references); and Aloys A. Michel, *The Indus Rivers* (New Haven, Conn.: Yale University Press, 1967).

14. For text of treaty, see Charles I. Bevans, *Treaties and other International Agreements of the United States of America, 1776–1949*, 12 vols. (Washington, D.C.: Department of State, 1968–74), vol. 12, pp. 319–27 (see Chapter 2, note 19). On Canada–United States environment-related agreements generally, see L. M. Bloomfield and Gerald F. Fitzgerald, *Boundary Waters Problems of Canada and the United States: The International Joint Commission, 1912–1958* (Toronto: Carswell, 1958); J. L. Callum, "The International Joint Commission," *Canadian Geographical Journal* 72 (March 1966): 76–87; Don C. Piper, *The International Law of the Great Lakes: A Study of Canadian–United States Cooperation* (Durham, N.C.: Duke University Press, 1967); Matthew E. Welsh, "The Work of the International Joint Commission," *Department of State Bulletin* (23 September 1968): 311–14; Richard B. Bilder, "Controlling Great Lakes Pollution," *Michigan Law Review* 70 (January 1972): 469–556; O. P. Dwivedi, "The International Joint Commission: Its Role in United States Canada Boundary Pollution Control," *International Review of Administrative Sciences* 40, no. 4 (1974): 369–76; Maxwell Cohen, "The International Joint Commission: Yesterday, Today, and Tomorrow," in *Proceedings of the Canada–United States Natural Resources and Environmental Symposium*, ed. John E. Carroll (Concord: University of New Hampshire Institute of Natural and Environmental Resources, 1978); Don Munton and Susan Eros, *Political and Legal Aspects of a Canada–United States Air Quality Accord* (Toronto: Canadian Institute of International Affairs, 1982); John E. Carroll and Newell B. Mark, "On Living Together in North America: Canada, the United States, and International Environmental Relations," *Denver Journal of International Law and Policy* 12 (Fall 1982): 35–50; John E. Carroll, *Environmental Diplomacy: An Examination and Prospective of Canadian-United States Transboundary Environmental Relations* (Ann Arbor: University of Michigan Press, 1983); Lynton K. Caldwell, ed., *Per-*

spectives on Ecosystem Management for the Great Lakes (Albany: State University of New York Press, 1988); Caldwell, "Emerging Boundary Environmental Challenges and Institutional Issues: Canada and the United States," *Natural Resources Journal* 33 (Winter 1993): 10–31; and Caldwell, "Disharmony in the Great Lakes Basin—Institutional Jurisdictions Frustrate the Ecosystem Approach," *Alternatives* 20, no. 3 (1994): 27–31.

15. U.S. Department of State, *United States Treaties and Other International Agreements*, vol. 30, pt. 2, pp. 1383–1487, TIAS no. 9257; see *Great Lakes Water Quality Agreement, 1978* (Windsor, Canada: International Joint Commission, 1980), p. 4; Great Lakes Research Advisory Board, "Special Report to the International Joint Commission," in *The Ecosystem Approach: Scope and Implications of an Ecosystem Approach to Transboundary Problems in the Great Lakes Basin* (n.p., July 1978); and J. F. Cartrilli and A. J. Dines, "Great Lakes Water Pollution Control: The Land Use Connection," *Environmental Policy and Law* 6 (February 1980): 9–16.

16. Maxwell Cohen, "Transboundary Attitudes and Policy—Some Canadian Perspectives," paper prepared for Harvard Center for International Affairs, 21 October 1980; and O. P. Dwivedi and John E. Carroll, "Issues in Canadian-American Environmental Relations," in *Resources and Environment Policy Perspectives for Canada*, ed. O. P. Dwivedi (Toronto: McClelland and Stewart, 1980), chap. 13.

17. U.S. Department of State, "Utilization of Waters of Colorado and Tijuana Rivers and of the Rio Grande," revised treaty in force, 8 November 1945, *Treaty Series* 994, 59 Stat. 1219. Albert E. Utton, ed., *Pollution and International Boundaries: United States–Mexican Environmental Problems* (Albuquerque: University of New Mexico Press, 1973); and Cesar Sepulveda, "Mexican-American International Water Quality Problems: Prospects and Perspectives," *Natural Resources Journal* 12 (October 1972): 487–95. See also Jerry E. Mueller, *Restless Rivers: International Law and the Behavior of the Rio Grande* (El Paso: Texas Western Press, 1975); Marshall E. Wilcher, "Transnational Environmental Problems—The United States, Canada, Mexico," *Environmentalist* 3 (Spring 1983): 45–54; Stephen P. Mumme, "Innovation and Reform in Transboundary Resource Management: A Critical Look at the International Boundary and Water Commission, United States and Mexico," *Natural Resources Journal* 33 (Winter 1993): 93–120; and Alberto Szekely, "Emerging Boundary Environmental Challenges and Institutional Issues: Mexico and the United States," *Natural Resources Journal* 33 (Winter 1993): 33–46. For current information on Mexican-U.S. regional environmental developments see the journal *Borderlands Research Monograph Series*, New Mexico State University; publications of the Texas Center for Policy Studies, Austin, Texas, and of the International Transboundary Resources Center, University of New Mexico School of Law.

18. *The NAFTA: Report on Environmental Issues* (Washington, D.C.: U.S. Government Printing Office, November, 1993); *Executive Order of the President 12916, Implementation of the Border Environment Cooperation Commission and the North American Development Bank*, May 23, 1994, vol. 30, no. 20, p. 1084; Annette Baker Fox, "Environment and Trade: The NAFTA Case," *Political Science Quarterly* 110, no. 1 (Spring 1995): 49–69. The North American Development Bank was also established as a side agreement to NAFTA.

19. *EFTA* (Geneva, Switzerland: EFTA Secretariat, Press and Information Service, January, 1994).

20. Emile Noel, *Working Together—The Institutions of the European Community* (Luxembourg: Official Publications of the EC, 1988), and P. L. DeReeder, *Environmental Programmes of Intergovernmental Organizations—With Special Reference to the Sphere of Interest of the Chemical Industry* (The Hague: Martinus Nijhoff, May 1977). For relationships with the Council of Europe and other European regional organizations see Sir Barnett Cocks, *The European Parliament: Structure, Procedure and Practice* (London: Her Majesty's Stationery Office, 1973); Ken Collins, "Plans and Prospects for the European Parliament in Shaping Future Environmental Policy," *European Environmental Law Review* 4, no. 3 (March 1995): 74–77. On the environmental programs of the EC, see Aguilar Fernández, "Differences and Dynamics in European Environmental Policy," *Science and Public Policy* 22, no. 3 (June 1995): 189–94; Stanley P. Johnson, and Guy Corcelle, *The Environmental Policy of the European Communi-*

ties, 2d ed. (London: Kluwer Law International, 1995); and Commission of the European Communities, "Environment Programme, 1977–1981," *Bulletin of the European Communities,* Supplement 6 (June 1976). See also J. McLaughlin and M. J. Forster, *The Law and Practice Relating to Pollution Control in the Member States of the European Communities: A Comparative Survey,* 2d ed. (London: Graham and Trotman, 1982); Marshall E. Wilcher, *Environmental Cooperation in the North Atlantic Area* (Washington, D.C.: University Press of America, 1980); Gianfranco Amendola and Peter H. Sand, "Transnational Environmental Cooperation between Different Legal Systems in Europe," *Earth Law Journal* 1 (1975): 189–94. For further reference, see *Endoc Directory: Environmental Information and Documentation Centres of the European Communities* (Hitchin, U.K.: Peter Peregrinus, 1978). This publication is to be periodically updated from a computer-based permanent inventory. For a comparison of U.S. and European Community law, see Turner T. Smith Jr., *Understanding US and European Environmental Law: A Practitioner's Guide* (London: Graham and Trotman, 1989). This publisher has issued numerous titles on European and international environmental law, e.g., Nigel Haigh et al., *European Community Environmental Law in Practice,* 4 vols. Note primarily vol. 1, *Comparative Report, Water and Waste in Four Countries: A Study of the Implementation of the EEC Directives in France, Germany, Netherlands, and United Kingdom,* 1986.

21. Homer C. Angelo, "Protection of the Human Environment—First Steps Toward Regional Cooperation in Europe," *International Lawyer* 5 (July 1971): 511–26; also see Todd Howland, "Chernobyl and Acid Deposition: An Analysis of the Failure of European Cooperation to Protect the Shared Environment," *Temple International and Comparative Law Journal* (Fall–Spring 1988): 11–37; Menno Kamminga, "Improving Integration of Environmental Requirements into Other EC Policies," *European Environmental Law Review* 3, no. 1 (January 1994): 23–25.

22. Marshall E. Wilcher, *Environmental Cooperation in the North Atlantic Area* (Washington, D.C.: University Press of America, 1979), pp. 21, 64, 128–29. Various decision categories within the EU have differing degrees of binding force. Regulations are binding on all bodies to which they apply, both public and private, in all member states. Directives are binding only on member states and not private enterprises. Discretion is left to individual member states concerning the methods of bringing about a specified result. Decisions are made in individual cases and are binding in every respect on the member states or private enterprise to which they apply.

23. "Declaration of the Council of the European Communities and of the Representatives of the Governments of the Member States Meeting in the Council of 22 November 1973 on the Programme of Action of the European Communities on the Environment," *International Legal Materials* 13 (1974): 164–216; see also Declaration of the Council of the European Communities and of the Representatives of the Member States Meeting in the Council of 22 November 1973 on the Programme of Action of the European Communities on the Environment, *Official Journal of the European Communities* 16, no. C112 (20 December 1973), entire issue.

24. *Information and Notices,* 20, (13 June 1977).

25. Commission of the European Communities, *State of the Environment, Second Report 1979,* p. 6. Articles by Peter Malanczuk currently updating reports on the environmental activities of the EU have appeared with regularity in issues of *Environmental Policy and Law,* e.g., "Developments to the End of 1979," vol. 6 (February 1980): 35–37. See also Nigel Haigh, "EC Success Story: Action Program on the Environment Works Well," *Europe,* no. 237 (May–June 1973): 16–17; also Athleen Ellington and Tom Burke, *Europe, Environment— The European Communities' Environmental Policy* (London: Ecobooks, 1981); and "Environment—A Decade of Cooperation within the Economic Commission for Europe," *Economic Bulletin for Europe* (Pergamon Press) 34, no. 1 (1982).

26. Geoffrey Wandesforde-Smith, "Environmental Impact Assessment in the European Community," *Zeitschrift für Umweltpolitik* (January 1979): 35–76.

27. Commission of the European Communities (CED), *Proposal for a Council Directive Con-*

cerning the Assessment of the Environmental Effects of Certain Public and Private Projects,
COM/80/313 final (Brussels: CED, 1980). See Wandesforde-Smith, "Environmental Impact
Assessment in the European Community," and the following articles by Norman Lee and
Christopher Wood, who served as consultants on environmental impact assessment procedures and methods to the Commission of the European Community: "Environmental
Impact Assessment of Projects in EEC Countries," *Journal of Environmental Management*
6, no. 1 (January 1978): 57–71; "The Assessment of Environmental Impacts in Project Appraisal in the European Communities," *Journal of Common Market Studies* 16 (March 1978):
101–10; "Environmental Impact Assessment in the European Economic Community," *Environmental Impact Assessment Review* 1 (September 1980): 287–300; and *Environmental Impact
Assessment: Proceedings of a Seminar of the United Nations Economic Commission for Europe—
Villach, Austria, September 1979* (Oxford: Pergamon Press, for the United Nations, 1979). See
also R. J. Cerny and W. R. Sheate, "Strategic Environmental Assessment in the European
Community: Amending the EIA Directive," *Environmental Policy and Law* 22, no. 3 (June
1992): 154–59; and W. R. Sheate, "Amending the EC Directive (85/337/EEC) on Environmental Impact Assessment," *European Environmental Law Review* 4, no. 3 (March 1995): 79.

28. *Official Journal of the European Communities,* vol. 26, no. C46 (17 February 1983): 1–16.

29. For background, see *Manual of the Council of Europe: Structure, Functions and Achievements,*
by a Group of Officials of the Secretariat (London: Stevens and Sons, 1970); and *The Europe
of the 21: Uniting for European Democracy* (Croton-on-Hudson, N.Y.: U.S. and World Publications, 1981). See also P. L. DeReeder, *Environmental Programmes of Intergovernmental
Organizations—With Special Reference to the Sphere of Interest of the Chemical Industry* (The
Hague: Martinus Nijhoff, May 1977). For a pessimistic report on the conservation of natural ecosystems in Europe, see Peter Baum, "No Progress Report," *Naturopa,* no. 38 (1981):
15–20. A somewhat more hopeful survey of twenty years of nature protection in Europe is
provided by Peter Gay and Georges Tendron in *Naturopa,* no. 41 (1982): entire issue. For
background on the environmental policies of the Council of Europe, the European Parliament, and the European Communities, see Edgar Faure, *For a European Environmental
Policy,* trans. Gordon Jenkins (Brussels: European Cooperation Fund, 1977).

30. *Explanatory Report Concerning the Convention on the Conservation of European Wildlife and
Natural Habitats* (Strasbourg: Council of Europe, 1979). For the Council-EC relationship
in this convention, see Gunnar Seidenfaden, "Berne: The Convention Open to All," *Naturopa* 39 (1981): 4–6. Institutional and legal aspects are outlined by Pierre-Henri Imbert in
"La Convention relative à la conservation de la vie sauvage et du milieu naturel de l'Europe,
exception au étape?" *Annuaire français du droit international* (1979).

31. *Man in a European Society* (Strasbourg: Council of Europe, 1966), pp. 33–35, 58. See issues
of *Forward in Europe: Bulletin of the Council of Europe,* which has a section on environment.

32. *Europe's Environment—The Dobris Assessment: The Report on the State of the Pan-European
Environment Requested by the Environment Ministers for the Whole of Europe at the Ministerial
Conference held at Dobris Castle, Czechoslovakia, June 1991,* ed. David Stanners and Philippe
Bordeau (Copenhagen: European Environment Agency, 1995).

33. Günter Weichart, "The North Sea," *Environment* (January–February 1974): 29–33; reprinted from *Ambio* 2, no. 4, under the title "Pollution of the North Sea." See also Konrad
von Moltke and Nigel Haigh, "Environmental Protection of the North Sea," *European Environment Review* 1, no. 3 (June 1987): 12–18; H. Peter Krosby, "Oil and the Environment:
Norway's Enlightened Policy," *Scandinavian Review* 64 (December 1976): 39–44; and Ian R.
Manners, *North Sea Oil and Environmental Planning: The United Kingdom Conference* (Austin:
University of Texas Press, 1982). See also *Report of the ICES Working Group on Pollution of
the North Sea,* Cooperative Research Report, series A, no. 13 (Charlottenlund, Denmark:
International Council for the Exploration of the Sea, July 1969); *Marine Pollution Bulletin,*
September 1973, p. 135; and *Marine Pollution Bulletin,* August 1973, p. 118.

34. *Environmental Impacts from Offshore Exploration and Production of Oil and Gas* (Paris: Environment Directorate, OECD, 1977), pp. 35–43.

35. Stanley Foundation, "What to Do with Derricks? Oil Drilling Platforms in the North Sea," *World Press Review*, September 1995; "Greenpeace Seeks Peace with Oil Industry," *Reuters World Service*, 24 August 1995; Brandon Mitchener, "Greenpeace Admits Slip on Oil Rig Risk," *International Herald Tribune*, 6 September 1995 and Brent Spar, "Greenpeace, Shell UK Discuss Disposal Issues," *American Political Network, Greenwire*, 12 September 1995 (worldview).

36. Peter Fotheringham and P. W. Birnie, "Regulation of North Sea Marine Pollution," in *The Effective Management of Marine Resources*, ed. C. M. Mason (New York: Nichols Publishing, 1979), pp. 168–223.

37. On this agreement and others related to it, see Jan Schneider, *World Public Order of the Environment Towards an International Ecological Law and Organization* (Toronto, Canada: University of Toronto Press, 1979), pp. 32–38. For texts of this and related agreements, see Barros and Johnson, *The International Law of Pollution*.

38. Bernard A. Dubois, "The 1976 London Convention on Civil Liability for Oil Pollution Damage from Offshore Operations," *Journal of Maritime Law and Commerce* 9 (October 1977): 61–77.

39. There are also technological proposals to combat pollution in the North Sea. The relative shallow depth of its southern portion located between major industrial and commercial centers has suggested the development of artificial islands for a variety of purposes, which might lead to the alleviation of the pollution control problems but which also might contribute to them. This proposition is not new, having been raised in 1954 and on several occasions thereafter by Patrick Horsbrugh, initiator of the Environic Foundation International. More recently, a letter by Michael Brophy in the London *Times* (31 January 1980) proposed "the first Euro-airport on a site equidistant from Holland, Belgium, France, and England." A set of papers, including copies of correspondence entitled "North Sea Islands as a Multipurpose Means of Reaching the Continent, Stimulating Advanced Technologies in Transportation, Construction and Industrial Processes while Creating Structures Designed in Accordance with Basic Ecological Disciplines," has been prepared by Environic Foundation International, P.O. Box 88, Notre Dame, Indiana 46556. On the need for a comprehensive multilateral agreement on environmental management for the North Sea, see Ton Igistra, "Regional Cooperation in the North Sea: An Inquiry," *International Journal of Estuarine and Coastal Law* 3, no. 2 (1988): 181–207; and D. Freestone and Ton Igistra, eds., *The North Sea: Basic Legal Documents on Regional Environmental Cooperation* (Dordrecht, Netherlands: Graham and Trotman/Martinus Nijhoff, 1991).

40. See *Man and the Baltic* (Helsinki: Finnish Baltic Sea Committee, Ministry of the Interior, 1977); Bengt Owe Jansson, *Ecosystem Approach to the Baltic Problem*, Bulletin from the Ecological Research Committee, no. 16 (Stockholm: Swedish Natural Science Research Council, 1972); and Interim Baltic Marine Environment Protection Commission, "Joint Activities of the Baltic Sea States within the Framework of the Convention on the Protection of the Marine Environment of the Baltic Sea Area, 1974–1978," *Baltic Sea Environment Proceedings*, no. 1 (Helsinki: Government Printing Centre, 1979). For general accounts and background, see Lennert J. Lundqvist, "Saving the Baltic," *Scandinavian Review* 64 (December 1976): 46–53; Joseph B. Board, "All in the Same Boat: The Scandinavian Example in Environmental Protection" and "Oresund: A Case of Planning," in *Scandinavian Review* 64 (December 1976): 4–9, 56–58; Toivo Miljan, "The Baltic Sea: Mare Clusum or Mare Liberum?" *Cooperation and Conflict* 9, no. 1 (1974): 19–28; and Gunnar Alexandersson, "The Baltic Straits," in *International Straits of the World*, vol. 6, ed. Gerard J. Mangone (The Hague: Martinus Nijhoff, 1982).

41. E.g., Janet Pawlak, "Land-Based Inputs of Some Major Pollutants to the Baltic Sea," *Ambio* 9, nos. 3–4 (1980): 163; and Erkki Leppäkoski, "Man's Impact on the Baltic Ecosystem," *Ambio* 9, nos. 3–4 (1980): 180.

42. Convention on the Protection of the Marine Environment of the Baltic Sea Area, "Final Act of the Diplomatic Conference on the Protection of the Marine Environment of the

Baltic Sea Area, Helsinki, 18-22 March 1974"; and Baltic Marine Environment Protection Commission, "Report of the Interim Commission (IC) to the Baltic Marine Environment Protection Commission," *Baltic Sea Environment Proceedings*, no. 2 (Helsinki: Government Printing Centre, 1981); Bertil Hägerhäll, "International Cooperation to Protect the Baltic," *Ambio* 9, nos. 3-4 (1980): 183; and Helena Rytovuori, "Structures of Détente and Ecological Interdependence: Cooperation in the Baltic Sea Area for the Protection of Marine Environment and Living Resources," *Cooperation and Conflict* 15 (1980): 95.

43. "Convention on Fishing and Conservation of the Living Resources in the Baltic Sea and the Belts" (Gdansk, 13 September 1973), in *Register of International Conventions and Protocols in the Field of the Environment* 2 (Nairobi: UNEP, 7 February 1977), p. 88; *International Legal Materials* 12 (1973): 1291-97 (A/C.1/1035/15, October 1973).

44. Hägerhäll, "International Cooperation to Protect the Baltic," p. 186; and Board, "Oresund: A Case of Planning," pp. 56-58.

45. See "Nordic Cooperation," *Annual Report*, 1974 (July 1975): 22; and *Cooperation Agreements between Nordic Countries*, 2d rev. ed. (1978), pp. 47-51.

46. Bernt I. Dybern, "The Organizational Pattern of Baltic Marine Science," *Ambio* 9, nos. 3-4 (1980): 188-93, describes the functions of organizations engaged in environmental research in the Baltic area. See also Jeannie E. Kenberger, "International Organizations at Nordic Council Conference—Environmental Policies Discussed," *Ambio* 4, no. 4 (1975): 175-76.

47. See the survey *Nordic Environment Protection*, comp. Nordforsk's Secretariat of Environmental Sciences (Helsinki, 1972), p. 7.

48. UNEP, *An Introduction to the UNEP Regional Seas Programme* (Geneva: Regional Seas Programme Activity Centre, n.d.); "Action on Regional Seas," *Uniterra* 1 (August 1976); Patricia A. Bliss-Guest and Stjepan Keckes, "The Regional Seas Programme of UNEP," *Environmental Conservation* 9 (Spring 1982): 43-49; Stjepan Keckes, "Regional Seas: An Emerging Marine Policy Approach," in *Comparative Marine Policy*, Rhode Island University, Center for Ocean Management Studies (New York: Praeger, 1981), pp. 17-20; Keckes, "The Regional Seas Programme—The Principal Elements . . . ," *Uniterra* 6 (March–April 1981): 12; Thomas Land, "The Regional Seas Approach: A Focal Point for Environmental Cooperation," *Ceres* 15 (March–April 1982): 43-46; and Peter Thacher, "A Master Plan for the Watery Planet," *Uniterra* 6 (March–April 1981): 6. See also UNEP Governing Council, *The Environment Programme: Medium Term Plan 1982-1983*, UNEP/GC.9/6 (Nairobi, 6 March 1981); and Peter Hulme, "The Regional Seas Program: What Fate for UNEP's Crown Jewels," *Ambio* 12, no. 1 (1983): 2-13. For detailed descriptions of the programme and the several Action Plans, see UNEP Regional Seas Reports and Studies, 1982-.

49. "Regional Seas Expanding," *The Siren*, no. 9 (Summer 1980): 12; Keckes, "The Regional Seas Programme—The Principal Elements," p. 12; Keckes, "Taking Stock," *The Siren*, no. 16 (Spring 1982): entire issue; Keckes, "Now we are 12," *The Siren* 43 (March–September 1989).

50. *Annual Report of the Executive Director*, 1986, pp. 66-68; ibid., 1988, pp. 40-44.

51. *Achievements and Planned Development of UNEP's Regional Seas Programme and Comparable Programmes Sponsored by Other Bodies*, UNEP Regional Seas Reports and Studies, no. 1 (Geneva: UNEP, 1982). See also Anathea L. Brooks, Wayne H. Bell, and John R. Greer, *Our Coastal Seas: What is their Future?* Proceedings of the 1993 Environmental Management of the Coastal Seas Conference (College Park: Maryland Sea Grant College, 1995).

52. See Scott C. Truver, *The Strait of Gibraltar and the Mediterranean* (The Hague: Martinus Nijhoff, 1980); Peter M. Haas, *Saving the Mediterranean: The Politics of International Environmental Cooperation* (New York: Columbia University Press, 1989). For the difficult and critical problem of national jurisdiction over territorial sea space, see Fabrizio Bastianelli, "Boundary Delimitation in the Mediterranean Sea," *Marine Policy Reports* 5 (February 1983): 1-6 (published at University of Delaware). For an informal but informative discussion of the environmental issues, see transcript of "Mediterranean Prospect," *Nova* (Boston: WCBH Transcripts, 1980).

53. "Mediterranean Pilot Study of Environmental Degradation and Pollution from Coastal Development" (Paris: OECD Environment Directorate, 1975), p. 20; Bonnie L. Knapp, "The
Action Plan for the Mediterranean: A Regional Arrangement to Control Marine Pollution,"
Marine Affairs Journal, no. 5 (1978): 121–22; Paul E. Ress, "International Pollution Treaties
in the Mediterranean," *Environmental Conservation* 5 (Summer 1978): 100; and Ress, "Mediterranean States Agree on Treaty to Control Land-Based Pollution," *Ambio* 9, nos. 3–4
(1980): 194.

54. "Mediterranean Pollution Treaties Enter into Force" and "The Seven Seas," *The Siren*, no. 1
(June 1978).

55. UNEP, *Mediterranean Action Plan and the Final Act of the Conference of Plenipotentiaries of the
Coastal States of the Mediterranean Region for the Protection of the Mediterranean Sea* (New York:
United Nations, 1978). Also Knapp, "The Action Plan for the Mediterranean," p. 131.

56. *Mediterranean Action Plan*, p. 3.

57. *Future for the Mediterranean Basin: The Blue Plan*, ed. Michel Grenon and Michel Batisse
(Oxford: Oxford University Press, 1989). See also Peter Haas, *Saving the Mediterranean: The
Politics of International Environmental Cooperation* (New York: Columbia University Press,
1989).

58. "Convention for the Protection of the Mediterranean Sea against Pollution," *Mediterranean
Action Plan*, pp. 26–33 and Annex A; "Protocol for the Prevention of Pollution of the Mediterranean Sea by Dumping from Ships and Aircraft," *Mediterranean Action Plan*, pp. 39–46;
"Protocol Concerning Cooperation in Combating Pollution of the Mediterranean Sea by
Oil and Other Harmful Substances in Cases of Emergency," *Mediterranean Action Plan*, pp.
47–52; Nicholas A. Robinson, "Convention for the Protection of the Mediterranean Sea
against Pollution," *Earth Law Journal* 2 (1976): 289–91.

59. "Land-based Pollution in Mediterranean," *Uniterra* 2 (March 1977): 1, 3; Paul E. Ress,
"Treaty Turns the Tide of Mediterranean Pollution," *Uniterra* 5 (July 1980): 3; Ress, "Mediterranean States Agree on Treaty," p. 194. Also see Jean-Pierre Dobbert, "Protocol to Control Pollution in the Mediterranean," *Environmental Policy and Law* 6 (June 1980): 10–14.

60. "Mediterranean Action Plan Enters New Phase," *The Siren*, no. 12 (Spring 1981): 8. On the
Regional Oil Combating Centre (ROCC), see *IMCO News*, no. 1 (1982): 6–7, 16.

61. See Council of Europe, *Newsletter—Nature*, no. 81-8/9; *Uniterra* 8 (September–October
1980): 12.

62. See R. K. Ramazani, *The Persian Gulf and the Strait of Hormuz* (The Hague: Martinus
Nijhoff, 1980); Margi Bryant, "The Fragile Persian Gulf," *World Press Review* (March 1981):
55; "Oil-Producers Make Waves," *The Siren*, no. 1 (June 1978): 4–5; and S. H. Amin, "Marine
Pollution Regulation in the Persian Gulf," *Marine Policy Reports* 5 (September 1982): 1–4.
The last includes discussion of a full range of marine antipollution measures.

63. *Oversight Hearing on the Ecoterrorism Inflicted as a Result of the Persian Gulf War: The Environmental Role of the National Oceanic and Atmospheric Administration of the U.S. Coast Guard
in the Persian Gulf Conflict*, Hearing in the U.S. House of Representatives, 17 October 1991,
Serial No. 102–441; and *The Environmental Aftermath of the Gulf War: A Report*, prepared for
the Committee on Environment and Public Works-Gulf Pollution Task Force, U.S. Senate,
March 1992.

64. On the development of the plan, see "Government Experts Plan Protection Moves," *Uniterra* 2 (July 1977): 1, 3; "Action Plan for the Gulf," *Uniterra* 2 (December 1976–January
1977): 1–2.

65. UNEP, *Final Act of Kuwait Regional Conference of Plenipotentiaries on the Protection and De-
velopment of the Marine Environment and the Coastal Areas* (n.p., 1978), pp. 13–27, including
"Action Plan for the Protection and Development of the Marine Environment and the
Coastal Areas of Bahrain, Iran, Iraq, Kuwait, Oman, Qatar, Saudi Arabia and the United
Arab Emirates." Hereinafter referred to as the *Kuwait Action Plan*.

66. "Kuwait Regional Convention for Co-operation on the Protection of the Marine Environ-

ment from Pollution," *Kuwait Action Plan*, pp. 28–48; and "Protocol Concerning Regional Co-operation in Combatting Pollution by Oil and Other Harmful Substances in Cases of Emergency," ibid., pp. 49–62.

67. "Iran-Iraq War Causes Worst Oil Spill in Persian Gulf," *World Environment Report* 9 (15 April 1983): 1–2; and "Gulf Oil Spills into Ecological Disaster," *World Environment Report* 9 (15 May 1983): 3.

68. "KAP Projects Move Ahead," *The Siren*, no. 7 (Winter 1980): 2; *Achievements and Planned Development of UNEP's Regional Seas Programme and Comparable Programmes Sponsored by Other Bodies*, pp. 12–13. See also Makram George, "Rehabilitating ROPME's Ecosystems," *The Siren*, no. 48 (March–September 1993): 1–4.

69. See Ruth Lapidoth-Eschelbacher, *The Red Sea and the Gulf of Aden* (The Hague: Martinus Nijhoff, 1980); and three articles by Lapidoth-Eschelbacher, "Red Sea Programme Remodeled," *The Siren*, no. 12 (Spring 1981): 2–3; "Experts Revise Red Sea Convention," *The Siren*, no. 11 (Winter 1981): 10; and "Red Sea Treaty," *Environmental Policy and Law* 10 (January 1983): 2, 28–30.

70. *Protecting the Gulf of Aquaba: A Regional Environmental Challenge*, ed. Adam Tevah V'Din, Emad Adly, and Mahmoud A. Al-Kohman (Washington, D.C.: Environmental Law Institute, 1993).

71. For background, see John D. Weaver, ed., *Geology, Geophysics and Resources of the Caribbean* (Paris: UNESCO, Intergovernmental Oceanographic Commission, 1975); Carl A. Carlozzi and Alice A. Carlozzi, *Conservation and Caribbean Regional Progress* (Antioch, Ohio: Antioch Press, for the Caribbean Research Institute, 1968); see also *Environmental News Letter* and other publications of the Caribbean Conservation Association and the Caribbean Research Institute; "Caribbean Experts Give Nod to Fund Projects," *The Siren*, no. 12 (Spring 1981): 1, 4.

72. William S. Beller, ed., *Transactions at the Conference on Environmental Management and Economic Growth in the Smaller Caribbean Islands, September 17–21, 1979*, International Organization and Conference Series 143, Department of State Publication 8996 (Washington, D.C.: U.S. Department of State, 1979).

73. See United Nations Press Release HE/584, 10 April 1981; "Report of the Meeting," UNEP/CE PAL/IC.27/2, 16 April 1982; "A Conservation Strategy for the Caribbean," *IUCN Bulletin*, n.s., 12 (May–June 1981): entire issue; "Caribbean States Act on Environment," *The Siren*, no. 8 (Spring 1980): 1, 4; Paul E. Ress, "Caribbean Countries Agree on Sweeping Environmental Plan" and "The Jamaica Accord," *Uniterra* 6 (March–April 1981): 8–9; and Lawrence Mosher, "Maybe Later, U.S. Says in Response to Bid for Caribbean Cleanup Funds," *National Journal* 13 (2 May 1981): 783–84.

74. "The Jamaica Accord" (excerpts from an address given by the Hon. Edward Seaga, Prime Minister of Jamaica, on the Caribbean Action Plan), *Uniterra* 6 (March–April 1981): 9.

75. "Caribbean Sea Treaty Signed in Colombia," *World Environment Report* 9 (15 April 1983): 2.

76. Paul E. Ress, "Seas Convention and Protocol Signed for West Africa Region," *Uniterra* 5 (November–December 1980): 3. See also Mohamed Tangi, "Discovering Oil in West and Central Africa," *The Siren* (July 1982): 9–16.

77. "Seaboard States Agree to 'Hot Pursuit' of Polluting Oil Tankers," *Uniterra* 6 (March–April 1981): 11.

78. For background, see Donald K. Emmerson, "The Case for a Maritime Perspective on Southeast Asia," *Journal of Southeast Asian Studies* 11 (March 1980): 139–45; Mark J. Valencia, "The South China Sea: Prospects for Marine Regionalism," *Marine Policy* (April 1978): 87–104; and Valencia, "Southeast Asia: National Marine Interests and Marine Regionalism," *Ocean Development and International Law Journal* 5, no. 4 (1978): 421–76.

79. UNEP, "Report of the Meeting of Experts to Review the Draft Action Plan for the East Asian Seas," Baguio, Philippines, 17–21 June 1980; "Report of the First ASEAN Working Group Meeting on Marine Sciences," Jakarta, Indonesia, 12–14 June 1979; and "Report of

the Second Meeting of the ASEAN Experts on the Environment," Penang, Malaysia, 17–20 September 1979. For the program, see "ASEAN Experts Consider Draft Plan," *The Siren*, no. 9 (Summer 1980): 4; and UNEP, *Asia-Pacific Report 1980* (Bangkok: UNEP Regional Office for Asia and the Pacific, 1981), p. 45; James Crawford and Donald R. Rothwell, *The Law of the Sea in the Asian Pacific Region* (Dordrecht, Netherlands: Martinus Nijhoff, 1995).

80. *IMCO News*, no. 2 (1980): 10.

81. UNEP, *Asia-Pacific Report 1980*, p. 78.

82. Annual Reports of the ASEAN Standing Committees, e.g. 1992–93, "Cooperation and Environment," and "Singapore Resolution on Environment and Development of 1992"; and *Environmental Management in ASEAN: Perspectives on Critical Regional Issues*, ed. Maria Sadas (Singapore: Institute of Southeast Asian Studies, 1993).

83. See *Yearbook of International Organizations*, 1994–1995, "South Pacific Commission," pp. 1424–25; and *Regional Cooperation in the South Pacific, The South Pacific Commission: History, Aims and Activities*, 3d ed. (Noumea, New Caledonia: South Pacific Commission, 1977).

84. Suzanne Murrell, "Conference on the Human Environment—What It Means to the South Pacific Region," *South Pacific Commission: Monthly News of Activities*, no. 33 (March 1982): 14–15, also pp. 8–9. The South Pacific Region geographically defined for the purpose of the action plan drafted at Rarotonga in March 1982 includes the area of responsibility of the South Pacific Commission, together with any associated national maritime resource management zones.

85. *The Siren*, no. 10 (Fall 1980): 1, 11. On the importance of coral reef protection in the Pacific, see Teny Topalian, "Protecting Marine Resources in Tropical Pacific Islands," *Seawind* 2, no. 3 (July–September 1988): 6–11.

86. "Paving the Way in the South-East Pacific," *The Siren*, no. 12 (Spring 1981): 11.

87. UNEP, *Achievements and Planned Development of UNEP's Regional Seas Programme and Comparable Programmes Sponsored by Other Bodies*, p. 18.

88. "Now We Are 12," *The Siren*, no. 43 (December 1991); and *The Siren*, no. 48 (March–September 1993).

89. "Protection of the Black Sea Ministerial Declaration, Odessa, 7 April 1993 and the Bucharest Convention on the Protection of the Black Sea Against Pollution," *International Journal of Marine and Coastal Law* 9, no. 1 (1994): 72–77.

90. Erik Brüel, *International Straits: A Treatise on International Law*, 2 vols. (Copenhagen: Nyt Nordisk Forelag, 1947), vol. 1, pp. 18–20. On particular straits see the International Straits of the World Series, edited by Gerard J. Mangone, published by Sijthoff and Noordhoff, Alphen aan den Rijn, Netherlands: William E. Butler, *Northeast Arctic Passage*, 1978; Michael Leifer, *Malacca, Singapore, and Indonesia*, 1978; R.K. Ramazani, *The Persian Gulf*, 1979; Scott C. Truver, *The Strait of Gibraltar and the Mediterranean*, 1980; (the following published by Martinus Nijhoff) Ruth Lapidoth-Eschelbacher, *The Red Sea and the Gulf of Aden*, 1982; Gunnar Alexandersson, *The Baltic Straits*, 1982; Donat Pharand, *The Northwest Arctic Passage: Arctic Straits*, 1984; Ruk Cuyvere, *The Strait of Dover*, 1986; Christos Rozakis, *The Turkish Straits*, 1987; Chi Young Pak, *The Korean Straits*, 1988.

91. William L. Schachte, "International Straits and Navigational Freedoms," *Ocean Development and International Law* 24, no. 2 (April–June 1982): 179–96; and *The Law of the Sea: Straits Used for International Navigation*, Legislative History of Part III of the United Nations Convention on the Law of the Sea, 2 vols. (New York: UN Office of Ocean Affairs and the Law of the Sea, 1992).

92. *International Environment Reporter* 12, no. 2 (February 1989): 57; and U.S. Arctic Research Commission, *The United States Arctic Research Commission* (leaflet) and *Statement of Goals and Objectives to Guide United States Arctic Research* (Washington, D.C.: U.S. Arctic Research Commission, December 1988). See also Elena N. Nikitina, "International Mechanisms and Arctic Environmental Research," in *Peace and Violence* 12, no. 7 (1989): 123–32; Alexei Roginka, "Arctic Environmental Cooperation: Prospects and Possibilities," paper presented at

the International Studies Convention, London, March 1989. See also articles in *International Studies Notes* 11, no. 3 (Spring 1985).

93. Canada, secretary of state for External Affairs, press release, 23 April 1990; and David D. Caron, "Toward an Arctic Environmental Regime," *Ocean Development and International Law* 24, no. 4 (October–December 1993): 377–92.

94. See *Goals and Priorities to Guide United States Arctic Research* Biennial Statement, Arctic Research Commission, January 1995, and Interagency Research Policy Committee, *Arctic Research of the United States* 5 (Fall 1991): 29–35; also "Recent U.S.-U.S.S.R. Agreements Relating to the Bering Region," *Arctic Research of the United States* 5 (Fall 1991): 37–47.

95. On Antarctica generally, see the following: Trevor Hatherton, ed., *Antarctica* (New York: Praeger, 1965), a New Zealand and Antarctic Society survey; H. C. R. King, *The Antarctic* (London: Blandford Press, 1969); Neal Potter, *Natural Resource Potentials of the Antarctic* (New York: American Geographical Society, 1969); J. F. Lovering and J. R. V. Prescott, *Last of Lands—Antarctica* (Melbourne, Australia: Melbourne University Press, 1979); *Antarctic* (a news bulletin published quarterly by the New Zealand Antarctic Society, P.O. Box 2110, Wellington, New Zealand), 1956–; *Antarctic Journal of the United States* (Washington, D.C.: National Science Foundation), 1966–.

For background and details on environmental policy, see Ross C. Peavey and Laurence M. Gould, "Antarctica—International Land of Science," *UNESCO Courier* 15 (January 1962): 9–14; Laurence M. Gould, "Antarctica—Continent of International Science," *Science* 150 (December 1965): 1775–81; Vivian Fuchs, "Antarctica: The International Laboratory," *Science Journal* 2 (November 1966): 48–54; Raymond F. Dasmann, "Conservation in the Antarctic," *Antarctic Journal* (January–February 1968): 1–6; articles by John Hanessian Jr. in *American Universities Field Staff Reports*, Polar Area Series; "Antarctica Since the ICY," *Bulletin of the Atomic Scientists* 26 (December 1970): entire issue; and Gerald S. Schatz, ed., *Science, Technology and Sovereignty in the Polar Regions* (Lexington, Mass.: Lexington Books, 1974), p. 215. Most pertinent to the subject of the book, with extensive treatments of environmental factors, living resources such as krill, minerals, and international agreements, are F. M. Auburn, *Antarctic Law and Politics* (Bloomington: Indiana University Press, 1982); and "The Antarctic," pt. 2 of Jonathan I. Charney, ed., *The New Nationalism and the Use of Common Spaces* (Totowa, N.J.: Allenhold, Osmun, 1982), pp. 115–332.

For more recent developments, including the Convention on the Regulation of Antarctic Mineral Resource Activities, see Susan A. Fletcher, *Antarctica: Environmental Protection Issues*, summary of a CRS workshop (Washington, D.C.: Congressional Research Service, Library of Congress, April 10, 1989); Malcolm W. Browne, "In Once Pristine Antarctica a Complicated Cleanup Begins," *New York Times* (19 December 1989): 21, 26; Jeffrey D. Myhre, "Exploiting Antarctica's Minerals: The Treaty System Completed," *Harvard International Review*, 2 (Summer 1989): 37–39; Lee A. Kimball and Scully R. Tucker, "Antarctica: Is There Life after Minerals? The Minerals Treaty and Beyond," *Marine Policy: The International Journal of Ocean Affairs* 13 (April 1989): 87–98; Ian Anderson, "Antarctic Minerals Deal Heads for Rocks: Regulating Future Mining and Oil Exploration in Antarctica," *New Scientist* (20 May 1989): 21; "Antarctic Antics: Australia Will Refuse to Sign Antarctic Minerals Convention," editorial in *New Scientist* (3 June 1989): 20; Malcolm W. Browne, "French and Australians Kill Accord on Antarctic," *New York Times* (25 September 1989): 10; Rick Pullen, "Treaty Said Needed to Avert Antarctica Gold Rush," *American Metal Market* (2 October 1989): 4; Deborah Shapley, "Polar Thinking on the Antarctic," *New York Times* (17 October 1989): 15, 27; Christopher C. Joyner, "The Antarctic Treaty System and the Law of the Sea—Competing Regimes in the Southern Ocean," *International Journal of Marine and Coastal Law* 10, no. 2 (May 1995): 301–31 (the respective regimes may be reinforcing). And on the Madrid Protocol of 1991 see Mahind Perera, *Change and Continuity in Antarctic Environmental Protection: Politic and Policy*, Ph.D. diss., Dalhousie University, 1995.

Relevant also for background are Jeffrey D. Myhre, *The Antarctic Treaty System: Politics, Law, and Diplomacy* (Boulder, Colo.: Westview Press, 1986); William D. Triggs, *The Antarc-*

tic Treaty Regime (Cambridge: Cambridge University Press, 1987); M. J. Peterson, *Managing the Frozen South: The Creation and Evolution of the Antarctic Treaty System* (Berkeley: University of California Press, 1988); William E. Westermeyer, *The Politics of Mineral Resource Development in Antarctica: Alternative Regimes for the Future* (Boulder, Colo.: Westview Press, 1984); Francisco Orrego Vicuna, *Antarctic Mineral Exploration: The Emerging Legal Framework* (Cambridge: Cambridge University Press, 1988); and Sir Arthur Watts, *International Law and the Antarctic Treaty System* (Cambridge: Grotius, 1992).

96. The text of the act appears in Barros and Johnston, *International Law of Pollution*, pp. 262–76. For further comment, see William V. O'Brien and Armando C. Chapelli, "The Law of the Sea in the 'Canadian' Arctic: The Pattern of Controversy," pt. 1, *McGill Law Journal* 19 (November 1973): 322–66; and pt. 2, *McGill Law Journal* (December 1973): 477–542; Jeffrey J. Sherrin, "International Law and Canadian Arctic Pollution Control," *Albany Law Review* 38, no. 4 (1974): 911–42; and Albert E. Utton, "The Arctic Waters Pollution Prevention Act, and the Right of Self-Protection," *University of British Columbia Law Review* 7, no. 2 (1972): 221–34.

97. "The Antarctic Treaty," *United Nations Treaty Series*, vol. 402 (1961), no. 5778, pp. 71–102. See "Antarctic Treaty Signed by IGY Nations: Polar Region Established as Neutral Science Reserve," *Science* 130 (11 December 1959): 1641–44. See also Auburn, *Antarctic Law and Politics*, chap. 4.

98. David F. Salisbury, "Antarctica: Geopolitical Football," *Technology Review* 79 (March–April 1977): 14, 80. See also Barbara Mitchell, "The Politics of Antarctica," *Environment* 22 (January–February 1980): 12–20; and Auburn, *Antarctic Law and Politics*.

99. Truls Hanevold, "The Antarctic Treaty Consultative Meetings: Form and Procedure," *Cooperation and Conflict* 6, no. 3 (1971): 183–99.

100. U.S. Department of State, "Antarctica: Measures in Furtherance of Principles and Objectives of the Antarctic Treaty," *Treaties and Other International Acts Series* 6058 (2–13 June 1964); David Anderson, "The Conservation of Wildlife under the Antarctic Treaty," *Polar Record* 14, no. 88 (1968): 25–32.

101. Auburn, *Antarctic Law and Politics*, p. 270. See also F. M. Auburn, "The Antarctic Environment," *Yearbook of World Affairs* 35 (Boulder, Colo.: Westview Press, for the London Institute of World Affairs, 1981), pp. 248–65; Auburn, "The Falkland Island Dispute and Antarctica," *Marine Policy Reports* 5 (December 1982): 1–4; and Brian Roberts, "International Cooperation for Antarctica Development: The Test for the Antarctic Treaty," *Polar Record* 19, no. 119 (1978): 107–20.

102. U.S. Department of State, "Conservation of Antarctic Seals," *Treaties and Other International Acts Series* 8826 (1 June 1972).

103. "Draft Convention on the Conservation of Antarctic Marine Living Resources," *Environmental Policy and Law* 5 (February 1979): 4–5, 58–62; and "Convention on the Conservation of Antarctic Marine Resources," *Environmental Policy and Law* 6 (September 1980): 135, 140. See also U.S. Congress, Senate, *Convention on the Conservation of Antarctic Marine Living Resources: Message from the President*, Executive Document X, 96th Cong., 2d sess., 1980. See also Ralph L. Harry, "The Antarctic Region and the Law of the Sea Convention," *Virginia Journal of International Law* 21, no. 4 (1981): 727–44.

104. Barbara Mitchell and Richard Sandbrook, with additional material by John Beddington and Seamus McElroy, *The Management of the Southern Ocean* (London: International Institute for Environment and Development, 1980), p. 4.

105. Personal communication from Sidney J. Holt, 29 November 1981. For some of the political factors involved, see Auburn, *Antarctic Law and Politics*, pp. 127–38.

106. W. Nigel Bonner, "The Krill Problem," *Oryx* (16 May 1981): 31–37.

107. On Biological Investigations of Marine and Antarctic Systems and Stocks (BIOMASS), see Auburn, *Antarctic Law and Politics*, indexed references and chap. 7, "Krill," pp. 205–67. Also Sayed Z. El-Sayed, "SCAR/SCOR Conference on Living Resources of the Southern Ocean," *Antarctic Journal* 12, nos. 1–2 (1977): 1–4; and "The Southern Ocean," *Oceanus* 18

(Summer 1975): entire issue, especially Gerald S. Schatz, "A Sea of Sensitivities," pp. 40–55.

108. For marine issues in Antarctic waters generally, including the krill issue, see Mitchell and Sandbrook, *Management of the Southern Ocean.* Specifically on krill, see Sayed Z. El-Sayed and Mary Alice McWhinnie, "Antarctic Krill: Protein of the Last Frontier," *Oceanus* 22 (Spring 1979): 13–20; *Mammals in the Seas* 1: 79, 233; Willis E. Pequenat, "Whales, Plankton, and Man," *Scientific American* 198 (January 1959): 84–90; Dayton L. Alverson, "Tug-of-War for the Antarctic Krill," *Ocean Development and International Law* 8, no. 2 (1980): 171–82; Gerald Bakus, Wendy Garling, and John E. Buchanan, *The Antarctic Krill Resource: Prospects for Commercial Exploitation* (Pasadena, Calif.: Tetra Tech, 1978); "Antarctica: Inching toward a Krill Convention," *IUCN Bulletin*, n.s., 10 (April 1978): 28–29; and "How Much Krill Can the Antarctic Spare?" *Environmental Law* 5 (February 1979): 5–6.

109. Auburn, *Antarctic Law and Politics*, p. 296.

110. See John Urquhart, "Northward Ho! Canada's Mining Firms Turn to Arctic Islands, Site of Rich Resources" and "Canada to Open Hearings on Pilot Project of $2.1 Billion to Ship LNG from Arctic," *Wall Street Journal*, 5 February 1982, pp. 1, 14, and 2 February 1982, p. 12. Many Antarctic geopolitical issues have Arctic precedents. See Finn Sollie et al., *The Challenge of New Territories* (Oslo: Universitetsforlaget, 1974).

111. Note New York Times News Services dispatches from Antarctica by Robert Reinhold, e.g., 13 January 1982; and Arlen J. Large, "Rules on Oil Exploration in Antarctica To Be Set at Meeting in Bonn Next Month," *Wall Street Journal*, 15 June 1983, p. 36; and Jane Naczynski-Phillips, "Antarctica Faces a Rush of Resource Exploitation," *World Environment Report* 9 (15 July 1983): 1–2. See also "SCAR on Research in Antarctica," *World Environment Report* 9 (15 November 1983): 3; and Protocol on Environmental Protection to the Antarctica Treaty in *International Legal Materials* 30 (1991), 1461–86.

112. Hedwig Hintze, "Regionalism," *Encyclopedia of the Social Sciences* 13 (1934): 208–18; Robert B. Vance, "Region," *International Encyclopedia of the Social Sciences* 13 (1968): 377–82.

113. See *World Development* 3, no. 2 (1990): 6–7; *National Geographic* 177, no. 2 (February 1990): 73–92; and Philip P. Micklin, "Dessication of the Aral Sea: A Water Management Disaster in the Soviet Union," *Science* 241 (1988): 1170–76.

9. International Commons: Air, Sea, Outer Space

1. Among the many general descriptions of the atmospheric environment are the following: Frederick Kenneth Hare, *The Restless Atmosphere* (New York: Harper and Row, 1978); Bert Bolin, ed., *The Atmosphere and the Sea in Motion* (New York: Oxford, 1959); H. S. W. Massey and R. L. F. Boyd, *The Upper Atmosphere* (London: Hutchinson, 1958); National Research Council, *The Atmospheric Sciences and Man's Needs* (Washington, D.C.: National Academy of Sciences, 1971); and *The Atmospheric Sciences: Problems and Applications* (Washington, D.C.: National Academy of Sciences, 1977).

2. See *International Environmental Reporter* 21: 3001–5.

3. UNEP, *The State of the Environment: 1979* (Nairobi: UNEP, 1979), pp. 16–21; P. F. Cunliffe, *Environmental Noise Pollution* (London: Wiley, 1977); George Bugliarello et al., *The Impact of Noise Pollution* (New York: Pergamon Press, 1976); Karl D. Kryter, "Sonic Booms from Supersonic Transport," *Science* 163 (24 January 1969): 359–67. See citations in Bernd Ruster, Bruno Simma, and Michael Bock, eds., *Air and Noise Pollution*, vol. 7 of *International Protection of the Environment: Treaties and Related Documents* (Dobbs Ferry, N.Y.: Oceana Publications, 1975).

4. On Fiji and New Zealand protest of French neutron bomb tests in Polynesia, see Alastair Matheson, "Pacific Fights N-Tests and Waste Dumping," *Uniterra* 5 (November–December 1980): 11; and "French Nuclear Testing and International Law," *Rutgers Law Review* 24, no. 1 (1969): 144–70. On the Chernobyl incident, see *Keesing's Contemporary Archives* (June 1986): 34460–62.

5. It appears that atmospheric dust, carried from deserts and semiarid regions disturbed by human misuse, may have measurable effects upon climate, weather, and man. See Troy L. Pewe, ed., *Desert Dust: Origin, Characteristics, and Effect on Man*, Special Paper no. 186, (Boulder, Colo.: Geological Society of America, 1981).

6. For a comprehensive summation of environmental problems in the atmosphere, see *World Environment—1982*, pp. 19–72; Jeremy Leggett, ed., *Global Warming: The Greenpeace Report* (Oxford: Oxford University Press, 1990); Francesca Lyman, *The Greenhouse Trap: What Are We Doing to the Atmosphere and How Can We Slow Global Warming?* (Boston, Mass.: Beacon Press, 1980). For detailed discussion, see the following: *Proposals for a National Policy and Program* and *The Role of Statistics in Weather Resources Management: Report of the Statistical Task Force to the Weather Modification Advisory Board*, vols. 1 and 2 of *The Management of Weather Resources* (Washington, D.C.: Government Printing Office, 1978); Charles F. Cooper, "What Might Man-Induced Climate Change Mean?" *Foreign Affairs* 56 (April 1978): 500–20; J. W. Samuels, "International Control of Weather Modification Activities: Peril or Policy?" *National Resources Journal* 13 (April 1973): 327–42; Edith Brown Weiss, "International Responses to Weather Modification," *International Organization* 29 (Summer 1975): 805–26; *Climate Change, Food Production and Interstate Conflict: Summary of an International Conference* (New York: Rockefeller Foundation, 1976), p. 71; Herman Pollock, "International Aspects of Weather Modification," *U.S. Department of State Bulletin* 67 (21 August 1972): 212–14; *Implications of Intentional Weather and Climate Modification on the Human Environment: Paper Prepared for the United Nations Conference on the Human Environment, held at Stockholm, Sweden, June 5–16, 1972* (Geneva: WMO, 1971); Ray Jay Davies and Lewis Grant, eds., *Weather Modification: Technology and Law* (Boulder, Colo.: Westview Press, 1978); and Crispin Tickell, *Climatic Change and World Affairs* (New York: Pergamon Press, 1977).

7. *The Atmospheric Sciences: Problems and Applications*, p. 88.

8. *The Atmospheric Sciences and Man's Needs*, Recommendation III-6, p. 79.

9. U.S. Congress, Senate Subcommittee on Oceans and International Environment of the Committee on Foreign Relations, *Prohibiting Military Weather Modification: Hearing on S. Res. 281*, 92d Cong., 2d sess., 26–27 July 1972; and U.S. Congress, Senate Subcommittee on Oceans and International Environment of the Committee on Foreign Relations, *Weather Modification*, 93d Cong., 2d sess., 25 January and 20 March 1974. The opening statement by Senator Claiborne Pell reviews the history of the Senate resolution. See also Louise A. Purrett, "Weather Modification as a Future Weapon," *Science News* 101 (15 April 1972): 254–55.

10. *Weekly Compilation of Presidential Documents*, vol. 10, no. 27 (6 July 1974), p. 753.

11. United Nations, *Multilateral Treaties in Respect of Which the Secretary-General Performs Depositary Functions*, vol. 13 (1979) (New York: United Nations, 1980), p. 613.

12. For an account of the political difficulties encountered, see Lawrence Juda, "Negotiating a Treaty on Environmental Warfare: The Convention on Environmental Warfare and Its Impact upon Arms Control Negotiations," *International Organization* 32 (Autumn 1978): 975–91; U.S. Congress, Senate Committee on Foreign Relations, *Environmental Modification Treaty*, 95th Cong., 2d sess., 3 October 1978; and U.S. Congress, Senate Committee on Foreign Relations, *Environmental Modification Techniques*, 96th Cong., 1st sess., 10 May 1979.

13. Reid A. Bryson, "Some Lessons of Climatic History," in *Atmospheric Quality and Climatic Change*, ed. Richard Kopec, University of North Carolina Studies in Geography, no. 9 (Chapel Hill: University of North Carolina Press, 1975), pp. 38–51; *The Atmospheric Sciences: Problems and Applications*; Frederick Kenneth Hare, "Climate: The Neglected Factor?" *International Journal* (Spring 1981): 371–87; William W. Kellogg, "Effects of Human Activities on Global Climate," pt. 1, *WMO Bulletin* 26 (1977): 229–40; pt. 2, *WMO Bulletin* 27 (1978): 3–10; William W. Kellogg and Robert Schware, *Climate Change and Society* (Boulder, Colo.: Westview Press, 1981); H. E. Landsberg, *Special Environmental Report No. 7: Weather, Climate, and Human Settlements*, World Meteorological Organization Publication, no. 448 (Geneva: WMO, 1976); William H. Matthews, William W. Kellogg, and C. D. Robinson, eds., *Man's Impact on the Climate* (Cambridge, Mass.: MIT Press, 1971); Ved P. Nanda,

ed., *World Climate Change: The Role of International Law and Institutions* (Boulder, Colo.: Westview Press, 1981); Stephen H. Schneider, with Lynn E. Mesirow, *The Genesis Strategy: Climate and Global Survival* (New York: Plenum Press, 1976); Carroll L. Wilson, ed., *Inadvertent Climate Modification* (Cambridge, Mass.: MIT Press, 1971); *Man's Impact on Global Environment: Assessment and Recommendations for Action: Report of the Study of Critical Environment Problems* (Cambridge, Mass.: MIT Press, 1970); *Life on a Warmer Earth: Possible Climatic Consequences of Man-Made Global Warming*, Executive Report 3, based on research by H. Flohn (Laxenburg, Austria: International Institute for Applied Systems Analysis, 1981); and Melinda Cain, "Carbon Dioxide and the Climate: Monitoring and Search for Understanding," in David A. Kay and Harold K. Jacobson, eds., *Environmental Protection: The International Dimension* (Totowa, N.J.: Allenheld, Osmun, 1983), pp. 75–100.

14. Alden McLellan, *The Upper Atmospheric Environment of the Supersonic Transport: A Bibliography*, 2d ed. (Madison: University of Wisconsin Press, 1973).

15. See Asit K. Biswas, ed., *The Ozone Layer, Environmental Sciences and Applications Series*, vol. 4 (Oxford: Pergamon Press, 1979); United Kingdom, Department of the Environment, *Chlorofluorocarbons and Their Effect on Stratospheric Ozone*, Pollution Paper no. 5 (London: Her Majesty's Stationery Office, 1976) has an extensive bibliography; National Research Council, *Protection against Depletion of Stratospheric Ozone by Chlorofluorocarbons* (Washington, D.C.: National Academy of Sciences, 1979); "New Assessments of Ozone Depletion," *Physics Today* 33 (February 1980): 21–22; Horst Siebert, ed., *Global Environmental Resources: The Ozone Problem* (Frankfurt am Main: Verlag Peter Lang, 1982); and Thomas B. Stoel Jr., "Fluorocarbons: Mobilizing Concern and Action," in Kay and Jacobson, *Environmental Protection: The International Dimension*, pp. 45–74. Thomas E. Downing and Robert W. Kates, "The International Response to the Threat of Chlorofluorocarbons to Atmospheric Ozone," *American Economic Review* 72 (May 1982): 267–72.

16. U.S. Congress, Senate Committee on Aeronautical and Space Science, *The International Legal and Institutional Aspects of the Stratospheric Ozone Problem*, 94th Cong., 1st sess., 15 August 1975; *Protection against Depletion of Stratospheric Ozone by Fluorocarbons*, Report Prepared by the Committee on Impact of Stratospheric Change, Assembly of Mathematical and Physical Sciences, Committee on the Alternatives for the Reduction of Fluorocarbon Emissions, and the Commission on Socio-Technical Systems (Washington, D.C.: National Academy of Sciences, 1979). "Scientists Issue Latest Views on Ozone Damage," *Uniterra* 5 (November–December 1980): 3–4; and Peter Huhne, "WMO Reiterates Risk of Ozone Depletion," *Ambio*, no. 1 (1982): 70.

17. Craig R. Whitney, "80 Nations Favor Ban to Help Ozone," *New York Times*, 3 May 1989, p. 13; Richard Benedick, *Ozone Diplomacy: New Directions in Safeguarding the Planet* (Cambridge, Mass.: Harvard University Press, 1991); and Karen T. Litfin, *Ozone Discourses: Science and Politics in Global Cooperation* (New York: Columbia University Press, 1994).

18. Stephen H. Schneider, "The CO_2 Problem: Are There Policy Implications—Yet? An Editorial," *Climatic Change* 2 (1980): 205. But see Eleanor Randolph, "Experts Find Possible Climatic Bomb," *Los Angeles Times*, 9 August 1981, sec. A3, p. 1; and James E. Hansen et al., "Climate Impact of Increasing Atmospheric Carbon Dioxide," *Science* 213 (28 August 1981): 957–66. Also see Charles C. Coutant, "Foreseeable Effects of CO_2-Induced Climatic Change: Freshwater Concerns," *Environmental Conservation* 8 (Winter 1981): 285–97.

19. George M. Woodwell et al., "The Biota and the World Carbon Budget," *Science* 199 (13 January 1978): 141–46; U.S. Congress, Senate Committee on Government Affairs, *Carbon Dioxide Accumulation in the Atmosphere, Synthetic Fuels and Energy Policy: A Symposium* (Washington, D.C.: Government Printing Office, 30 July 1978); National Research Council, Climate Research Board, *Carbon Dioxide and Climate: A Scientific Assessment*, Report of an Ad Hoc Study Group on Carbon Dioxide and Climate, Woods Hole, Mass., 23–27 July 1979, to the Climate Research Board (Washington, D.C.: National Academy of Sciences, 1979); U.S. Department of Energy, *Carbon Dioxide Effects Research and Assessment Program* (Springfield,

Va.: National Technical Information Service, 1979–81), as of January 1981, fourteen reports were published; Boyd R. Strain and Thomas V. Armentano, *Position Paper on Environmental and Societal Consequences of CO₂ Induced Climatic Change: Response of "Unmanaged Ecosystems,"* (Durham, N.C.: Duke University Phytotron Laboratory, 11 October 1980), paper produced for the Climate Program of the American Association for the Advancement of Science; U.S. Council on Environmental Quality, *Global Energy Futures and the Carbon Dioxide Problem* (Washington, D.C.: Government Printing Office, January 1981); C. D. Keeling and R. B. Bacastow, "Impact of Industrial Cases on Climate," in *Energy and Climate: Studies in Geophysics* 14 (Washington, D.C.: National Academy of Sciences, 1977), pp. 72–95.

20. Jill Williams, ed., *Carbon Dioxide, Climate and Society: Proceedings of an IIASA Workshop, February 21–24, 1978* (New York: Pergamon Press, 1978), p. 1. See also B. Bolin et al., eds., *The Global Carbon Cycle*, Scope 13 (New York: John Wiley, 1979); and the two following reports: National Research Council, *Changing Climate* (Washington, D.C.: National Academy Press, 1983), reviewed in *Science* 222 (4 November 1983): 491; and Stephen Seidel and Dale Keyes, *Can We Delay a Greenhouse Warming? The Effectiveness and Feasibility of Options to Slow the Build-up of Carbon Dioxide in the Atmosphere* (Washington, D.C.: U.S. Environmental Protection Agency, 1983). In addition, see C. M. Woodwell et al., "Global Deforestation: Contribution to Atmospheric Carbon Dioxide," *Science* 222 (9 December 1983): 1081–86; and Richard A. Kerr, "The Carbon Cycle and Climate Warming," *Science* 222 (9 December 1983): 1107–8.

21. U.S. Environmental Protection Agency, *Acid Rain*, EPA-600/9-79-036 (Washington, D.C.: EPA, July 1980); EPA, *Research Summary*, EPA-600/8-79-028 (Washington, D.C.: EPA, October 1979); "How Many More Lakes Have to Die?" *Canada Today* 12 (February 1981): 2–11; D. M. Whelpdale, *Atmospheric Pathways of Sulphur Compounds: A General Report* (London: Monitoring and Assessment Research Centre [MARC], Chelsea College, University of London, 1978); W. B. Clapham, "Acid Precipitation," *Mazingira* 5, no. 3 (1981): 8–19; John E. Carroll, *Acid Rain: An Issue in Canadian-American Relations* (Toronto and Washington: C.D. Howe Institute and National Planning Association, 1982); Ross Howard and Michael Perley, *Acid Rain: The North American Forecast* (Toronto, Canada: Anansi, 1980); Karen A. Mingst, "Evaluating Public and Private Approaches to International Solutions to Acid Rain," *Natural Resources Journal* 22 (January 1982): 5–20; Irene H. van Lier, *Acid Rain and International Law* (Toronto and the Netherlands: Bunsel Environmental Consultants and Sijthoff and Noordhoff at Alphen Aan Den Rijn, 1980); and Sweden, Ministry of Agriculture, Environment '82 Committee, *Acidification Today and Tomorrow*, trans. Simon Harper (Stockholm: Ministry of Agriculture, n.d.). See also *Acid News*, newsletter published by the Swedish NGO Secretariat on Acid Rain, Swedish Society for the Conservation of Nature.

22. Frank Fraser Darling, "Forestry, the Environment, and Man's Needs," *Unasylva* 27 (Winter 1974–75): 2–8; and A. Gomez-Pompa, C. Vasquez-Yomes, and S. Guevara, "The Tropical Rain Forest: A Non-Renewable Resource," *Science* 177 (1 September 1972): 762–75. Intergovernmental Panel on Climate Change, *Scientific Assessment of Climate Change: Report Prepared for IPCC by Working Group One* (June 1990); *Climate Change: The IPCC Response Strategies* (World Meterological Organization and United Nations Environment Programme, Intergovernmental Panel on Climate Change, Geneva, 1990); W. J. M. Tegart, G. W. Sheldon and D. C. Griffiths, *Climate Change: The IPCC Impact Assessment* (Canberra, Australia: Australian Government Publishing Services, 1990). U.S. Congress, Senate Committee on Energy and Natural Resources, *Global Climate Change, Hearings before the Committee on Energy and Natural Resources*, 102d Cong., 2d sess., 6 and 12 May 1992; U.S. Congress, House Subcommittee on Health and the Environment of the Committee on Energy and Commerce, *Global Climate Change and Greenhouse Emissions*, Hearings before the House Subcommittee on Health and the Environment, 102d Cong., 1st sess., 21 February and 1 August 1991; David D. Kemp, *Global Environmental Issues–A Climatological Approach*, 2d ed. (London: Routledge, 1994). See also notes under ocean levels.

23. LaMont C. Cole, "Can the World Be Saved?" *BioScience* 18 (July 1978): 679–84; Cole, "Are We Running Out of Oxygen?" *Catalyst for Environmental Quality* 1 (1970): 2–4; Paul Ehrlich, "Eco-Catastrophe," in *Eco-Catastrophe* (San Francisco: Canfield Press, 1970), pp. 1–14, a cautionary science fiction scenario by the editors of *Ramparts*. But see Wallace S. Broecker, "Man's Oxygen Reserve: Claims That This Important Resource Is in Danger of Serious Depletion Are Not at All Valid," *Science* 168 (26 June 1970): 1537–38; and the chapter entitled, "The Contemporary Atmosphere," in James E. Lovelock, *Gaia: A New Look at Life on Earth* (New York: Oxford University Press, 1979).

24. See Reid A. Bryson, "Climatic Modification by Air Pollution," in Nicholas Polunin, ed., *The Environmental Future: Proceedings of the First International Conference on the Environmental Future, Held in Finland from 27 June to 3 July 1971* (New York: Barnes and Noble, 1972), pp. 133–73. On effects of dust, see Pewe, *Desert Dust*; and J. M. Prospero, "Mineral and Sea Salt Aerosol Concentrations in Various Ocean Regions," *Journal of Geophysical Research* (20 February 1979): 725–31. See also *WMO Bulletin* 29 (October 1980): 250–55; and 30 (January 1981): 3–9.

25. Bernard Cosset, "First GARP Global Experiment," *WMO Bulletin* 28 (January 1979): 5–17. See also "WMO-ICSA Global Atmospheric Research Programme (GARP)," *Annual Report of Intergovernmental Oceanographic Commission, 1976–1977* (Paris: UNESCO, 1977), pp. 8–9. At least eighteen national governments, the Intergovernmental Oceanographic Commission, and the ICSU Scientific Committee on Oceanographic Research (SCOR) cooperated with the Joint Organizing Committee (JOC) of GARP in managing the complex experiment.

26. *International Perspectives on the Study of Climate and Society: Report of the International Workshop on Climate Issues, Schloss Laxenbourg, Austria, April 1978, to the Climate Research Board, National Research Council* (Washington, D.C.: National Academy of Sciences, 1978).

27. Cited in *World Climate Conference, February 12–23, 1979, Proceedings*, WMO, no. 557 (Geneva: WMO, 1979), p. 6, and note bibliographic references; *Outline Plan and Basis for the World Climate Programme, 1980–1983*, WMO, no. 540 (Geneva: WMO, 1980); and B. W. Bovine and B. R. Doos, "Why a World Climate Programme?" *Nature and Resources* 17 (January–March 1981): 2–7. See articles in journal *Climate Change* (Dordrecht, Netherlands: D. Reidel, 1977–). See also David D. Kemp, *Global Environmental Issues: A Climatological Approach*, 2d ed. (London and New York: Routledge, 1994). On a possible international convention on the atmosphere, see Toufiq A. Siddiqi, "A Comprehensive Law of the Atmosphere as a Framework for Addressing Carbon Dioxide and Climate Change Issues," in *Steps toward an International Convention Stabilizing the Composition of the Atmosphere* (Woods Hole, Mass.: Woods Hole Research Center, 1989): 59–65.

28. "Declaration of Legal Principles Governing the Activities of States in the Exploration and Use of Outer Space," Resolution 1962 (XVIII), in UN General Assembly, *Official Records*, Supplement 15, 1963. For the text of "Treaty on Principles Governing the Activities of States in the Exploration and Use of Outer Space, Including the Moon and Other Celestial Bodies," 27 January 1967, see U.S. Department of State, *United States Treaties and Other International Agreements* (hereinafter, *US Treaties*), vol. 18, pt. 3, p. 2410, TIAS no. 6347; or *United Nations Treaty Series*, vol. 610 (1970), no. 8843, p. 205; also Carl Q. Christol, "Protection of Space from Environmental Harms," *Annals of Air and Space Law* 4 (1979): 433–58; and U.S. Congress, Senate Committee on Aeronautical and Space Sciences, *International Cooperation in Outer Space: A Symposium*, 92d Cong., 1st sess., 9 December 1971, Document No. 92-57.

29. See reports of the UN Committee on the Peaceful Uses of Outer Space; U.S. Congress, Senate Committee on Commerce, Science, and Transportation, *Space Law: Selected Basic Documents*, 2d ed. (December 1978); Delbert D. Smith, *Space Stations: International Law and Policy* (Boulder, Colo.: Westview Press, 1979); and "The International Regime of Outer Space," Appendix F of *Civilian Space Policy and Applications* (Washington, D.C.: Office of Technology Assessment, June 1982).

30. UN Conference on the Exploration and Peaceful Uses of Outer Space (Vienna, 1968), *Space Exploration and Applications: Papers Presented at the . . . Conference, Vienna, 14–27 August 1968*, 2 vols. (New York: United Nations, 1969), A/CONF.34/2, vols. 1–2; and *Practical Benefits of Space Exploration: A Digest of Papers presented at the . . . Conference* (New York: United Nations, 1969).

31. Ralph Chipman, ed., *The World in Space: A Survey of Space Activities and Issues* (Englewood Cliffs, N.J.: Prentice-Hall, 1982), papers prepared for UNISPACE '82; and Dag Hammerskjold Library, *Outer Space: A Selected Bibliography*, ST/LIB/Ser.B33 (New York: United Nations, 1982); and *Space Science Research in the United States: A Technical Memorandum* (Washington, D.C.: Office of Technology Assessment, September 1982).

32. Report of the Second United Nations Conference on the Exploration and Peaceful Uses of Outer Space, Vienna, 9–21 August 1982, A/CONF.101/10 (New York: United Nations, 1982).

33. For a U.S. evaluation, see Office of Technology Assessment, *UNISPACE '82: A Context for International Cooperation and Competition* (Washington: OTA, March 1983). See also Harvey Brooks, "Managing the Enterprise in Space," *Technology Review* 86 (April 1983): 38–47.

34. Michael Crichton, *The Andromeda Strain* (New York: A. A. Knopf, 1969). See also P. H. A. Sneath, "Dangers of Contamination of Planet and Earth," in *The Biology of Space Travel*, ed. N. W. Pirie (London: Institute of Biology, 1961), pp. 95–106.

35. For accounts of these experiments, see "The Integrity of Science: A Report by the AAAS Committee on Science in the Promotion of Human Welfare," *American Scientist* 53 (June 1965): 174–98; and Edward W. Lawless, "Project West Ford—Orbital Belt of Needles," in *Technology and Social Shock* (New Brunswick, N.J.: Rutgers University Press, 1977), pp. 375–79.

36. Donald M. Waltz, "The Promise of the Space Factory," *Technology Review* 79 (May 1977): 38–49. The author provides an extensive bibliography.

37. Smith, *Space Stations*; and Jean-Claude Picker, *Space Observatories*, trans. Janet Rountree Lisk (Dordrecht, Netherlands: D. Reidel, 1970).

38. Michael J. Caffey and Thomas B. McCord, "Mining Outer Space," *Technology Review* 79 (June 1977): 50–59; and William K. Holtmann, "Mines in the Sky Promise Riches, a Greener Earth," *Smithsonian* 13 (September 1982): 70–76. See also Audrey V. Irvin (Associated Press), "Lunar Mining Rights Sought by Earthmen," *Pasadena Star-News*, 1 March 1981; and Robert J. Neumann and Harvey W. Herring, *Materials Processing in Space: Early Experiments* (Washington, D.C.: National Aeronautics and Space Administration, 1980).

39. R. D. Johnson and C. Holbrow, "Space Settlements: A Study Design," NASA Publication no. Sp-413, 1977. See also Eugene B. Konecci, "Space Ecological Systems," in *Bioastronautics*, ed. Karl Ernst Schaefer (New York: Macmillan, 1964), pp. 274–304.

40. M. Mitchell Waldrop, "Citizens for Space," *Science* 211 (9 January 1981): 152. The British Interplanetary Society (founded 1933) has been more concerned with space travel.

41. Michael Modell, "Sustaining Life in a Space Colony," *Technology Review* 79 (July-August 1977): 36–43; "Colonies in Space," *Newsweek* (27 November 1978): 95–101; Edward F. Crawley, "Designing the Space Colony," *Technology Review* 79 (July-August 1977): 45–50; Gerard K. O'Neill, "The Colonization of Space," *Physics Today* 32 (September 1974): 32–40; and O'Neill, *The High Frontier: Human Colonies in Space* (New York: Morrow, 1977). See also George S. Robinson, *Living in Outer Space* (Washington, D.C.: Public Affairs Press, 1975); and Robinson, *Space Trek* (Harrisburg, Pa.: Stackpole Books, 1978).

42. For questions regarding space colonization, see John Holt, "Outpost of Progress," *Technology Review* 79 (July-August 1977): 12–13, 80. But see also J. Peter Vojk, "The Impact of Space Colonization on World Dynamics," *Technological Forecasting and Social Change* 9 (1976): 361–99; T. Stephen Cheston, "Space Technology—A Subject for Social Scientists and Humanity Scholars," *Technology Tomorrow* 3 (October 1980): 3–5; and Arthur C. Clarke, "Beyond Centaurus," in *Voices from the Sky* (New York: Pyramid Books, 1967), pp. 40–51.

43. David M. Leive, *International Telecommunications and International Law: The Regulations of the*

Radio Spectrum (Leyden: Sijthoff, 1970); and *The Future of the International Telecommunications Union* (Washington, D.C.: American Society for International Law, 1972). For a more detailed account of international management of the electromagnetic spectrum, see chap. 10, "Telecommunications: Managing a Technological Revolution," in Marvin S. Soroos, *Beyond Sovereignty: The Challenge of Global Policy* (Columbia: University of South Carolina Press, 1986), pp. 323–49 (bibliographical references).

44. Thomas M. Donahue, "The Upper Atmosphere," in *The Atmospheric Sciences: Problems and Applications*, pp. 199–205. Note also pp. 104–5.

45. H. M. Assenheim, ed., *The Biological Effects of Radio-Frequency and Microwave Radiation* (Ottawa: Division of Biological Sciences, National Research Council of Canada, 1979); and Joseph H. Battochetti, *Electromagnetism, Man and the Environment* (London: Paul Elek, 1976); and John Salter, "Environmental Impact of Telecommunications," *European Environmental Law Review* 3 (February 1994): 46–53.

46. *Satellite Situation Report*, vol. 29, no. 2, Goddard Space Flight Center, NASA, 30 June 1989. For current information in a rapidly developing technology, see *Satellite Communications*, 1977– ; *Satellite News*, 1978– ; *Satellite Telecommunications Newsletter*, 1978– ; and *Satellite Week*, 1979–. Including active satellites, an estimated total of 20,119 artificial objects were reported as orbiting the earth. International controversies may be developing over space allocation for future satellites; see "Communications: A Special Report," *IEE Spectrum* 16 (October 1979): 2881.

47. Nicholas N. Matte and Hamilton De Saussure, eds., *Legal Implications of Remote Sensing from Outer Space* (Leyden, Netherlands: A. W. Sijthoff, 1976); and Hamilton De Saussure, "Remote Sensing by Satellite: What Future for an International Regime?" *American Journal of International Law* 71 (October 1977): 707–24. See also M. Griggs, "Legal Aspects of Remote Sensing and Air Enforcement," *Journal of Air Pollution* 28, no. 2 (1978): 519–22.

48. Gerard K. O'Neill, "Space Colonies and Energy Supply to the Earth," *Science* 190 (5 December 1975): 943–47; and Carl Q. Christol, *Satellite Power Systems (SPS): International Agreements*, sponsored by NASA and the Department of Energy and prepared for PRC Energy Analysis Company, October 1978, N79-23497/7GA (Springfield, Va.: National Technology Information Service, 1978). See also Thomas L. Neff, *The Social Costs of Solar Energy: A Study of Photovoltaic Energy Systems* (New York: Pergamon Press, 1981).

49. For the text of the treaty, in effect 16 July 1979, see "Convention on the International Maritime Satellite Organization (INMARSAT) and Operating Agreements," in *Space Law*, pp. 409–47; and "INMARSAT: A New Communications System for Shipping," *IMCO News*, no. 2 (1979): 9–11. Also see U.S. Congress, House Subcommittee on Communications of the Committee on Interstate and Foreign Commerce, *International Communications Services, on the Need for an Improved and Expanded System of International Telecommunications*, 95th Cong., 1st sess., 15, 16, 22, 23 March 1977, Serial No. 95-56, p. 408; on the United States Marine Satellite System (MARISAT), see ibid., pp. 406–7.

50. Joseph N. Pelton, *Global Communications Satellite Policy: INTELSAT Politics and Functionalism* (Mt. Airy, Md.: Lamond, 1974). For the text of the International Telecommunications Satellite Organization (INTELSAT) multilateral treaty and operating agreements, see *Space Law*, pp. 173–304. For commentary, see "Statement by Santiago Astrain, Director General, International Telecommunications Satellite Organization," *International Communications Services*, pp. 256–63; and U.S. Congress, Senate Subcommittee on Communications of the Committee on Science and Transportation, *Oversight Hearing on International Telecommunications Policies*, 95th Cong., 1st sess., 13 July 1977, no. 95-54, pp. 136–40.

51. *International Telecommunications Policies*, pp. 148–49 (see "Communications: A Special Report"). For text-relevant portions of the Spoleto document, see "Europe-North America Telecommunications Service Principles and Policy Development 1980–1990: The European Views," in U.S. Congress, House Subcommittee on Communications of the Committee on Interstate and Foreign Commerce, *The Communications Act of 1978*, vol. 2, pt. 2, 95th Cong., 2d sess., no. 95-196.

52. Marvin S. Soroos, "The Commons in the Sky: The Radio Spectrum and Geosynchronous Orbit as Issues in Global Policy," *International Organization* 36 (Summer 1982): 665–77. For a review of issues, see Abram Chayes, Paul Laskin, and Monroe Price, *Direct Broadcasting from Satellites: Policies and Problems* (Washington, D.C.: West Publishing, for the American Society of International Law, 1975). Note especially pp. 80–84 and 89–98.

53. For a comprehensive treatment of environmental remote sensing and its applications, see Joseph Lintz Jr. and David S. Simonett, *Remote Sensing of Environment* (Reading, Mass.: Addison-Wesley, 1976); Donald J. Clough and Lawrence W. Morley, eds., *Earth Observation Systems for Resource Management and Control, NATO Conference Series* (New York: Plenum Press, 1977); and Michel Lenco, "Remote Sensing and Natural Resources," *Nature and Resources* 18 (April–June 1982): 2–9; and the journal *Remote Sensing of Environment* (Amsterdam: Elsevier Science Publisher, 1968–). See also M. I. Dyer and D. A. Crossley Jr., *Coupling of Ecological Studies with Remote Sensing: Potentials at Four Biosphere Reserves in the United States* (Washington, D.C.: U.S. Department of State Publication 9504, 1986); V. E. Sokolov and B. V. Vinogradov, "Man and the Biosphere: The View from Above," *Nature and Resources* 22, nos. 1–2 (January 1986): 13–23 (Remote Sensing for Ecology in the USSR).

54. U.S. Congress, Senate Committee on Aeronautical and Space Sciences, *An Analysis of the Future Landsat Effort*, 94th Cong., 2d sess., 10 August 1976. See also U.S. Congress, House Subcommittee on Space Science and Applications and the Subcommittee on Natural Resources and Environment of the Committee on Science and Technology, *Operational Civil Remote Sensing Systems*, 96th Cong., 2d sess., 24–25 June and 29 July 1980, no. 131. Parallel hearings were held in the Senate: *Civil Remote Sensing Satellite Systems*, 26 June and 24 July 1980, no. 96–111. In March 1983 President Ronald Reagan announced a decision to sell the government's weather satellites and land remote sensing satellites (Landsat) to the private sector. The announcement was followed by vigorous objection from scientific and international communities. See M. Mitchell Waldrop, "Landsat Plan Hits Stormy Weather," *Science* 219 (25 March 1983): 1410.

55. Peter Jankowitsch, "The U.N.: Framework for a Consensus on Remote Sensing"; and Martin Mentor, "The United Nations' Contribution towards an International Agreement on Remote Sensing," in Nicholas N. Matte and Hamilton De Saussure, eds., *Legal Implications of Remote Sensing from Outer Space* (Leyden, Netherlands: A. W. Sijthoff, 1976), pp. 159–66.

56. Ricardo Umali, "Landsat: Uninvited Eye," *East–West Perspectives* 1 (Winter 1980): 12–21.

57. *Remote Sensing from Space: Prospects for Developing Countries: Report of the Ad Hoc Committee on Remote Sensing for Development* (Washington, D.C.: National Academy of Sciences, 1977). This report contains extensive bibliographical references and surveys the principal issues and problems. See also Chris J. Johannsen and James L. Sanders, eds., *Remote Sensing for Resource Management* (Ankeny, Iowa: Soil Conservation Society of America, 1982).

58. International Commission for the Study of Communication Problems, *The New World Information Order* (Paris: UNESCO, 1978); *The New World Information Order: A Selective Bibliography* (New York: Dag Hammarskjöld Library, United Nations, 1984); and *Communication and Society: A Documentary History of a New World Information and Communication Order Seen as an Evolving and Continuous Process, 1975–1986* (Paris: UNESCO, 1986).

59. See John M. Van Dyke, Durwood Zaelke, and Grant Hewson, eds., *Freedom of the Seas in the 21st Century* (Washington, D.C.: Island Press, 1994); Richard A. Davis, *Principles of Oceanography* (Reading, Mass.: Addison-Wesley, 1977); John C. Harvey, *Atmosphere and Ocean* (Sussex, United Kingdom: Artemis, 1976); and Mykola M. Volekvakha, *Water and Air of Our Planet* (Kiev, now Ukraine: Nankova Dumka Publishers, 1975). There are as well numerous journals and annuals concerned specifically with the oceans and relevant to international environmental policies. Among them are: *Marine Policy Reports* (Newark, Del.), 1977– ; *Journal of Marine Research* (New Haven, Conn.), 1937– ; *Ocean, Marine Technology Society, Annual Conference Proceedings* (Washington, D.C.), 1965– ; *Ocean Development and International Law* (New York), 1973– ; *Ocean Management* (Amsterdam), 1973– ; *Ocean Science News* (Washing-

ton, D.C.), 1958–; *Oceans* (San Diego, Calif.), 1969–; *Oceanus* (Woods Hole, Mass.), 1952–; and Harold Barnes and Margaret Barnes, eds., *Oceanography and Marine Biology: Annual Review* (London: George Allen and Unwin), 1963–. More specifically on governance of the oceans see Elisabeth Mann Borgese and David Krieger, eds., *The Tides of Change: Peace, Pollution and Potential of the Oceans* (New York: Mason/Charter, 1975), p. 357; Elisabeth Mann Borgese, *The Ocean Regime* (Santa Barbara, Calif.: Center for the Study of Democratic Institutions, 1968); and Borgese, *The New International Economic Order and the Law of the Sea* (Valletta, Malta: International Ocean Institute, 1975). See also Robert L. Friedheim, *Understanding the Debate on Ocean Resources* (Denver, Colo.: University of Denver Press, 1969); Edmund A. Gullion, ed., *Uses of the Seas* (Englewood Cliffs, N.J.: Prentice-Hall, 1968); John Robert Victor Prescott, *The Political Geography of the Ocean* (New York: John Wiley, 1975); "The UN and the Sea," *UNITAR News* 6 (5 November 1974): 2–27; Edward Wenk Jr., *The Politics of the Oceans* (Seattle: University of Washington Press, 1972); James C. F. Wang, *Handbook on Ocean Politics and Law* (Westport, Conn.: Greenwood Press, 1992); and *Symposium on the International Regime of the Sea-bed*, ed. J. Sztucki (Rome: Accademia Nazionale dei Lincei, 1970). For a further perspective, see Leonid M. Brekhovskikh, "Foreword: World Ocean and International Organizations," in Bekiashev and Serebriakov, *International Marine Organizations*.

60. John King Gamble Jr., "Status of the UN Convention on the Law of the Sea," *Marine Policy Reports* 10, no. 1 (July 1988): 105; James K. Sebenius, *Negotiating the Law of the Sea* (Cambridge, Mass.: Harvard University Press, 1984); and Jacques C. Richardson, *Managing the Ocean: Resources, Research, and Law* (Mt. Airy, Md.: Lomond Publications, 1985).

61. Joint Group of Experts on the Scientific Aspens of Marine Pollution (CESAMP), *Report of the Tenth Session held at UNESCO Headquarters, Paris, 29 May–2 June 1978*, Rep. Stud. CESAMP, no. 9. See also John Lawrence Hargrove, ed., *Who Protects the Oceans? Environment and the Development of the Law of the Sea* (St. Paul, Minn.: West Publishing Co., for the American Society of International Law, 1975), pp. 96–98; and Velimir Pravdic, *CESAMP: The First Dozen Years* (Geneva: UNEP, Regional Seas Programme Activities Centre, 1981).

62. For a brief description of the Intergovernmental Oceanographic Commission (IOC), its membership organization, and activities, see its undated leaflet entitled *Intergovernmental Oceanographic Commission* (Paris: UNESCO, [1977]). Note also reports of the sessions of the IOC (e.g., *Report of the Eleventh Session of Intergovernmental Oceanographic Commission, Paris, 15 October–3 November 1979* [Paris: UNESCO]). The IOC also issues *Biennial Reports*.

63. See National Academy of Engineering, *An Oceanic Quest: The International Decade of Ocean Exploration* (Washington, D.C.: National Academy of Sciences, 1969), p. 115; "National Science Foundation Launches Ten IDOE Projects," *International Marine Science Newsletter*, no. 1 (February 1973); and *Report of the Decade: The International Decade of Ocean Exploration* (Washington, D.C.: National Science Foundation, Division of Ocean Sciences, n.d.); Margaret E. Galey, *The Intergovernmental Oceanographic Commission: Its Capacity to Implement an International Decade of Ocean Exploration*, Occasional Paper, no. 2 (Kingston: University of Rhode Island, Law of Sea Institute, 1973), p. 38; and Lauriston King, "The International Decade of Ocean Exploration in Marine Science Affairs," *Marine Technology Journal* 11 (1978): 10–15. On the Long-Term and Expanded Programme of Oceanic Exploration and Research (LEPOR), see *Biennial Report of Intergovernmental Oceanographic Commission, 1976–1977*, p. 5.

64. "National Oceanographic Programmes — NOP's DNPs — Declared National Programmes," IOC/INF/4 Rev. (Paris: UNESCO, n.d.).

65. C. F. Humphrey, "The Unknown Indian Ocean: An International Investigation," *New Scientist* 9 (5 January 1961): 36–38. For a description of the purpose and organization of the International Cooperative Investigation of the Tropical Atlantic (ICITA), see the preface to A. C. Kolesnikov (director of the Marine Hydrophysical Institute, Ukrainian Academy of Science, U.S.S.R.), ed., *Equalant I and Equalant II, Oceanographic Atlas* (Washington, D.C.: Government Printing Office, 1973).

66. See "Global Investigation of Pollution in the Marine Environment (GIPME)," *Biennial Report of Intergovernmental Oceanographic Commission, 1976–1977*, pp. 9ff. Also Edward D. Goldberg, *The Health of the Oceans* (Paris: UNESCO, 1976).

67. "UNESCO Resumes Publication of IMS," *IMCO Newsletter* 1 (February 1979). From April 1963 to February 1970 *International Marine Science* was published as a quarterly, that is, through vol. 7, no. 4.

68. M. A. Kohler, "The International Hydrological Decade," *UNO Bulletin* 12 (October 1963): 193–97; and Michel Batisse, "Launching the Hydrological Decade," *New Scientist* 25 (7 January 1965): 38–40. Also, Andrew Jamison, "IHD: International Symbol or 'National Embarrassment'?" *Science* 161 (13 September 1968): 1118–19.

69. See UN, FAO, International Technical Conference (Rome, 18 April–10 May 1955), "Papers Presented at the International Technical Conference on the Conservation of the Living Resources of the Sea," A/Conf.10/7, 1956.

70. See Mario Ruvio, ed., *Marine Pollution and Sea Life* (London: Fishing News [Books], 1972) for edited contributions and summaries of discussions at the FAO Technical Conference on Marine Pollution and Its Effects on Living Resources and Fishing, Rome, 9–18 December 1970.

71. Angelantonia D. M. Forte, "Nuclear Pollution of the Marine Environment: Some Recent Controls," in *The Impact of Marine Pollution*, ed. Douglas J. Cusine and John P. Grant (London: Croom Helm, 1980), pp. 241–50; IAEA, "Disposal of Radioactive Wastes in Seas, Oceans, and Surface Water," Proceedings Series of IAEA (New York: UNIPUB, 1966); *Disposal of Radioactive Wastes into Marine and Fresh Waters*, IAEA Bibliographical Series, no. 5 (New York: UNIPUB, 1962); and Thomas C. Jackson, *Nuclear Waste Management: The Ocean Alternative*, Proceedings of the Public Policy Forum Sponsored by the Ocean Society in the Georgetown University Law Center, February 1980 (New York: Pergamon Press, 1981).

72. International Environmental Programs Committee, *Early Action on the Global Environmental Monitoring System [GEMS]* (Washington, D.C.: National Academy of Sciences, 1976); UNEP Governing Council, "The Global Environmental Monitoring System: Report of the Executive Director," Item 7(b) of Provisional Agenda, UNEP/GC/31/Add.2, Nairobi, 25 February 1975. Illustrative of the GEMS collaboration is the report *Air Monitoring Programme Design for Urban and Industrial Areas*, published under the joint sponsorship of UNEP, WHO, and WMO (Geneva: WHO, 1977).

73. R. S. Wimpenny, "The International Council for the Exploration of the Sea," *Nature* 170 (29 November 1952): 906–8; "International Council for the Exploration of the Sea," *Yearbook of International Organizations*, 1978, A1733; U.S. Congress, Senate, *International Council for the Exploration of the Sea*, Senate Report No. 93-31, 93d Cong., 2d sess., 22 August 1974, pp. 1–3. For continuing accounts of council activities, see *Journal du Conseil*, 1926–, and *Procès-Verbaux* of council meetings. Although the titles of these publications are in French, the greater part of the text is in English.

74. "General Fisheries Council for the Mediterranean," in *International Governmental Organizations*, comp. Dorothy Peaslee Xydis, pt. 2, 3d ed. (The Hague: Martinus Nijhoff, 1975), pp. 121–22; and "Agreement for the Establishment of a General Fisheries Council for the Mediterranean (CFCM), Resolution No. 39-63," *Resolutions Adopted by the FAO Conference at the Twelfth Session*, 1963 (Rome: FAO, 1964), pp. 44–45.

75. John P. Wise, "Food from the Sea: Myth or Reality," in *Drugs and Food from the Sea: Myth or Reality?* ed. Pushker N. Kaul and Carl J. Sinderman (Norman: University of Oklahoma Press, 1978), pp. 405–13; Frederick W. Bell, *Food from the Sea: The Economics and Politics of Ocean Fisheries* (Boulder, Colo.: Westview Press, 1978); and "Food from the Sea," *Oceanus* 18 (Winter 1975): entire issue.

76. Bell, *Food from the Sea*, pp. 339–40.

77. See Wilbert McLeod Chapman, "The Theory and Practice of International Fisheries Commissions and Bodies," in *Gulf and Caribbean Fisheries Institute, Twentieth Annual Ses-*

sion (Miami, Fla.: University of Miami, November 1967), pp. 77-105; *Principles for a Global Fisheries Management Regime*, prepared by an Expert Interdisciplinary Working Group (Washington, D.C.: American Society of International Law, 1974); Francis T. Christy Jr. and Anthony Scott, *The Common Wealth in Ocean Fisheries* (Baltimore, Md.: Johns Hopkins University Press, 1966); Edward Miles, *Organizational Arrangements to Facilitate Global Management of Fisheries* (Washington, D.C.: Resources for the Future, June 1974); Charles B. Heck, Collective Arrangements for Managing Ocean Fisheries," *International Organizations* 29 (Summer 1975): 711-44; Brian J. Rothschild, ed., *World Fisheries Policy: Multidisciplinary Views* (Seattle: University of Washington Press, 1972); and Oran Young, *Resource Management at the International Level: The Case of the North Pacific* (London: Frances Pinter, 1977). For a table listing international bodies concerned with fisheries management with acronyms and 1972 membership, see J. A. Gulland, *The Management of Marine Fisheries* (Bristol, United Kingdom: Scientechnica, 1974).

78. R. E. Kearney, "The Law of the Sea and Regional Fisheries Policy," *Ocean Development and International Law* 5, nos. 2-3 (1978): 249-86; and George Kent, *The Politics of Pacific Island Fisheries* (Boulder, Colo.: Westview Press, 1980).

79. James Joseph and Joseph W. Greenough, *International Management of Tuna, Porpoise, and Billfish: Biological, Legal, and Political Aspects* (Seattle: University of Washington Press, 1979), p. 190.

80. Ibid., p. 181.

81. "UN Conference on Straddling and Highly Migratory Fish Stocks," in Biliana Cincin-Sam and Robert W. Krecht, "Implications of the Earth Summit for Ocean and Coastal Governance," 148-49. Barbara Kwiatkowska, "The High Seas Fishing Regime: At a Point of No Return?" *International Journal of Marine and Coastal Law* 8, no. 3 (1993): 327.

82. See David W. Ehrenfeld, "The Fate of the Blue Whales," in *Conserving Life on Earth* (New York: Oxford University Press, 1972), pp. 205-13; Noel Simon, "Of Whales and Whaling," *Science* 149 (27 August 1965): 943-46; Ray Gambell and Sidney G. Brown, "Status and Conservation of the Great Whales," *IUCN Bulletin*, n.s., 2 (October-December 1971): 185-89; Jamff H. W. Hain, "The International Regulation of Whaling," *Marine Affairs Journal* (3 September 1975): 28-48; Larry L. Booda, "Of Whales and Men, Editorial," *Sea Technology* 21 (September 1980): 7; Maxine McCloskey, "Will Commercial Whaling End This Year?" *Sierra* 64 (July-August 1979): 41; W. E. Schevill, ed., *The Whale Problem: A Status Report* (Cambridge, Mass.: Harvard University Press, 1974); George L. Small, *The Blue Whale* (New York: Columbia University Press, 1971); Carl Q. Christol, John R. Schmidhauser, and George O. Totten, *Working Paper on the Law and the Whale: Current Developments in the International Whaling Controversy* (Washington, D.C.: Seventh Conference on the Law of the World, 12-17 October 1975); and John R. Schmidhauser and George O. Totten, eds., *The Whaling Issue in U.S.-Japan Relations* (Boulder, Colo.: Westview Press, 1978).

83. See Edward Mitchell, *Porpoise, Dolphin, and Small Whale Fisheries of the World: Status and Problems*, Monograph no. 3 (Morges, Switzerland: IUCN, 1975), p. 129. Also Karen Pryor and Kenneth Harris, "The Tuna-Porpoise Problem: Behavioral Aspects," *Oceanus* 21 (Spring 1978): 33-37. For effects of competitive overfishing, see Michael J. A. Butler, "Plight of the Bluefin Tuna," *National Geographic* 162 (August 1982): 220-39.

84. E.g., Malcolm J. Forster, "IWC Makes Some Progress," *Environmental Policy and Law* 5 (October 1979): 170-74; Anthony J. Mence, "Thoughts about the International Whaling Commission after Its Thirty-First Meeting," *Environmental Conservation* 6 (Winter 1979): 255-56; and Richard Fitter, "Whaling Bonus, Another Inch Forward," *Oryx* 16 (October 1981): 123-24. For background, see J. N. Tonnessen and A. O. Johnson, *The History of Modern Whaling* (Berkeley: University of California Press, 1982).

85. "Bowhead and Sperm Whaling Quotas See-saw," *Environmental Conservation* 5 (Spring 1978): 68; and Charles D. Evans and Larry S. Underwood, "How Many Bowheads?" *Oceanus* 21 (Spring 1978): 17-23.

86. *IUCN Bulletin*, n.s., 10 (April 1979): 25; *IUCN Bulletin*, n.s., 10 (June 1979): 49; and "Pirate Whaling," *IUCN Bulletin*, n.s., 10 (June 1979): 45-48. See also U.S. Congress, Senate Committee on Commerce, Science, and Transportation, *Outlaw Whaling—Hearing on Whaling Operations Conducted Outside the Control of the International Whaling Commission*, 96th Cong., 1st sess., 22 June 1979.

87. *International Environment Reporter 5*, no. 8, Current Report (11 August 1982): 329.

88. Lee Talbot, *IUCN Bulletin*, n.s., 13 (July, August, September 1982): 51.

89. Council of Europe, *Newsletter-Nature*, no. 82-12, p. 1; and *World Environment Report 9* (14 March 1983): 7.

90. See *Oceanus* 21 (Spring 1978): entire issue, especially Sidney J. Holt, "Changing Attitudes toward Marine Mammals," and Rodney V. Salm, "Strategies for Protecting Marine Mammal Habitats"; Michel J. Savini, *Report on International and National Legislation for the Conservation of Marine Mammals*, FAO Fisheries Circular, no. 326 (Rome: FAO, June 1974), p. 80; Michael J. Bean, "Marine Mammal Protection under Scrutiny," *Environmental Law* (Summer 1981): 1-2; Patricia Birnie, *Legal Measures for the Conservation of Marine Mammals*, Environmental Policy and Law Paper 19 (Gland, Switzerland: IUCN, 1982), looseleaf compilation; and *National Marine Sanctuary Program* (Washington, D.C.: National Oceanic and Atmospheric Administration, 1981). See also monthly issues of *Marine Mammal News*, 1975– (Washington, D.C.: Nautilus Press).

91. *Mammals in the Sea*, 4 vols. (Rome: FAO, 1978). On the FAO/UNEP Plan of Action, see UNEP Governing Council, *The Environment Programme: Programme Performance Report, January–April 1981, Report of the Secretary-General*, Annex 11, pp. 23-29; and *Draft Global Plan of Action for the Conservation, Management, and Utilization of Marine Mammals*, FAO/UNEP Project, no. 0502-78/02 (Rome: FAO, 1981).

92. See Daniel Navid, "World Conference on Sea Turtles," *Environmental Law and Policy* 6 (February 1980): 17-18; and Karen A. Bjorndal, ed., *Biology and Conservation of Sea Turtles: Proceedings of the World Conference on Sea Turtle Conservation* (Washington, D.C.: Smithsonian Institution Press, 1982). See also Archie Carr, *So Excellent a Fishe: A Natural History of Sea Turtles* (Garden City, N.Y.: Natural History Press, 1967); and *IUCN Bulletin*, n.s., 13 (July, August, September, 1982): 52-53.

93. See "Marine Biomedicine," *Oceanus* 19 (Winter 1976): entire issue. For detailed information, see M. H. Baslow, *Marine Pharmacology* (Huntington, N.Y.: Krieger, 1977); D. J. Faulkner and W. F. Fenical, eds., *Marine Natural Product Chemistry* (New York: Plenum Press, 1977); D. J. Faulkner, "The Search for Drugs from the Sea," *Oceanus* 22 (Summer 1979): 44-50; and Pushker N. Kaul and Carl J. Sinderman, eds., *Drugs and Food from the Sea* (Norman: University of Oklahoma Press). See also Richard A. Fralick and John H. Ryther, "Uses and Cultivation of Seaweeds," *Oceanus* 19 (Summer 1976): 32-39; Lewis Thomas, "Marine Models in Modern Medicine," *Oceanus* 19 (Winter 1976): 2-5; and Goran Michanek, *Seaweed Resources of the Ocean*, FAO Fisheries Paper, no. 138 (Rome: FAO, 1975). See also Nicholas V. C. Polunin, "Marine 'Genetic Resources' and the Potential Role of Protected Areas in Conserving Them," *Environmental Conservation* 10, no. 1 (Spring 1983): 31-41.

94. Kaul and Sinderman, *Drugs and Food from the Sea*, p. 226.

95. *Organizational Arrangements to Facilitate Global Management of Fisheries*, p. 22. See also William T. Burke, *The New International Law of Fisheries: UNCLOS 1982 and Beyond* (Oxford: Clarendon Press, 1994) (includes marine mammals); Evelyne Meltzer, "Global Overview of Straddling and Highly Migratory Fish Stocks: The Nonsustainable Nature of High Seas Fisheries," *Ocean Development and International Law* 25, no. 3 (July–September 1994): 255-344; Gulland, *The Management of Marine Fisheries*; and H. Gary Knight, *Managing the Sea's Living Resources: Legal and Political Aspects of High Seas Fisheries* (Lexington, Mass.: D. C. Heath, 1977).

96. Barbara Hastings, "Pacific Isles Try to Put Hook on Illegal Fishing," *Honolulu Star Bulletin*, 18 January 1981.

97. Santiago Astrain, "A New Decade for Intelsat in the Pacific," *Satellite Communications* 41 (April 1980): 16–17.

98. Robert B. Biggs, "Offshore Industrial-Port Islands," *Oceanus* 19 (Fall 1975): 56–66; and Satoshi Hayashi, "Offshore Structures: Their Present Situation and Problems," in *Marine Technology and Law: Development of Hydrocarbon Resources and Offshore Structures*, Proceedings of the Second International Ocean Symposium, 13-15 December 1977 (Tokyo: Japan Shipping Club, 1978). Note also in this publication Ben C. Gerwick Jr., "Current Projects in Offshore Structures," pp. 109–14; Kiyonore Kukutaki, "Offshore Structures and Human Environment," pp. 114–20; John P. Craven, "The Effect of Technology on the Ultimate Regime for Coastal Waters," pp. 128–29. See also Fred S. Ellers, "Advanced Offshore Oil Platforms," *Scientific American* 246 (April 1982): 39–49; John P. Craven, "Present and Future Uses of Floating Platforms," *Oceanus* 19 (Fall 1975): 67–71 (entire issue devoted to "Seaward Expansion"); and letter by Michael Brophy, *Times* (London), 31 January 1980, proposing an island Euro-airport.

99. Michael J. Herz, ed., *Palau and the Superport: The Development of an Ocean Ethic*, Oceanic Society Symposium, 1977 (San Francisco, Calif.: Ocean Society, 1977); and "Petition of the People of Palau to the Trusteeship Council of the United Nations," T/Pet., 10/121, 5 July 1977.

100. See Jack N. Barkenbus, *Deep Seabed Resources: Politics and Technology* (New York: Macmillan and Free Press, 1979); and *Oceanus* 25 (Fall 1982): entire issue.

101. The involvement of national and multinational enterprise in ocean mining is described by Jonathan Bartlett, ed., *The Ocean Environment*, The Reference Shelf (New York: Wilson, 1977), especially pp. 118 and 127–28. For a survey of U.S. marine science engineering capabilities as of 1968, see Otto F. Kline and Gibson M. Wolf, "The Oceans: Unexploited Opportunities," *Harvard Business Review* 46 (March–April 1978): 140–56, especially the table on pp. 150–51 that shows the activities of ninety-eight companies. Although the data are not recent, they indicate the scope of business involvement in marine research and development. See also Elaine H. Burnell and Piers von Simson, *Pacem in Maribus: Ocean Enterprises; A Summary of the Prospects and Hazards of Man's Impending Commercial Exploitation of the Underseas* (Santa Barbara, Calif.: Center for the Study of Democratic Institutions, 1970); and Kurt Michael Shusterich, *Resource Management and the Oceans: The Political Economy of Deep Seabed Mining* (Boulder, Colo.: Westview Press, 1982).

102. UN Ocean Economics and Technology Office, *Manganese Nodules: Dimensions and Perspectives* (Dordrecht, Netherlands: D. Reidel, 1979), p. 1; *Ocean Manganese Nodules*, United States Congressional Research Service, 94th Cong., 2d sess., 1976.

103. Wolfrum Rüdiger, "Common Heritage of Mankind," *Encyclopedia of Public International Law*, vol. 11, *Law of the Sea, Air and Space* (Amsterdam, Netherlands: Elsevier Science, 1989), pp. 65–69 (references); David Hunter, Julia Sommer, and Scott Vaughan, "Legal Status of Natural Resources and Common Areas," in *Concepts and Principles of International Environmental Law* (Environment and Trade Series) UNEP 1994, 35–40; Rudolph Preston Arnold, "The Common Heritage of Mankind as a Legal Concept," *International Lawyer* 9 (1975): 153–58. Also, from a marine science perspective, David A. Ross, "Whose Common Heritage?" *Oceanus* (Summer 1973): 2–13.

104. See Scott Allen and John P. Craven, eds., *Alternatives in Deepsea Mining* (Honolulu: University of Hawaii, Law of the Sea Institute, 1979).

105. Vincent J. Negrelli, "Ocean Mineral Revenue Sharing," *Ocean Development and International Law Journal* 5, nos. 2-3 (1978): 152–80.

106. See U.S. Department of Interior, Ocean Mining Administration, *Ocean Mining—An Economic Evaluation*, 1976; and Francis T. Christy Jr., "Marigenous Minerals: Wealth, Regimes, and Factors of Decision," in *Symposium on the International Regime of the Seabed*, pp. 113–53.

107. "Law of the Sea Forum: The 1994 Agreement on Implementation of the Seabed Provisions of the Convention on the Law of the Sea," *American Journal of International Law* 88, no. 4 (October 1994): 687–714.

108. For historical background on international marine pollution issues, see James Barros and Douglas M. Johnston, *The International Law of Pollution* (New York: Free Press, 1974), pp. 200–293; Douglas J. Cusine and John P. Grant, eds., *The Impact of Marine Pollution* (London: Croom Helm, 1980); Donald W. Hood, ed., *The Impingement of Man on the Oceans* (New York: Wiley Interscience, 1971); William H. Matthews, Frederick E. Smith, and Edward D. Goldberg, eds., *Man's Impact on Terrestrial and Oceanic Ecosystems* (Cambridge, Mass.: MIT Press, 1971); and "Marine Pollution," pt. 1 of *The New Nationalism and the Use of Common Spaces*, ed. Jonathan I. Charney (Totowa, N.J.: Allanheld, Osmun, 1982), pp. 7–111. For current information, see *Marine Pollution Bulletin*, 1970–. For a survey of the antipollution and marine protection treaties sponsored by IMCO, see "Development, Ratification, and Entry Into Force," *IMCO News*, no. 1 (1979) and subsequent issues. For a broader social perspective, see Virginia K. Tippie and Dana Kester, *Impact of Marine Pollution on Society* (South Hadley, MA: Bergin, 1982); Janet Pawlak, "Land-Based Inputs of Some Major Pollutants to the Baltic Sea," *Ambio* 9, nos. 3–4 (1980): 163; and General Fisheries Council for the Mediterranean, *The State of Marine Pollution in the Mediterranean and Legislative Controls* (Rome: FAO, 1972). See also references to regional seas, Chapter 8.

109. See "High Level Nuclear Wastes in the Seabed?" *Oceanus* 20 (Winter 1977): entire issue, especially Robert A. Frosch, "Disposing of High-Level Radioactive Waste," pp. 4–7. Also see notes and discussions under section entitled "Energy and Environment" in Chapter 10 of this volume; Daniel P. Finn, "International Cooperation to Protect the Marine Environment: The Case of Radioactive Waste Disposal," *Oceans 81*, pp. 601–4; and, "Ocean Disposal of Radioactive Wastes: The Obligation of International Cooperation to Protect the Marine Environment," *Virginia Journal of International Law* 21, no. 4 (1981): 621–90; and S. Jacob Scherr, "Radioactive Waste Disposal: The Quest for a Solution," in Kay and Jacobson, *Environmental Protection: The International Dimension*, pp. 101–18.

110. "International Convention for the Prevention of Pollution of the Sea by Oil, 1954," *United Nations Treaty Series*, vol. 327 (1959), no. 4714, pp. 4–33. For discussions of the 1954 London Convention and subsequent negotiations regarding prevention of marine pollution from ships, see Robert A. Shinn, *The International Politics of Marine Pollution Control* (New York: Praeger, 1974); R. Michael M'Gonigle and Mark W. Zacher, *Pollution, Politics, and International Law: Tankers at Sea* (Berkeley: University of California Press, 1979); Louis Henkin et al., *International Law: Cases and Materials* (St. Paul, Minn.: West Publishing Co., 1980), especially Section 4A, "Protection of the Marine Environment," pp. 398–407; and Alan B. Sielen and Robert J. McManus, "IMCO and the Politics of Ship Pollution," in Kay and Jacobson, *Environmental Protection: The International Dimension*, pp. 140–83.

111. "International Convention for the Prevention of Pollution from Ships (MARPOL), 1973," *International Legal Materials* 12 (1973): 1319–35; *International Environment Reporter*, Reference File, vol. 1, sec. 21, pp. 2301, 2346.

112. Barbara Maslam, "Global Marine Pollution Treaty Has Been Ratified," *World Environment Report* 8 (30 November 1982): 1–2. For a status summary of IMCO-sponsored treaties as of 1 May 1980, see *IMCO News*, no. 2 (1980), p. 2.

113. See J. E. Smith, ed., *'Torrey Canyon' Pollution and Marine Life: A Report by the Plymouth Laboratory of the Marine Biological Association of the United Kingdom* (Cambridge: Cambridge University Press, 1968). On the Brussels conference of 1969, see M'Gonigle and Zacher, *Pollution, Politics, and International Law*, pp. 152–58.

114. "International Convention Relating to Intervention on the High Seas in Cases of Oil Pollution Casualties, 29 November 1969," *UST*, vol. 26, pt. 1, p. 765, TIAS no. 8068; in effect 6 May 1975 (IMCO Public Law Convention).

115. "International Convention on Civil Liability for Oil Pollution Damage, 29 November 1969" (CLC or "Private Law" Convention), *International Legal Materials* 9 (1970): 45; and "Protocol" of 19 November 1976, *International Legal Materials* 16 (1977): 606–17.

116. "Resolution on Establishment of an International Compensation Fund for Oil Pollution Damage," *International Legal Materials* 9 (1970): 66–67.

117. "Convention on the Prevention of Marine Pollution by Dumping of Wastes and Other Matter," 29 December 1972, *UST*, vol. 26, pt. 2, p. 2403, TIAS no. 8165. See also U.S. Congress, House Committee on Foreign Affairs, Subcommittee on International Organizations and Movements, *International Implications of Dumping Poisonous Gas and Waste into the Oceans*, 91st Cong., 1st sess., 8–15 May 1969.

118. "Convention for the Prevention of Marine Pollution by Dumping from Ships and Aircraft (1972)," *International Legal Materials* 12 (1973): 251–89; see also *International Environment Reporter*, Reference File, vol. 2, sec. 121, pp. 0101–4 (Oslo Convention).

119. "Convention for the Prevention of Marine Pollution from Land-Based Sources (1974), *International Legal Materials* 13 (1974): 546–85; and *International Environment Reporter*, Reference File, vol. 2, sec. 121, pp. 201–6 (Paris Convention). See also Richard Tucker Scully, "International Regulation of Pollution from Land-Based Sources," *Marine Affairs Journal*, no. 3 (September 1975): 84–107.

120. Greg Batey, "Regulation and Assessment of Ocean Dumping," *Sea Technology* 20 (October 1979): 17–20, provides a concise account of the oceanic dumping problem. See also Robert J. McManus, "Ocean Dumping: Standards in Action," in Kay and Jacobson, *Environmental Protection: The International Dimension*, pp. 119–39; and Barbara Massam, "Sea Dumping Treaty Has Legal Problems," *World Environment Report* 9 (15 February 1983): 4.

121. "Belgium, Denmark, European Communities, Finland, France, Germany, Iceland, Ireland, Luxembourg, Netherlands, Norway, Portugal, Spain, Sweden, Switzerland, United Kingdom: Convention for the Protection of the Marine Environment of the North-East Atlantic," 32 *International Legal Materials* 32 (1993): 1069–71. For analysis of the convention see Ellen Hey, Jon Ijlstra, and Andrea Nollkaemper, "The 1992 Paris Convention for the Protection of the Marine Environment of the North-East Atlantic: A Critical Analysis," *International Journal of Marine and Coastal Law* 8, no. 1 (1993): 1–49.

122. United Nations Environment Programme, *Environment Data Report 1993–94* (Oceans and Sea Level), London: Blackwell, pp. 120–22; U.S. Senate Committee on Commerce, Science, and Transportation, *Global Change Research, Global Warming, and the Oceans, Hearing before the Committee on Commerce, Science, and Transportation*, 102d Cong., 2d sess., 20 May 1992; U.S. Senate Committee on Energy and Natural Resources, *Role of Oceans in Global Climate Change, Hearing before the Committee on Energy and Natural Resources*, 103d Cong., 2d sess., 8 March 1994.

123. U.S. Senate Committee on Energy and Natural Resources, *Global Climate Change and the Pacific Islands, Hearing before the Committee on Energy and Natural Resources*, 102d Cong., 2d sess., 26 May 1992.

124. Fred Hoyle, *Ice: The Ultimate Human Catastrophe* (New York: Continuum, 1981); S. G. Dobrovalski, *Global Climatic Changes in Water and Heat Transfer Accumulation Processes* (New York: Elsevier, 1992).

125. Garrett Hardin, "The Tragedy of the Commons," *Science* 162 (13 December 1968): 1243–48. See also Julian J. Edney, "The Commons Problem: Alternative Perspectives," *American Psychologist* 35 (February 1980): 131–50 (provides an extensive bibliography on the commons issue); Per Magnus Wijkman, "Managing the Global Commons," *International Organization* 36 (Summer 1982): 511–16; and Soedjatmoko, "Managing the Global Commons," *Mazingira* 6, no. 2 (1982): 32–39. For an economic perspective, see Ralph C. d'Arge, *On Managing the Global Commons* (Chapel Hill: Institute for Environmental Studies, University of North Carolina, 1981).

126. See John Kish, *The Law of International Spaces* (Leyden, Netherlands: A. W. Sijthoff, 1973). For historical background on relationships in which changes are impending, see Irvin L. White, *Decision-Making for Space: Law and Politics in Air, Sea and Outer Space* (West Lafayette, Ind.: Purdue University Studies, 1970).

127. Phil Kuntz, "Pie in the Sky: Big Science Is Ready for Blastoff," *Congressional Quarterly Weekly Report* 48, no. 17 (28 April 1990): 1254–60.

128. See *Washington Post*, 12 May 1990, A6; and *Weekly Compilation of Presidential Documents*,

vol. 26, no. 10, pp. 381–82, and vol. 26, no. 20, pp. 748–49. Also William K. Stevens, "Huge Space Stations Seen as Distorting Studies of Earth," *New York Times*, 19 June 1990, B5, B8.

10. Sustainability: Population, Resources, Development

1. Clive Ponting, *A Green History of the World: The Environment and the Collapse of Great Civilizations* (New York: St. Martin's Press, 1992).

2. Kenneth E. Boulding, *Ecodynamics: A New Theory of Societal Evolution* (Beverly Hills, Calif.: Sage Publications, 1987), and *The World as a Total System* (Beverly Hills, Calif.: Sage Publications, 1985).

3. Jay W. Forrester, *World Dynamics* (Cambridge, Mass.: Wright-Allen, 1971).

4. "World Conservation Strategy: Managing for the Future," *IUCN Bulletin* 20, nos. 4–6 (April–June 1989): 3. See also *Caring for the Earth: A Strategy for Sustainable Living* (successor to the World Conservation Strategy) (London: Earthscan, 1991).

5. World Commission on Environment and Development, *Our Common Future* (Oxford: Oxford University Press, 1987). For a report of the Experts' Group on Environmental Law established to assist the commission, see R. D. Munro and J. C. Lammers, eds., *Environmental Protection and Sustainable Development: Legal Principles and Recommendations* (London: Graham and Trotman, 1987).

6. See *Sourcebook on Sustainable Development* (Winnipeg, Canada: International Institute for Sustainable Development, 1992). Also L. C. Trzyna, ed., *A Sustainable World: Defining and Measuring Sustainable Development* (Sacramento, Calif.: published for IUCN by the California Institute of Public Affairs, 1995). See also *Strategies for National Sustainable Development: A Handbook for their Implementation* (London: Earthscan, 1994); and *What Works: An Annotated Bibliography of Case Studies of Sustainable Development*, ed. D. Scott Slocombe et al. (Sacramento, Calif.: International Center for the Environment and Public Policy, 1993).

7. It is not feasible to do more than cite a few of the more pertinent sources regarding population and environment. Following are a few that complement this text. Publications of the United Nations Population Fund are the best generally available source: Expert Group on Population, Resources, Environment, and Development, *Proceedings of the Expert Group*, Geneva, 25–29 April 1983 (New York: United Nations, 1984); United Nations Population Fund, "Links between Population, Environment and Resources" and "UNFPA in 1988," *1988 Report of the Executive Director of the United Nations Population Fund State of World Population* (New York: UNFPA, 1988); *Report of the International Conference on Population, Mexico City, 6–14 August 1984* (New York: United Nations Department of Technical Cooperation for Development, 1984), E/Conf.76/19; see also Paul R. Ehrlich and Anne H. Ehrlich, *Population, Resources, Environment: Issues in Human Ecology* (San Francisco: W. H. Freeman, 1970; republished 1977 as *Ecoscience*); Garrett Hardin, *Living within Limits: Ecology, Economics, and Population Taboos* (Oxford: Oxford University Press, 1993); Don Hinrichsen, "Critical Links between Population and Resources," *Populi* 15, no. 1 (1988): 15–25; R. Paul Shaw, "Rapid Population Growth and Environmental Degradation: Ultimate versus Proximate Factors," *Environmental Conservation* 16, no. 1 (Autumn 1989): 119–208; UN Economic and Social Council, *Resolution 1763 (LIV) on UN Fund for Population Activities*, 1858th Plenary Session, 18 May 1973.

8. International Conference on Population and Development, *Report of the Conference . . .* (Cairo, 5–13 September 1994), New York: United Nations, A/Conf. 171/13, 18 October 1994; General Assembly, Preparatory Committee for the International Conference on Population and Development. *Draft Final Document for the Conference — Draft Programme of Action of the Conference — Note by the Secretary-General* (New York: United Nations, February 1994), A/Conf. 171/PC/5, 18. The conference was initially announced as "Population, Environment and Development." Sometime after January 1992 "Environment" was dropped from the name of the conference.

9. For example, see JoAnn Fagot Azeil, *Resource Shortages and World Politics* (Washington,

D.C.: University Press of America, 1977); Thomas Bertelman et al., *Resources, Society, and the Future*, trans. Roger C. Tanner (New York: Pergamon, 1980); Wallace D. Bowman, *A Brief Summary of Theories Relating to Natural Resources Development* (Washington, D.C.: Legislative Reference Service, Library of Congress, 1968); Lester R. Brown, *The Global Politics of Resource Scarcity* (Washington, D.C.: Overseas Development Council, 1974); J. A. Butlin, ed., *Economics of Environmental and Natural Resources Policy* (Boulder, Colo.: Westview Press, 1981); Marion Clawson, *Natural Resources and International Development* (Baltimore, Md.: Johns Hopkins University Press, 1964); Anthony J. Dolman, *Resources, Regimes, World Order* (New York: Pergamon, published in cooperation with Foundation Reshaping the International Order, Rio, 1981); D. Labor and U. Colombo, *Beyond the Age of Waste: A Report to the Club of Rome* (New York: Pergamon Press, 1981); Joseph L. Fisher and Neal Potter, *World Prospects for Natural Resources* (Baltimore, Md.: Johns Hopkins University Press, 1964); Gerald Garvey and Lou Ann Garvey, eds., *International Resource Flows* (Lexington, Mass.: Lexington Books, 1977); Dick Kischten, "The Earth May Be Running Out of Earth as Demands for the Use of Land Increase," *National Journal* (19 January 1980): 93–98; Gary Klee, ed., *World Systems of Traditional Resource Management* (New York and Silver Spring, Md.: John Wiley and V. H. Winston, 1980); Dennis L. Little, Robert E. Dils, and John Gray, eds., *Renewable Natural Resources: A Management Handbook for the 1980s* (Boulder, Colo.: Westview Press, 1982); C. M. Mason, ed., *The Effective Management of Resources: The International Politics of the North Sea* (New York: Nichols, 1979); Zuhayr Mikdashi, *The International Politics of Natural Resources* (Ithaca, N.Y.: Cornell University Press, 1976); Peter N. Nemetz, ed., *Resource Policy: International Perspectives* (Montreal: Institute for Research on Public Policy, 1980); Kenneth Ruddle and Walther Manshard, eds., *Renewable Natural Resources and the Environment: Pressing Problems in the Developing World* (Dublin: Tycooly International Publications, 1982); J. E. S. Fawcett and Audrey Parry, *Law and International Resources Conflicts* (Oxford: Clarendon Press, 1981); and Albert E. Utton and Ludwik A. Teclaff, eds., *Transboundary Resources Law* (Boulder, Colo.: Westview Press, 1987).

10. See Bruce Lord Bandurski, "Ecology and Economics: Partners for Productivity," *Annals of the American Academy of Political and Social Science* 405 (January 1975): 75. See also J. A. Hanson, "Towards an Ecologically-Based Economic Philosophy," *Environmental Conservation* 4 (Spring 1977): 3–10.

11. E.g., H. J. Barnett and C. Morse, *Scarcity and Growth: The Economy of Natural Resource Availability* (Baltimore, Md.: Johns Hopkins University Press, 1963); P. T. Bauer, *Equality, the Third World and Economic Delusion* (Cambridge, Mass.: Harvard University Press, 1981), pp. 49–51; C. Angers, W. P. Gramm, and S. C. Maurice, *Does Resource Conservation Pay?* IIER Original Paper, no. 14 (Los Angeles: International Institute of Economic Research, n.d.); Julian L. Simon, "Resources, Population, Environment: An Oversupply of False Bad News," *Science* 208 (27 June 1980): 1431–37. For an objective economic analysis of the resource adequacy issue, see V. Kerry Smith, "The Evaluation of Natural Resource Adequacy: Elusive Quest or Frontier of Economic Analysis?" *Land Economics* 56 (August 1980): 257–98, references. See also notes 12 and 13 following.

12. Harrison Brown, James S. Bonner, and John Weir, "What Is a Resource?" in *The Next Hundred Years: Man's Natural and Technological Resources* (New York: Viking Press, 1970), pp. 89–94.

13. See Nicholas Georgescu-Roegen, *The Entropy Law and the Economic Process* (Cambridge, Mass.: Harvard University Press, 1971); and "The Crisis of Resources: Its Nature and Its Unfolding," in *Energy, Economics, and the Environment*, ed. Gregory A. Daneke (Lexington, Mass.: Lexington Books, 1982), pp. 9–24. On materials policy generally, see *Material Needs and the Environment: Proceedings of a Joint Meeting of the National Academy of Sciences, National Academy of Engineers, October 25–26* (Washington, D.C.: National Academy of Sciences, 1973). But see also Dieter C. Altenpohl, *Materials in World Perspective* (Berlin: Springer-Verlag, 1980).

14. Note "American Indian Religious Freedom," Joint Resolution 102 of the 95th Congress, Public Law 95-341, 42 United States Code, p. 1996.
15. E.g., Wilfred Beckerman, "The Myth of 'Finite' Resources," *Business and Society Review* 12 (Winter 1974-75): 21-25; Julian L. Simon, *The Ultimate Resource* (Princeton, N.J.: Princeton University Press, 1981); and, "The Scarcity of Raw Materials," *Atlantic Monthly* (June 1981): 33-41 (a chapter in *The Ultimate Resource*); and William Tucker, *Progress and Privilege: America in the Age of Environmentalism* (Garden City, N.Y.: Doubleday, 1982). P.T. Bauer's contention that "mineral deposits do not represent exhaustible resources" may technically be true, but practically misleading. This statement and the assertion that "there is no danger of population growth posing a threat to long term energy supplies" disregards the social effects of shortages, the problems associated with access to resources, and the costs of recycling and recovery and the artificial development of basic materials. See *Equality, the Third World and Economic Delusion*, p. 50.
16. E.g., Wilfred Beckerman, *Two Cheers for the Affluent Society: A Spirited Defense of Economic Growth* (New York: St. Martin's Press, 1974), and *In Defense of Economic Growth* (London: Cape, 1974); Herman Kahn and Ernest Schneider, "Globaloney 2000," *Policy Review* 16 (Spring 1981): 129-47; Simon, "Resources, Population, Environment."
17. See issues of *Conservation and Recycling: An International Journal*, published by Pergamon Press, 1976-.
18. National Research Council, Committee on Remote Sensing for Agricultural Purposes, *Remote Sensing with Special Reference to Agriculture and Forestry* (Washington, D.C.: National Academy of Sciences, 1970), includes bibliography; and E. C. Barrett and L. F. Curtis, *Introduction to Environmental Remote Sensing*, 2d ed. (London: Chapman and Hall, 1982).
19. Norman Myers, "Preservation and Production: Multinational Timber Corporations and Tropical Moist Forests," *Council on Economic Priorities Newsletter*, no. 5 (September 1980): n.p.; and *UNCTAD Bulletin*, no. 185 (July 1982) and no. 192 (April 1983). Susan R. Fletcher, *Tropical Deforestation: International Implications*, CRS Issue Brief (Washington, D.C.: Congressional Research Service, Library of Congress, July 1989); and Nicholas Guppy, "Tropical Deforestation: A Global View," *Foreign Affairs* 62 (Spring 1984): 928-65.
20. Betsy Cody, *Debt-for-Nature Swaps in Developing Countries*, CRS Report for Congress (Washington, D.C.: Congressional Research Service, Library of Congress, 26 September 1988).
21. James H. Cobbe, *Government and Mining Companies in Developing Countries* (Boulder, Colo.: Westview Press, 1979), p. 280. For further discussion of international problems in minerals development, see comments by James H. Cobbe, Ronald T. Libby, and David A. Jodice, *International Organization* 35 (Autumn 1981): 725-54.
22. Raymond F. Mikesell, *New Patterns of World Mineral Development* (London and Washington: British-North American Committee and National Planning Association, 1979), pp. 72-75. See also David N. Smith and Louis T. Wells Jr., "Mineral Agreements in Developing Countries; Structure and Substances," *American Journal of International Law* 69 (July 1975): 560-90; for background, James F. McDivitt, *Minerals and Men* (Baltimore, Md.: Johns Hopkins University Press, for Resources for the Future, 1965); and Leonard L. Fischman, *World Mineral Trends and U.S. Supply Problems* (Study Project Director) (Washington, D.C.: Resources for the Future, 1980).
23. L. K. Caldwell, "Transboundary Conflicts: Resources and Environment" in *Canada–United States Relationships: The Politics of Energy and the Environment*, ed. Jonathan Lemco (Westport, Conn.: Praeger, 1992), pp. 24-26.
24. United Nations, *Natural Resources and Energy Newsletter* 5 (May 1981): 7-8.
25. Rudolph W. Knoepfel, "Cooperation, Greater Sensitivity Needed to Improve Economic Stature of LDCs," *AMA Forum* 67 (June 1978): 30. For further discussion of the multinational corporation and the environment, see Walter Ingo, "A Survey of International Economic Repercussions of Environmental Policy," in J. A. Butlin, ed., *The Economics of Environmental and Natural Resources Policy* (Boulder, Colo.: Westview Press, 1981), pp. 163-81;

Thomas A. Gladwin, *Environment, Planning and the Multinational Corporation* (Greenwich, Conn.: JAI Press, 1977); United Nations, Department of Economic and Social Affairs, *The Impact of Multinational Corporations on Development and on International Relations*, E/5500/Rev. 1, ST/ESA/6 (New York, 1974); and Walter Ingo, ed., *Studies in International Environmental Economics* (New York: Wiley, 1976).

26. In Roy S. Rauschkolb, ed., *Land Degradation—Soils Bulletin* 13 (Rome: FAO, 1971), p. iii. The work was prepared for the United Nations Conference on the Human Environment as an "interagency focal point," with contributions from IAEA, UNESCO, and WHO. John Lawrence Hargrove and Janie Callison, "Soil Degradation: New Concerns but Uncertain Prospects," in David A. Kay and Harold K. Jacobson, eds., *Environmental Protection: The International Dimension* (Totowa, N.J.: Allenheld, Osmun), pp. 217–39. See also J. Riquier, "A World Assessment of Soil Degradation," *Nature and Resources* 18 (April–June 1982): 18–21. For historical perspective on soil as a resource, see Vernon Gill Carter and Tom Dale, *Topsoil and Civilization*, rev. ed. (Norman: University of Oklahoma Press, 1974). Largely technical in treatment are Milos Holy, *Erosion and Environment: Environmental Sciences and Applications*, trans. Jana Ondrackova, vol. 9 (Oxford: Pergamon Press, 1980); and *Soil Conservation and Erosion in the Tropics* (Madison, Wis.: American Society of Agronomy, ASA Special Publication no. 43, 1982). Most directly relating food and environment is Erik P. Eckholm, *Losing Ground: Environmental Stress and World Food Prospects* (New York: W. W. Norton, 1976). For failure of international response to the issue, see Libby Bassett, "Governments Ignore World Soils Policy," *World Environment Report* 9 (30 May 1983): 3.

27. Amory B. Lovins, *Soft Energy Paths: Toward a Durable Peace* (San Francisco: Friends of the Earth International, 1977; distributed by Ballinger Publishing, Cambridge, Mass.); and Lovins, *World Energy Strategies: Facts, Issues, and Options* (Cambridge, Mass.: Ballinger Publishing, 1975).

28. Public Affairs Department of Exxon, "World Energy Outlook" (April 1978), p. 5.

29. *Energy: Global Prospects 1985–2000—Report of the Workshop on Alternative Energy Strategies* (New York: McGraw-Hill, 1977).

30. See James Howe and James Tarrant, *An Alternative Road to the Post-Petroleum Era: North–South Cooperation* (Washington, D.C.: Overseas Development Council, July 1980). On prospects for renewable energy resources, see Wilfrid Bach et al., "Renewable Energy Prospects," *Energy: The International Journal* 4 (October 1979): entire issue; and Essam El-Hinnawi and Asit K. Biswas, eds., *Renewable Resources of Energy and the Environment* (Dublin: Tycooly International Publications, 1981).

31. See Essam El-Hinnawi, ed., *The Environmental Impacts of the Production and Use of Energy* (Dublin: Tycooly International Publications, 1981); Toufiq A. Siddiqi, "Environmental Considerations in Energy Policies," in *Ecology and Development*, ed. Desh Bandu and Veena Bhardwaj, East–West Environment and Policy Institute Reprint, no. 10 (New Delhi: Indian Environmental Society, 1979); and "Energy Policy Issues and the Environmental Agenda of the 1980s," *Current Issues IV: The Yearbook of Environmental Education and Environmental Studies* (Troy, Ohio: National Association for Environmental Education, 1980), pp. 91–104, references. See also David James and Toufiq A. Siddiqi, *The Increased Use of Coal in the Asia-Pacific Region: Achieving Energy and Environmental Goals*, Working Paper (Honolulu, Hawaii: East-West Center, Environment and Policy Institute, January 1981).

32. Massachusetts Institute of Technology, *Coal—Bridge to the Future and Future Coal Prospects: Country and Regional Assessments*, vols. 1 and 2, Carroll L. Wilson, Project Director (Cambridge, Mass.: Ballinger Publishing Co., 1980).

33. Melvin A. Conant and Fern Racine Gold, *The Geopolitics of Energy* (Boulder, Colo.: Westview Press, 1978), p. 89.

34. Ibid., p. 116.

35. Herbert Inhaber, "Is Solar Power More Dangerous Than Nuclear?" *IAEA Bulletin* 23 (February 1981): 11–17; Thomas L. Neff, *The Social Costs of Solar Energy: A Study of Photovoltaic*

Energy Systems (New York: Pergamon Press, 1981). For other aspects of solar energy, see Stephen W. Sawyer and Stephen L. Feldman, "Technocracy versus Reality: Perceptions in Solar Policy," *Policy Sciences* 4 (September 1981): 459–72; U.S. Council on Environmental Quality (CEQ), *Solar Energy: Progress and Promise* (Washington, D.C.: CEQ, April 1978); and United Nations Conference on New Sources of Energy (Rome, 21–31 August 1961), "Proceedings of the United Nations Conference on New Sources of Energy," vols. 4, 5, and 6, "Solar Energy," (New York: United Nations, 1964) E/Conf.35.5, E/Conf.35.6, and E/Conf.35.7; and Preparatory Committee for the United Nations Conference on New and Renewable Sources of Energy, "Report of the Technical Panel on Solar Energy in its Second Session," A/Conf.100/PC/27, 6 April 1981.

36. See Jacques Constans, *Marine Sources of Energy* (New York: Pergamon Press, 1979), chap. 2, pp. 35–77; also William F. Whitmore, "OTEC: Electricity from the Ocean," *Technology Review* 81 (October 1978): 58–63; and Eric Bender, "The OTEC Gamble," *Sea Technology* 20 (August 1970): 16.

37. *Alcohol Production from Bio-Mass in the Developing Countries* (Washington, D.C.: World Bank, September 1980).

38. See Colin Norman, "U.N. Grapples with Renewable Energy," *Science* 213 (24 July 1981): 419; "Wood Energy," *Unasylva* 33, no. 131 (1981): entire issue; and W. B. Morgan and R. P. Moss, eds., *Fuelwood and Rural Energy Production and Supply in the Humid Tropics* (Dublin: Tycooly International Publications, 1982).

39. See articles and references in *Nasopharyngeal Carcinoma: Etiology and Control: Proceedings of an International Symposium in Kyoto, Japan, 4–6 April 1977*, ed. G. de-The and Y. Ito, IARC Scientific Publications, no. 20 (Lyon: International Agency for Cancer Research, 1978). See also Peter Clifford, "Carcinogens in the Nose and Throat: Nasopharyngeal Carcinoma in Kenya," *Proceedings of the Royal Society of Medicine* 65 (August 1972): 682–85.

40. "Law of the Sea Forum: The 1994 Agreement on the Seabed Provisions of the Convention on the Law of the Sea," *American Journal of International Law* 88, no. 4 (October 1994): 687–714.

41. *The International Energy Agency of OECD* (Paris: OECD, 1975); and *Annual Report on Energy Research, Development and Demonstration: Activities of the IEA 1979–1980* (Paris: OECD, 1980). See also Mason Willrich and Melvin A. Conant, "The International Energy Agency: An Interpretation and Assessment," *American Journal of International Law* 71 (April 1977): 199–223.

42. *Workshop on Energy Data in Developing Countries—Proceedings*, 2 vols. (Paris: OECD, 1979).

43. Gunther Handl, "Managing Nuclear Wastes: The International Connection," *Natural Resources Journal* 21 (April 1981): 268–314. See also notes 44 and 45 following.

44. Gisela Dreschhoff, D. F. Saunders, and E. J. Zeller, "International High Level Nuclear Waste Management," *Science and Public Affairs* 30 (January 1974): 28–33; and Thomas C. Jackson, ed., *Nuclear Waste Management: The Ocean Alternative* [Proceedings of the Public Policy Forum], sponsored by the Oceanic Society in cooperation with the Georgetown University Law Center (New York: Pergamon Press, 1981).

45. See E. J. Zeller, D. F. Saunders, and E. E. Angino, "Putting Radioactive Wastes on Ice: A Proposal for an International Radionuclide Depository in Antarctica," *Science and Public Affairs* 29 (January 1973): 4ff.

46. Among numerous treatments of Canadian Arctic policy, see the following: William V. O'Brien and Armando C. Chapelli, "The Law of the Sea in the Canadian Arctic: The Pattern of Controversy," pt. 1, *McGill-Law Journal* 19 (November 1973): 322–66; ibid., pt. 2 (December 1973): 477–542; Albert Utton, "The Arctic Waters Pollution Prevention Act, and the Right of Self Protection," *University of British Columbia Law Review* 7, no. 2 (1972): 221–34; John C. Klatz, "Are Ocean Polluters Subject to Universal Jurisdiction—Canada Breaks the Ice," *International Lawyer* 6, no. 4 (1972): 706–17.

47. See chapters on "Military Activity and the Environment" and "Energy and the Environment," in Essam El-Hinnawi and Manzur Hashmi, *Global Environmental Issues* (Dublin:

Tycooly International Publishing, 1982), note references; Eric Chivian et al., eds., *Last Aid: The Medical Dimensions of Nuclear War* (San Francisco, Calif.: W. H. Freeman, 1982). For what is widely regarded as an optimistic assessment of consequences, see Herman Kahn, *On Thermonuclear War*, 2d ed. (Princeton, N.J.: Princeton University Press, 1961).

48. For example, the Fourth Session of the Governing Council of UNEP (30 March–14 April 1976) considered environmental aspects of energy production. See "Review of the Impact of Production and Use of Energy on the Environment: Report of the Executive Director," UNEP/GC/61/Add. 1, Nairobi, January 1976. See also Peter Auer, ed., *Energy and the Developing Nations* (Washington, D.C.: World Bank, August 1980); G. A. Daneke, ed., *Energy, Economics, and the Environment: Toward a Comprehensive Perspective* (Lexington, Mass.: D. C. Heath, 1982); Kirk R. Smith and Harrison Brown, "From Pakistan to Japan: The Energy Problems of Asia," *OPEC Review* 4 (Autumn 1980): 19–54; Bach et al., "Renewable Energy Prospects"; William F. Martin and Frank J. P. Pinto, "Energy for the Third World," *Technology Review* 80 (June/July 1978): 48–56; and Toufiq A. Siddiqi and Gerald F. Hein, "Energy Resources of the Developing Countries and Some Priority Markets for the Use of Solar Energy," *Journal of Energy and Development* 3 (Autumn 1977): 164–89.

49. From the fourth (1971) conference, see E. E. Larson and H. C. Brown Jr., "The Energy-Environmental Equation," in *Proceedings of the Fourth International Conference on the Peaceful Uses of Atomic Energy—Under the Co-Sponsorship of the United Nations and the International Atomic Energy Agency, held at Geneva, Switzerland, September 6–16, 1971* (Vienna: IAEA, 1972), vol. 2, pp. 529–45.

50. UN Conference on New Sources of Energy (Rome, 21–31 August 1961), "Report of the United Nations Conference on New Sources of Energy," E/3577/Rev. 1, 1962; Conference on New Sources of Energy, "Proceedings: General Sessions," vol. 1, E/Conf.35/2, 1963; and Preparatory Committee for the UN Conference on New and Renewable Sources of Energy (10–22 August 1981, Nairobi), "Report of the A/Conf.100/PC/41, 10 March 1981. See also reports of technical panels and ad hoc groups of experts (A/Conf.100/PC 23–34, 36–39, 42). For a summary of the 1961 conference, see *Yearbook of the United Nations, 1961* (New York: United Nations, 1963), pp. 268–69. For the 1981 conference, see *UNDP and Energy: Exploration, Conservation, Innovation* (New York: UN Development Programme, 1981), C-39, in 17 parts; and Erik P. Eckholm, "U.N. Conference on New and Renewable Energy," *Science* 212 (5 June 1981): editorial, p. 1089.

51. Irene Tinker, "U.N. Energy Conference: Substance and Politics," *Science* 214 (4 December 1981), editorial, p. 1079.

52. Cited by Colin Norman, "An Empty Plan for Renewable Energy," *Science* 213 (11 September 1981): 1235. See also Betty Bassan, *Report on the 1981 United Nations Energy Conference* (New York: Sierra Club International Earthcare Center, 1981).

53. See "U.N. Grapples with Renewable Energy," *Science* 213 (24 July 1981): 419.

54. Eckholm, "U.N. Conference on New and Renewable Energy," p. 1089.

55. Peter H. Freeman, *Large Dams and the Environment: Recommendations for Development Planning, A Report Prepared for the United Nations Water Conference, Mar del Plata, Argentina, March 1977* (Washington, D.C.: International Institute for Environment and Development, 1977), bibliography, pp. 51–55. See also Hussein M. Fahim, *Dams, People and Development: The Aswan High Dam Case* (New York: Pergamon, 1981); and International Commission on Large Dams, United States Committee, Committee on Environmental Effects, *Environmental Effects of Large Dams, Report by the Committee on Environmental Effects of the United States Committee on Large Dams* (New York: American Society of Civil Engineers, 1978).

56. It is hardly feasible to reference the voluminous literature on development here. The following items, however, provide various insights into problems of reconciling environmental and developmental objectives: Mostafa Kamal Tolba, *Development without Destruction: Evolving Environmental Perceptions* (Dublin: Tycooly International Publications, 1982); for case studies in environmental consequences of development projects, see M. Taghi Far-

var and John P. Milton, eds., *The Careless Technology: Ecology and International Development* (Garden City, N.Y.: Natural History Press, 1972): for a note on perspectives of French development economists, see pp. 929-31. See also Kim Q. Hill, ed., *Toward a New Strategy for Development: A Set of Papers Commissioned by the Ralhko Chapel and Presented at Ralhko Chapel Colloquium held at Houston, Texas, on February 3-5, 1977* (Elmsford, N.Y.: Pergamon Press, 1979); and Lisa Peattie, *Thinking about Development* (New York: Plenum Press, 1981). For a sharply critical view of development concepts, see P. T. Bauer, *Dissent on Development: Studies and Debates in Development Economics* (Cambridge, Mass.: Harvard University Press, 1972); for development concepts at the time of the Stockholm Conference, see Allen Lowell Doud, "International Environmental Development: Perceptions of Developing and Developed Countries," *Natural Resources Journal* 12 (October 1972): 520-29; and for a later perspective, Brian Harvey and Gary W. Knamiller, "Development and Conservation: A Global Dilemma," *Environmental Conservation* 8 (Autumn 1981): 199-205.

57. Shadia Schneider-Sawiris, *The Concept of Compensation in the Field of Trade and Environment,* IUCN Environmental Policy and Law Paper, no. 4 (Morges, Switzerland, 1973); and Yvonne I. Nicholls, *Source Book: Emergence of Proposals for Recompensing Developing Countries for Maintaining Environmental Quality,* IUCN Environmental Policy and Law Paper, no. 5 (Morges, Switzerland, 1974).

58. Relationships between world trade and international environmental policy are of concern to all nations. The issues are complex and the consequences of particular policies often difficult to trace. The American Society for International Law has sponsored a study of the relation of international trade to environmental policy, including discussion of a possible environmental code of conduct for multinational companies. See Seymour J. Rubin and Thomas A. Graham, *Environment and Trade: The Relation of International Trade and Environmental Policy* (Totowa, N.J.: Allanheld, Osmun, 1982); Durwood Zaelke, Paul Orbuch, and Robert F. Hausman, *Trade and the Environment: Law, Economics, and Policy* (Washington, D.C.: Island Press, 1993); *The Environmental Effects of Trade* (Paris: OECD, 1994); and Thomas Anderson, Carl Folke, and Sefan Andersson, *Trading with the Environment: Ecology, Economics, Institutions, and Policy* (London: Earthscan, 1995); and *International Agreements to Protect the Environment and Wildlife,* Report to the U.S. Senate Committee on Finance, on Investigation, no. 332-287 under sec. 332 of the Tariff Act of 1930, Washington, D.C.: U.S. International Trade Commission, Publication 2351, January 1991.

59. See Maynard M. Hufschmidt and Eric L. Hyman, eds., *Economic Approaches to Natural Resource and Environmental Quality Analysis: Proceedings and Papers of a Conference on Extended Benefit-Cost Analysis held at the Environmental and Policy Institute, East-West Center, Honolulu, Hawaii, September 19-26, 1979* (Dublin: Tycooly International Publications, 1982); and Hufschmidt et al., *Environment, Natural Systems, and Development: An Economic Evaluation Guide* (Baltimore, Md.: Johns Hopkins University Press, 1983).

60. Raymond F. Dasmann, John P. Milton, and Peter Freeman, *Ecological Principles for Economic Development* (New York: John Wiley, for IUCN and the Conservation Foundation, 1973).

61. Duncan Poore, "Ecological Guidelines," *Unasylva* 27, no. 110 (1975): 16-20. IUCN issued a series of separate guidebooks for various ecological regions.

62. See M. Taghi Farvar and John P. Milton, eds., *The Careless Technology: Ecology and International Development* (Garden City, N.Y.: Natural History Press, 1972).

63. E. F. Schumacher, "The Work of the Intermediate Technology Development Group in Africa," *International Labour Review* 106 (July 1972): 75-92; Schumacher, *Small is Beautiful: Economics as If People Mattered* (New York: Harper and Row, 1973); and Schumacher, *A Guide for the Perplexed* (New York: Harper and Row, 1979). Also see Amulya Kumar N. Reddy, *Technology, Development, and the Environment: A Reappraisal* (Nairobi: UNEP, 1979), and Wade Greene, "An Introduction to AT—The Appropriate Technology Movement" (New York: Alicia Patterson Foundation, 1978).

64. Daniel M. Dworkin, ed., *Environment and Development: Collected Papers, Summary Reports,*

and Recommendations—SCOPE/UNEP, Symposium of Environmental Sciences in Developing Countries, Nairobi, February 11-23, 1974 (Indianapolis, Ind.: SCOPE Miscellaneous Publications, 1974).

65. See Richard L. Clinton, "Ecodevelopment," *World Affairs* 10 (Fall 1977): III-26; Ignacy Sachs, *Stratégies de l'ecodeveloppement* (Paris: Editions Economist Humanisma, 1980); Clinton, "Ecodevelopment," *Ceres* 42 (November–December 1974): 8-12; and Clinton, "Environment and Styles of Development," *African Environment* 1 (December 1974): 9-33. See also UNEP Governing Council, "Environmental Impact of Irrational and Wasteful Use of Natural Resources," Item 15(a) of the Provisional Agenda, UNEP/GC/79, Nairobi, January 1976; and Governing Council, "Environment and Development: Report of the Director General," Item 14 of the Provisional Agenda, UNEP/GC/76, Nairobi, 29 January 1976. For further articles and reports, see publications of *Cahiers de l'ecodeveloppement*, Paris, in English and French including the periodical *Ecodevelopment News;* and Robert Riddell, *Ecodevelopment: Economics, Ecology and Development* (New York: St. Martin's Press, 1981). For a critique of ecodevelopment ideas, see review by Armelle Braun, "The Earth as a Garden," *Ceres* 14 (November–December 1981): 46-47.

66. See Kenneth A. Dahlberg, *Beyond the Green Revolution: The Ecology and Politics of Global Agricultural Development* (New York: Plenum Press, 1979), especially chaps. 4, 5, and 6. On ecologically sustainable development, see Richard Carpenter, "Using Ecological Knowledge for Development Planning," *Environmental Management* 4, no. 1 (1980): 13-20; Peter Hendy, "The Case for Economic Ecology," *Ceres* 13 (March–April 1982): 15-18; note earlier work by Jaro Mayda, *Environment and Resources: From Conservation to Ecomanagement* (San Juan: University of Puerto Rico, School of Law, 1968); and *Legal, Regulatory, and Institutional Aspects of Environmental and Natural Resource Management in Developing Countries* (Washington, D.C.: Agency for International Development and National Park Service, November 1981). On the closely related items of training for ecomanagement, see *NSMD: Natural Systems Management for Development—Report from the East-West Environment and Policy Institute Workshop on Training for Natural Systems Management, 22 October–2 November 1979* (Honolulu, Hawaii: East-West Center, 1980). See also "Technologies for Ecodevelopment," *African Environment* 3 (November 1977): 251-95.

67. UN General Assembly, "United Nations Environment Programme—The Cocoyoc Declaration Adopted by the Participants in the UNEP/UNCTAD Symposium on 'Patterns of Resource Use, Environment and Development Strategies,' held at Cocoyoc, Mexico, from 8-12 October 1974," A/C.2/292, 1 November 1974. Declaration reprinted in full in *International Organization* 29 (Summer 1975): 893-901, and in UNEP Executive Series, *In Defense of the Earth: The Basic Texts on Environment.*

68. The principal documents of the New International Economic Order (NIEO) include the following: "Declaration on the Establishment of a New International Economic Order" (1974), "Programme of Action on the Establishment of a New International Economic Order" (1974), "Charter of Economic Rights and Duties of States" (1974). A comprehensive collection has been brought together by Alfred George Moss and Harry N. M. Winton, comps., *A New Inter-National Economic Order: Selected Documents, 1945-1974*, 2 vols., UNITAR Document Service, no. 1 (New York: United Nations Institute for Training and Research, n.d.). See also Ervin Laszlo et al., *The Objectives of the New International Economic Order* (New York: Pergamon Press, 1978); and Jan Tinbergen, Anthony J. Dolman, and Jon Van Ettinger, eds., *Reshaping the International Order: A Report to the Club of Rome* (New York: E. P. Dutton, 1976). For the official text of the principal declarations, see "Resolutions Adopted by the General Assembly During its Sixth Special Session, 9 April-2 May 1974," and Resolutions 3201 (S-VI) "Declaration on the Establishment of a New International Economic Order" and 3202 (S-VI) "Programme of Action on the Establishment of a New International Economic Order," in UN General Assembly, *Official Records*, Sixth Special Session, Supplement no. 1, A/9559. For an NIEO perspective, see Kamal Hossain and

Subrata Roy Chowdhury, eds., *Permanent Sovereignty over Natural Resources in International Law: Principle and Practice* (New York: St. Martin's Press, 1984). See also "Legal Aspects of a New International Economic Order," *The International Law Association—Report of the Sixtieth Conference Held at Montreal, August 29th, 1982, to September 4th, 1982* (London: International Law Association, 1983).

69. "Charter of Economic Rights and Duties of States" (1974), Resolution 3281 (XXIX), reprinted in Laszlo et al., *The Objectives of the New International Economic Order.*

70. Farvar and Milton, *The Careless Technology.*

71. B. J. Brown et al., "Global Sustainability: Towards Definition," *Environmental Management* 11 (November 1987): 713-19; Lynton K. Caldwell, "Political Aspects of Ecologically Sustainable Development," *Environmental Conservation* 11 (Winter 1986): 299-308; and "Sustainable Development: Viable Concept and Attainable Goal?" *Environmental Conservation* 21, no. 3 (Autumn 1994): 193-95; William C. Clark and R. E. Munn, eds., *Sustainable Development of the Biosphere* (Cambridge, Mass.: Cambridge University Press, 1986) (published on behalf of the IIASA); Michael Redclift, *Sustainable Development: Exploring the Contradictions* (London: Methuen, 1987); Mostafa Kamal Tolba, *Sustainable Development: Constraints and Opportunities* (London: Butterworth, 1987); R. K. Turner, *Sustainable Environmental Management* (Boulder, Colo.: Westview Press, 1988). See also Paul Hawken, *The Ecology of Commerce: A Declaration of Sustainability* (New York: Harper Business, 1993); Stephen Viederman, "Sustainable Development: What Is It and How Do We Get There?" *Current History* (April 1993): 180-85; *Implementing Sustainable Development: Experiences in Sustainable Development Administration*, DDSMS/SEM.95/1, INT-91-R71, reports from 11 countries (New York: UN Department for Development Support and Management Services, 1995).

72. Tolba, *Sustainable Development*, vii.

11. Enhancing the Quality of Life: Natural and Cultural Environments

1. There have been numerous publications dealing with quality of life and environment and their measurement. Following is a selective listing: Frank M. Andrews and Stephen B. Withey, "Developing Measures of Perceived Life Quality: Results from Several National Surveys," *Social Indicators Research* 1, no. 1 (1974): 1-26; and *Social Indicators of Well-Being* (New York: Plenum Press, 1976); Mihai C. Botez et al., *Preliminaries on a Comparative Analysis of the Various Viewpoints on the Quality of Life* (Tokyo: United Nations University, 1979), a Marxist analysis; Gerardo Budowski, "The Quantity-Quality Relationship in Environmental Management," *Impact of Science on Society* (20 July-September 1970): 235-46; Uriel G. Foa and Edna B. Foa, "Measuring Quality of Life: Can It Help Solve The Ecological Crisis?" *International Journal of Environmental Studies* 5 (July 1973): 21-26; George Gallup, "Human Needs and Satisfactions: A Global Survey," *Public Opinion Quarterly* 40 (Winter 1976-77): 459-67; E. Hankiss, R. Manchin, and L. Fuster, *Cross-Cultural Quality of Life Research: An Outline for a Conceptual Framework and Some Methodological Issues* (Budapest: Hungarian Academy of Sciences, Center for Quality of Life Research, 1978); Peter W. House and Edward R. Williams, *The Carrying Capacity of a Nation: Growth and the Quality of Life* (Lexington, Mass.: Lexington Books, 1975); Sylvan J. Kaplan and Evelyn Kivy Rosenberg, eds., *Ecology and the Quality of Life* (Springfield, Ill.: Charles C. Thomas, 1973); G. E. Lasker, ed., *Quality of Life: Systems Approaches*, vol. 1 of *Applied Systems and Cybernetics: Proceedings of an International Congress* (New York: Pergamon Press, 1981); Lester W. Milbrath, "A Conceptualization and Research Strategy for the Study of Ecological Aspects of the Quality of Life," *Social Indicators Research* 10 (1982): 133-57, note bibliographical references; and "Policy Relevant Quality of Life Research," *Annals of the American Academy of Political and Social Sciences* 444 (July 1979): 32-45; David Morris and Florizelle B. Liser, "The PQLI [Physical Quality of Life]: Measuring Progress in Meeting Human Needs," *Commu-*

nique No. 32 (Washington, D.C.: Overseas Development Council, August 1977); National Research Council, Division of Behavioral Sciences, *Environmental Quality and Social Behavior: Strategies for Research, Report on a Study Conference on Research Strategies in the Social and Behavioral Sciences on Environmental Problems and Policies, held in Irvington, Virginia, June 22–27, 1969* (Washington, D.C.: National Academy of Sciences, 1973); U.S. Department of State, *World Environmental Quality: A Challenge to the International Community* (Washington, D.C.: Government Printing Office, 1973); U.S. Environmental Protection Agency, *Quality of Life Indicators: A Review of the State-of-the-Art and Guidelines as Derived to Assist in Developing Environmental Indicators* (Washington, D.C.: Office of Research and Monitoring, 1972); C. P. Wolf, *Quality of Life, Concept and Measurement: A Preliminary Bibliography*, Public Administration Series, P-249 (Monticello, Ill.: Vance Bibliographies, June 1979); Marvin E. Wolfgang, ed., "The Environment and the Quality of Life: A World View," *Annals of the American Academy of Political and Social Sciences* 444 (July 1979): entire issue; and Wolfgang Zapf, "The Polity as a Monitor of the Quality of Life," in *The Politics of Environmental Policy*, eds. Lester W. Milbrath and Frederick R. Inscho (Beverly Hills, Calif.: Sage, 1975), pp. 35–39. Readers should be aware of a large number of relevant reports issued through the UNESCO-sponsored Man and the Biosphere Programme (MAB).

2. See *Indicators of Environmental Quality and Quality of Life*, Reports and Papers in the Social Sciences, no. 38 (Paris: UNESCO, 1978).

3. George E. Lasker, ed., *Applied Systems and Cybernetics: Proceedings of the International Congress of Applied Systems Research and Cybernetics, December 12–16, Acapulco, Mexico*, 6 vols. (New York: Pergamon Press, 1981).

4. E.g., Lou Clapper, " 'Life Quality' Tops Priorities at UNEP Geneva Meet," *Center Magazine* (5 October 1973): 3–6. For one interpretation by implication, see Georges Fradier, *About the Quality of Life: UNESCO and Its Programme* (Paris: UNESCO, 1976).

5. See Larry Kohl, "Pere David's Deer Saved from Extinction," *National Geographic* 162 (October 1982): 478–85; Dale J. Osborn, "Pere David's Deer," *Field Museum of Natural History Bulletin* 48 (October 1977): 10–13; Janet Newlan Bower, "Pere David's Deer: The Trek from Extinction," *National Parks and Conservation Magazine* (April 1979): 14–18; and Willy Ley, "The Story of the Milhu," in *Dragons in Amber: Further Adventures of a Romantic Naturalist* (New York: Viking, 1951), pp. 120–37. For the European bison, see Heinz-Georg Klör and Arnfried Wunschman, "How the European Bison Survived," in *Grzimek's Animal Life Encyclopedia*, vol. 13, *Mammals*, pp. 394–98; for Przewalski's horse, see Jere Wolf, "Other Wild Horses," *Grzimek's Animal Life Encyclopedia*, vol. 12, *Mammals*, pp. 565–68; for the Franklinia Gordonia altamaha, see Peter Jenkins, "The Historical Background of Franklin's Tree," *The Pennsylvania Magazine of History and Biography* 57 (July 1933): 270–93. See also Donald Culrose Peattie, "Wilderness Plantsmen: Bartram and Michaux," in *Green Laurels: The Lives and Achievements of the Great Naturalists* (New York: Literary Guild, 1936), pp. 186–200.

6. Bernard Grzimek and Michael Grzimek, *Serengeti Shall Not Die* (London: Hamish Hamilton, 1960; New York: Fontana Books, 1964). See also Harold T. P. Hayes, "A Reporter at Large: The Last Place," *New Yorker* (6 December 1976): 51–133.

7. Paul Ehrlich and Anne Ehrlich, *Extinction: The Causes and Consequences of the Disappearance of Species* (New York: Random House, May 1981); Paul S. Martin and H. E. Wright, Jr., eds., *Pleistocene Extinctions: The Search for a Cause* (New Haven, Conn.: Yale University Press, 1967); Carl O. Sauer, "Extinction of Pleistocene Mammals," in "A Geographic Sketch of Early Man in America," *Geographical Review* 34 (October 1944): 529–73. See also H. A. Goodwin and J. M. Goodwin, comps., *List of Mammals Which Have Become Extinct or Are Possibly Extinct since 1600*, Occasional Paper, no. 18 (Morges, Switzerland: IUCN, 1973).

8. "International Biological Programme: Report of the Planning Committee, November 15, 1963," *BioScience* 14 (April 1964): 43–49; and Special Committee on the International Biological Programme, *Environment: The Quest for Quality Mobilizing Science, Industry, and Government* (Morges, Switzerland: Special Committee on the International Biological Pro-

gramme and IUCN, 1970), p. 26. See also Nigel Calder, "World Biology Project Takes Shape," *New Scientist* 23 (30 July 1964): 260-62; and U.S. Congress, House Committee on Science and Astronautics, Subcommittee on Science, Research, and Development, *The International Biological Program: Its Meaning and Needs, Report of the Subcommittee on Science, Research, and Development of the Committee on Science and Astronautics*, 90th Cong., 2d sess., 1968.

9. E. Barton Worthington, ed., *The Evolution of IBP: The International Biological Programme* (Cambridge: Cambridge University Press, 1975). See also W. Frank Blair, *Big Biology*, US./ IBP Synthesis Series (Stroudsburg, Pa.: Dowden, Hutchinson, and Ross, 1977); and Gina Douglas, "Synthesis of the International Biological Programme," *Environmental Conservation* 5 (Summer 1978): 125-26; and for a retrospective appraisal, see E. Barton Worthington, "The Twentieth Anniversary of I.B.P.," *Environmental Conservation* 9 (Spring 1982): 70.

10. See Michel Batisse, "The Biosphere Reserve: A Tool for Environmental Conservation and Management," *Environmental Conservation* 9 (Summer 1982): 101-11. For a description and analysis of biosphere reserves in the United States, see Paul G. Risser and Kathy D. Cornelison, *Man and the Biosphere* (Norman: University of Oklahoma Press, 1979). Twenty-eight reserves are described. The totals reported in this book are from *World Resources 1994-95: A Guide to the Global Environment* (Oxford: Oxford University Press, prepared by the World Resources Institute in collaboration with the United Nations Environment Programme and the United Nations Development Programme, 1994), pp. 316-19.

11. Batisse, "The Biosphere Reserve," p. 105. See also *The Biosphere Reserve and Its Relationship to Other Protected Areas* (Gland, Switzerland: IUCN, 1979); and Programme on Man and the Biosphere (MAB), Task Force on Criteria and Guidelines for the Choice and Establishment of Biosphere Reserves—Organized Jointly by Unesco and UNEP, Paris, 20-24 May 1974, MAB Report Series, no. 22, "Final Report," SC.74/Conf.203/2. For an assessment see Batisse, "Developing and focusing the biosphere reserve concept," *Nature and Resources* 22, no. 3 (July–September 1986): 2-11.

12. U.S. Department of State, "Convention on International Trade in Endangered Species of Wild Fauna and Flora, with Appendices," 3 March 1973, *United States Treaties and Other International Agreements*, vol. 27, pt. 2, p. 1087, TIAS no. 8249; and *IUCN Bulletin*, n.s., 4 (March 1973): 1, and twelve-part supplement. See also T. Inskipp and Sue Welle, *International Trade in Wildlife* (London: Institute for Environment and Development and the Fauna Preservation Society, 1979); and David S. Favre, *International Trade in Endangered Species: A Guide to CITES* (London: Graham and Trotman, 1989).

13. Peter H. Sand (secretary-general of Convention on International Trade in Endangered Species of Wild Fauna and Flora [CITES]), "Controlling the International Wildlife Trade," in UNEP, *Reports to Government*, no. 31 (May 1981), pp. 1-2; Gerhard Emonds, *Guidelines for National Implementation of the Convention on International Trade in Endangered Species of Wild Fauna and Flora—CITES*, Environmental Policy and Law Paper 17 (Gland, Switzerland: IUCN, 1981); "A Desperate Struggle to Stamp Out Illegal Wildlife Trade: Hong Kong at the Hub," *Ceres* 15 (January–February 1982): 3-5; Winston Harrington, "Endangered Species— A Global Threat," *Resources*, no. 71 (October 1982): 24; and Norman Boucher, "The Wildlife Trade," *Atlantic Monthly* (March 1983): 10-11.

14. James Fisher, Noel Simon, and Jack Vincent, *The Red Data Book: Wildlife in Danger* (London: Collins, 1969); Bjorn Berglund, *Noah's Ark Is Stranded* (New York: Dell, 1976); Jean Dorst, *Before Nature Dies* (Boston: Houghton Mifflin, 1970); Norman Myers, *The Sinking Ark: A New Look at the Problem of Disappearing Species* (New York: Pergamon Press, 1979). See also Patrick Marnham, "Rescuing the Ark: Conservation and Paternalism in Madagascar," *Harper's Magazine* (October 1980): 86-89; Roger M. Williams, "The Politics of Wild Animals," *World* (13 February 1973): 25-27; and Colin Norman, "The Threat to One Million Species," *Science* 214 (4 December 1981): 1105-7. For a valuable account of the international movement to protect endangered species, see Harold J. Coolidge, "An Outline of the Origins and Growth of the IUCN Survival Service Commission," in *Transactions of*

the Thirty-Third North American Wildlife and Natural Resources Conference, 11–13 March 1968 (Washington, D.C.: Wildlife Management Institute, 1978), pp. 407–17. See also David Lee, *The Sinking Ark: Environmental Problems in Malaysia and Southeast Asia* (Kuala Lumpur, Malaysia: Heinemann, 1981); and Jean Fauchon, "The Dreams and Delusions of Wildlife Conservation," *Ceres* 15 (March–April 1982): 33–39. On the Invertebrate Red Data Book, see *IUCN Bulletin*, n.s., 14 (July–September 1983): 80.

15. For text of treaty, see Special Supplement to *IUCN Bulletin* (January–February 1980). See also "A Convention Is Born," *IUCN Bulletin*, n.s., 10 (June 1979): 41; and "The Bonn Convention Concluded," *Environmental Policy and Law* 5 (June 1979): 135.

16. Draft Global Plan of Action, FAO/UNEP project no. 0502-78/02 (Rome: FAO, 1981). This publication provides an excellent account of the marine mammals problem and the history of efforts to deal with it. Appendix B lists and describes "International Organizations, Agreements and Programmes Concerned with Marine Mammals and their Environment," pp. 131–39. For the World Wildlife Fund's Marine Programme, see "The Seas Must Live," *IUCN Bulletin*, n.s., 7 (December 1976): entire issue.

17. "Agreement on Conservation of Polar Bears," 15 November 1973, *UST*, vol. 24, pt. 4, pp. 3918–23, TIAS no. 8409. For background, see *Bears—Their Biology and Management: Papers and Proceedings of the International Conference on Bear Research and Management, Calgary, 6–9 November 1970*, IUCN Publications, n.s., no. 23, Panel 3, Polar Bear Studies (Morges, Switzerland: IUCN, 1972), pp. 138–98.

18. For "Final Act of the Conference on the Conservation of Antarctic Marine Living Resources and the Convention" as adopted, see *International Legal Materials* 19 (July 1980): 837–62. For a critique of the draft convention, see Barbara Mitchell and Richard Sandbrook, *The Management of the Southern Ocean* (London: International Institute for Environment and Development, 1980), sec. 11.

19. *IUCN Bulletin*, n.s., 10 (February 1979): 9. See also Grenville Lucas and A. H. M. Synge, "The IUCN Threatened Plants Committee and Its Work Throughout the World," *Environmental Conservation* 4 (Autumn 1977): 179–87; and Lucas, "Threatened Plants—How to Save Them," *Oryx* 13 (February 1976): 257–58.

20. *United Nations Treaty Series*, vol. 149 (1952), no. 1963, p. 68. See also *FAO Plant Protection Bulletin* for reports on plant diseases and pests and phytosanitary legislation and regulations. See also William B. Hewitt and Luigi Chiarappon, *Plant Health and Quarantine in International Transfer of Genetic Resources* (Cleveland, Ohio: CRC Press, 1977).

21. *United Nations Treaty Series*, vol. 247 (1956), no. 1963, p. 400, concluded as a supplementary agreement under the International Plant Protection Convention of 1951. For illustration of the operation of this agreement, see *Report of the Tenth Session of the Plant Protection Committee for the South East Asia and Pacific Region* (Bangkok, Thailand: Food and Agriculture Organization, 1976).

22. "Agreement Concerning Cooperation in the Quarantine of Plants and Their Protection Against Pests and Diseases" (Sofia, Bulgaria, 14 December 1959), *United Nations Treaty Series*, vol. 422 (1962), no. 6067, pp. 33 and 42.

23. E.g., see William C. Brice, ed., *The Environmental History of the Near and Middle East Since the Last Ice Age* (New York: Academic Press, 1978); Rhoads Murphy, "The Decline of North Africa Since the Roman Occupation: Climatic or Human?" *Annals of the Association of American Geographers* 41 (June 1951): 116–32; and George Perkins Marsh, "The Woods," chap. 6 of *Man and Nature; or Physical Geography as Modified by Human Action* (New York: Scribners, 1864; reprint ed., Cambridge, Mass.: Belknap Press of Harvard University Press, 1965). See also Arnold Toynbee, *Mankind and Mother Earth: A Narrative History of the World* (London: Oxford University Press, 1976).

24. "The Disappearing Forest," *Uniterra* 5 (January 1980): 1–2; "The Tropics," *Unasylva* 27, no. 110 (1975): entire issue. See also Norman Myers, *Conversion of Tropical Moist Forests: A Report Prepared by Norman Myers for the Committee on Research Priorities in Tropical Biology of*

the National Research Council (Washington, D.C.: National Academy of Sciences, 1980); and Myers, "The Present Status and Future Prospects of Tropical Moist Forests," *Environmental Conservation* 7 (Summer 1980): 101–14. See review of Myers's book and author's response: Ariel E. Lugo and Sandra Brown, "Tropical Ecosystems and the Human Factors," *Unasylva* 33, no. 133 (1981): 45–49, 49–52. See also *Forestry Activities and Deforestation Problems in Developing Countries*, Report by the Forest Products Laboratory, PASA no. AG/TAB-1080-10-78; and "Vanishing Forests," *Newsweek* (24 November 1980), pp. 117–22; U.S. Interagency Task Force on Tropical Forests, *The World's Tropical Forests: A Policy, Strategy, and Program for the United States* (Washington, D.C.: Department of State Publication 9117, May 1980); National Research Council, (U.S.) Committee on Selected Biological Problems in the Humid Tropics, *Ecological Aspects of Development in the Humid Tropics* (Washington, D.C.: National Academy Press, 1982); and U.S. Congress, House of Representatives, *Deforestation: Environmental Impact and Research Needs, Joint Hearing before the Subcommittee on Natural Resources, Agriculture Research and Environment of the Committee on Science and Technology and the Subcommittee on Human Rights and International Organizations of the Committee on Foreign Affairs*, 97th Cong., 2d sess., 16 September 1982, no. 161. For an opinion dissenting from claims of adverse effects of cutting tropical forests, see Ariel E. Lugo and Sandra Brown, "Tropical Lands: Popular Misconceptions," *Mazingira* 5, no. 2 (1981): 10–19.

25. Dennis Richardson, "A Faustian Dilemma," *Unasylva* 29, no. 117 (1977): 12–14.

26. Loren McIntyre, "Jari: A Billion Dollar Gamble," *National Geographic* 157 (May 1980): 686–711. This project encountered political complications; see "Jari—Driblando o aval: Governo empresta a Antunes, que paga por Ludwig," *Veja* (1 July 1981): 99 (a Brazilian account). For an account of the economic difficulties of the Jari project, see Given Kinkead, "Trouble in D. K. Ludwig's Jungle," *Fortune* 103 (21 April 1981): 102–17, "Jari: First Phase Ends in Attrition," Latin American Regional Reports—Brazil, R.B.-81-01 (2 January 1981): 7; "Daniel Ludwig is Giving up Amazon Dream," *Wall Street Journal*, 11 January 1982, p. 25; and "End of an Amazon Dream," *Latin American Weekly Report*, WR-82-03 (15 January 1982): 6. For a more inclusive perspective, see Robert J. A. Goodland, "Brazil's Environmental Progress in Amazonian Development," in J. Heming, ed., *Change in the Amazon Basin* (Manchester, U.K.: Manchester University Press, 1983).

27. Allen Keast and Eugene S. Morton, eds., *Migrant Birds in the Neotropics: Ecology, Behavior, Distribution and Conservation—Proceedings of a Symposium held at the Conservation and Research Center, National Zoological Park, Smithsonian Institution, October 27–29, 1977* (Washington, D.C.: Smithsonian Institution Press, 1980). Also Peter Steinhart, "Trouble in the Tropics," *National Wildlife* 22 (December–January 1984): 16–20.

28. Bayard Webster, New York Times News Service, in the *Chicago Tribune*, 17 August 1980, sec. 3, p. 8.

29. Gary S. Hartshorn, "El Salvador: An Ecological Disaster," letter to Peter Martin, executive director, Institute of Current World Affairs, Hanover, N.H., 7 June 1961, GSH-1.

30. R. J. W. Aluma, "Uganda: A Damage Report," *Unasylva* 31 (1979): 20–24. Among numerous reports of deforestation, see Robert J. A. Goodland and H. S. Irwin, *The Amazon Jungle: Green Hell to Red Desert?* (Amsterdam: Elsevier, 1975); and numerous articles in *Unasylva* and *Uniterra*. Also Robert J. A. Goodland, "Environmental Ranking of Amazonian Development Projects in Brazil," *Environmental Conservation* 7 (Spring 1980): 6–26; and references under note 34 below.

31. Alison Jolly, *A World Like Our Own: Man and Nature in Madagascar* (New Haven, Conn.: Yale University Press, 1980), p. 25.

32. *Unasylva* 30, no. 121 (1978): 35–38. For an account of the Eighth Forestry Congress (1978) and its "Forests for People Declaration," see *Uniterra* 3 (November 1978): 1–2.

33. R. H. Kemp, "Exploration, Utilization and Conservation of Genetic Resources," *Unasylva* 30, no. 119-20 (1978): 10–16, extensive references. See also Gabor Vida, "Genetic Diversity and the Environmental Future," *Environmental Conservation* 5 (Summer 1978): 127–32; and

"Germplasm Conservation and Genetic Resources" in *Ecological Aspects of Development in the Humid Tropics*, pp. 77–92.

34. "The Disappearing Forests," *Uniterra* 5 (January 1980): 1–2; and Chiek Dia, "African States Agree Forests Convention," *Uniterra* 5 (July 1980): 2. The status of the Yaounde center is uncertain.

35. James Sholto Douglas, "Forest Farming: An Ecological Approach to Increase in Nature's Food Productivity," *Impact of Science on Society* 23, no. 2 (1973): 117–32. Also Richard A. Carpenter, ed., *Assessing Tropical Forest Lands: Their Suitability for Sustainable Uses* (Dublin: Tycooly International Publications, 1981); and W. R. H. Perera, *Man—Forests—and Environment: A New Forest Policy?* (Honolulu, Hawaii: Environment and Policy Institute, East-West Center, 1980). Examples of destructive policies cited from Sri Lanka, Indonesia, and Malaysia. Cf. Duncan Poore, comp., *Ecological Guidelines for Development in Tropical Forest Areas of South East Asia*, Occasional Paper, no. 10 (Morges, Switzerland: IUCN, 1974); and H. Sutlive et al., eds., *Where Have All the Flowers Gone? Deforestation in the Third World*, Studies in Third World Societies, no. 13 (Williamsburg, Va.: College of William and Mary, 1981). For a review of forestry in rural development, see *Ceres* 14 (July–August 1981).

36. J. P. Lanly and J. Clement, "Present and Future Natural Forest and Plantation Areas in the Tropics," *Unasylva* 31, no. 123 (1979): 20.

37. For a comprehensive review of the tropical deforestation problem see Nicholas Guppy, "Tropical Deforestation: A Global View," *Foreign Affairs* 62 (Spring 1984): 928–65. For background see Tatsuro Kunugi, "Consensus on Elements of International Timber Agreement—Negotiating Conference Expected Early Next Year," *Monthly Bulletin* [UNCTAD], no. 185 (July 1982): n.p.; *UNCTAD Bulletin*, no. 192 (April 1983): 19. See also UNEP, *1989 The State of the World Environment*, Nairobi (April 1989): iv.

38. International Conference on Wetlands, *Project Mar: The Conservation and Management of Temperate Marshes, Bogs and Other Wetlands, Proceedings of the MAR Conference organized by IUCN, ICBP, and IWRB at Les Saintes-Maries-de-la-Mer, 12–16 November 1962*, IUCN Publications, n.s., no. 13. See also Richard H. Goodwin, "The Project Mar Conference," *The Nature Conservancy News* 12 (Winter 1962): 3–4. See the later report of an international working group: UNESCO, MAB, "Expert Panel on Project 5: Ecological Effects of Human Activities on the Value and Resources of Lakes, Marshes, Rivers, Deltas, Estuaries and Coastal Zones—Final Report," MAB Report Series, no. 5, SC.72/Conf.142/3, 1972.

39. E. Capr, ed., *International Conference on the Conservation of Wetlands and Water Fowl, Ramsar, Iran. 30 January–3 February 1971* (Slimbridge, U.K.: International Wild Fowl Research Bureau, 1972), p. 5.

40. Ibid., "Final Act and Summary Record."

41. *IUCN Bulletin*, n.s., 14 (January–March 1983): 3; United Kingdom, Parliament, *Sessional Papers* (Commons), Treaty Series 34 (1981), CMND 8276. See also "Governments Urged to Protect National Wetlands," *Uniterra* 5 (November–December 1980): 12, which reports on the first meeting of signatories, at Cagliari, Sardinia. For later assessment, see "Special Report: Wetlands," *IUCN Bulletin* 20, nos. 4–6 (April–June 1989).

42. Thomas Land, "Ecological 'Disaster' in China," *Toronto Star*, 14 October 1981, p. A6; James Walls, ed., *Combatting Desertification in China* (Nairobi: UNEP, 1982). See also Vaclav Smil, "Ecological Mismanagement in China," *Bulletin of the Atomic Scientists* 38 (October 1982): 18–23; Smil, *The Peoples Republic of China: Environmental Aspects of Economic Development* (Washington, D.C.: World Bank, Office of Environmental Affairs, September 1982); Paul and Anne Ehrlich, "China on the Brink," *Mother Earth News* (July–August 1983): 146–47; and Wolfgang K. H. Kinzelbach, "China: Energy and Environment," *Environmental Management* 7 (July 1983): 303–10. Since 15 August 1989 the People's Republic of China has published *China Environment News*, a monthly English language newspaper, a source of information about environmental problems and policies in China hitherto not readily obtained.

43. See Asim I. el Maghraby, "The Jonglei Canal—Needed Development or Potential Eco-

disaster?" *Environmental Conservation* 9 (Summer 1982): 114–48. Also see Roger M. Berthe-
lot, "UNDP/Fact-Finding Mission—Sudan: Jonglei Canal," SUD/GEN, February 1976;
Peter Hayes, "Jonglei Canal: Risking Social and Economic Disaster?" n.d., unpublished;
and John M. Bradley, "Controversial African Canal Has Eco-Impact," *World Environment
Report* 9 (30 June 1983): 6; and Robert Eshman, "The Jonglei Canal: A Ditch Too Big?"
Environment 25 (June 1983): 15–20, 32.

44. Clare Oxby, *Pastoral Nomad and Development* (London: International African Institute, 1975),
with select annotated bibliography; Rada Dyson-Hudson and Neville Dyson-Hudson, "Nomadic Pastoralism," *Annual Review of Anthropology* 9 (1980): 1561; Laurence Krader, "Ecology
of Central Asian Pastoralism," *Southwestern Journal of Anthropology* 11 (Winter 1955): 301–
26; Denis Sinor, "Horse and Pasture in Inner Asian History," *Oriens Extremus* 19, nos. 1–2
(1972): 171–83.

45. "Conservation Rediscovered: Learning from the Bedouin," *IUCN Bulletin*, n.s., 12 (March–
April 1981): 14. See also Harold F. Heady, "Ecological Consequences of Bedouin Settlements
in Saudi Arabia," Frank Fraser Darling and Mary A. Farvar, "Ecological Consequences of
Sedentarization of Nomads," and Lee M. Talbot, "Ecological Consequences of Rangeland
Development in Masailand, East Africa," in M. Taghi Farvar and John P. Milton, *The Care-
less Technology: Ecology and International Development* (Garden City, N.Y.: Natural History
Press, 1972), pp. 671–82, 683–93, and 694–711; and Daniel N. Stiles, "Camel Pastoralism and
Desertification in North Kenya," *Desertification Control* 8 (June 1983): 2–8.

46. "Conservation Rediscovered," p. 14.

47. Z. A. Konezacki, *The Economics of Pastoralism: A Case Study of the Sub-Sahara Africa* (London: Frank Cass, 1978). See also H. J. Cook, "The Struggle Against Environmental Degradation—Botswana's Experience," *Desertification Control* 8 (June 1983): 9–15.

48. Callisto Eneas Madavo, "Uncontrolled Settlements," in *Finance and Development* 13 (March
1976): 16–19. See also Charles Abrams, *Man's Struggle for Shelter in an Urbanizing World*
(Cambridge, Mass.: MIT Press, 1964).

49. United Nations Centre for Human Settlements (Habitat), *Survey of Slum and Squatter Settlements* (Dublin: Tycooly International Publications, 1982).

50. Barbara Ward, "The Poor World's Cities," *Economist* (6 December 1969): 56–62. For descriptions of urban environmental problems in Africa, see *African Environment* 2, nos. 12
(1976). For a "worst case" example, see Geoffrey Moorhouse, *Calcutta* (New York: Harcourt
Brace Jovanovich, 1971); and Sumil K. Munsi, *Calcutta Metropolitan Explosion: Its Nature and
Roots* (New Delhi: Peoples' Publishing House, 1975). But lest it be thought that the socio-
environmental dereliction of cities in the Third World is only a contemporary phenomenon,
see Rudyard Kipling, *City of the Dreadful Night*, 16, pt. 2: 287–363, in *The Writings in Prose of
Rudyard Kipling* (New York: Scribner's, 1916).

51. Hartshorn to Martin, 7 June 1978, GSH-1.

52. The Tepoztlan Seminar was sponsored by three nongovernmental organizations: the Committee on Urban and Regional Studies of the Latin American Social Science Council
(CLACSON), the International Institute for Environment and Development (IIED), and
the Inter-American Planning Society (SIAP). See *Habitat News* 3, no. 2 (November 1980): 18.

53. *Human Settlements in the Arctic—An ECE Symposium on Human Settlements, Planning and
Development in the Arctic* (New York: Pergamon Press, for the United Nations, 1980); and
George C. West, "Environmental Problems Associated with Arctic Development Especially
in Alaska," *Environmental Conservation* 3 (Autumn 1976): 218–24.

54. *Report of Habitat: United Nations Conference on Human Settlements, Vancouver, 31 May–11 June
1976* (New York: United Nations, 1976). A list of supplementary papers and reports was
published by the conference secretariat. Eloquently summing up the issues underlying the
Habitat Conference is the book by Barbara Ward, *The Home of Man*, introduction by
Enrique Penalosa, secretary-general, United Nations Conference on Human Settlements
(New York: W. W. Norton, 1976). See also M. R. Biswas, "Habitat in Retrospect," *Inter-*

national Journal of Environmental Studies 11, no. 4 (1978): 267–79; and Enrique Penalosa, "The United Nations Conference on Human Settlements: What Happened at Vancouver?" *Mazingira* 6, no. 1 (1982): 40–45.

55. UN, HABITAT, Ad Hoc Intergovernmental Working Groups (Geneva, 22–26 September 1975), "Report of the Meeting of Experts held at Dubrovnik from 20–23 May 1975," A/Conf.70/WG14, 18 July 1975. See also papers published under MAB Project 11: "Ecological Aspects of Urban Systems and Other Human Settlements." Note Stephen Boyden, *An Integrative Ecological Approach to the Study of Human Settlements*, MAB Technical Notes 12 (Paris: UNESCO, 1979); and the journal *Habitat International: The International Journal on All Aspects of Human Settlements both Urban and Rural* (Elmsford, N.Y.: Pergamon Press, 1976–).

56. Chaplin B. Barnes, "Habitat and the Human Environment: Challenge to the Non-Governmental Company," *Earth Law* 2 (1976): 107–9.

57. Cesar Quintana, remark reported in *Habitat Foundation News*, no. 6 (November 1978): 1.

58. United Nations, *Report of the Preparatory Committee for the United Nations Conference on Human Settlements* (HABITAT II) General Assembly, Official Records, 49th Session, Supplement No. 37 (A/49/37).

59. *IULA Newsletter* 11 (May 1977): 4.

60. Raimund O. A. Becker-Ritterspach, "The Restoration of Bhaktapur," *Unasylva* 30, no. 121 (1978): 2.

61. Peter Graham, "The Double-Edged Sword of Tourism: A UNESCO Study of the Visitor Industry," *Honolulu Star Bulletin*, 31 January 1981, p. A17, commenting on a report by Emmanuel de Kadt, ed., *Tourism: Passport to Development? Perspectives on the Social and Cultural Effects of Tourism on Developing Countries* (Paris: UNESCO, 1976). See also *Monumentum* 6 (1970), a special issue devoted to the Conference on Monuments and Tourism, Oxford, 7–11 July 1969; and *The Impact of Tourism on the Environment: General Report* (Paris: OECD, 1980), extensive references pp. 96–148.

62. Gerardo Budowski, "Tourism and Environmental Conservation: Conflict, Coexistence, or Symbiosis?" *Environmental Conservation* 3 (Spring 1976): 27–31. Also *Ecological Impact of Recreation and Tourism on Temperate Environments*, pt. 1 of *Towards a New Relationship of Man and Nature in Temperate Lands*, Proceedings and Papers of the IUCN Tenth Technical Meeting, Lucerne, Switzerland, June 1966 (Morges, Switzerland: IUCN, 1967).

63. Paul Hoffmann, "Saving Italy's Ruins from Ruin," *Art News* 80 (March 1981): 78–85.

64. Raymond H. M. Goy, "The International Protection of the Cultural and Natural Heritage," *Netherlands Yearbook of International Law* 1973, p. 122.

65. See *Protection of Mankind's Cultural Heritage: Sites and Monuments* (Paris: UNESCO, 1970), which describes legal and scientific cooperation, international campaigns for monuments, for Florence, and for Venice, the campaign to save the monuments of Nubia, and cultural tourism.

66. "Recommendation Concerning the Safeguarding of the Beauty and Character of Landscapes and Sites," *Records of the General Conference 12th Session — Resolutions, Paris, 1962* (Paris: UNESCO, 1963), pp. 139–42.

67. For descriptions of the International Council of Monuments and Sites (ICOMOS), see *Yearbook of International Organizations, 1978*, 17th ed., A 1748, and *Monumentum* 12 (1976): 4.

68. Ernest Allen Connally, "The International Organization to Link Public Authority, Institutions, and Individuals Interested in the Study and Preservation of Monuments and Sites," *Monumentum* 12 (1976): 4.

69. In addition to *Monumentum* see, for example, the following publications of ICOMOS: *Conference on the Conservation, Restoration and Revival of Areas and Groups of Buildings of Historic Interest*, Caceres, Spain, 15–19 March 1967; and *Second Conference on the Conservation, Restoration and Revival of Areas and Groups of Buildings of Historic Interest*, Tunis, 1968.

70. *Yearbook of International Organizations, 1978*, 17th ed., A174g. For illustration of its activities,

see *The Museum in the Service of Man, Today and Tomorrow: The Museum's Educational and Cultural Roles*, Papers from the Ninth Geneva Conference of the International Council of Museums (ICOM) (Paris: ICOM, International Museum Documentation Center, 1972).

71. See leaflet entitled *International Centre for the Study of the Preservation and Restoration of Cultural Property* (Washington, D.C.: International Centre Committee, n.d.).

72. "The Race to Save Abu Simbel," *Life* 61 (2 December 1966): 32–39; and issues of the *Unesco Courier* 13 (February 1960); 15 (October 1961); and 17 (December 1964).

73. *Unesco Chronicle* 13 (June 1967): 259; (December 1967): 486–87; and 15 (March 1969): 116.

74. R. Soekmono, *Chandi Borobudur: A Monument of Mankind* (Paris: UNESCO Press; Assen/ Amsterdam: Van Gorcum, 1976); Soekmono, "Borobudur," *Unesco Courier* 36 (February 1983): 8–15; and R. Soekmono and Caesar Voute, "How Borobudur Was Saved," *Unesco Courier* 36 (February 1983): 16–23.

75. Becker-Ritterspach, "The Restoration of Bhaktapur," pp. 2–10.

76. Dogan Kurban in *Monumentum* 6 (1970): 76.

77. U.S. Council on Environmental Quality, *Environmental Quality: The Second Annual Report of the Council on Environmental Quality* (Washington, D.C.: Government Printing Office, August 1971), p. ix.

78. Robert L. Meyer, "Travaux Préparatoires for the UNESCO World Heritage Convention," *Earth Law Journal* 2 (1976): 45–79; *Ambio* 12, nos. 3–4 (1983): entire issue. For the text of the convention, see any of the following: UNESCO General Conference, Seventeenth Session, Paris, 16 November 1972, "Convention Concerning the Protection of the World Cultural and Natural Heritage" (Paris: Office of International Standards and Legal Affairs, UNESCO, 1972); *International Legal Materials* 2 (1972): 1538; and *U.S. Treaties*, vol. 27, pt. 1, pp. 37–71, TIAS no. 8226.

79. World Heritage Convention: World Heritage Reference Map and List of Recorded Sites (Paris: UNESCO, 1987); *World Heritage Newsletter*, no. 7 (March 1995).

80. Patrick J. O'Keefe and James A. R. Napziger, "Report: The Draft Convention on Protection of the Underwater Cultural Heritage," *Ocean Development and International Law* 25, no. 4 (October–December 1994): 391–418.

81. "The Caracas Convention," *IUCN Bulletin* 23, no. 2 (June 1992): 16–17.

82. See articles by Gerard Bolla, "The International Convention" and by Michel Batisse, "A New Partnership in the Making," in *Unesco Courier* 33 (August 1980). The issue lists and describes its first fifty-seven entries in the World Heritage List; text is by Georges Fradier. See also *The World's Greatest Natural Areas: An Indicative Inventory of Natural Sites of World Heritage Quality* (Gland, Switzerland: IUCN, 1982).

83. Meyer, "Travaux Préparatoires." See also Donatus de Silva, "In Defence of the Wonders of the World," *Uniterra* 5 (September–October 1980): 7.

84. "World Heritage Committee," *Yearbook of International Organizations, 1978*, 17th ed., B2358g.

85. Alexander B. Adams, ed., *First World Conference on National Parks: Proceedings . . . Seattle, Washington, 30 June–7 July 1962* (Washington, D.C.: United States Department of the Interior, National Park Service, n.d.). See also Philip Quigg, *Protecting Natural Areas: An Introduction to the Creation of National Parks and Reserves* (New York: National Audubon Society, 1978); Edmund A. Schofield, ed., *Earthcare: Global Protection of Natural Areas, Proceedings of the Fourteenth Biennial Wilderness Conference* (Boulder, Colo.: Westview Press, 1978); Norman Myers, "National Parks in Savannah Africa," *Science* 178 (22 December 1972): 1255–63; "The World's Protected Areas," *Ambio* 11, no. 5 (1982): entire issue; and Jeffrey A. McNeely and Kenton R. Miller, "IUCN, National Parks, and Protected Areas: Priorities for Action," *Environmental Conservation* 10, no. 1 (Spring 1983): 13–21.

86. Hugh Elliott, ed., *Second World Conference on National Parks, Yellowstone and Grand Teton Parks, U.S.A., September 18–27, 1972* (Morges, Switzerland: IUCN, 1974).

87. *IUCN Bulletin*, n.s., 13 (July, August, September 1982): 57; (October, November, December 1982): 77–79, and *The World National Parks Congress* (Gland, Switzerland: IUCN, 1982).

88. The list is revised periodically. For one including descriptions of parks, see *United Nations List of National Parks and Equivalent Reserves*, 2d ed. (Morges, Switzerland: IUCN, 1971).
89. Harold J. Coolidge, "Highlights in International Developments Bearing on National Parks History," presented at "A Short Course in Administration of National Parks and Equivalent Reserves," School of Natural Resources, University of Michigan, Ann Arbor, 10 May 1965.
90. "Parks for Life: A New Beginning," *IUCN Bulletin* 23, no. 2 (June 1992): 10–12.
91. See Ian Burton, Robert W. Kates, and Gilbert E. White, *The Environment as Hazard* (New York: Oxford University Press, 1978).
92. U.S. Congress, House Committee on Foreign Affairs Subcommittee on National Security Policy and Scientific Developments, *The Politics of Global Health*, prepared by Freeman H. Quimby for Science Policy Research Division, Congressional Research Service, Library of Congress as part of an extended study of the interactions of science and technology with United States foreign policy (Washington, D.C.: Government Printing Office, May 1971), p. 3.
93. WHO, "The Changing Scope of Environmental Health," *WHO Chronicle* 22 (March 1968): 95–99; and Evelyn E. Meyer and Peter Sainsbury, eds., *Promoting Health in the Human Environment: A Review Based on the Technical Discussions Held during the Twenty-Seventh World Health Assembly, 1974* (Geneva: WHO, 1975); WHO, *Environmental Change and Resulting Impacts on Health*, Report of a WHO Expert Committee, WHO Technical Report Series 292 (Geneva: WHO, 1964). For background, see Neville M. Goodman, *International Health Organizations and Their Work* (Philadelphia, Pa.: Blakiston, 1952); and *The First Ten Years of the World Health Organization* (Geneva: WHO, 1958). For general, nontechnical treatment of environmental health, see Erik P. Eckholm, *The Picture of Health: Environmental Sources of Disease* (New York: W. W. Norton, 1977); and John Lenihan and William W. Fletcher, eds., *Health and the Environment* (New York: Academic Press, 1976).
94. *The Global Eradication of Smallpox: Final Report of the Global Commission for the Certification of Smallpox Eradication, Geneva, December 1979* (Geneva: WHO, 1980). See also relevant passages in *Building the Health Bridge: Selections from the Works of Fred L. Soper, M.D.*, ed. J. Austin Kerr (Bloomington: Indiana University Press, 1970). In addition to this account of the Rockefeller Foundation International Health Division in the eradication of infectious diseases in various parts of the world, see Jack W. Hopkins, *The Eradication of Smallpox: Organizational Learning and Innovation in International Health* (Boulder, Colo.: Westview Press, 1989).
95. On the ecology of disease, see the following: "Considerations of Human Ecology in Environmental Health Programmes," *Report on an Interregional Seminar held in Geneva, 31 July–9 August 1972*, DIS/73.2 (Geneva: WHO, 1972); Jacques M. May, "Medical Geography: Its Methods and Objectives," *Geographical Review* 40 (January 1950): 941; May, "Science on the March: The Geography of Pathology," *Scientific Monthly* 72 (February 1951): 128–31; May, *The Ecology of Human Disease* (New York: MD Publications, 1958); May, ed., *Studies in Disease Ecology* (New York: Hafner, 1961); Douglas H. K. Lee, "Environmental Health and Human Ecology," *American Journal of Public Health* 54 (Supplement to January 1964): 7–10; Howard C. Hopps, "Geographic Pathology," in *Pathology*, eds. W. A. D. Anderson and John M. Kissane (St. Louis: C. V. Mosby, 1977), pp. 692–736; and Hopps, "The Ecology of Disease," *Pfizer Spectrum International* 19 (Fall 1976): 43–46. See also Evgeny N. Pavlovsky, *Natural Nidality of Transmittable Diseases*, trans. Frederick K. Plaus, Jr., and ed. Norman D. Levine (Urbana: University of Illinois Press, 1966); and Evgeny N. Pavlovsky, ed., *Human Diseases with Natural Foci*, trans. D. Rottenburg (Moscow: Foreign Languages Publishing House, n.d.). For interesting case studies, see "The Cancer Detectives of Lin Xian," *Nova* (Boston: WGBH Transcripts, 1980); and "Lassa Fever," ibid. (1982).
96. Michael J. Sharpston, "Health and the Human Environment," *Finance and Development* 13 (March 1976): 26. Cf. John Bryant, ed., *Health and the Developing World* (Ithaca, N.Y.: Cornell University Press, 1969); and James E. Banta, "Effecting Changes in Health Behavior in Developing Countries," *Archives of Environmental Health* 18 (February 1969): 265–68.

97. *Environmental Health and Human Ecological Considerations in Economic Development Projects* (Washington, D.C.: World Bank, 1974), extensive references.

98. For historical background on the campaign against malaria, see: F. L. Hoffman, *A Plea and a Plan for the Eradication of Malaria throughout the Western Hemisphere* (Newark, N.J.: Prudential Press, 1917); League of Nations Health Organization, *Malaria Commission Reports 1926–1931* (Geneva: League of Nations, 1926–1931); Forest Ray Moulton, ed., *Human Malaria* (Washington, D.C.: American Association for the Advancement of Science, 1941); Sir Malcolm Watson, *African Highway, the Battle for Health in Central Africa* (London: Murray, 1953); *First Ten Years of the World Health Organization*, pp. 172–87; Maurice Candau, "World War against Malaria," *International Development Review* 1 (October 1959): 7–11; Peter Newman, *Malaria Eradication and Population Growth* (Ann Arbor: University of Michigan Press, 1965); and Fred L. Soper, "The Eradication of Malaria: Local, Continental, Global," pt. 2 of *Building the Health Bridge* (1970). For a summary statement, see "Environmental Disease: Malaria," *The State of the Environment 1978: Selected Topics* (Nairobi: UNEP, 1978), pp. 13–17.

99. Akilu Lemma, "Schistosomiasis: The Social Challenge of Controlling a Man-Made Disease," *Impact of Science on Society* 23 (April–June 1973): 133–42; J. Gus Liebenow, "Bilharzia Control in Swaziland—The Dilemma of Development," Africa, no. 2, *American Universities Field Staff Reports* (1981); F. Eugene McJunkin, "Schistosomiasis: Limiting Adverse Health Consequences of Development Projects," in David A. Kay and Harold K. Jacobson, eds., *Environmental Protection: The International Dimension* (Totowa, N.J.: Allenheld, Osmun), pp. 200–216; and the following chapters in Milton and Farvar, *The Careless Technology:* Charles C. Hughes and John M. Hunter, "The Role of Technological Development in Promoting Disease in Africa," pp. 69–101 (references); C. J. Shiff, "The Impact of Agricultural Development on Aquatic Systems and Its Effect on the Epidemiology of Schistosomes in Rhodesia," pp. 102–15; and Henry Van der Schalie, "World Health Organization Project Egypt 10: A Case History of a Schistosomiasis Control Project," pp. 116–36. At its fourth session the Governing Council of UNEP reviewed an "Action Plan on Ecological and Habitat Management of Schistosomiasis," UNEP/GC (IV) Inf.1, 22 March 1976.

100. *Intersectoral Coordination and Health in Environmental Management: An Examination of National Expenence,* Public Health Papers Series, no. 74 (Geneva: WHO, 1980), pp. 13–14. See also *National Environmental Health Programmes: Their Planning, Organization and Administration,* Report of a WHO Expert Committee, WHO Technical Report Series, no. 439 (Geneva: WHO, 1970); *Health Hazards of the Human Environment* (Geneva: WHO, 1972); and Meyer and Sainsbury, *Promoting Health in the Human Environment.* For examination of some environmental health problems of the industrialized countries, see Asher J. Finkel, ed., *Energy, the Environment and Human Health* (Acton, Mass.: Publishing Sciences Group, for the American Medical Association, 1977).

101. On the history of the World Health Organization specifically, see Norman Howard Jones, "The World Health Organization in Historical Perspective," *Perspectives in Biology and Medicine* 24 (Spring 1981): 467–82; Sir John Charles, "Origins, History, and Achievements of the World Health Organization," *British Medical Journal* 2 (4 May 1968): 293–96; Howard B. Calderwood, "The World Health Organization and Its Regional Organizations," *Temple Law Review* 37 (Fall 1963): 15–25; and *First Ten Years of the World Health Organization.* For background, see Goodman, *International Health Organizations and Their Work.*

102. See United Nations, International Conference on Primary Health Care (Alma Ata, U.S.S.R., 6–12 September 1978), "Primary Health Care: A Joint Report by the Director-General of the World Health Organization and the Executive Director of the United Nations Children's Fund" (Geneva: WHO and UNICEF, 1978); and "Primary Health Care: Report of the International Conference on Primary Health Care" (Geneva: WHO and UNICEF, 1978). Note Declaration of Alma Ata adopted by the conference.

103. "Alma Ata Meeting," *Uniterra* 3 (November 1978): 3.

104. Peter G. Bourne, "The United Nations International Drinking Water Supply and Sanitation Decade," *Mazingira* 5, no. 4 (1981): 14–24; Philip H. Jones, "International Drinking

Water and Supply Decade 1981–1990," *Environmentalist* 1 (Summer 1981): 101–3; Anil Agari-
val et al., *Water, Sanitation, Health—For All? Prospects for the International Drinking Water
Supply and Sanitation Decade 1981–1990* (London: Earthscan, 1981); and F. Eugene McJunkin,
"Water Supply and Sanitation: Improving Life for the Rural Majority," in Kay and Jacobson, *Environmental Protection: The International Dimension*, pp. 184–99. See also *UN Chronicle*
18 (January 1981): 28–30.

105. On the International Hydrological Programme (IHP), see *Unesco Chronicle* 21 (January-
February 1975): 30, 175–76. Reports of the sessions of the IHP Intergovernmental Council are published by UNESCO. For background, see "International Hydrological Decade,"
Nature and Resources 1 (June 1965): 6–8; Michel Batisse, "Launching the Hydrological Decade," *New Scientist* 25 (7 January 1965): 38–40; and "Taking Stock of the International
Hydrological Decade: Achievements and Prospects," *Unesco Chronicle* 20 (May–June 1974):
189–94. See also the journal *Water International* (Lausanne: Elsevier Sequoia, 1975-).

106. Richard Helmer, "Water-Quality Monitoring: A Global Approach," *Nature and Resources* 17
(January–March 1981): 7–12; and Silvio Barabas, "GEMS/WATER—Global Water-quality
Monitoring," *Impact of Science on Society*, no. 1 (1983), pp. 105–15 (see entire issue on fresh
water resources).

107. See Richard W. Franks and Barbara H. Chasin, *Seeds of Famine: Ecological Destruction and
the Development Dilemma in the West African Sahel* (New York: Universe, 1980); Michael H.
Glantz, ed., *The Politics of Natural Disaster: The Case of the Sahel Drought* (New York: Praeger,
1976); Noel V. Lateef, *Crisis in the Sahel: A Case Study in Development Cooperation* (Boulder,
Colo.: Westview Press, 1980).

108. Robert W. Kates, "Drought in the Sahel: Competing Views as to What Really Happened
in 1910–14 and 1968–74," *Mazingira* 5, no. 2 (1981): 72–83. For an ecological assessment,
see UNESCO, MAB, "Regional Meeting on Integrated Ecological Research and Training
Needs in the Sahelian Region Final Report," MAB Report Series, no. 18, SC.74/Conf.670/
Niamey, 9–15 March 1974. See also *Sahel: A Guide to the Microfiche Collection of Documents
and Dissertations* (Ann Arbor, MI: University Microfilms International, 1980); Claire Ster-
ling, "The Making of the Sub-Saharan Wasteland," *Atlantic Monthly* 283 (May 1974): 98–
105; *African Environment* 1 (April 1975): special issue on the Sahelian drought; Jean Dresch,
"Reflections on the Future of the Semi-Arid Regions," in *African Environment: Problems and
Perspectives*, ed. Paul Richards, Special Report 1, (London: International African Institute,
1975); and David Dalby, R. J. Harrison Church, and Fatima Bezzas, *Drought in Africa*, 2d
ed. (London: International African Institute, 1977).

109. See *L'Economie des pays du Sahel: L'eau et L'irrigation* (Paris: Ediafric, 1976), pp. 167–97, and
"The Sahel Programme—Blueprint for Change," *IUCN Bulletin* 19, nos. 4–6 (April–June
1988): 4–5.

110. See Secretariat of the United Nations Conference on Desertification (Nairobi, Kenya,
29 August–9 September 1977), ed. and comp., *Desertification: Its Causes and Consequences*
(Oxford: Pergamon Press, 1977); *Alternative Strategies for Desert Development and Manage-
ment*, papers submitted to the UNITAR—State of California Conference, vols. 1–4 (Oxford:
Pergamon Press, 1982); UNEP, Governing Council, "Implementation of General Assembly
Resolution 3337 (XXIX): International Cooperation to Combat Desertification—Report of
the Executive Director," UNEP/GC/67, Nairobi, 28 January 1976; "Report of the United
Nations Conference on Desertification," A/Conf.74/36, Nairobi, n.d.; and Mostafa Kamal
Tolba, "The United Nations Conference on Desertification: A Review," *Mazingira* 6, no. 1
(1982): 14–23. See also "Positive Support for Desert Control," *Uniterra* 5 (April 1980): 1–7;
"An Implementation of the Plan of Action to Control Desertification," *Uniterra* 5 (May–
June 1980): 10; Margaret R. Biswas, "U.N. Conference on Desertification in Retrospect,"
Environmental Conservation 5 (Winter 1978): 247–62; *Environmental Conservation* 5 (Spring
1978): 69–70; *Environmental Conservation* 6 (Summer 1979): 160–61; Amin S. Hassanyar,
"Restoration of Arid and Semi-Arid Ecosystems in Afghanistan," *Environmental Conserva-
tion* 4 (Winter 1977): 297–301; J. L. Cloudsley-Thompson, "The Expanding Sahara," *Envi-*

ronmental Conservation 1 (Spring 1974): 5–13; and James Walls, *Land, Man and Sand: Desertification and Its Solution* (New York: Macmillan, 1980), fifteen case studies based on the UN conference. Note also UNEP's biannual publication, *Desertification Control*, 1976–.

111. "The United Nations and Water," *UNITAR News* 9 (1977): entire issue; A. K. Biswas, ed., *United Nations Water Conference: Summary and Main Documents* (Oxford: Pergamon Press, 1978); and Yahia Abdel Mageed, "The UN Water Conference: The Scramble for Resolutions and the Implementation Cap," *Mazingira* 6, no. 1 (1982): 2–13.

112. Troy L. Pewe, ed., *Desert Dust: Origin, Characteristics, and Effect on Man* (Boulder, Colo.: Geological Society of America, 1981). See also additional references under Chapter 9, n. 5, of this book. See also Joseph M. Prospero and Ruby T. Nees, "Dust Concentrations in the Atmosphere of the Sahelian North Atlantic: Possible Relationship to the Sahelian Drought," *Science* 196 (10 June 1977): 1196–98.

113. David S. G. Thomas and Nicholas J. Middleton, *Desertification: Exploding the Myth* (New York: John Wiley, 1994).

114. See Glenn E. Schweitzer, "Toxic Chemicals: Steps toward Their Evaluation and Control," in Kay and Jacobson, *Environmental Protection: The International Dimension*, pp. 22–44.

115. Reported in *Newsweek*, 17 August 1981, p. 53.

116. E.g., Barry I. Castleman, "The Export of Hazardous Factories to Developing Countries," *International Journal of Health Services* 9, no. 4 (1979): 569–606.

117. U.S. Congress, House Committee on Foreign Affairs, *Export of Hazardous Products, Hearings before the Subcommittee on International Economic Policy and Trade of the Committee on Foreign Affairs*, 92d Cong., 2d sess., 5, 12 June and 9 September 1980. See also Jane H. Ives, *International Occupational Safety and Health Resource Catalogue* (New York: Praeger, 1981). An International Conference on the Exportation of Hazardous Industries, Technologies, and Products to Developing Countries was held at Hunter College, New York City, 2–3 November 1979, under the sponsorship of the University of Connecticut Health Center and in conjunction with the annual meeting of the American Public Health Association.

118. UN General Assembly, "Report of the Economic and Social Council on Protection Against Products Harmful to Health and Environment," A/C./37/L.65/Rev.1, 7 December 1982; and Libby Bassett, "U.N. Votes to Protect People from Harmful Imports," *World Environment Report* 9 (15 January 1983): 1–2. On the Basel Convention see "UN Conference Supports Curbs on Exporting of Hazardous Waste: 34 Nations Sign Pact and U.S. Plans to Review It," *New York Times* (23 March 1989): A1, B11; Iwona Rummel Bulska, "The Road to Basle: Taking the International Hazards out of Hazardous Wastes," *Our Planet* 1 (March 1989): 3, 14–15; and Diane Brady, "New Curbs on the Commerce in Poison," *Our Planet* 1, nos. 2–3 (1989): 18–20.

119. For more detailed treatment and references, see Norman Myers, *A Wealth of Wild Species* (Boulder, Colo.: Westview Press, 1983); Kenton R. Miller, "The Earth's Living Terrestrial Resources: Managing their Conservation," in Kay and Jacobson, *Environmental Protection: The International Dimension*, pp. 240–66; and Bryan C. Norton, *The Preservation of Species: The Value of Biodiversity* (Princeton, N.J.: Princeton University Press, 1986).

120. UN General Assembly, Resolution Adopted by the General Assembly, *World Charter for Nature*, A/RES/37/7, 9 November 1982, p. 3.

121. E.g., see Kenneth A. Dahlberg, "Plant Germplasm Conservation: Emerging Problems and Issues," *Mazingira* 7, no. 1 (1983): 14–25; Dahlberg, *Beyond the Green Revolution*; Kenneth O. Rachie and Judith M. Lyman, *Genetic Engineering for Crop Development: A Rockefeller Foundation Conference, May 12–15, 1980* (New York: Rockefeller Foundation, August 1981); FAO, *Report of the FAO/UNEP/IBPGR International Conference on Plant Genetic Resources, Rome, 6–10 April 1981* (Rome: FAO, Plant Production Protection Division, 1981); Jane Robertson Vernhes, "US-MAB Symposium: Learning to Conserve Genetic Resources," *Ambio* 12, no. 1 (1983): 34–37. See also Mark Shapiro, "Seeds of Disaster," *Mother Jones* (12 December 1982), pp. 12ff.

122. Lynton K. Caldwell, "International Aspects of Biotechnology: Guest Editorial and Re-

view," *MIRCEN Journal* 4, no. 4 (1988): 245–58 (see references and subsequent issues of the *Journal*).

123. "The European Networks of Biogenetic Reserves," *Naturopa*, no. 41 (1982): 24; and see notes 11–12 above, and United Nations Development Organization, "The Role of the International Centre for Genetic Engineering and Technology in Fostering Development through Applied Microbiology" prepared by UNIDO Secretariat UNIDO/15.521, 28 February 1985. See *Annual Report of UNIDO*, 1993. IDB. 12/2 PBC. 10/2, Subprogramme 453.

124. *Proceedings of the U.S. Strategy Conference on Biological Diversity, November 16–18, 1981* (Washington, D.C.: U.S. Department of State, Publication 9262, April 1982).

125. Gregory Jensen (United Press International), "Rare Breeds Trust Saving Endangered Farm Animals," in *Bloomington-Bedford (Indiana) Herald-Times*, 2 May 1976, p. 10; David Blackwell, "Breeding Livestocks for Farming Posterity," *Financial Times*, 10 September 1991, p. 20.

126. Alessandra Stanley, "Ashkabad Journal," *New York Times*, 10 November 1995, A4; Jonathan Evan Maslow, *Sacred Horses: The Memoirs of a Turkmen Cowboy* (New York: Random House, 1994).

127. UNEP, *1989 State of the World Environment* (Nairobi: April 1989), p. 11; Jeffrey A. McNeely, "Dollars and Sense about Biological Diversity," *IUCN Bulletin* 19, no. 46 (April–June 1988): 10; *Economics and Biological Diversity: Developing and Using Economic Incentives to Conserve Biological Resources* (Cambridge: IUCN Publications Unit, 1989); *Biodiversity Prospecting: Using Genetic Resources for Sustainable Development* (Washington, D.C.: World Resources Institute, 1993).

12. Strategies for Global Environmental Protection

1. Andronico O. Adede, *International Environmental Law Digest: Instruments for International Responses to Problems of Environment and Development, 1972–1992* (Amsterdam: Elsevier Science, 1993).

2. See Lester W. Milbrath, "Environmental Values and Beliefs of the General Public and Leaders in the United States, England and Germany," *Occasional Papers* (Buffalo: Environmental Studies Center, State University of New York, August 1980), also found in *Environmental Policy Formation: The Impact of Values, Ideology, and Standards*, ed. Dean E. Mann (Lexington, Mass.: Lexington Books, 1981), pp. 43–61. See also Peter Ester, "Environmental Concern in the Netherlands," in *Progress on Resources Management and Environmental Planning*, ed. T. O'Riordan and R. Kerry Turner (New York: John Wiley, 1981), 3: 81–108; and *Public Opinion on Environmental Issues: Results of a National Public Opinion Survey* (Washington, D.C.: Council on Environmental Quality, 1980); Ronald Inglehart, *Cultural Shift in Advanced Industrial Societies* (Princeton, N.J.: Princeton University Press, 1989); also *Our Planet* 1.

3. Gregg Easterbrook, *A Moment on the Earth: The Coming Age of Environmental Optimism* (New York: Viking, 1995). Union of Concerned Scientists, *World Scientists Warning to Humanity*, 26 Church St., Cambridge, Mass., 1995 (more than 1,600 signatories from 70 countries including 104 Nobel prize winners).

4. *Man's Role in Changing the Face of the Earth*, ed. William L. Thomas Jr. (Chicago: University of Chicago Press, 1956).

5. Decline in political control in the conventional sense is of course relative and does not imply decline in all areas, e.g., science, medicine, technology, etc., in which Western society continues to advance. The phenomenon may be more of transition than of decline, and there is widespread agreement among historians that the end of the twentieth century marks the end of a distinctive era of human expansion and transformation of the earth. The environmental consequences of this era strongly affect what will (or can) follow it. For different interpretations of this transition, see Daniel Bell, *The Coming of Post-Industrial Society* (New York: Basic Books, 1973); Kenneth E. Boulding, *The Meaning of the Twentieth Century: The*

Great Transition (New York: Harper and Row, 1964); Karl Polanyi, *The Great Transforma-tion* (Boston: Beacon Press, 1944); and Walter Prescott Webb, *The Great Frontier* (Boston: Houghton Mifflin, 1952). I have summarized some of the historical trends during the "mod-ern" era in "1992: Threshold to the Postmodern World," in *A Time to Hear and Answer: Essays for the Bicentennial Season*, Franklin Lectures in the Sciences and Humanities, 4th series (University: University of Alabama Press, for Auburn University, 1977), pp. 177-218, and in *Between Two Worlds: Science, the Environmental Movement, and Policy Choice* (Cam-bridge: Cambridge University Press, 1990). Also Fritjhof Capra, *The Turning Point: Science, Society, and the Rising Culture* (New York: Simon and Schuster, 1984). For a Russian perspec-tive, see I. P. Gerasimov, *Geography and Ecology* (Moscow: Progress Publishers, 1983).

6. See William Ophuls, *Ecology and the Politics of Scarcity* (San Francisco, Calif.: Freeman, 1977), and comments by Carl Cook in the *Social Science Quarterly* 62 (March 1981): 23-25. The fol-lowing books describe the social and economic consequences of human disregard for the requirements for environmental renewal, considerations that are as valid today as when they were written: Fairfield Osborn, *Our Plundered Planet* (New York: Pyramid Books, 1968); Paul Sears, *Deserts on the March* (Norman: University of Oklahoma Press, 1935); William Vogt, *The Road to Survival* (New York: William Sloan, 1960).

7. See Gerald Garvey and Lou Ann Garvey, eds., *International Resource Flows* (Lexington, Mass.: Lexington Books, 1977). Also see Ruth W. Arad and Uzi B. Arad, "Scarce Natu-ral Resources and Potential Conflict," in *Sharing Global Resources* (New York: McGraw-Hill, 1979); Michael T. Klare, "Resource Wars," *Harper's Magazine* (January 1981): 20-23; James A. Miller, Daniel I. Fine, and R. Daniel McMichael, *The Resource War in 3-D—De-pendency, Diplomacy, Defense* (Pittsburgh, Pa.: Pittsburgh World Affairs Council, 1981).

8. The conceptual issue is developed by P. T. Bauer, *Equality, The Third World and Economic Delusion* (Cambridge, Mass.: Harvard University Press, 1981). For a more extensive analysis of semantic problems with environmental terms and concepts, see Harold Sprout and Mar-garet Sprout, "Focal Terms and Concepts," in *The Ecological Perspective on Human Affairs: With Special Reference to International Politics* (Princeton, N.J.: Princeton University Press, 1965).

9. Russell E. Train, *Corporate Use of Information Regarding Natural Resources and Environmental Quality* (Washington, D.C.: World Wildlife Fund-U.S. for the Council on Environmental Quality, 1984).

10. *Environmental Activities of Non-Government Organizations (NGOs) Related to UNEP Pro-grammes: Report and Directory* (Nairobi: Environment Liaison Centre, 1980), p. 11. Among the countries in which three or more environmental NGOs were listed were Kenya, Ghana, Nigeria, India, Malaysia, Philippines, Sri Lanka, Colombia, Costa Rica, Mexico, and Vene-zuela. Also *The Environment in 1982: Retrospect and Prospect*, UNEP/GC(SSC)/2, Nairobi, 29 January 1982, pp. 15-16.

11. UNEP, *1988 Annual Report of the Executive Director*, paragraph 41, p. 14.

12. Michel Batisse, "The Relevance of MAB," *Environmental Conservation* 7 (Winter 1980): 180; see also Batisse, "The Future of MAB," *Environmental Conservation* 9 (Spring 1982): 71-72, and "The Silver Jubilee of the MAB and Its Revival," *Environmental Conservation* 20, no. 2 (Summer 1993): 107-112.

13. For the origin and objectives of MAB, see UNESCO, "Final Report of the Intergovern-mental Conference of Experts on the Scientific Basis for Rational Use and Conservation of the Resources of the Biosphere, held at UNESCO House, Paris, France, 4-13 September, 1968," SC/MD/9, Paris, 1969; UNESCO, *Use and Conservation of the Biosphere: Proceedings of the Intergovernmental Conferences of Experts on the Scientific Basis for Rational Use and Conser-vation of the Resources of the Biosphere*, UNESCO Natural Resources Research Series, no. 10, Paris, 1970; and UNESCO, General Conference, Sixteenth Session, 6 October 1970, "Plan for a Long-term Intergovernmental and Interdisciplinary Programme on Man and the Bio-sphere," Paris, 1970; Michel Batisse, "Can We Keep Our Planet Habitable?" *Unesco Courier*

22 (January 1969): 4-5; Harold J. Coolidge, "World Biosphere Conference: A Challenge to Mankind," *IUCN Bulletin*, n.s., 2 (October–December 1968): 65-66; Raymond Dasmann, "Conservation and Rational Uses of the Environment," *Nature and Resources* 4 (June 1968): 2-5; and William P. Gregg, Jr., "MAB and Its Biosphere Reserves Project: A New Dimension in Global Conservation," *The George Wright Forum* (Spring 1983), pp. 17-31.

14. Batisse, "The Relevance of MAB," p. 181.

15. For the substance of this aspect of MAB, see *Expert Panel on Project 13: Perception of Environmental Quality; Final Report*, MAB Report Series, no. 9 (Paris: UNESCO, March 1973), pp. 26-29.

16. See "Ecology in Practice: A Conference/Exhibition for MAB's First Ten Years," *Nature and Resources* 17 (October–December 1981): 29-30; and Francisco di Castri, Malcolm Hadley, and Jeanne Damiamian, "Ecology in Action—An exhibit: An experiment in communicating scientific information," *Nature and Resources* 18 (April–June 1982): 10-17.

17. See *Connect: UNESCO-UNEP Environmental Education Newsletter* 1, no. 1 (January 1976).

18. See Intergovernmental Conference on Environmental Education, Tbilisi (U.S.S.R.), 14-26 October 1977, "Final Report," ED/MD/49, April 1978, and report in *Uniterra* 9 (October–November 1977): 1-7. Preceding the conference a number of preliminary documents were issued by UNESCO and UNEP. See publications under the UN code ED-77/Conf. 203.

19. *Connect* 7 (March 1982): entire issue. See also *Environmental Education in the Light of the Tbilisi Conference* (Paris: UNESCO, 1980); and *Connect* 7 (December 1982) dealing with environmental education and UNESCO's Medium Term Plan (1984-89).

20. *Connect* 12, no. 3 (September 1987): entire issue.

21. "A World Strategy for Conservation," *Uniterra* 5 (March 1980): 11; Robert Wazeka, "A World Conservation Strategy Is Launched," *Unasylva* 31 (1979): 39-41; and Thomas E. Lovejoy, "A Strategy for Survival," *Uniterra* 8 (September–October 1980): 6. The World Conservation Strategy was issued initially in packet form. Subsequently it appeared (unofficially) in book form, suitable for general readers. See Robert Allen, *How to Save the World: Strategy for World Conservation* (Totowa, N.J.: Barnes and Noble, 1980).

22. *World Conservation Strategy: Living Resource Conservation for Sustainable Development* (Gland, Switzerland: IUCN, UNEP, WWF, 1980). Preceding publication of the World Conservation Strategy was the prospectus of A Conservation Programme for Sustainable Development 1980-1982: Prepared by the International Union for Conservation of Nature and Natural Resources as a framework for the activities of IUCN, the World Wildlife Fund (WWF) and their associates (Gland, Switzerland: IUCN, 1 December 1979).

23. Lee M. Talbot, "The World's Conservation Strategy," *Environmental Conservation* 7 (Winter 1980): 268.

24. Thomas E. Lovejoy, *Uniterra* 8 (September–October 1980): 6.

25. See UNEP, "Meeting with Multilateral Development Financing Agencies to Adopt a Draft Declaration of Principles on the Incorporation of Environmental Considerations in Development Policies, Programmes and Projects, Paris, 12-13 September 1979" (Nairobi: UNEP, 1 November 1979), UNEP/GC.8/Int.; *World Bank Annual Report 1980* (Washington, D.C.: World Bank, n.d.), p. 78; and Yusuf J. Ahmad, "Environmental Concerns of the Least Developed Countries," *Mazingira* 5, no. 3 (1981): 29. For final text, see *International Legal Materials* 19 (July 1980): 837-38. For more recent activities, see UNEP, *1987 Annual Report of the Executive Director*, p. 20, and *1988 Annual Report*, pp. 16-17.

26. Mohamed T. El-Ashry, "The World Bank's Post-Rio Strategy," *EPA Journal* (April–June 1993): 22-25.

27. "In Surinam Rainforests, A Fight over Trees vs. Jobs," *New York Times*, 4 September 1994. The headline misleads. The issue is money vs. ecology.

28. Lynton K. Caldwell, "An Ecological Approach to International Development: Problems of Policy and Administration," in M. Taghi Farvar and John P. Milton, eds., *The Careless Technology: Ecology and International Development* (Garden City, N.Y.: Natural History Press, 1972), pp. 941-42.

29. *UNEP North America News*, 4, no. 4 (August 1989): 6.
30. Note, for example, *The World Environment—1982*, chap. 16; Essam El-Hinnawi, "The Arms Race and the Environment," *Mazingira* 6, no. 2 (1982): 4-15; Arthur H. Westing, "Environmental Impact of Nuclear Warfare," *Environmental Conservation* 8 (Winter 1981): 269-73, references, and "Environmental Consequences of Nuclear War," *Environmental Conservation* 9 (Winter 1982): 269-72; Mostafa K. Tolba, Arthur H. Westing, and Nicholas Polunin, "The Environmental Imperative of Nuclear Disarmament," *Environmental Conservation* 9 (Summer 1983): 91-95; also "Nuclear War: The Aftermath," *Ambio* 11, no. 23 (1982): entire issue; and Libby Bassett, "Nuclear Aftermath: Biological Havoc," *World Environmental Report* 9 (15 November 1983): 1-2; J. P. Robinson, *The Effects of Weapons on Ecosystems*, UNEP Studies, vol. 1 (New York: Pergamon Press, for UNEP, 1979); *Warfare in a Fragile World: Military Impact on the Human Environment* (Stockholm: Stockholm International Peace Research Institute; New York: Crane, Russak, 1980); also *Threat of Modern Warfare to Man and the Environment: an Annotated Bibliography*, prepared under the auspices of the International Peace Research Association (Paris: UNESCO, 1979); and U.S. Congress, House Committee on Foreign Affairs, *Chemical-Biological Warfare: U.S. Policies and International Effects: Hearings before the Subcommittee on National Security Policy and Scientific Developments of the Committee on Foreign Affairs*, 91st Cong., 1st sess., November and December 1969; and "War, Law, and the Environment," *IUCN Bulletin* 22, no. 3 (September 1991): entire issue.

13. A Changing World Order: Into the Twenty-First Century

1. See Joan Martin-Brown, "Converging Worlds: The Implications of Environmental Events for the Free Market and Foreign Policy Development," *Environmentalist* 4, no. 2 (1984): 139-42.
2. For a list of world models and references, see UNEP, *The Environment in 1982: Retrospect and Prospect*, p. 43. See also *Global Models, World Futures, and Public Policy: A Critique* (Washington, D.C.: Office of Technology Assessment, 1982); and U.S. Council on Environmental Quality and Department of State, *The Global 2000 Report to the President: Entering the Twenty-First Century*, 3 vols. (Washington, D.C.: Government Printing Office, 1980). For a review of models see Asit K. Biswas, "Global Future Studies: Review of Past Decade," *Mazingira* 6, no. 1 (1982): 68-75.
3. Robert D. Munro, in *Mazingira* 6, no. 1 (1982): 56; and "World Conservation Strategy: Managing for the Future," *IUCN Bulletin* 20, nos. 4-6 (April-June 1989): 3.
4. See *World Environment Handbook: A Directory of Government Natural Resource Management Agencies in 144 Countries* (New York: World Environment Center, 1982). See also Raymonde Gour-Tanguay, ed., *Environmental Policies in Developing Countries* (Berlin: Erich Schmidt Verlag, 1977); Essam El-Hinnawi and Asit K. Biswas, eds., *Third World and the Environment* (Dublin: Tycooly International Publications, 1982); Whitman Bassow, "The Third World: Changing Attitudes toward Environmental Protection," *Annals of the American Academy of Political and Social Science* 444 (July 1979): 112-20; and Yusuf J. Ahmad, "Environmental Concerns of the Least Developed Countries," *Mazingira* 5, no. 3 (1981): 20-29. But see also B. Bowonder, "Environmental Management Conflicts in Developing Countries: An Analysis," *Environmental Management* 7, no. 3 (1983): 211-21.
5. *Ronald Reagan and the American Environment: An Indictment* (San Francisco: Friends of the Earth, 1982). For a European perspective, see Erwan Fouere, "Clashing on the Environment," *Europe*, no. 237 (May-June 1983): 12-15; and for indicators of a change in U.S. policy, see Libby Bassett, "From Inertia to Impetus: New Look in U.S. Leadership," *World Environment Report* 9 (30 June 1983): 1-2.
6. Russell J. Dalton, *The Green Revolution: Environmental Groups in Western Europe* (New Haven, Conn.: Yale University Press, 1994). See also Fritjhof Capra and Charlene Spretnak, *Green Politics* (New York: Dutton, 1984; rev. ed., Santa Fe, N. Mex.: Bear, 1986). Also Richard J. Roddewig, *Green Bans: The Birth of Australian Environmental Politics: A Study in*

Public Opinion and Participation (Montclair, N.J.: Allanheld, Osmun, 1978); Richard J. Rod-dewig and John S. Rosenburg, "In Australia Unions Strike for Environment," *Conservation Foundation News Letter* (November 1975): 8; Wellington Long, "Green Power Environmentalists Show Growing Influence in German Politics," *Europe*, no. 233 (September–October 1982): 20–21; David H. Handley, "The Environmental Protection Issue and Value Change in Western Europe," paper presented at the meeting of the International Society of Political Psychology, 2–4 September 1978, New York; Ronald Koven, "Polls Give Left Lead in France as Vote Nears," *Washington Post*, 8 February 1978, pp. A1, A16; "West German Green Parties," *New York Times*, 16 August 1978, p. 94; "The 'Green Party': A New Political Party," *The Week in Germany* 10 (8 November 1979): 2; Stephen Levine and Juliet Lodge, "The New Zealand General Election of 1975," *Parliamentary Affairs* 29 (Summer 1976: 310–26; Anna Lubinska, "An Overview of the European 'Greens,'" *World Environment Report* 9 (15 April 1983): 7; and Piero Valsecchi, "Italy Forms a 'Green Wave,'" *World Environment Report* 9 (15 April 1983): 7–8; and for environmental parties in Austria, see report by J. M. Bradley in *World Environment Report* 9 (30 May 1983): 1–3.

7. David Brand, "Polish Ecology Club Risks Government Ire by Battling Pollution," *Wall Street Journal*, 24 July 1981.

8. Norman Myers and Dorothy Myers, "How the Global Community Can Respond to International Environmental Problems," *Ambio* 12, no. 1 (1983): 20–26; and Peter M. Haas, Robert O. Keohane, and Marc Levy, eds., *Institutions for the Earth: Sources of Effective Environmental Protection* (Cambridge, Mass.: MIT Press, 1993).

9. Buckminster Fuller, *Operating Manual for Spaceship Earth* (New York: Simon and Schuster, 1970), p. 492, n. 3.

10. Max Nicholson, *The Big Change: After the Environmental Revolution* (New York: McGraw-Hill, 1973). See also Nicholson, *The Environmental Revolution: A Guide for the New Masters of the World* (New York: McGraw-Hill, 1970).

11. For example, see Philip Shabecoff, "The Environment as a Diplomatic Issue," *New York Times* (25 December 1987), and "Traditional Definitions of National Security Are Shaken by Environmental Threats," *New York Times* (29 May 1989).

12. *International Legal Materials* 28, no. 5 (September 1989): 1311–13.

13. Norman, Myers, "Environmental Unknowns," *Science* 269 (21 July 1995): 358–60.

14. For additional aspects of security-environment relationships, see Arthur F. Westing, ed., *Global Resources and International Conflict: Environmental Factors in Strategic Policy and Action* (Oxford: Oxford University Press, 1986), and *Cultural Norms, War, and the Environment* (Oxford: Oxford University Press, 1988). Also Lester R. Brown, *Redefining National Security*, Worldwatch Paper 14 (Washington, D.C.: Worldwatch Institute, 1977); also Thomas F. Homer-Dixon, *Environmental Scarcity and Global Security* (New York: Foreign Policy Assoc., 1993); and Al Gedicks, *The New Resource Wars: Struggle against Multinational Corporations* (Boston: South End Press, 1993); also W. Ophuls, *Ecology and the Politics of Scarcity: Prologue to a Political Theory of a Steady State* (San Francisco: W. H. Freeman, 1977).

15. Forecasts vary. See Jacques Theys, "21st Century: Environment and Resources," *European Environment Review* 1, no. 5 (December 1987): 2–11 (a generally pessimistic outlook—little hope of preserving natural areas). For a contrasting prognosis see Gregg Easterbrook, *A Moment on the Earth: The Coming Age of Environmental Optimism* (New York: Viking, 1995).

Index to Topics and Institutions

In general, institutions, conferences, and conventions are listed under the topic to which they pertain.

Acid precipitation, 209–210
Acid rain. *See* Acid precipitation; Atmospheric pollution
Additionality, concept of, 70, 73, 74, 77, 78, 81, 271
Aden, Gulf of: as component of UNEP's Oceans and Coastal Areas Programme, 186–187
Africa: African Convention for Conservation of Nature and Natural Resources, 40; African Convention Relative to the Preservation of Flora and Fauna in their Natural State, 40; College of Wildlife Management, 124; International African Migratory Locust Organization (OICMA), 159; Anti-Locust and Anti-Avarian Organization, 134; Organization of African Unity (OAU), 124. *See also individual countries of East Africa*
Agenda 21. *See* United Nations Conference on Environment and Development
Agriculture: applications of biotechnology to, 319; Consultative Group on Agricultural Research (CGIAR), 320; erosion of top soil, 258; National Agricultural Chemical Association (U.S.), 317; plant breeding, 318–320. *See also* Food and Agriculture Organization
Air pollution. *See* Atmospheric pollution
Amazon basin, 238; Treaty of Amazonian Cooperation, 99. *See also* Amazon Declaration; Forests, tropical; Jari Project
Amazon Declaration (Declaration of Manaus), 101, 357

Antarctica, 195–200; Agreed Measures for Antarctic Fauna and Flora, 197; Antarctic Treaty of 1959, 18, 196, 197, 238, 347; Biological Investigations of Marine and Antarctic Systems and Stocks (BIOMASS), 198; Convention on the Conservation of Antarctic Marine Living Resources, 197, 286; Convention for the Conservation of Antarctic Seals, 197; Convention on the Regulation of Antarctic Mineral Resource Activities, 93; krill, 197, 226; marine ecosystems, 226, 235–236; mineral development, 263–264; Madrid Protocol to the Antarctic Treaty, 199; national claims, 196; physical factors, 196; Scientific Committee on Antarctic Research (SCAR), 142, 198, 199
Apollo XI, 14, 30
Arab League Educational, Cultural, and Scientific Organization (ALECSO), 187
Arbitration. *See* Law, international
Arctic, 194–195; Declaration on the Protection of the Arctic Environment, 195; mineral exploitation, oil pollution and exploration, 194–195; settlement planning, 297
Arctic Research and Policy Act (U.S.), 194
Association of Southeast Asian Nations (ASEAN), 4, 159, 190; Workshop on Nature Conservation (Bali, 1980), 191
Atmosphere, 203–213; Global Atmospheric Research Programme (GARP), 137, 211; International Civil Aviation Organization (ICAO), 205; international policy for, 210–213; National Center for Atmospheric Research (U.S.), National Oceanic and Atmospheric Administration (NOAA)(U.S.), 211, 220; upper atmosphere, experiments in, 214, 215

Index to Authors

Lynton Keith Caldwell is Arthur F. Bentley
Professor of Political Science Emeritus and Professor
of Public and Environmental Affairs at Indiana
University.

Library of Congress Cataloging-in-Publication Data
Caldwell, Lynton Keith, 1913–
International environmental policy : from the
twentieth to the twenty-first century / Lynton Keith
Caldwell. — 3rd ed. / rev. and updated with the
assistance of Paul Stanley Weiland.
Includes indexes.
ISBN 0-8223-1861-X (cloth : alk. paper). —
ISBN 0-8223-1866-0 (paper : alk. paper)
1. Environmental law, International.
2. Environmental policy.   I. Weiland, Paul Stanley.
II. Title.
K3585.4.C34   1996
341.7'62—dc20   96-20958   CIP